RESOLVING THE RISE AND FALL OF STAR FORMATION IN GALAXIES

IAU SYMPOSIUM 373

COVER ILLUSTRATION:

The nearby barred galaxy NGC 3627 in its optical light as seen by VLT/MUSE (top) and in its CO(2-1) line emission as captured by ALMA (bottom).

Credit: NASA/HST, ESO/VLT, ALMA, PHANGS team & J. Neidel/MPIA.

IAU SYMPOSIUM PROCEEDINGS SERIES

INTERNATIONAL ASTRONOMICAL UNION

UNION ASTRONOMIQUE INTERNATIONALE

International Astronomical Union

RESOLVING THE RISE AND FALL OF STAR FORMATION IN GALAXIES

PROCEEDINGS OF THE 373rd SYMPOSIUM OF THE INTERNATIONAL ASTRONOMICAL UNION HELD IN BUSAN, REPUBLIC OF KOREA 9–11 AUGUST, 2022

Edited by

TONY WONG
University of Illinois, USA

and

WOONG-TAE KIM
Seoul National University, Republic of Korea

CAMBRIDGE
UNIVERSITY PRESS

CAMBRIDGE UNIVERSITY PRESS
University Printing House, Cambridge CB2 8BS, United Kingdom
1 Liberty Plaza, Floor 20, New York, NY 10006, USA
10 Stamford Road, Oakleigh, Melbourne 3166, Australia

First published 2023

Printed in the UK by Bell & Bain, Glasgow, UK

Typeset in System LaTeX 2ε

A catalogue record for this book is available from the British Library Library of Congress Cataloguing in Publication data

This journal issue has been printed on FSC™-certified paper and cover board. FSC is an independent, non-governmental, not-for-profit organization established to promote the responsible management of the world's forests. Please see www.fsc.org for information.

ISBN 9781009352956 hardback
ISSN 1743-9213

Table of Contents

Chapter 1: Scales of Star Formation: From Molecular Cores to Galaxies

Chapter 4: The Rise and Fall of Star Formation Across Cosmic Time

Preface

Star formation is relevant to nearly every area in astrophysics, from planetary science to galaxy evolution. Yet the physical processes that determine the rate of star formation and its spatial and temporal distribution are still poorly understood. This Symposium focused on the impact that resolved studies of galaxies, both observational and theoretical, are having on the understanding of star formation on all scales. Its goal was to highlight the latest advances in understanding star formation in its galactic context (via resolved studies) and how it drives galaxy evolution.

The general picture of galaxy evolution is one of steady, inside-out growth and maturation of galaxy disks accompanied by rapid aging of the highest mass galaxies due to internal and/or external processes that inhibit star formation. Yet the roles of various factors in the aging process – e.g., galaxy mergers, gas consumption, environmental refueling (or lack thereof), and nuclear activity – remain poorly understood. Integrated over the galaxy population, the effect of galactic aging can be seen as a dramatic decline in the cosmic star formation rate since the epoch of "cosmic noon" at a redshift $z \sim 2$. This is the "rise and fall of star formation" as measured across the galaxy population. For galaxies that are still forming stars today, the star formation rate is strongly correlated with the stellar mass and the supply of molecular gas. However, the exact form of these star formation (scaling) relations, their universality, and the role of additional physical parameters (including galaxy conditions and environment) remain important topics of discussion. Moreover, these relations have provided limited insight into the cessation of star formation, commonly termed "quenching", which can occur locally well before the star formation of an entire galaxy shuts down. This is the "rise and fall of star formation" as measured within individual galaxies.

Until fairly recently, the communities that studied star formation in galaxies were divided into those who studied small scale processes at high resolution in very nearby galaxies (including our own), and those who treated star formation as a galaxy-scale process, studied out to high redshifts. In the last decade, the study of star formation has been undergoing a revolution that has connected these communities in both theory and observations. The key advance has been the ability to spatially resolve the sub-kiloparsec scales on which star formation relations are established, bridging the gap between resolved studies in the local neighborhood and large-scale galaxy surveys. Helping to interpret these new data are a new generation of cosmological simulations, fully resolved ISM simulations, and new techniques for confronting them with observations.

IAU Symposium 373, running from 9–11 August 2022 at the XXXI General Assembly meeting in Busan, Republic of Korea, provided an opportunity to share the latest findings in the field of star formation on sub-galactic scales. It included 21 invited talks, 36 contributed talks, 78 e-posters, and 78 e-talks. Owing to the COVID-19 pandemic, 20 of the talks were given remotely. It is a great pleasure to acknowledge the support of the NOC of the General Assembly in Busan.

Tony Wong (co-chair, SOC) and Woong-Tae Kim
November 2022

Editors

Tony Wong
University of Illinois, USA

Woong-Tae Kim
Seoul National University, Republic of Korea

Organizing Committee

Scientific Organizing Committee

Tony Wong (Co-chair) , University of Illinois, USA
Eva Schinnerer (Co-chair), MPIA, Germany
Guillermo Blanc, University of Chile, Chile
Sara Ellison, University of Victoria, Canada
Robert C. Kennicutt Jr., University of Arizona, USA
Woong-Tae Kim, Seoul National University, Republic of Korea
Johan Knapen, IAC, Spain
Kotaro Kohno, University of Tokyo, Japan
Claudia Lagos, University of Western Australia, Australia
Janice C. Lee, NOIRLab, USA
Karín Menéndez-Delmestre, UFRJ, Brazil
Yingjie Peng, Peking University, China
Amelie Saintonge, University College London, UK

National and Local Organizing Committees of the XXXIth General Assembly of the IAU

Hyesung Kang (Chair), Pusan National University, Republic of Korea
Byeong-Gon Park (Vice-Chair), Korea Astronomy & Space Science Institute, Republic of Korea
Deokkeun An, Ewha Womans University, Republic of Korea
Jungyeon Cho, Chungnam National University, Republic of Korea
Joon-Young Choi, Busan National Science Museum, Republic of Korea
Aeree Chung, Yonsei University, Republic of Korea
Junga Hwang, Korea Astronomy & Space Science Institute, Republic of Korea
Ho-Seong Hwang, Seoul National University, Republic of Korea
Chunglee Kim, Ewha Womans University, Republic of Korea
Dohyeong Kim, Pusan National University, Republic of Korea
Ji-hoon Kim, Seoul National University, Republic of Korea
Jongsoo Kim, Korea Astronomy & Space Science Institute, Republic of Korea
Minjin Kim, Kyungpook National University, Republic of Korea
Sungsoo S. Kim, Kyung Hee University, Republic of Korea
Woong-Tae Kim, Seoul National University, Republic of Korea
Woojin Kwon, Seoul National University, Republic of Korea
Jeong-Eun Lee, Kyung Hee University, Republic of Korea
Kang Hwan Lee, Institut Pasteur Korea, Republic of Korea
Sang-Sung Lee, Korea Astronomy & Space Science Institute, Republic of Korea
Seo-gu Lee, Korea Astronomy & Space Science Institute, Republic of Korea
Soo-Chang Rey, Chungnam National University, Republic of Korea
Hyunjin Shim, Kyungpook National University, Republic of Korea
In-Ok Song, Korea Science Academy of KAIST, Republic of Korea
Hong-Jin Yang, Korea Astronomy & Space Science Institute, Republic of Korea
Suk-Jin Yoon, Yonsei University, Republic of Korea
Sung-Chul Yoon, Seoul National University, Republic of Korea

Participants

Hedieh **Abdollahi**, ALzahra university — abdollahi.hedieh2013@gmail.com
Biruk **Abrham**, Ethiopia space science technology institute — bab3590@gmail.com
Dominic **Adams**, University of Minnesota — adams900@umn.edu
Binod **Adhikari**, St. Xavier's College — binod.adhi@gmail.com
José Feliciano **Agüí Fernández**, IAA-CSIC — feli@iaa.es
Hojae **Ahn**, Kyunghee University — hojaeahn@khu.ac.kr
Kyungjin **Ahn**, Chosun University — kjahn@chosun.ac.kr
Yuri **Aikawa**, University of Tokyo — aikawa@astron.s.u-tokyo.ac.jp
Bhairab **Ale**, Tribhuvan University — alebhairab0@gmail.com
Hiddo **Algera**, Hiroshima University — algera@hiroshima-u.ac.jp
Dana **Alina**, Nazarbayev University — dana.alina@nu.edu.kz
Rommy **Aliste Castillo**, Seoul National University — rommy@astro.snu.ac.kr
Myriam **Alqassab**, Bahrain Stargazers Astronomy Club — stargazers.bahrain@gmail.com
Mashhoor **Alwardat**, University of Sharjah — malwardat@sharjah.ac.ae
Dhanushka **Amaradasa**, University of Ruhuna — subathamaradasa@gmail.com
Deokkeun **An**, Ewha Womans University — deokkeun.an@gmail.com
Irham Taufik **Andika**, Max Planck Institute for Astronomy — andika@mpia.de
Hong Bae **Ann**, Pusan National University — hbann@pusan.ac.kr
Susana Beatriz **Araujo Furlan**, Argentine Institute of Radioastronomy — saraujo@iar.unlp.edu.ar
Manuel **Aravena**, Universidad Diego Portales — manuel.aravenaa@mail.udp.cl
Haylee **Archer**, Arizona State University — harcher@lowell.edu
Brent **Archinal**, U. S. Geological Survey — barchinal@usgs.gov
Itziar **Aretxaga**, INAOE — itziar@inaoep.mx
Aayush **Arya**, Lovely Professional University — aayusharya71@gmail.com
Yusuke **Aso**, Korea Astronomy and Space Science Institute — yaso@kasi.re.kr
Cecilia **Bacchini**, INAF — cecilia.bacchini@inaf.it
Yoonkwan **Bae**, Yonsei University — yoonkwan95@yonsei.ac.kr
Giseon **Baek**, Kyung Hee University — giseon8871@gmail.com
Junhyun **Baek**, Yonsei University — jhbaek1001@gmail.com
Hyeonguk **Bahk**, Seoul National University — bahkhyeonguk@gmail.com
Marc **Balcells**, Isaac Newton Group — balcells@ing.iac.es
Stefania **Barsanti**, Australian National University — stefania.barsanti@anu.edu.au
Andrew **Battisti**, Australian National University — andrew.battisti@anu.edu.au
Rachael **Beaton**, Princeton University — rachael.l.beaton@gmail.com
Laura **Becerra Bayona**, Pontificia Universidad Católica de Chile — laura.marcela.becerra@gmail.com
Sara **Beck**, Tel Aviv University — becksarac@gmail.com
Enrica **Bellocchi**, Universidad Complutense de Madrid — enricbel@ucm.es
Vardha N. **Bennert**, California Polytechnic State University San Luis Obispo — vbennert@calpoly.edu
Apurba **Bera**, Inter-university Centre for Astronomy and Astrophysics — apurba.bera@iucaa.in
Danielle **Berg**, The University of Texas at Austin — daberg@austin.utexas.edu
Matheus **Bernini Peron**, Universtaet Heidelberg — matheus.bernini@uni-heidelberg.de
Shankar **Bhattarai**, Korea Astronomy and Space Science Institute — shankar@kasi.re.kr

Alya **BinAshour**, University of Sharjah abinashoor@gmail.com
Mark **Bishop**, Victoria University of Wellington mark@bishop.kiwi.nz
Guillermo **Blanc**, Carnegie Observatories gblancm@carnegiescience.edu
Stefan **Bode**, LJMU bode_stefan@gmx.de
Alberto **Bolatto**, University of Maryland at College Park bolatto@umd.edu
Médéric **Boquien**, Universidad de Antofagasta mederic.boquien@uantof.cl
Connor **Bottrell**, University of Tokyo connor.bottrell@ipmu.jp
Richard **Boyle**, Vatican Observatory rpboyle@mac.com
Vinicius **Branco**, Geofísica e Ciências Atmosféricas vbranco@usp.br
Gustavo **Bruzual**, UNAM g.bruzual@irya.unam.mx
Tomasz **Bulik**, University of Warsaw tb@astrouw.edu.pl
Martin **Bureau**, University of Oxford martin.bureau@physics.ox.ac.uk
Denis **Burgarella**, Laboratoire d'Astrophysique de Marseille denis.burgarella@lam.fr
Maria Luisa **Buzzo**, Swinburne University of Technology lgomesbuzzo@swin.edu.au
Woowon **Byun**, Korea Astronomy and Space Science Institute wbyun87@gmail.com
Luz Marina **Cairós**, Institute for Astrophysics and Geophysics luzma@astro.physik.uni-goettingen.de
Artemi **Camps Fariña**, Universidad Complutense de Madrid arcamps@ucm.es
Lina **Canas**, IAU/NAOJ lina.canas@nao.ac.jp
Remziye **Canbay**, Istanbul University rmzycnby@gmail.com
Luisa **Cardona Torres**, Instituto Nacional de Astrofísica Óptica y Electrónica lucardona@inaoep.mx
Leticia **Carigi**, Instituto de Astronomia UNAM carigi@astro.unam.mx
Claude **Carignan**, University of Cape Town claude.carignan@gmail.com
Yvelice-Soraya **Castillo-Rosales**, National Autonomous University of Honduras yvelice.castillo@unah.edu.hn
Barbara **Catinella**, ICRAR, University of Western Australia barbara.catinella@uwa.edu.au
Alagie **Ceesay**, Gambia National Space Research Institute amiesannehahgpi@yahoo.com
Vitor **Cernic**, Universidade de São Paulo vitorcernic@gmail.com
Solène **Chabanier**, Lawrence Berkeley National Laboratory schabanier@lbl.gov
Jennifer **Chacón**, Escuela Politécnica Nacional jenniferx12@hotmail.com
Jihyo **Chae**, Yonsei University antares0116@gmail.com
Nushkia **Chamba**, Stockholm University/Oskar Klein Center nushkia.chamba@astro.su.se
Katie **Chamberlain**, University of Arizona katiechambe@email.arizona.edu
Eleni **Chatzichristou**, European research Council Executive Agency elthchatz@gmail.com
Judy **Chebly**, Leibniz Institute for Astrophysics Potsdam jchebly@aip.de
Jin-Hong **Chen**, Sun Yat-Sen University chenjh258@mail2.sysu.edu.cn
Qingxiang **Chen**, Yale-NUS qingxiang.chen@yale-nus.edu.sg
Xiaopeng **Cheng**, Korea Astronomy and Space Science Institute xcheng@kasi.re.kr
Daya Nidhi **Chhatkuli**, Tribhuvan University chhatkulidn@gmail.com
Brian S. **Cho**, Seoul National University brian@astro.snu.ac.kr
Hyejeon **Cho**, Yonsei University hyejeon@yonsei.ac.kr

Jungyeon **Cho**, Chungnam National University jcho@cnu.ac.kr
Youngjun **Cho**, UNIST youngjun@unist.ac.kr
Gwangson **Choe**, Kyung Hee University gchoe@khu.ac.kr
Bo-Eun **Choi**, University of Washington bechoi@uw.edu
Ena **Choi**, Korea Institute of Advanced Study enachoi@kias.re.kr
Hyeonho **Choi**, Seoul National University hhchoi1022@gmail.com
Jeong yoon **Choi**, UNIST jychoi@unist.ac.kr
Jeong Yun **Choi**, Yonsei University jychoi8.32@yonsei.ac.kr
Woorak **Choi**, Yonsei University woorak.c@yonsei.ac.kr
Youngwoo **Choi**, Seoul National University cyw3614@naver.com
Yumi **Choi**, UC Berkeley ychoi@stsci.edu
Jørgen **Christensen-Dalsgaard**, Aarhus University jcd@phys.au.dk
Martyna **Chruslinska**, Max-Planck Institute for Astrophysics martich@mpa-garching.mpg.de
Kyungwon **Chun**, Korea Astronomy and Space Science Institute kwchun87@gmail.com
Sang-Hyun **Chun**, Korea Astronomy and Space Science Institute shyunc@kasi.re.kr
Aeree **Chung**, Yonsei University achung@yonsei.ac.kr
Haeun **Chung**, University of Arizona haeunchung@arizona.edu
Milan **Cirkovic**, Astronomical Observatory of Belgrade mcirkovic@aob.rs
Cressida **Cleland**, University of Birmingham cressidac@star.sr.bham.ac.uk
Rachel **Cochrane**, Center for Computational Astrophysics rcochrane@flatironinstitute.org
Matthew **Colless**, Australian National University matthew.colless@anu.edu.au
Francoise **Combes**, Observatoire de Paris francoise.combes@obspm.fr
Rodney Seayne **Cooper**, Environmental Protection Agency of Liberia drdiahl452@gmail.com
Mirko **Curti**, University of Cambridge mc2041@cam.ac.uk
Elisabete **Da Cunha**, University of Western Australia elisabete.dacunha@uwa.edu.au
Daniel **Dale**, University of Wyoming ddale@uwyo.edu
Ashkbiz **Danehkar**, University of Michigan ashkbiz.danehkar@gmail.com
Mousumi **Das**, Indian Institute of Astrophysics mousumi@iiap.res.in
Roger **Davies**, University of Oxford roger.davies@physics.ox.ac.uk
Loreany **De Araújo**, University of Sao Paulo loreanyfa@usp.br
Erwin **De Blok**, ASTRON blok@astron.nl
Elisabete M. **De Gouveia Dal Pino**, IAG-USP dalpino@iag.usp.br
Richard **de Grijs**, Macquarie University richard.de-grijs@mq.edu.au
Roelof **De Jong**, Leibniz-Institut für Astrophysik Potsdam rdejong@aip.de
Camila **De Sá Freitas**, ESO camila.desafreitas@eso.org
Antonio **De Ugarte Postigo**, Observatoire de la Côte D'Azur deugarte@oca.eu
Tim **De Zeeuw**, Leiden University tim@strw.leidenuniv.nl
Camilo **Delgado-Correal**, Francisco José de Caldas District University of Bogotá m.camilo.d@gmail.com
Caro **Derkenne**, Macquarie University caro.derkenne@hdr.mq.edu.au
Tyler **Desjardins**, STScI desjard@stsci.edu
Miroslava **Dessauges-Zavadsky**, University of Geneva miroslava.dessauges@unige.ch
Subhrata **Dey**, Astronomical observatory of Jagiellonian University sdey@oa.uj.edu.pl
María Alejandra **Díaz Teodori**, IAU madiazteo@gmail.com
Helene **Dickel**, University of Illinois hdickel2@gmail.com

Adam **Dong**, University of British Columbia adamdong@phas.ubc.ca
Cintia **Durán**, Tlaloque carrillo.du@gmail.com
Eva **Durán Camacho**, Cardiff University durancamachoe@cardiff.ac.uk
Saili **Dutta**, NCRA-TIFR, Pune sailidutta23@gmail.com
Mayssa **El Yazidi**, CISAS University of Padova mayssaelyazidi@gmail.com
Bruce **Elmegreen**, IBM Research bge@us.ibm.com
Debra **Elmegreen**, IAU elmegreen@vassar.edu
Juan **Espejo**, Swinburne University of Technology juancho9303@gmail.com
Ahmed **Estiak**, Shahjalal University of Science and Technology ahmedestiak14@gmail.com
Jakob **Faber**, McGill University/Caltech jakob.faber@mcgill.ca
Olayinka **Fagbemiro**, AWB Nigeria/African Astronomical Society yinkaojay@gmail.com
Xiaohui **Fan**, University of Arizona xfan@email.arizona.edu
Hira **Fatima**, University of Karachi hirafatima1407@gmail.com
Alieu **Faye**, Gambia National Space Research Institute camarasambakuru@yahoo.com
Laura **Ferrarese**, National Research Council of Canada lauraferrarese1@gmail.com
Ignacio **Ferreras**, Instituto de Astrofísica de Canarias iferreras@iac.es
Dana **Ficut-Vicas**, IAU Office of Astronomy for Development dana.vicas@gmail.com
Maccagni **Filippo**, ASTRON maccagni@astron.nl
Molly **Finn**, University of Virginia mf4yu@virginia.edu
Deanne **Fisher**, Swinburne University dfisher@swin.edu.au
Francesca **Fragkoudi**, ESO/Durham University francesca.fragkoudi@gmail.com
Didier **Fraix-Burnet**, CNRS didier.fraix-burnet@univ-grenoble-alpes.fr
Michiko **Fujii**, University of Tokyo fujii@astron.s.u-tokyo.ac.jp
Yutaka **Fujita**, Tokyo Metropolitan University y-fujita@tmu.ac.jp
Bryan **Gaensler**, University of Toronto bryan.gaensler@utoronto.ca
Yu **Gao**, Xiamen University yugao@xmu.edu.cn
Luz **García**, Universidad ECCI lgarciap@ecci.edu.co
Iveth Adaena **Gaspar Gorostieta**, Instituto de Astrofísica de Canarias igaspar@iac.es
Tobias **Géron**, University of Oxford tobias.geron@physics.ox.ac.uk
Etsegenet **Getachew**, Ethiopia Space Sceince and Technology Instititute tsegiastro@gmail.com
Sepideh **Ghaziasgar**, Institute for Research in Fundamental Science sepide.ghaziasgar@gmail.com
Laya **Ghodsi**, The University of British Columbia layaghodsi@phas.ubc.ca
Mahtab **Gholami**, Institute for research in fundamental sciences mhtab.gholami@gmail.com
Aishwarya **Girdhar**, European Southern Observatory, Garching bei Munich aishwarya.girdhar@eso.org
Oleksiy **Golubov**, V. N. Karazin Kharkiv National University oleksiy.golubov@karazin.ua
Carlos **Gomez-Guijarro**, CEA Saclay carlos.gomezguijarro@cea.fr
Denise R. **Gonçalves**, Federal University of Rio de Janeiro denise@ov.ufrj.br
Geraldo **Gonçalves dos Santos Junior**, Universidade de São Paulo geraldo.goncalves.santos@usp.br
Rosa Amelia **Gonzalez-Lopezlira**, Universidad Nacional Autonoma de Mexico r.gonzalez@irya.unam.mx

Kathryn **Grasha**, Australian National University kathryn.grasha@anu.edu.au
Richard **Green**, University of Arizona rgreen@arizona.edu
Isabelle **Grenier**, Université de Paris and CEA Saclay isabelle.grenier@cea.fr
Preben **Grosboel**, European Southern Observatory pgrosbol@eso.org
Marco **Grossi**, Universidade Federal do Rio de Janeiro grossi@astro.ufrj.br
Steven **Gullberg**, University of Oklahoma gullberg439@earthlink.net
Fulin **Gürsoy**, Istanbul University fulingursoy@gmail.com
Ji-Hoon **Ha**, Ulsan National Institute of Science and Technology hjhspace223@gmail.com
Massissilia **Hamadouche**, University of Edinburgh mham@roe.ac.uk
Rania **Hamdani**, Youth And Science Association of Tunisia AJST raniahamdeni28.11.00@gmail.com
Mahmoud **Hamed**, National Centre for Nuclear Research mahmoud.hamed@ncbj.gov.pl
Chris **Hamilton**, Institute for Advanced Study chamilton@ias.edu
Daniel **Han**, Yonsei University daniel.han@yonsei.ac.kr
Ilseung **Han**, Astronomy and Space Science Institute ishan@kasi.re.kr
Jimin **Han**, Kyunghee University jimin@khu.ac.kr
Jiwon **Han**, Ewha Woman University jiwon9603@naver.com
San **Han**, Yonsei University sn1994a@gmail.com
Toshihiro **Handa**, Kagoshima University handa@sci.kagoshima-u.ac.jp
Shyam **Harimohan Menon**, Australian National University shyam.menon@anu.edu.au
Syed Najamul **Hasan**, Maulana Azad National Urdu University hasan.najam@gmail.com
Martha **Haynes**, Cornell University haynes@astro.cornell.edu
Hao **He**, Mcmaster University heh15@mcmaster.ca
Julia **Healy**, ASTRON healy@astron.nl
Jakob **Helton**, University of Arizona jakobhelton@email.arizona.edu
Gerhard **Hensler**, University of Vienna gerhard.hensler@univie.ac.at
Miftahul **Hilmi,** The University of Melbourne mhilmi@student.unimelb.edu.au
Yutaka **Hirai**, University of Notre Dame/Tohoku University yhirai2@nd.edu
Kelly **Holley-Bockelma**nn, Vanderbilt University k.holley@vanderbilt.edu
Ayami **Hotta**, University of Tsukuba hotta@ccs.tsukuba.ac.jp
Marc **Huertas-Company**, IAC - Paris Observatory mhuertas@iac.es
Lucas **Hunt**, United States Naval Observatory/CPI lrhunt87@gmail.com
Deidre **Hunter**, Lowell Observatory dah@lowell.edu
Ho Seong **Hwang**, Seoul National University hwang.ho.seong@gmail.com
Jeeun **Hwang**, Seoul National University jehwang@snu.ac.kr
Narae **Hwang**, Korea Astronomy and Space Science Institute nhwang@kasi.re.kr
Sungyong **Hwang**, Seoul National University viacruxiel@snu.ac.kr
Minhee **Hyun**, Korea Astronomy and Space Science Institute minhee20400@gmail.com
Adaeze Lorreta **Ibik**, University of Toronto adaeze.ibik@mail.utoronto.ca
Kohei **Ichikawa**, Tohoku University k.ichikawa@astr.tohoku.ac.jp
Ryota **Ikeda**, NAOJ/SOKENDAI ryota.ikeda@grad.nao.ac.jp
Elizabeth J. **Iles**, University of Hokkaido iles@astro1.sci.hokudai.ac.jp
Myungshin **Im**, Seoul National University myungshin.im@gmail.com
Sang Hyeok **Im**, Seoul National University tkdgur0117@gmail.com
Susumu **Inoue**, Bunkyo University/RIKEN sinoue@bunkyo.ac.jp
Nur Hidayah **Ismail**, ISMAIL, University of Malaya hidayahismail14@gmail.com

Nariman **Ismailov**, Shamakhy Astrophysical Observatory of ANAS ismailovnshao@gmail.com
Naoki **Isobe**, Japan Aerospace Exploration Agency n-isobe@ir.isas.jaxa.jp
Pascale **Jablonka**, EPFL pascale.jablonka@epfl.ch
Faezeh **Jahediparizi**, Shahid Bahonar University of Kerman faezejhdi@yahoo.com
Joscha **Jahns**, Max-Planck-Institute for Radio Astronomy jjahns@mpifr-bonn.mpg.de
Agnieszka **Janiuk**, Center for Theoretical Physics agnes@cft.edu.pl
Atefeh **Javadi**, Institute for Research in Fundamental Sciences atefehj@gmail.com
Bangally **Jawara**, Gambia National Space Research Institute gnsribrikama@gmail.com
Sarah **Jeffreson**, Center for Astrophysics sarah.jeffreson@cfa.harvard.edu
Seyoung **Jeon**, Yonsei University syj3514@yonsei.ac.kr
Young-Beom **Jeon**, Korea Astronomy and Space Science Institute ybjeon@kasi.re.kr
Ha Yeong **Jeong**, Pusan National University gkdud1163@gmail.com
Jaehong **Jeong**, Seoul National University mre942@gmail.com
Mankeun **Jeong**, Seoul National University jmk5040@gmail.com
Taebong **Jeong**, Kyunghee University csvwnb@khu.ac.kr
Woong-Seob **Jeong**, Korea Astronomy and Space Science Institute jeongws@kasi.re.kr
Yejin **Jeong**, Kyungpook National University yeijin0585@naver.com
Yongje **Jeong**, Seoul National University stiria@snu.ac.kr
Hannah **Jhee**, University of Seoul phynah15@gmail.com
Tae-Geun **Ji,** Kyung Hee University jtg7285@gmail.com
Yongseok **Jo**, Seoul National University g.kerex@gmail.com
Yun-A **Jo**, Kyungpook National University jyajyajya@hanmail.net
Gyula I. G. **Józsa**, Max-Planck-Institut für Radioastronomie gjozsa@mpifr.de
Eunsoo **Jun**, Korea Astronomy and Space Science Institute esjun@kasi.re.kr
Dooseok **Jung**, University of Massachusetts Amherst djung@umass.edu
Minyong **Jung**, Seoul National University wispedia@snu.ac.kr
S. Lyla **Jung**, Australian National University lyla.jung@anu.edu.au
Edward **Jurua**, Mbarara University of Science and Technology ejurua@must.ac.ug
Poonam **K.C.**, Tribhuvan University punu0504@gmail.com
Melanie **Kaasinen**, European Southern Observatory melanie.kaasinen@eso.org
Hamza **Kabba**, Gambia National Space Research Institute haddyndure1983@yahoo.com
Boris Sindhu **Kalita**, CEA-Saclay, France boris.kalita@cea.fr
Joshua **Kalognia**, NRA jkalognia@st.ug.edu.gh
Da Eun **Kang**, Heidelberg University kang@uni-heidelberg.de
Gungwon **Kang**, Chung-Ang University gwkang@cau.ac.kr
Gwibong **Kang**, Chungnam National University gwibongkang@gmail.com
Hyunwoo **Kang**, Korea Astronomy and Space Science Institute orionkhw@kasi.re.kr
Michele **Kaufman**, retired kaufmanrallis@icloud.com
Balpreet **Kaur**, National Centre for Radio Astrophysics bkaur@ncra.tifr.res.in
Fatoumata **Kebe**, Association Ephemerides fatoumata.kebe@gmail.com
Ryan **Keenan**, University of Arizona rpkeenan@email.arizona.edu

Elahe **Khalouei**, Seoul national university e.khalouei1991@gmail.com
Habib **Khosroshahi**, IPM habib@ipm.ir
Chang-Goo **Kim**, Princeton University cgkim@astro.princeton.edu
Changseok **Kim**, Seoul National University kcs1996kcs@snu.ac.kr
Chulhwan **Kim**, Kyung-Hee university ch_kim@khu.ac.kr
Chun-Hwey **Kim**, Chungbuk National University kimch@chungbuk.ac.kr
Dongkok **Kim**, Seoul National University cruithne33@gmail.com
Helen **Kim**, University of California, Los Angeles helenkkim@ucla.edu
Hyowon **Kim**, UST/KASI hwkim1011@gmail.com
Jaeeun **Kim**, Kyung Hee University blue41@khu.ac.kr
Jae-Woo **Kim**, Korea Astronomy and Space Science Institute kjw0704@kasi.re.kr
Jaeyeon **Kim**, Heidelberg University kim@uni-heidelberg.de
Jaeyeong **Kim**, Korea Astronomy and Space Science Institute jaeyeong@kasi.re.kr
Jeein **Kim,** Yonsei University jeeinkim@yonsei.ac.kr
JeongCho **Kim**, Inje University jeongcho.kim@gmail.com
Jeong-Gyu **Kim**, Korea Astronomy and Space Science Institute jeonggyu.astro@gmail.com
Ji Hoon **Kim**, Seoul National University jhkim.astrosnu@gmail.com
Ji-hoon **Kim**, Seoul National University mornkr@snu.ac.kr
Jinhyub **Kim**, University of Oxford jinhyubkim89@gmail.com
Jongsoo **Kim**, Korea Astronomy and Space Science Institute jskim@kasi.re.kr
Juhan **Kim**, Korea Institute for Advanced Study kjhan0606@gmail.com
Jungha **Kim**, Korea Astronomy and Space science Institute junghakim@kasi.re.kr
Junhan **Kim**, Caltech junhank@caltech.edu
Kee-Tae **Kim**, Korea Astronomy and Space Science Institute ktkim@kasi.re.kr
Keunho **Kim**, University of Cincinnati kim2k8@ucmail.uc.edu
Minjin **Kim**, Kyungpook National University mkim.astro@gmail.com
Minsu **Kim**, Sejong University mandu447@gmail.com
Sang Chul **Kim**, Korea Astronomy and Space Science Institute sckim@kasi.re.kr
Seongjae **Kim**, Korea Astronomy and Space Science Institute seongjkim@kasi.re.kr
Seonho **Kim**, Ulsan National Institute of Science and Technology shkim0707@unist.ac.kr
Seoyoung **Kim**, Seoul National University mintlux@snu.ac.kr
Seungmin **Kim**, Pusan National University sbaru1998@gmail.com
Shinna **Kim**, Sejong University kimsn9711@gmail.com
Shinyoung **Kim**, Korea Astronomy and Space science Institute syberith@gmail.com
Soon-Wook **Kim**, Korea Astronomy and Space Science Institute xrnovae@gmail.com
Sungeun **Kim**, Sejong University sek@sejong.ac.kr
Woong-Tae **Kim**, Seoul National University unitree@snu.ac.kr
Wooseok **Kim**, Chungbuk University robin7639@gmail.com
Yeong Ill **Kim**, ChungNam University ghetro07@naver.com
Yeonsik **Kim**, Kyungpook National University wave6563@gmail.com
Yonghwi **Kim**, Korea Institute for Advanced Study yonghwi.kim@gmail.com
Yongjung **Kim**, Kyungpook National University yjkim.ast@gmail.com
Jonas **Klevas**, Vilnius University jonas.klevas@tfai.vu.lt

Johan **Knapen**, Instituto de Astrofisica de Canarias johan.knapen@iac.es
Eunhee **Ko**, Seoul National University eunhee.ko.astro@gmail.com
Jongwan **Ko**, Korea Astronomy and Space Science Institute jwko@kasi.re.kr
Jin **Koda**, Stony Brook University jin.koda@stonybrook.edu
Keiichi **Kodaira**, Sokendai kodaira_keiichi0815@soken.ac.jp
Donghyeok **Koh**, Korea Astronomy and Space Science Institute dhkoh@kasi.re.kr
Kotaro **Kohno**, University of Tokyo kkohno@ioa.s.u-tokyo.ac.jp
Lena **Komarova**, University of Michigan komarova@umich.edu
Toktarkhan **Komesh**, Al-Farabi Kazakh National University toktarkhan.komesh@nu.edu.kz
Ayu **Konishi**, Osaka prefecture university ayu.kori818@gmail.com
Bon-Chul **Koo**, Seoul National University koo@astro.snu.ac.kr
Hyeonmo **Koo**, Universe of Seoul mike1919@naver.com
Timea **Kovacs**, MPIfR tkovacs@mpifr-bonn.mpg.de
Renée C. **Kraan-Korteweg**, University of Cape Town kraan@ast.uct.ac.za
Diederik **Kruijssen**, Heidelberg University kruijssen@uni-heidelberg.de
Gajanan **Kulkarni**, Korea Astronomy and Space Science Institute kpgajanan@kasi.re.kr
Kyujin **Kwak**, Ulsan National Institute of Science and Technology kkwak@unist.ac.kr
Woojin **Kwon**, Seoul National University wkwon@snu.ac.kr
Mathieu **Langer**, Université Paris-Saclay mathieu.langer@universite-paris-saclay.fr
John **Lattanzio**, Monash University john.lattanzio@monash.edu
Lauren **Laufman**, University of Minnesota Twin Cities laufm008@umn.edu
Marco **Laversveiler**, Federal University of Rio de Janeiro marcoaurelio18@astro.ufrj.br
Blake **Ledger**, McMaster University ledgeb1@mcmaster.ca
Nicolas **Ledos**, Osaka University ledos@astro-osaka.jp
Bomee **Lee**, Korea Astronomy and Space Science Institute bomee@kasi.re.kr
Chang Won **Lee**, Korea Astronomy and Space Science Institute cwl@kasi.re.kr
Gain **Lee**, Seoul National University wag.ur.coke@gmail.com
Jaehyun **Lee**, Korea Institute of Advanced Study jaehyun@kias.re.kr
Jeong Hwan **Lee**, Seoul National University joungh93@snu.ac.kr
Jeong-Eun **Lee**, Kyung Hee University jeongeun.lee@khu.ac.kr
Jong Chul **Lee**, Korea Astronomy and Space Science Institute jclee@kasi.re.kr
Joohyun **Lee**, University of Texas at Austin joohyun.lee@austin.utexas.edu
Minseon **Lee**, Kyung Hee University mslee@khu.ac.kr
Min-Young **Lee**, Korea Astronomy and Space Science Institute mlee@kasi.re.kr
Myung Gyoon **Lee**, Seoul National University mglee@astro.snu.ac.kr
Sang Hyun **Lee**, Korea Astronomy and Space Science Institute shlee@kasi.re.kr
Seokho **Lee**, Korea Astronomy and Space Science Institute seokholee@kasi.re.kr
Seong-Kook **Lee**, Seoul National University s.joshualee@gmail.com
Seungjae **Lee**, Seoul National University balgun1004@gmail.com
Sieun **Lee**, Kyunghee University dltldms0113@gmail.com
TaeYong **Lee**, KyungHee University tylee@khu.ac.kr

Woojin **Lee**, UNIST woojoowj123@unist.ac.kr
Yun Hee **Lee**, Korea Astronomy and Space Science Institute yhinjesus@kasi.re.kr
Lerothodi **Leeuw**, University of the Western Cape Lerothodi@alum.mit.edu
Claus **Leitherer**, Space Telescope Science Institute leitherer@stsci.edu
Adam **Leory**, Ohio State University leroy.42@osu.edu
Chun Sing **Leung**, Hong Kong Polytechnic University csinleung@polyu.edu.hk
Ho-Hin **Leung**, University of St Andrews hhl1@st-andrews.ac.uk
James **Leung**, University of Sydney jleu9465@uni.sydney.edu.au
Benjamin **L'Huillier**, Sejong University benjamin@sejong.ac.kr
Kunyang **Li**, Institut d'Astrophysique de Paris kunyangli356@outlook.com
Shanghuo **Li**, KASI shanghuo.li@gmail.com
Fuheng **Liang**, University of Oxford fuheng.liang@physics.ox.ac.uk
Li-Wen **Liao,** National Tsing Hua University liwen@gapp.nthu.edu.tw
Seunghwan **Lim,** CITA, Univ. of Toronto shlim1206@gmail.com
Sungsoon **Lim**, Yonsei University sslim00@gmail.com
Wanggi **Lim**, USRA wanggi.lim.astro@gmail.com
Chia-Lung **Lin**, National Central University m1059006@gm.astro.ncu.edu.tw
Lihwai **Lin**, ASIAA lihwailin@asiaa.sinica.edu.tw
Yen-Ting **Lin**, Academia Sinica ytl@asiaa.sinica.edu.tw
Maria Luiza **Linhares Dantas**, Nicolaus Copernicus Astronomical Center mlldantas@protonmail.com
Carlos **Lopez-Coba**, ASIAA calopez@asiaa.sinica.edu.tw
Elismar **Lösch**, University of São Paulo elismar.l@usp.br
James **Lowenthal**, Smith College jlowenth@smith.edu
Aran **Lyo,** Korea Astronomy and Space Science Institute arl@kasi.re.kr
Yik Ki **Ma**, Australian National University yikki.ma@anu.edu.au
Fumiya **Maeda**, University of Tokyo fmaeda@ioa.s.u-tokyo.ac.jp
Vincenzo **Mainieri**, ESO vmainier@eso.org
Katarzyna **Malek**, National Centre for Nuclear Research Katarzyna.Malek@ncbj.gov.pl
Bhupendra **Malvi,** Barkatullah University Bhup1201@gmail.com
Amit Kumar **Mandal**, Seoul National University amitastro.am@gmail.com
Yunus **Manjoo**, African Astronomical Society yunus.manjoo@afasociety.org
Lachlan **Marnoch**, Macquarie University lachlan.marnoch@hdr.mq.edu.au
Isabel **Marquez**, IAA-CSIC isabel@iaa.es
Garreth **Martin**, KASI, University of Arizona garreth.martin@ntlworld.com
Nicolas **Martinet**, Aix Marseille Université nicolas.martinet@lam.fr
Josefa **Masegosa**, IAA-CSIC pepa@iaa.es
Davide **Massari**, INAF davide.massari@inaf.it
Karen **Masters**, Haverford College klmasters@haverford.edu
Demetrios **Matsakis**, Masterclock, Inc. dnmyiasou@yahoo.com
Hidenori **Matsui**, Asahikawa College matsui@asahikawa-nct.ac.jp
Ren **Matsusaka**, Kagoshima University k9435354@kadai.jp
Jorryt **Matthee**, ETH Zurich jorryt@gmail.com
Giancarlo **Mattia**, Max Planck Institute for Astronomy mattia@mpia.de
Karin **Menendez-Delmestre**, Federal University of Rio de Janeiro kmd@astro.ufrj.br
Alejandra **Meza**, University of Chile ale.mravest@gmail.com
Areg **Mickaelian**, Byurakan Astrophysical Observatory aregmick@yahoo.com
Gor **Mikayelyan**, NAS RA Byurakan Astrophysical Observatory gormick@mail.ru

Daniel **Mikkola**, Lund University mikkola@astro.lu.se
Amy **Miller**, Macquarie University amyelizmiller@gmail.com
Bryan **Miller**, Gemini Observatory/NOIRLab bryan.miller@noirlab.edu
Elena **Milyute**, theoretical physicist litavem3@gmail.com
Ikki **Mitsuhashi**, the University of Tokyo ikki.mitsuhashi@grad.nao.ac.jp
Yusuke **Miyamoto**, Fukui University of Technology yusuke.miyamot@gmail.com
Rajsekhar **Mohapatra**, Australian National University rajsekhar.mohapatra@anu.edu.au
Jun-Sung **Moon**, Seoul National University jsmoon.astro@gmail.com
Sanghyuk **Moon**, Seoul National University s.moon@snu.ac.kr
Christopher **Moore**, Harvard-Smithsonian CfA Christopher.s.moore@cfa.harvard.edu
Jorge **Moreno**, Pomona College jorge.moreno@pomona.edu
Masao **Mori,** University of Tsukuba mmori@ccs.tsukuba.ac.jp
Skarleth **Motino-Flores**, USRA/NASA Ames Research Center skarlethgm@gmail.com
Marcie **Mun**, Australian National University jaeyeon.mun@anu.edu.au
Jeff **Munn**, U.S. Naval Observatory jeffrey.a.munn2.civ@us.navy.mil
Takeru **Murase**, Kagoshima University takerun.charvel@gmail.com
Chandrashekar **Murugeshan**, CSIRO chandrashekar.murugeshan@csiro.au
Rahul **Musale**, University of Pune rahulmusale1100@gmail.com
Kentaro **Nagamine**, Osaka University kn@astro-osaka.jp
Masahiro **Nagashima**, Bunkyo University masahiro@bunkyo.ac.jp
David **Nagy**, University of Geneva nagydavid@hotmail.ch
Roman **Nagy**, Comenius University in Bratislava roman.nagy@fmph.uniba.sk
Lilianne **Nakazono**, Universidade de São Paulo lilianne.nakazono@usp.br
Themiya **Nanayakkara**, Swinburne University of Technology wnanayakkara@swin.edu.au
Norma Araceli **Nava Moreno**, Instituto Nacional de Astrofísica, Óptica y Electrónica nava14@inaoe.mx
Alexander **Navarre**, University of Cincinnati navarrae@mail.uc.edu
Patrick **Neunteufel**, Max Planck Institute for Astrophysics pneun@mpa-garching.mpg.de
Ngoc **Nguyen**, Vietnam National Space Center capi37capi@gmail.com
Thi Phuong **Nguyen**, Korea Astronomy and Space Science Institute tpnguyen@kasi.re.kr
Giang **Nguyen Chau**, Korea University of Science and Technology chaugiang@kasi.re.kr
Nathalie **Nguyen-Quoc Ouellette**, Université de Montréal nathalie.ouellette.2@umontreal.ca
Anna **Niemiec**, Durham University anna.niemiec@durham.ac.uk
Masafumi **Noguchi**, Tohoku University noguchi@astr.tohoku.ac.jp
Joseph **Nuth**, NASA Goddard Space Flight Center joseph.a.nuth@nasa.gov
Ikechukwu **Obi**, National Space Research and Development Agency tonykassidy_z@yahoo.com
Caitlin **O'Brien**, The Ohio State University obrien.847@osu.edu
Sally **Oey**, University of Michigan msoey@umich.edu
Onuche **Ogu**, NASRDA/AWBNigeria onucheogu@gmail.com
Boon Kiat **Oh**, Seoul National University ohboonkiat@gmail.com
Heeyoung **Oh**, Korea Astronomy and Space Science Institute hyoh@kasi.re.kr
Junghwan **Oh**, Sejong University joh@sejong.ac.kr
Kyuseok **Oh**, Korea Astronomy and Space Science Institute oh@kasi.re.kr

Se-Heon **Oh**, Sejong University seheon.oh@sejong.ac.kr
Katsuya **Okoshi**, Tokyo University of Science okoshi@rs.tus.ac.jp
Yuri **Oku**, Osaka University oku@astro-osaka.jp
Natanael **Oliveira**, Federal University of Rio de Janeiro natanael18@astro.ufrj.br
Taira **Oogi**, Ehime University oogi@cosmos.phys.sci.ehime-u.ac.jp
Yasna **Ordenes Briceño**, Pontifical Catholic University of Chile yordenes@astro.puc.cl
John **Orlowski-Scherer**, University of Pennsylvania jorlo@sas.upenn.edu
Mabel **Osorio**, UNAM jmosorio@astro.unam.mx
Eve **Ostriker**, Princeton University eco@astro.princeton.edu
Koki **Otaki**, University of Tsukuba otaki@ccs.tsukuba.ac.jp
Juergen **Ott**, National Radio Astronomy Observatory jott@nrao.edu
Justin **Otter**, Johns Hopkins University jotter2@jhu.edu
Riley **Owens**, University of Cincinnati m.riley.owens@gmail.com
Hamsa **Padmanabhan**, Université de Genève hamsa.padmanabhan@unige.ch
Gregory **Paek**, Seoul National University gregorypaek94@gmail.com
Insu **Paek**, Seoul National University insupaek@snu.ac.kr
Mina **Pak**, Korea Astronomy and Space Science Institute minapak@kasi.re.kr
Soojong **Pak**, Kyung Hee University soojong@khu.ac.kr
Hsi-An **Pan**, Tamkang University hapan@gms.tku.edu.tw
Zhizheng **Pan**, Purple Mountain Observatory panzz@pmo.ac.cn
Divya **Pandey**, National Institute of Technology Rourkela divyapandey1212@gmail.com
Ayush **Pandhi**, University of Toronto ayush.pandhi@mail.utoronto.ca
Byeong-Gon **Park**, Korea Astronomy and Space Science Institute bgpark@kasi.re.kr
Changbom **Park**, Korea Institute for Advanced Study cbp@kias.re.kr
Gyueun **Park**, University of Science and Technology KASI gpark@kasi.re.kr
Hye-Jin **Park**, The Australian National University hyejin.park@anu.edu.au
Jaehong **Park**, Korea Institute for Advanced Study jaehongpark@kias.re.kr
So-Myoung **Park**, Korea Astronomy and Space Science Institute smpark@kasi.re.kr
Woojin **Park**, Korea Astronomy and Space Science Institute wjpark@kasi.re.kr
Kaelee **Parker**, University of Texas at Austin kaelee.parker@utexas.edu
Bogdan Adrian **Pastrav**, Institute of Space Science bapastrav@spacescience.ro
Atharva **Patil**, National Central University, Taiwan d1079002@gm.astro.ncu.edu.tw
Amber **Patten**, Global Family Alp-love@gmx.com
Sanjaya **Paudel**, Yonsei University sanjpaudel@gmail.com
William **Pearson**, National Center for Nuclear Research william.pearson@ncbj.gov.pl
Zhiyuan **Pei**, Guangzhou University matt.pui.astro@gmail.com
Yingjie **Peng**, Kavli Institute for Astronomy and Astrophysics, Peking University yjpeng@pku.edu.cn
Michele **Perna**, Centro de Astrobiología michele.perna@cab.inta-csic.es
Annalisa **Pillepich**, Max Planck Institute for Astronomy pillepich@mpia.de
Francesca **Pinna**, Max Planck Institute for Astronomy francescapinna9@gmail.com
Sandor **Pinter**, University of Public Service pinter.sandor@uni-nke.hu
Julia **Piotrowska**, Jagiellonian University jpiotrowska@byk.oa.uj.edu.pl
Bianca **Poggianti**, INAF-Osservatorio Astronomico di Padova bianca.poggianti@inaf.it

Hom Bahadur **Pokhrel,** St. Xavier's College pokhrelsanjay3@gmail.com
Mamta **Pommier**, CNRS mamtapan@gmail.com
Emanuela **Pompei**, ESO epompei@eso.org
Cristina **Popescu**, University of Central Lancashire cpopescu@uclan.ac.uk
Sergey **Popov**, Lomonosov Moscow State University polar@sai.msu.ru
Antonio J. **Porras-Valverde**, Vanderbilt University antonio.j.porras@vanderbilt.edu
Ana Carolina **Posses Nascimento**, Universidad Diego Portales ana.posses@mail.udp.cl
Anna **Queiroz**, Leibniz Institute for Astrophysics Potsdam -AIP aqueiroz@aip.de
Miora **Rakototafika**, University of Antananarivo minomrak@gmail.com
Varsha **Ramachandran**, Heidelberg University vramachandran@uni-heidelberg.de
Jenny **Ramos Lázaro**, National University of San Marcos jenny.ramos@unmsm.edu.pe
Zara **Randriamanakoto**, South African Astronomical Observatory zara@saao.ac.za
Adarsh **Ranjan**, Korea Astronomy and Space Science Institute ranjan@kasi.re.kr
Tim **Rawle**, European Space Agency tim.rawle@esa.int
Soo-Chang **Rey**, Chungnam National University screy86@gmail.com
Jinsu **Rhee**, Yonsei University jinsu.rhee@gmail.com
Mateus **Ribeiro**, Universidade Federal do Rio de Janeiro mribeiro@astro.ufrj.br
Gabriele **Riccio**, National Centre of Nuclear Research, Warsaw gabriele.riccio@ncbj.gov.pl
Dominik **Riechers**, Universitaet zu Köln riechers@ph1.uni-koeln.de
Joshua **Roberson**, University of Cincinnati robersju@mail.uc.edu
Victor **Robles**, Yale University victoresfm@hotmail.com
Erik **Rodrigues de Lima**, Universidade de São Paulo erik.vini@usp.br
Jimena **Rodríguez**, CONICET – UNLP jimerod@gmail.com
Joel **Roediger**, National Research Council Canada Joel.Roediger@nrc-cnrc.gc.ca
Ciaran **Rogers**, Leiden University ciaran-rogers@hotmail.com
Duk-Gyoo **Roh**, KASI dgroh@kasi.re.kr
Anish **Roshi**, Arecibo Observatory anish.roshi@gmail.com
Arpita **Roy**, Scuola Normale Superiore di Pisa, Italy arpita.roy@sns.it
Monica **Rubio**, Universidad de Chile mrubio@das.uchile.cl
Raquel **Ruiz Valença**, Universidade de São Paulo vruizraquel@usp.br
Michael **Rutkowski**, Minnesota State University michael.rutkowski@mnsu.edu
Stuart **Ryder**, Macquarie University Stuart.Ryder@mq.edu.au
Bohyun **Ryou**, Kyung Hee University rbohyun@khu.ac.kr
Taeho **Ryu**, The Max Planck Institute for Astrophysics tryu@slogin.mpa-garching.mpg.de
Elaine **Sadler**, University of Sydney elaine.sadler@sydney.edu.au
Teymoor **Saifollahi**, University of Groningen teymur.saif@gmail.com
Gh. **Saleh**, Saleh Research Centre postmaster@saleh-theory.com
Mambo Abdallah **Salum**, Mbeya University of Science and Technology salummambo7@gmail.com
Anahit **Samsonyan**, Byurakan Astrophysical Observatory anahit.sam@gmail.com
Pablo Manuel **Sánchez Alarcón**, Instituto de Astrofísica de Canarias pablo.manuel.sanchez.alarcon@iac.es
Sebastian Francisco **Sanchez Sanchez**, IA - UNAM sfsanchez@astro.unam.mx
Divita **Saraogi**, IIT BOMBAY divitasaraogi@iitb.ac.in
Craig **Sarazin**, University of Virginia sarazin@virginia.edu

Clifford Alanmayer **Sarfo**, BOZBEC MAYERS alancash777@gmail.com
Purnasha **Sarkar**, Indira Gandhi National Open University purnasha31@gmail.com
Asako **Sato**, Kyushu University sato.asako.322@s.kyushu-u.ac.jp
Genta **Sato**, Tohoku University g.sato@astr.tohoku.ac.jp
Christoph **Saulder**, KASI csaulder@kasi.re.kr
Eva **Schinnerer**, MPIA schinner@mpia.de
Jennifer **Schober**, EPFL Schober.Jen@gmail.com
Anshuraj **Sedai**, Nepal Astronomical Society sedaianshuraj@gmail.com
Peter **Senchyna**, Carnegie Institution for Science psenchyna@carnegiescience.edu
Mira **Seo**, Pusan National University mrseo@pusan.ac.kr
Ji Yeon **Seok**, Korea Astronomy and Space Science Institute jyseok@kasi.re.kr
Marta **Sewilo**, NASA/University of Maryland mmsewilo@gmail.com
Priya **Shah**, Maulana Azad National Urdu University priya.hasan@gmail.com
Suruchi **Shahi**, St. Xavier's College suruchishahi99@gmail.com
Samaneh **Shamyati**, Institute for research in fundamental sciences samaneh.shamyati93@gmail.com
Piyush **Sharda**, Australian National University piyush.sharda@anu.edu.au
Chelsea **Sharon**, Yale-NUS College chelsea.sharon@yale-nus.edu.sg
Yun-Kyeong **Sheen**, Korea Astronomy and Space Science Institute yksheen@kasi.re.kr
Hyunjin **Shim**, Kyungpook National University hjshim@knu.ac.kr
Takashi **Shimonishi**, Niigata University shimonishi@env.sc.niigata-u.ac.jp
Eun-jin **Shin**, Seoul National University shinej816@snu.ac.kr
Jihye **Shin**, KASI jhshin.jhshin@gmail.com
Zihey **Shin**, Korea Astronomy and Space Science Institute zhshin@kasi.re.kr
Yutaka **Shiratori**, Tokai University shiratori.star@gmail.com
Bekdaulet **Shukirgaliyev**, Nazarbayev University bekdaulet.shukirgaliyev@nu.edu.kz
Mattia **Sirressi**, Stockholm University mattia.sirressi@astro.su.se
J. Allyn **Smith**, Austin Peay State University smithj@apsu.edu
Tony **Sohn**, Space Telescope Science Institute tsohn@stsci.edu
Manuel **Solimano**, Universidad Diego Portales manuel.solimano@mail.udp.cl
Rachel **Somerville**, Flatiron Institute rsomerville@flatironinstitute.org
Changhee **Son**, UNIST ckdgml0424@unist.ac.kr
Alessandro **Sonnenfeld**, Leiden University sonnenfeld@strw.leidenuniv.nl
Kazuo **Sorai**, Hokkaido University sorai@phys.sci.hokudai.ac.jp
Roberto **Soria**, University of the Chinese Academy of Sciences rsoria@nao.cas.cn
Essa **Sowe**, Gambia National Space Research Institute abdoukadrydrame@yahoo.com
Essa **Sowe**, Gambia National Research Institute writegdy2007@yahoo.com
Navin **Sridhar**, Columbia University navinsridhar@gmail.com
Hannah **Stacey**, Max Planck Institute for Astrophysics hrstacey@icloud.com
Benjamin **Stappers**, University of Manchester ben.stappers@manchester.ac.uk
Tjitske **Starkenburg**, Northwestern University t.starkenburg@gmail.com
Eckhard **Sturm**, Max Planck Institute for Extraterrestrial Physics sturm@mpe.mpg.de
Akshaya **Subbanna**, Korea Astronomy and Space Science Institute akshayams@kasi.re.kr
Janette **Suherli**, University of Manitoba suherlij@myumanitoba.ca
Ivana **Sulaver**, University of Nova Gorica sullaweri@gmail.com
Nofoz **Suleiman**, ELTE university n.suleiman@ttk.elte.hu

Danny **Summers**, Memorial University of Newfoundland dsummers@mun.ca
Jiayi **Sun**, McMaster University sun208@mcmaster.ca
Yang **Sun**, University of Arizona sunyang@email.arizona.edu
Unnikrishnan Potty **Sureshkumar**, Jagiellonian University uk.potty.s@doctoral.uj.edu.pl
Kseniia **Sysoliatina**, Astronomisches Rechen-Institut Sysoliatina@uni-heidelberg.de
Sima **Taefi Aghdam**, Institute for Research in Fundamental Sciences sima.t.aghdam@gmail.com
Charles **Takalana**, African Astronomical Society charles.takalana@afasociety.org
Norimi **Takeba**, Kagoshima University n34.bamu.lss@gmail.com
Peng Kian **Tan**, National University of Singapore cqttpk@nus.edu.sg
Nahathai **Tanakul**, National Astronomical Research Institute of Thailand ntanakul@gmail.com
Elizabeth **Taylor**, University of Nottingham elizabeth.taylor@nottingham.ac.uk
Jerusalem Tamirat **Teklu**, Ariel University jerusalemt@ariel.ac.il
Grace **Telford**, Rutgers University grace.telford@rutgers.edu
Yu-Hsuan **Teng**, University of California at San Diego elthateng@gmail.com
Solomon **Tessema**, Ethiopian Space Science and Technology Institute researchspace816@gmail.com
Christina **Thöne**, Astronomical Institute of the Czech Academy of Sciences christina.thoene@gmail.com
Yong **Tian**, Natioanl Central University yongtian@astro.ncu.edu.tw
Roland **Timmerman**, Leiden University rtimmerman@strw.leidenuniv.nl
Yuan-Sen **Ting**, Australian National University yuan-sen.ting@anu.edu.au
Imen **Titouhi**, Tunis science city imen.titouhi@cst.rnu.tn
Kazuki **Tokuda**, Kyushu University tokuda@p.s.osakafu-u.ac.jp
Maryam **Torki**, IPM maryamtorki84@gmail.com
Silvia **Torres-Peimbert**, Universidad Nacional Autonoma de Mexico silvia@astro.unam.mx
L. Viktor **Tóth**, Eötvös University Budapest l.v.toth@astro.elte.hu
Scott **Trager**, University of Groningen sctrager@astro.rug.nl
Kim-Vy **Tran**, University of New South Wales kimvy.tran@gmail.com
Tony **Travouillon**, The Australian National University tony.travouillon@anu.edu.au
Patrick **Treuthardt**, North Carolina Museum of Natural Sciences patrick.treuthardt@naturalsciences.org
Dian **Triani**, Australian National University dian.triani@anu.edu.au
Masato **Tsuboi**, Japan Aerospace Exploration Agency tsuboi@vsop.isas.jaxa.jp
Takuji **Tsujimoto**, National Astronomical Observatory of Japan taku.tsujimoto@nao.ac.jp
Devendra Raj **Upadhyay**, Tribhuvan University devendra.upadhyay@ac.tu.edu.np
Sheona **Urquhart**, The Open University sheona.urquhart@open.ac.uk
David **Valcin**, Ohio University d.valcin@ohio.edu
Thijs **van der Hulst**, University of Groningen vdhulst@astro.rug.nl
Joshiwa **Van Marrewijk**, European Southern Observatory joshiwa.vanmarrewijk@eso.org
Carlos **Vargas**, University of Arizona cjvargas@email.arizona.edu
Lourdes **Verdes-Montenegro**, Instituto de Astrofísica de Andalucía lourdes@iaa.es
Marc **Verheijen**, University of Groningen verheyen@astro.rug.nl
Simone **Veronese**, ASTRON veronese@astron.nl
Marcelo **Vicentin**, University of São Paulo marcelo.vicentin@usp.br
Frederick **Vrba**, US Naval Observatory frederick.j.vrba.civ@us.navy.mil
Mike **Walmsley**, University of Manchester michael.walmsley@manchester.ac.uk

Di **Wang**, University of Sydney di.wang@sydney.edu.au
Zixian **Wang**, University of Sydney zwan0382@uni.sydney.edu.au
Jacob **Ward**, Univeristy of Maryland jacob.wolfgang.ward@gmail.com
Benjamin **Wehmeyer**, Konkoly Obs/Univ of Hertfordshire benjamin.wehmeyer@csfk.org
Patricia **Whitelock**, SAAO and UCT paw@saao.ac.za
Antonia **Wilmot**, H M Nautical Almanac Office toni.wilmot@ukho.gov.uk
Dejene Zewdie **Woldeyes**, Universidad Diego Portales dejene.woldeyes@mail.udp.cl
O. Ivy **Wong**, CSIRO ivy.wong@csiro.au
Tony **Wong**, University of Illinois wongt@illinois.edu
Jong-Hak **Woo**, Seoul National University woo@astro.snu.ack.kr
Angus **Wright**, Ruhr-University Bochum awright@astro.rub.de
Po-Feng **Wu**, National Taiwan University winder9255@gmail.com
Zeto **Xia**, Masaryk University, Czech Republic zeto.xia@gmail.com
Dandan **Xu**, Tsinghua University dandanxu@tsinghua.edu.cn
Jyoti **Yadav**, Indian Institute Of Astrophysics yadavjyoti636@gmail.com
Yana **Yakushina**, University of Ghent ya.yakushina@gmail.com
Yoshiyuki **Yamada**, Kyoto University yamada@scphys.kyoto-u.ac.jp
Soung-Chul **Yang**, Korea Astronomy and Space Science Institute sczoo@kasi.re.kr
Chikako **Yasui**, NAOJ ck.yasui@nao.ac.jp
Sibaek **Yi**, Kyung Hee University sibaekyi@khu.ac.kr
Sukyoung **Yi**, Yonsei University yi@yonsei.ac.kr
Kijeong **Yim**, Chungnam National University kijeong.yim@gmail.com
Yoshichika **Yokoe**, University of Tokyo y yokoe@icrr.u-tokyo.ac.jp
Suk Yee **Yong**, CSIRO yongsukyee@gmail.com
Jaewon **Yoo**, KASI/UST jwyoo@kasi.re.kr
Heesun **Yoon**, Ulsan National Institute of Science and Technology gmltjs315@gmail.com
Hyein **Yoon**, University of Sydney hyein.yoon@sydney.edu.au
Jeongkwan **Yoon**, Ulsan National Institute of Science and Technology logicyoon@unist.ac.kr
Suk-Jin **Yoon**, Yonsei University sjyoon0691@yonsei.ac.kr
Michitoshi **Yoshida**, National Astronomical Observatory of Japan michitoshi.yoshida@nao.ac.jp
Hyun Jung **Youn**, Chungnam National University 17171771@o.cnu.ac.kr
Qingzheng **Yu**, Xiamen University yuqingzheng@stu.xmu.edu.cn
Javier **Zaragoza Cardiel**, INAOE javier.zaragoza@inaoep.mx
Jorge **Zavala**, NAOJ jorgea.zavalas@gmail.com
Woong-bae **Zee**, Yonsei Univ galaxy.wb.zi@gmail.com
Haiyan **Zhang**, National Astronomical Observatories of CAS hyzhang@bao.ac.cn
Jielai **Zhang**, Swinburne University of Technology zhang.jielai@gmail.com
Jun **Zhang**, Shanghai Jiao Tong University betajzhang@sjtu.edu.cn
Shuang **Zhou**, University of Nottingham Shuang.Zhou@nottingham.ac.uk
Irina **Zhuravleva**, University of Chicago zhuravleva@astro.uchicago.edu
Tuila **Ziliotto**, Pontificia Universidad Catolica de Chile ziliottotuila@gmail.com
Catherine **Zucker**, Space Telescope Science Institute czucker@stsci.edu

Chapter 1:
Scales of Star Formation:
From Molecular Cores to Galaxies

Resolving the Rise and Fall of Star Formation in Galaxies
Proceedings IAU Symposium No. 373, 2023
T. Wong & W.-T. Kim, eds.
doi:10.1017/S1743921322003830

The local and global relations between $\Sigma_\star$, $\Sigma_{\rm SFR}$ and $\Sigma_{\rm mol}$ that regulate star-formation

Sebastián F. Sánchez[1], Daysi C. Gómez Medina[2], J.K. Barrera-Ballesteros[1], L. Galbany[3], A. Bolatto[4] and T. Wong[5]

[1]Instituto de Astronomía,
Universidad Nacional Autónoma de México,
A. P. 70-264, C.P. 04510, Ciudad de México, Mexico
email: sfsanchez@astro.unam.mx

[2]Facultad de Ciencias Espaciales,
Universidad Nacional Autónoma de Honduras

[3]Institute of Space Sciences (ICE, CSIC), Campus UAB, Carrer de Can Magrans, s/n,
E-08193 Barcelona, Spain.

[4]Department of Astronomy, University of Maryland,
College Park, MD 20742, USA

[5]Department of Astronomy, University of Illinois,
Urbana, IL 61801, USA

Abstract. Star-formation is one of the main processes that shape galaxies, defining its stellar population and metallicity production and enrichment. It is nowadays known that this process is ruled by a set of relations that connect three parameters: the molecular gas mass, the stellar mass and the star-formation rate itself. These relations are fulfilled at a wide range of scales in galaxies, from galaxy wide to kpc-scales. At which scales they are broken, and how universal they are (i.e., if they change at different scales or for different galaxy types) it is still an open question. We explore here how those relations compare at different scales using as proxy the new analysis done using Integral Field Spectroscopy data and CO observations data from the EDGE-CALIFA survey and the AMUSSING++ compilation.

Keywords. galaxies: evolution, galaxies: ISM, galaxies: TBW, techniques: spectroscopic

1. Introduction

One of the main processes that shape a galaxy is indeed the star-formation, defining its very existence. A certain set of physical conditions are required to trigger the thermonuclear reactions that define a star, starting from the fueling of atomic gas, its cooling and condensation to generate molecular gas clouds, and the fragmentation of those clouds and its collapse to reach the required physical conditions to ignite them. All those conditions shape a set of relations between the star-formation rage itself, the gas mass and the stellar mass content. Those relation are nowadays evident that regulates the star-formation process.

Schmidt (1959) first proposed a relation between the SFR and the density of gas in a certain volume, based on theoretical considerations. This relation was later analyzed in Schmidt (1968), but it was not until Kennicutt (1998a) that it was not expressed in its current form: a log-log (or power law) relation between ($\Sigma_{\rm SFR}$ and $\Sigma_{\rm mol}$), i.e., two intensive parameters that do not depend on the size of galaxies. In Kennicutt (1998b) it was predicted an slope of ∼1.4 for this relation, based on free-fall time considerations for a self-collapsion cloud. Although it was first derived as a relation between global intensive parameters, it is now known to hold above the typical scale of large molecular clouds (∼500 pc), i.e., described as a relation for spatially resolved sub-galactic structures (e.g., the rSK-law Wong & Blitz 2002; Kennicutt et al. 2007). Both the global and the resolved relations present a similar dispersion, of the order of ∼0.2 dex (Bigiel et al. 2008; Leroy et al. 2013). However, there is discrepancy in the slope with respect to the theoretical considerations, being nowadays more near to a slope ∼1 (e.g. Sánchez 2020; Sánchez et al. 2021a).

The relation between the star-formation rate (SFR) and the stellar mass (M_*) has been explored uncovered more recently, based on the exploration of large galaxy surveys such as the SDSS (York et al. 2000). This relation, known as the Star-Formation Main Sequence (e.g. Brinchmann et al. 2004; Renzini & Peng 2015) is a tight correlation (σ ∼0.25 dex) between the logarithm of both quantities, with a slope near to one, observed only for star-forming galaxies (i.e., those which ionization is clearly dominated by the effect of young massive OB stars). It has been observed in a wide range of redshifts, explored in more detail in the nearby Universe ($z \sim 0$), but present up to redshift 1-2 (e.g. Speagle et al. 2014; Rodríguez-Puebla et al. 2016). It presents a clear evolution, at least in its zero-point, as galaxies increase their stellar masses and decreases their SFR from early cosmological times to the present epoch. Two almost simultaneous studies, Sánchez et al. (2013) and Wuyts et al. (2013), proposed the existence of resolved version of this relation that holds down to ∼1kpc scales, the rSFMS, observed as a tight relation between the logarithms of the $\Sigma_{\rm SFR}$ and Σ_*. Detailed explorations, like the one presented by Cano-Díaz et al. (2016a), using data from the CALIFA survey (Sánchez et al. 2012), has estabilished definetively its shape and characteristics. A possible dependence of that relation with other properties of the galaxies, such as the morphology, has been explored by different authors (e.g. González Delgado et al. 2016; Catalán-Torrecilla et al. 2017; Cano-Díaz et al. 2019; Méndez-Abreu et al. 2019).

A third relation between the molecular gas and stellar masses has been described for SFGs too (e.g. Saintonge et al. 2016; Calette et al. 2018). Like the previous two it is a tight relation with a slope near to one. This relation, known as the Molecular Gas Main Sequence (MGMS), has been less frequently explored in the literature. It has not been until recent times that its resolved counterpart has been explored at a kpc scale, (rMGMS Lin et al. 2019), as a relation between the corresponding intensive parameters: $\Sigma_{\rm mol}$ vs. Σ_*, using a combination of IFS data provided by the MaNGA survey (Bundy et al. 2015), and CO-mapping provided by the ALMAQUEST compilatoin (Lin et al. 2020).

The connection between the global (galaxy-wide) and resolved (kpc-scale) relations, its universality, the possible dependence with additional parameters, and the interconnection or hierarchy between the three relations is a topic of study that it is still open. Bolatto et al. (2017), first shown that a simple parametrization of the global intensive SK-relation follows the same distribution observed in the rSK between $\Sigma_{\rm SFR}$ and $\Sigma_{\rm mol}$. More detailed explorations have shown the same correspondence between the global extensive SFMS and the local/resolved rSFMS (Pan et al. 2018; Cano-Díaz et al. 2019). Finally, Sánchez et al. (2021a) and Sánchez et al. (2021c), have shown that the global intensive relations (i.e., when $\Sigma_{\rm SFR}$, $\Sigma_{\rm mol}$ and Σ_* are measured as average quantities galaxy wide) and the local/resolved relations (i.e., when those parameters are measured in star-forming regions at a kpc-scale), present the same distributions, the same slopes and similar zero-points. These explorations were possible due to the unique combination of spatially resolved observations of large sample of galaxies using both optical IFS and CO millimetric data, and the use of recent tracers of the gas content based on dust attenuation values (e.g. Barrera-Ballesteros et al. 2020).

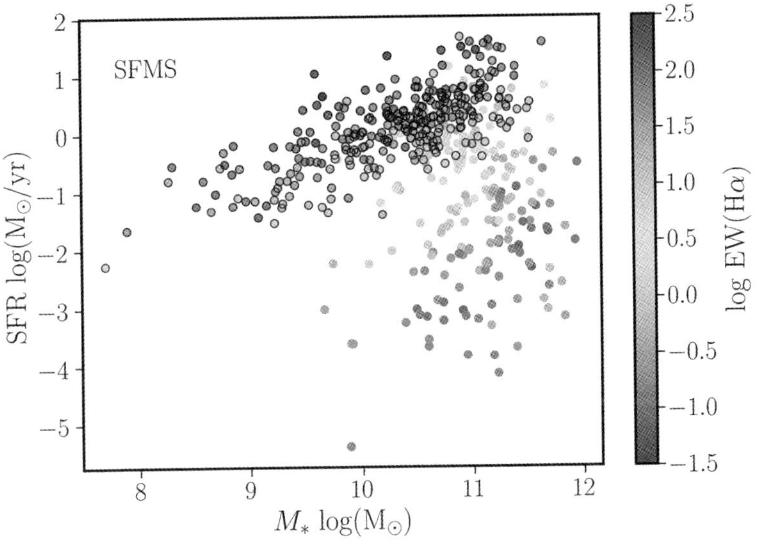

Figure 1. Distribution of the SFR along M_* for AMUSING++ galaxy compilation. Each solid-circle corresponds to one particular galaxy, color-coded by the EW(Hα) at the effective radius. Black open-circles indicate that SFGs selected from the original sample.

In this manuscript we present new explorations of the connection between the global and local relations using recent results obtained using IFS data on an unique compilation of MUSE (Bacon et al. 2001) data for a sample of galaxies in the nearby Universe. We use these results to discuss on the nature of these relations and the physical reason for their existence. Throughout this article we assume the standard Λ Cold Dark Matter cosmology with the parameters: $H_0 = 71$ km/s/Mpc, $\Omega_M = 0.27$, $\Omega_\Lambda = 0.73$.

2. Sample and Data

We made use of the data provided by AMUSING++ compilation (López-Cobá et al. 2020), a set of observations using MUSE on galaxies in the nearby Universe ($z \sim 0.03$), that comprises ~600 galaxies (in the current compilation). These observations comprises the galaxies included in the AMUSING survey (PI: J. Anderson), and data retrieved from the ESO archive. All galaxies were selected to match their optical extension with the FoV of the MUSE instrument (~1 arcmin^2), in a way that it covers between 1.0 and 2.0 effective radius. We exclude edge-on galaxies, i.e., with inclination larger than 70°, to avoid possible issues related with highly inclined galaxies. All data were analyzed using the Pipe3D pipeline (Sánchez et al. 2016; Lacerda et al. 2022), a tool aimed to separate the stellar population and emission line properties of the galaxies. Using this tool we retrieve for each galaxy: (i) the integrated stellar mass, M_*, derived from via the decomposition of the stellar population spectra in a set of single-stellar populations spectra, which allows to estimate the average mass-to-light ratio; (ii) the integrated the star-formation rate, estimated using the dust-corrected Hα luminosity, adopting the calibrator proposed by Kennicutt (1998a); and finally (iii) an estimation of the molecular gas mass (M_{gas}) based on the dust attenuation, using a modification of the calibrator proposed by Barrera-Ballesteros et al. (2020). All those quantities were derived using the procedures described in detail in Sánchez et al. (2021b) and Sánchez et al. (2022). In addition we retrieve the [OIII]/Hβ, [NII]/Hα line ratios and the EW(Hα) measured at the effective radius. Using those data we define a sample of star-forming galaxies (SFGs) following

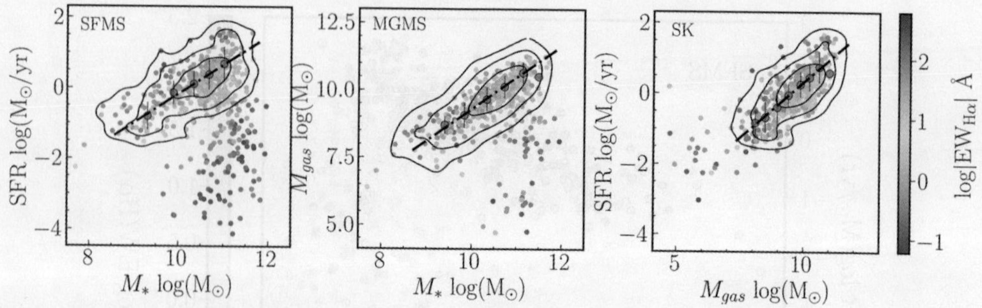

Figure 2. Distribution of the global extensive parameters: (i) SFR as a function of M_* (left panel), (ii) M_{gas} as a function of M_* (central panel) and (ii) SFR as a function of M_{gas} (right panel). In each panel each solid-circle corresponds to one AMUSING++ galaxy, color-coded by the EW(Hα) at the effective radius. The contours trace the density distribution of the SFGs (encircling 95%, 80% and 40% of this subsample, respectively). Grey solid-circles corresponds to the average values in bins of 0.2 dex in the x-axis, with the errorbars tracing the 1σ distribution. Finally the dashed-lines correspond the best fitted relations: SFMS, MGMS and SK, in each case.

Sánchez (2020), i.e., galaxies that are located below the Kewley et al. (2001) demarcation line in the BPT (Baldwin et al. 1981) diagram with an EW(Hα)>6Å. Figure 1 shows the distribution along the SFR-M_* plane of the 485 (non highly inclined) galaxies selected from the AMUSING++ compilation, together with the SFGs sub-sample, that comprises 297 galaxies. It is clearly seen that this compilation comprises galaxies of any star-formation stage, covering a wide range of stellar masses, and presumely a wide range of galaxy morphologies (as shown in López-Cobá et al. 2020). Finally, it is clear that our selected sub-sample of SFGs follow the expected trend along this diagram, i.e., the SFMS.

3. Analysis and Results

Following Cano-Díaz et al. (2019) we first explore the global extensive relations between the three retrieved parameters (SFR, M_* and M_{gas}), analyzing the distribution and relation between the SFR-M_* (SFMS), M_{gas}-M_* (MGMS) and SFR-M_{gas} (SK). Figure 2 show such distributions for the full sample of galaxies, highlighting the location of SFGs. As expected, in the three cases it is observed clear linear trends that define the well known relations. To parametrize them we perform a binning of the data in the x-axis, deriving the average and standard deviation in bins of 0.2 dex. The location of the binned data is shown in the different panels too. Then, we perform a simple linear regression for the binned data to derive the best fitted relation between both pairs of parameters. Equations 3.1, 3.2, and 3.3 show the results of this analysis. Errors were derived by performing a simple Monte-Carlo iteration on the original data, propagating the errors of each derived parameter.

$$log(SFR) = (0.83 \pm 0.22) log(M_*) - (8.43 \mp 0.93) \tag{3.1}$$

$$log(M_*) = (1.18 \pm 0.21) log(M_{gas}) - (2.52 \mp 0.83) \tag{3.2}$$

$$log(SFR) = (0.63 \pm 0.19) log(M_{gas}) - (5.97 \mp 0.67) \tag{3.3}$$

As expected we found three clear tight correlations between the analyzed extensive parameters. In all cases the slopes are near one, or slightly lower, by compatible within $\sim 2\sigma$ with this value. The largest deviation from this value is reported for the SK-law, most probably due to the more limited dynamical range of the two parameters involved in this relation.

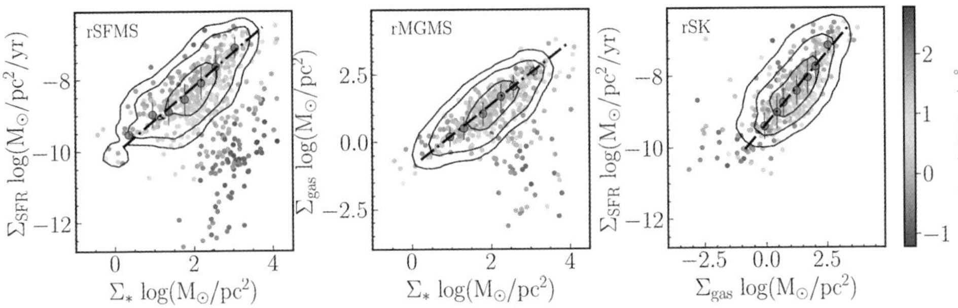

Figure 3. Distribution of the global intensive parameters: (i) Σ_{SFR} as a function of $\Sigma_\star$ (left panel), (ii) Σ_{gas} as a function of $\Sigma_\star$ (central panel) and (ii) Σ_{SFR} as a function of Σ_{gas} (right panel). In each panel each solid-circle corresponds to one AMUSING++ galaxy, color-coded by the EW(Hα) at the effective radius. The contours trace the density distribution of the SFGs (encircling 95%, 80% and 40% of this subsample, respectively). Grey solid-circles corresponds to the average values in bins of 0.2 dex in the x-axis, with the errorbars tracing the 1σ distribution. Finally the dashed-lines correspond the best fitted global intensive relations relations: rSFMS, rMGMS and rSK, in each case.

Once derived the extensive relations, we repeat the analysis using the global intensive relations (Σ_{SFR}, Σ_* and Σ_{mol}). To derive those parameters we follow Sánchez et al. (2021b), dividing the corresponding extensive parameter by the effective area covered by each galaxy (defined as $4\pi R_e^2$, i.e., the area encircled in 2 effective radius). Figure 3 show the distribution of parameters and the results of the binning and fitting analysis, with the best fitted linear relations listed in Equations 3.4, 3.5, and 3.6.

$$log(\Sigma_{SFR}) = (0.94 \pm 0.22) log(\Sigma_*) - (10.08 \mp 0.35) \tag{3.4}$$

$$log(\Sigma_{gas}) = (1.15 \pm 0.19) log(\Sigma_*) - (0.92 \mp 0.30) \tag{3.5}$$

$$log(\Sigma_{SFR}) = (0.80 \pm 0.18) log(\Sigma_{gas}) - (9.34 \mp 0.22) \tag{3.6}$$

As expected the three distributions show a clear tight relation between the involved pair of parameters, correlations that a tighter than the extensive ones, with a lower error for both the slopes and zero-points. Furthermore, the slopes are in all cases more near to one, within less than 1σ in the case of the rSFMS and rMGMS relations and just at 1σ for the rSK one.

4. Conclusions

Table 1 present a summary of the results together with similar values reported in the literature for both global intensive relations (Fig. 3), and local/resolved ones. From this comparison it is seen that:

- Global intensive parameters present a niche of opportunity to explore the physics of star-formation in a more coherent way than extensive ones, providing tighter and better defined relations (as the galaxy extension and aperture effects are limited).
- Global intensive relations can be directly compared with local/resolved ones (at least down to 1 kpc), showing similar trends both qualitatively and quantitatively, suggesting that they are originated by the same physical processes.
- There are hints of the evolution of the local/resolved relations that are evident when comparing results from sample at different redshits (e.g. Wuyts et al. (2013) vs. the rest of the data)
- If there is any secondary relation or a hierarchy among the relations, it is required to explore them on the tighter and best defined set of data.

Table 1. Results of the analysis of global intensive relations rSFMS, rSK y rMGMS.

References	β	α	r_c	σ_{obs}	σ_{exp}	# Gal.	# SFG
			rSFMS				
AMUSING++	-10.08 ± 0.35	0.94±0.22	0.79	0.466	0.290	485	297
Sa21a(EDGE)	-10.10 ± 0.22	1.02±0.16	0.68	0.266	0.190	126	12667
Sa21a(CALIFA)	-10.27 ± 0.22	1.01±0.15	0.85	0.244	0.192	941	533
Sa21a(APEX)	-9.78 ± 0.30	0.74±0.21	0.76	0.226	0.211	512	251
Wu13	-8.4^{1}	0.95					
CD16	-10.19 ± 0.33	0.72±0.04	0.63	0.16			
Li19	-10.54 ± 0.11	1.19±0.01		0.25		14	5383*
CD19	-10.48 ± 0.69	0.94±0.08	0.62	0.27		2737	~500K
Sa21	-10.35 ± 0.03	0.98±0.02	0.96	0.17		1512	~3M
El20	−10.07±1.44	1.03±0.17	0.57	0.28–0.39		28	~15035*
			rSK				
AMUSING++	-9.34 ± 0.22	0.80±0.18	0.8	0.478	0.294	485	297
Sa21a(EDGE)	-9.01 ± 0.14	0.98±0.14	0.73	0.249	0.216	126	12667
Sa21a(CALIFA)	-9.01 ± 0.16	0.95±0.21	0.77	0.293	0.297	941	533
Sa21a(APEX)	-8.84 ± 0.24	0.76±0.27	0.70	0.294	0.228	512	351
Bo17	-9.22	1.00				104	~5000
Li19	-9.03 ± 0.06	1.05±0.01		0.19		14	5383*
El20	-8.87 ± 0.66	1.05±0.19	0.74	0.22–0.32		28	~15035*
			rMGMS				
AMUSING++	-0.92 ± 0.30	1.15±0.19	0.85	0.528	0.283	485	297
Sa21a(EDGE)	-0.91 ± 0.16	0.93±0.11	0.68	0.218	0.209	126	12667
Sa21a(CALIFA)	-1.12 ± 0.27	0.93±0.18	0.74	0.276	0.288	941	533
Sa21a(APEX)	-0.70 ± 0.37	0.73±0.24	0.73	0.234	0.212	512	251
19	-0.59 ± 0.08	1.10±0.01		0.20		14	5383*
BB20	-0.95	0.93		0.20		93	~5000
El20	-0.99 ± 0.13	0.88±0.15	0.72	0.21–0.28		28	~15035*

Zero-point (β) and slope (α) of the three intensive global relations explored along this study (rSFMS, rSK and rMGMS), together with the values reported for a selection of similar relations extracted from the literature: Sánchez et al. (2021c)(Sa21a), Wuyts et al. (2013)(Wu13), Cano-Díaz et al. (2016b)(CD16), Lin et al. (2019)(Li19), Cano-Díaz et al. (2019)(CD19), Sánchez et al. (2021b)(Sa21) y Ellison et al. (2020a)(El20), Sánchez et al. (2021b)(Sa21), Bolatto et al. (2017)(Bo17), Lin et al. (2019)(Li19) and Barrera-Ballesteros et al. (2020)(BB20) y Ellison et al. (2020a)(El20). In addition we include the correlation coefficient (r_c), the standard deviation of the distribution of data before ($\sigma_{\rm obs}$) and after applying the best fitted linear relation ($\sigma_{\rm exp}$), together with the number of galaxies or number of individual spatial elements considered in the analysis. In italics we highlight the results from spatially resolved data, and in non-italics are indicated the explorations based on global intensive parameters.

In this regards, we need to highlight that it is still an open question which is of the explored relations is more fundamental than the previous one (if any). The nature of the scatter described in the three relations and the existence of possible secondary relations with other parameters, that may drive this scatter, is indeed and important topic of study. As indicated before, it could be the case that these relations depends on the morphology, that may affect the star-formation efficiency (SFE). Some authors have reported that the rSFMS preent a secondary relation that depends on the SFE (Ellison et al. 2020b, using the ALMAQUEST data) or the gas fraction (Colombo et al. 2020). This has open the seek for a more fundamental relation that involves pairs of parameters to trace the SFR (e.g. Shi et al. 2018; Dey et al. 2019; Barrera-Ballesteros et al. 2021). However, other results indicate that most probably those secondary relations are pure mathematical artifact introduced by correlation between errors (Sánchez et al. 2021c). The use of a larger samples, the tighter and better defined relations (i.e., the intensive ones), and better methods to derive the parameters (in particular for the molecular gas), seem to be required approach to perform future explorations.

References

Bacon, R., Copin, Y., Monnet, G., et al. 2001, MNRAS, 326, 23, doi: 10.1046/j.1365-8711.2001.04612.x

Baldwin, J. A., Phillips, M. M., & Terlevich, R. 1981, PASP, 93, 5, doi: 10.1086/130766
Barrera-Ballesteros, J. K., Utomo, D., Bolatto, A. D., et al. 2020, MNRAS, 492, 2651, doi: 10.1093/mnras/stz3553
Barrera-Ballesteros, J. K., Heckman, T., Sanchez, S. F., et al. 2021, arXiv e-prints, arXiv:2101.02711. https://arxiv.org/abs/2101.02711
Bigiel, F., Leroy, A., Walter, F., et al. 2008, AJ, 136, 2846, doi: 10.1088/0004-6256/136/6/2846
Bolatto, A. D., Wong, T., Utomo, D., et al. 2017, ApJ, 846, 159, doi: 10.3847/1538-4357/aa86aa
Brinchmann, J., Charlot, S., White, S. D. M., et al. 2004, MNRAS, 351, 1151, doi: 10.1111/j.1365-2966.2004.07881.x
Bundy, K., Bershady, M. A., Law, D. R., et al. 2015, ApJ, 798, 7, doi: 10.1088/0004-637X/798/1/7
Calette, A. R., Avila-Reese, V., Rodríguez-Puebla, A., Hernández-Toledo, H., & Papastergis, E. 2018, RMxAA, 54, 443. https://arxiv.org/abs/1803.07692
Cano-Díaz, M., Ávila-Reese, V., Sánchez, S. F., et al. 2019, MNRAS, 488, 3929, doi: 10.1093/mnras/stz1894
Cano-Díaz, M., Sánchez, S. F., Zibetti, S., et al. 2016a, ApJL, 821, L26, doi: 10.3847/2041-8205/821/2/L26
—. 2016b, ApJL, 821, L26, doi: 10.3847/2041-8205/821/2/L26
Catalán-Torrecilla, C., Gil de Paz, A., Castillo-Morales, A., et al. 2017, ApJ, 848, 87, doi: 10.3847/1538-4357/aa8a6d
Colombo, D., Sanchez, S. F., Bolatto, A. D., et al. 2020, arXiv e-prints, arXiv:2009.08383. https://arxiv.org/abs/2009.08383
Dey, B., Rosolowsky, E., Cao, Y., et al. 2019, MNRAS, 488, 1926, doi: 10.1093/mnras/stz1777
Ellison, S. L., Lin, L., Thorp, M. D., et al. 2020a, MNRAS, doi: 10.1093/mnras/staa3822
Ellison, S. L., Thorp, M. D., Lin, L., et al. 2020b, MNRAS, 493, L39, doi: 10.1093/mnrasl/slz179
González Delgado, R. M., Cid Fernandes, R., Pérez, E., et al. 2016, A&A, 590, A44, doi: 10.1051/0004-6361/201628174
Kennicutt, Jr., R. C. 1998a, ApJ, 498, 541, doi: 10.1086/305588
—. 1998b, ARAA, 36, 189, doi: 10.1146/annurev.astro.36.1.189
Kennicutt, Jr., R. C., Calzetti, D., Walter, F., et al. 2007, ApJ, 671, 333, doi: 10.1086/522300
Kewley, L. J., Dopita, M. A., Sutherland, R. S., Heisler, C. A., & Trevena, J. 2001, ApJ, 556, 121, doi: 10.1086/321545
Lacerda, E. A. D., Sánchez, S. F., Mejía-Narváez, A., et al. 2022, New Astronomy, 97, 101895, doi: 10.1016/j.newast.2022.101895
Leroy, A. K., Walter, F., Sandstrom, K., et al. 2013, AJ, 146, 19, doi: 10.1088/0004-6256/146/2/19
Lin, L., Pan, H.-A., Ellison, S. L., et al. 2019, arXiv e-prints, arXiv:1909.11243. https://arxiv.org/abs/1909.11243
Lin, L., Ellison, S. L., Pan, H.-A., et al. 2020, arXiv e-prints, arXiv:2010.01751. https://arxiv.org/abs/2010.01751
López-Cobá, C., Sánchez, S. F., Anderson, J. P., et al. 2020, arXiv e-prints, arXiv:2002.09328. https://arxiv.org/abs/2002.09328
Méndez-Abreu, J., Sánchez, S. F., & de Lorenzo-Cáceres, A. 2019, MNRAS, 488, L80, doi: 10.1093/mnrasl/slz103
Pan, H.-A., Lin, L., Hsieh, B.-C., et al. 2018, ApJ, 854, 159, doi: 10.3847/1538-4357/aaa9bc
Renzini, A., & Peng, Y.-j. 2015, ApJL, 801, L29, doi: 10.1088/2041-8205/801/2/L29
Rodríguez-Puebla, A., Primack, J. R., Behroozi, P., & Faber, S. M. 2016, MNRAS, 455, 2592, doi: 10.1093/mnras/stv2513
Saintonge, A., Catinella, B., Cortese, L., et al. 2016, MNRAS, 462, 1749, doi: 10.1093/mnras/stw1715
Sánchez, S. F. 2020, ARAA, 58, 99, doi: 10.1146/annurev-astro-012120-013326
Sánchez, S. F., Walcher, C. J., Lopez-Cobá, C., et al. 2021a, RMxAA, 57, 3, doi: 10.22201/ia.01851101p.2021.57.01.01

—. 2021b, RMxAA, 57, 3, doi: 10.22201/ia.01851101p.2021.57.01.01
Sánchez, S. F., Kennicutt, R. C., Gil de Paz, A., et al. 2012, A&A, 538, A8, doi: 10.1051/0004-6361/201117353
Sánchez, S. F., Rosales-Ortega, F. F., Jungwiert, B., et al. 2013, A&A, 554, A58, doi: 10.1051/0004-6361/201220669
Sánchez, S. F., Pérez, E., Sánchez-Blázquez, P., et al. 2016, RMxAA, 52, 21. https://arxiv.org/abs/1509.08552
Sánchez, S. F., Barrera-Ballesteros, J. K., Colombo, D., et al. 2021c, MNRAS, 503, 1615, doi: 10.1093/mnras/stab442
Sánchez, S. F., Barrera-Ballesteros, J. K., Lacerda, E., et al. 2022, ApJS, 262, 36, doi: 10.3847/1538-4365/ac7b8f
Schmidt, M. 1959, ApJ, 129, 243, doi: 10.1086/146614
—. 1968, ApJ, 151, 393, doi: 10.1086/149446
Shi, Y., Yan, L., Armus, L., et al. 2018, ApJ, 853, 149, doi: 10.3847/1538-4357/aaa3e6
Speagle, J. S., Steinhardt, C. L., Capak, P. L., & Silverman, J. D. 2014, ApJS, 214, 15, doi: 10.1088/0067-0049/214/2/15
Wong, T., & Blitz, L. 2002, ApJ, 569, 157, doi: 10.1086/339287
Wuyts, S., Förster Schreiber, N. M., Nelson, E. J., et al. 2013, ApJ, 779, 135, doi: 10.1088/0004-637X/779/2/135
York, D. G., Adelman, J., Anderson, Jr., J. E., Anderson, S. F., & et al. 2000, AJ, 120, 1579, doi: 10.1086/301513

Resolving the Rise and Fall of Star Formation in Galaxies
Proceedings IAU Symposium No. 373, 2023
T. Wong & W.-T. Kim, eds.
doi:10.1017/S1743921322005038

Hierarchical star formation in the Magellanic Clouds with the VMC survey

Amy E. Miller[1,2,3,4], Maria-Rosa L. Cioni[3], Richard de Grijs[1,2], Ning-Chen Sun[5] and The VMC Team

[1]School of Mathematical and Physical Sciences, Macquarie University, Balaclava Road, Sydney, NSW 2109, Australia

[2]Research Centre for Astronomy, Astrophysics and Astrophotonics, Macquarie University, Balaclava Road, Sydney, NSW 2109, Australia

[3]Leibniz-Institut für Astrophysik Potsdam, An der Sternwarte 16, D-14482 Potsdam, Germany

[4]Institut für Physik und Astronomie, Universität Potsdam, Haus 28, Karl-Liebknecht-Str. 24/25, D-14476 Golm (Potsdam), Germany

[5]Department of Physics and Astronomy, University of Sheffield, Hicks Building, Hounsfield Road, Sheffield S3 7RH, UK

Abstract. The VISTA Magellanic Clouds Survey (VMC) is a near-infrared survey of the Magellanic system. The VMC data has been exploited to detect and study statistically correlated young groups of stars — also known as "young stellar structures" — in the Large and Small Magellanic Clouds (LMC and SMC). We showcase the ~3000 recently detected young stellar structures in the LMC and their similarity to the fractal interstellar medium. We discuss how their properties indicate their formation mechanisms and that there are no preferred scales of star formation in the LMC.

Keywords. stars: formation - galaxies: Magellanic Clouds - stars: early-type - methods: statistical

1. Introduction

Within star-forming galaxies we generally observe a young stellar hierarchy composed of (listed in decreasing density) stars, clusters ($\sim 10^0$ pc), associations ($\sim 10^1 - 10^2$ pc), and complexes ($\sim 10^3$ pc). This hierarchical distribution of young stars is found to disperse rapidly, on timescales of ≥100 Myr (Bastian *et al.* 2009; Sun *et al.* 2017b). There is evidence that most stars form in clusters, but there is also evidence that stars form in more isolated environments, leading to two distinct star formation channels being proposed: 'clustered' and 'distributed' star formation (Lada & Lada 2003). However, there is emerging evidence that star forming structures exist in a continuous range of sizes, masses, and densities (Bressert *et al.* 2010). Further analysis of the young stellar hierarchy is necessary to improve understanding of the star formation distribution and to search for characteristic sizes that may correspond to physical star formation mechanisms.

When discussing the star formation hierarchy, most studies focus on the densest regions, clusters, or the loosest regions, galaxies. These two disparate spatial scales must be unified; therefore, we focus on the intermediate scale, from 10–1000 pc. To study this intermediate scale, Sun *et al.* (2017a,b, 2018) employed a contour–based clustering analysis of young stars in two small regions covering the Large Magellanic Cloud (LMC) and

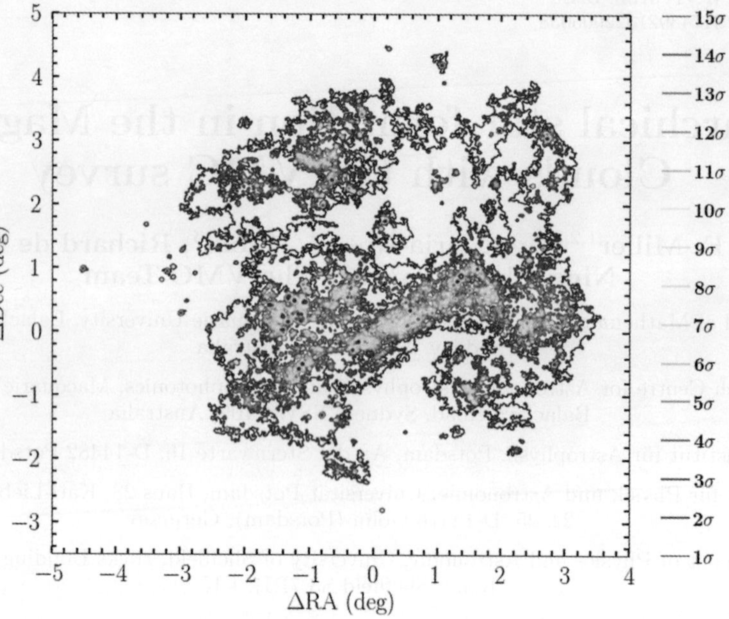

Figure 1. The identified young stellar structures coloured by 1σ to 15σ. After Miller *et al.* (2022).

the entire Small Magellanic Cloud (SMC). Their methodology was inspired by and incorporated work done by Gouliermis *et al.* (2010, 2015, 2017). We applied their methodology to the entire LMC in Miller *et al.* (2022). Our goals were to check for preferred scales of star formation, probe the dominant physical properties behind star formation, and to compare to the interstellar medium. In the following sections, we describe the data used, the detection methodology, results, and the implications of our work.

2. VMC data and young stellar structure identification

The Visible and Infrared Survey Telescope for Astronomy (VISTA) Magellanic Clouds Survey (VMC; Cioni *et al.* 2011), is a deep near-infrared YJK_s survey of the Magellanic system with spatial resolution of $\leq$1 arcsec. The VMC covers $\sim$120 square degrees on the LMC. We use point spread function photometry catalogues and deredden the magnitudes based on the extinction map provided by Skowron *et al.* (2021). We construct $(J - K_s)$ versus K_s colour–magnitude diagrams and use them to select $\sim 400,000$ young upper main-sequence stars (see Figure 2 of Miller *et al.* 2022). After binning the stars, we apply kernel density estimation (KDE) and make a surface density map with a resolution of 10 pc (see left panel of Figure 3 from Miller *et al.* 2022). From this map, we identify structures at 1 to 15σ above the median background density. After applying selection criteria that each structure on the $1 - 2\sigma$ levels have at least one structure at a higher significance level and that each structure have at least 5 stars, we identify about 3000 young stellar structures (see Figure 1).

3. Results and comparison to gas structures

The main result from our work that we want to emphasize is the size (radius) distribution of the young stellar structures (see left panel of Figure 3). For such a size distribution, it is important to assess if the distribution is lognormal or a power-law. In

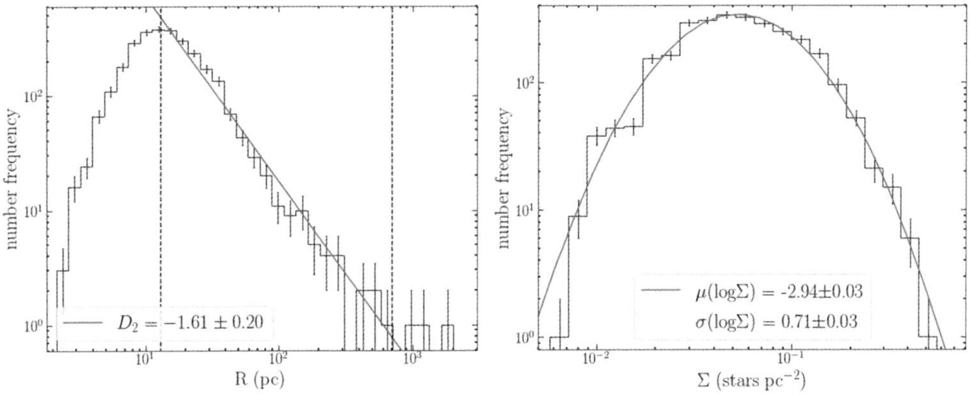

Figure 2. (Left) The size distribution of the young stellar structures. In red, the power-law we fit between 10–700 pc. (Right) The surface density distribution of the structures. In red, the lognormal distribution we fit. After Miller *et al.* (2022).

our size distribution, there is clearly an excess of large structures on the right-hand side. We fit a power-law between the dashed lines at 10 pc and 700 pc. Mathematically, this power–law corresponds to the 2D fractal dimension (D_2). This dimension encodes information about the spatial distribution of the young stellar structures. The one we derive, ∼1.6, suggests our structures have a fractal morphology and that there is no preferred scale from 10–700 pc.

The peak of the size distribution is ∼13 pc. The peak of a size distribution could be representative of a characteristic size of star formation. However, our structures are incomplete at $R < 10$ pc due to the resolution of our KDE map and due to our two selection criteria. Therefore, it seems the peak is related to the detection limit of our study and is not evidence of a characteristic scale of star formation.

The mass–size relation and the perimeter–area relation follow a power–law (see Figures 7 and 8 in Miller *et al.* 2022). The surface density distribution follows a lognormal shape (see right panel of Figure 3). We summarize LMC results from Miller *et al.* (2022) along with results from Sun *et al.* (2018) in the SMC and compare them to results obtained in studies of gas structures in Table 1. Due to the striking similarities, the young stellar structures in the LMC and SMC might inherit the hierarchical, fractal properties from the interstellar medium from which they form. A further important highlight from this comparison is that the density distribution of the young stellar structures follows the same lognormal shape as probability density functions that are characteristic of supersonic, non–gravitating turbulent gas (Vazquez-Semadeni 1994; Padoan *et al.* 1997; Federrath *et al.* 2010).

4. Conclusion

In Miller *et al.* (2022), we identified ∼3000 young stellar structures at 15 significance levels in the LMC using the near–infrared VMC survey. In this article we summarize our findings, and touch on those found in the SMC in Sun *et al.* (2018). We show the young stellar structures have:

- Irregular morphology and hierarchical organization
- Power–law mass–size relation, size distribution, and perimeter–area relation
- Lognormal surface density distribution
- Strong similarities with results in the interstellar medium and simulations of supersonic turbulence

Table 1. Comparison with results from young stellar structures in the LMC (Miller *et al.* 2022) and SMC (Sun *et al.* 2018) with gas structures in the interstellar medium (references therein).

property	young stellar structures	gas structures
morphology	irregular	irregular[1],[2]
structure	hierarchical	hierarchical[1],[2]
size distribution	power–law	power–law[1]
mass–size relation	power–law	power–law[3]
density distribution	lognormal	lognormal[4]
perimeter-area relation	power–law	power–law[5]
2D fractal dimension	$D_2 \approx 1.6$	$D_2 \approx 1.7$[2]

References: [1] Elmegreen & Falgarone (1996); [2] Sánchez & Alfaro (2008); [3] Sánchez *et al.* (2005); [4] Padoan & Nordlund (2002); [5] Sánchez *et al.* (2007)

Our findings imply that star formation creates a continuum of star–forming structures in the LMC from 10 pc to 700 pc and is scale–free. The findings are consistent with the hypothesis of turbulence–driven hierarchical star formation.

Acknowledgements

The author is supported by the International Cotutelle Macquarie University Research Excellence Scholarship. Under the European Union's Horizon 2020 research and innovation programme, the European Research Council (ERC) has provided funding for this project (grant agreement no. 682115). The Australian Research Council's Centre of Excellence for All Sky Astrophysics in Three Dimensions (ASTRO 3D), under the auspices of project CE170100013, also contributed to the funding of this study. We appreciate the support of the Science and Technology Facilities Council (STFC) in the UK, which enabled the Cambridge Astronomy Survey Unit (CASU) and the Wide Field Astronomy Unit (WFAU) in Edinburgh to provide the necessary data products. This investigation was conducted based on observations collected with VISTA at the ESO/La Silla Paranal Observatory with programme ID 179.B-2003.

References

Bastian, N., Gieles, M., Erolanco, B., *et al.* 2009, *MNRAS*, 392, 868
Bressert, E., Bastian, N., Gutermuth, R., *et al.* 2010, *MNRAS*, 409, L54
Cioni, M.-R. L., Clementini, G., Girardi, L., *et al.* 2011, *A&A*, 527, A116
Elmegreen, B. G., Falgarone, E., 1996, *ApJ*, 471, 816
Fenderrath, C., Roman-Duval, J., Klessen, R. S., *et al.* 2010, *A&A*, 512, 81
Gouliermis, D., Schmeja, S., Klessen, R. S., *et al.* 2010, *ApJ*, 725, 1717
Gouliermis, D., Thilker, D., Elmegreen, B. G., *et al.* 2015, *MNRAS*, 452, 3508
Gouliermis, D., Elmegreen, B. G., Elmegreen, D. M., *et al.* 2015, *MNRAS*, 468, 509
Lada, C. J., Lada, E. A. 2003, *ARA&A*, 41, 57
Miller, A. E., Cioni, M.-R. L., de Grijs, R., *et al.* 2022, *MNRAS*, 521, 1196
Padoan,P.,Nordlunk,A., 2002, *ApJ*, 576, 870
Padoan,P.,Jones,B. J. T., Nordlund, A., 1997, *ApJ*, 474, 730
Sánchez, N., Alfaro, E. J., Pérez, E., *et al.* 2005, *ApJ*, 625, 849
Sánchez, N., Alfaro, E. J., Pérez, E., *et al.* 2005, *ApJ*, 656, 222
Sánchez, N., Alfaro, E. J., *et al.* 2008, *ApJS*, 178, 1
Skowron, D.-M., Skowron, J., Udalski, A., *et al.* 2021, *ApJS*, 252, 23S
Sun, N.-C., de Grijs, R., Subramanian, S., *et al.* 2017a, *ApJ*, 835, 171
Sun, N.-C., de Grijs, R., Subramanian, S., *et al.* 2017b, *ApJ*, 849, 149
Sun, N.-C., de Grijs, R., Cioni, M.-R. L., *et al.* 2018, *ApJ*, 858, 31
Vazquez-Semadeni, E., 1994, *ApJ*, 423, 681

Resolving the Rise and Fall of Star Formation in Galaxies
Proceedings IAU Symposium No. 373, 2023
T. Wong & W.-T. Kim, eds.
doi:10.1017/S1743921322003684

Molecular Clouds with CO-dark Envelopes in the Extended Ultraviolet (XUV) Disk of M83

Jin Koda

Department of Physics and Astronomy, Stony Brook University, Stony Brook, NY 11794-3800

Abstract. We report a CO(3-2) detection of 23 molecular clouds in the extended ultraviolet (XUV) disk of the spiral galaxy M83 with ALMA. The observed 1 kpc^2 region is at about $1.24R_{25}$ from the disk center, where CO(2-1) was previously not detected. The detection and non-detection, as well as the level of star formation (SF) activity in the region, can be explained consistently if the clouds have the mass distribution common among Galactic clouds, such as Orion A – with star-forming dense clumps embedded in thick layers of bulk molecular gas, but in a low-metallicity regime where their outer layers are CO-deficient and CO-dark. The cloud masses, estimated from CO(3-2), range from 8.2×10^2 to $2.3 \times 10^4 M_\odot$. The most massive clouds appear similar to Orion A in SF activity as well as in gas mass. The common cloud mass structure also justifies the use of high-J CO transitions to trace the total gas mass of clouds, or galaxies, even in the high-z universe. This study is the first demonstration that CO(3-2) is an efficient tracer of molecular clouds even in low-metallicity environments. This study is published in the Astronomical Journal, entitled "First Detection of the Molecular Cloud Population in the Extended Ultraviolet (XUV) Disk of M83" by J. Koda, L. Watson, F. Combes, M. Rubio, S. Boissier, M. Yagi, D. Thilker, A. M Lee, Y. Komiyama, K. Morokuma-Matsui, and C. Verdugo.

Keywords. Interstellar medium (847), Molecular clouds (1072), Star formation (1569), Galaxy evolution (594), Spiral galaxies (1560), Galaxy disks (589)

1. Introduction

The GALEX satellite found massive star formation (SF) in the far outskirts of galactic disks (Gil de Paz et al. 2005; Thilker et al. 2005). Bright ultraviolet (UV) sources are distributed beyond the optical radius R_{25}. They reveal abundant and recent SF. These extended UV disks, dubbed XUV disks, are fairly common among local disk galaxies (Thilker et al. 2007). They offer an opportunity to study SF in extreme conditions, in particular, at a low average gas density and molecular fraction.

Molecular clouds host virtually all SF within the optical disks of local galaxies. It is crucial to study whether the same is true in XUV disks. Efforts have been made to detect CO emission there, however, they have rarely succeeded (Watson & Koda 2017, for review). Only four galaxies permitted CO detection at a few positions in their outskirts (Braine & Herpin 2004, Braine et al. 2007, Braine et al. 2010, Dessauges-Zavadsky et al. 2014). Most detections did not reveal individual clouds due to low sensitivity and resolution. Only M33, the closest to the Milky Way, allowed detection down to $\lesssim 4 \times 10^4 M_\odot$ in two regions beyond the R_{25} radius, and only one of them is identified as a single cloud (Braine et al. 2012). The rarity of the CO detection is at odds with the abundance of the UV sources across the XUV disks.

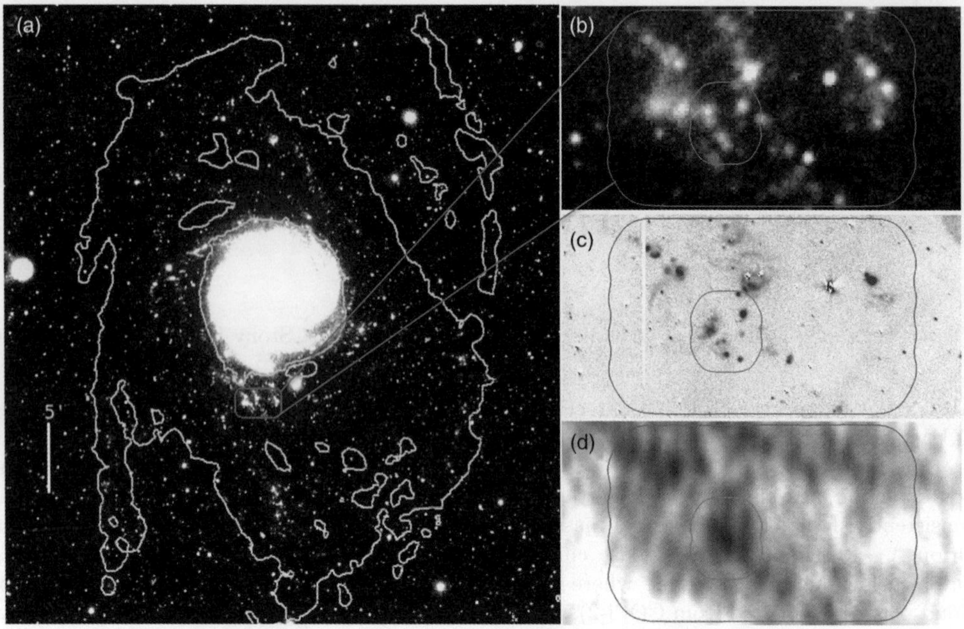

Figure 1. (a) GALEX FUV & NUV-band color composite image. The inner (green) and outer (cyan) contours are the edge of the optical disk and the extent of HI gas at an HI surface density of 1.5×10^{20} cm^{-2}. The red box is the $1.5' \times 3'$ ($2.0 \times 3.9\,\mathrm{kpc}^2$) field from the ALMA CO(J=2-1) observations (Bicalho et al. 2019). The three panels on the right are zoom-in to the CO(2-1) field: (b) the GALEX color composite, as in Panel a; (c) the continuum-subtracted Hα image from the Subaru telescope, and (d) the HI 21cm emission image from Walter et al. (2008). The area of our CO(3-2) observations is indicated by the inner magenta box of $\sim 0.75' \times 0.85'$ ($\sim 0.98 \times 1.11\,\mathrm{kpc}^2$).

The previous ALMA observations of CO(2-1) in the XUV disk of M83 also resulted in non-detection even at a high mass sensitivity of $2.2 \times 10^4 \mathrm{M}_\odot$ (3σ; Bicalho et al. 2019). The observed field is a relatively large area of $\sim 1.5' \times 3'$ ($2 \times 4\,\mathrm{kpc}^2$) at a galactic radius of $r_{\mathrm{gal}} \sim 1.24 R_{25}$ with $R_{25} = 6.44'$ (8.4 kpc; Figure 1 red box).

2. ALMA CO(3-2) Observations and Detection of 23 Clouds

We revisit a smaller area of $\sim 0.75' \times 0.85'$ ($\sim 1\,\mathrm{kpc}^2$; Figure 1 magenta) within the CO(2-1) field (red) at a higher mass sensitivity by a factor of about 10 in CO(3-2) with ALMA. The target region is selected around the maximum HI 21 cm intensity position within the CO(2-1) observation. It includes bright UV peaks located at $\sim 1.24 R_{25}$ and is outside the traditional optical disk (green contour).

The new ALMA data clearly shows the detection of CO(3-2) emission (Figure 2). In the data cube, we identify pixels with $> 5\sigma$ peaks and find their envelopes by expanding their volumes to adjacent pixels down to 3σ significance. In some cases the edges of the 3σ envelopes of two peaks (two clouds) are connected. We use the CLUMPFIND algorithm (Williams et al. 1994) to split them – the partition of the two clouds is determined simply by assigning each pixel to the nearest of the two peaks. We visually inspect the results and manually merge them back to a single cloud when their intensity-weighted positions are closer than the beam size of 0.96" × 0.82". Figure 2 shows the IDs of the detected clouds.

The observed region has several HII regions (Figure 3). We visually identified significant Hα peaks in the Subaru Hα image. Panel a shows their locations. The Hα luminosity

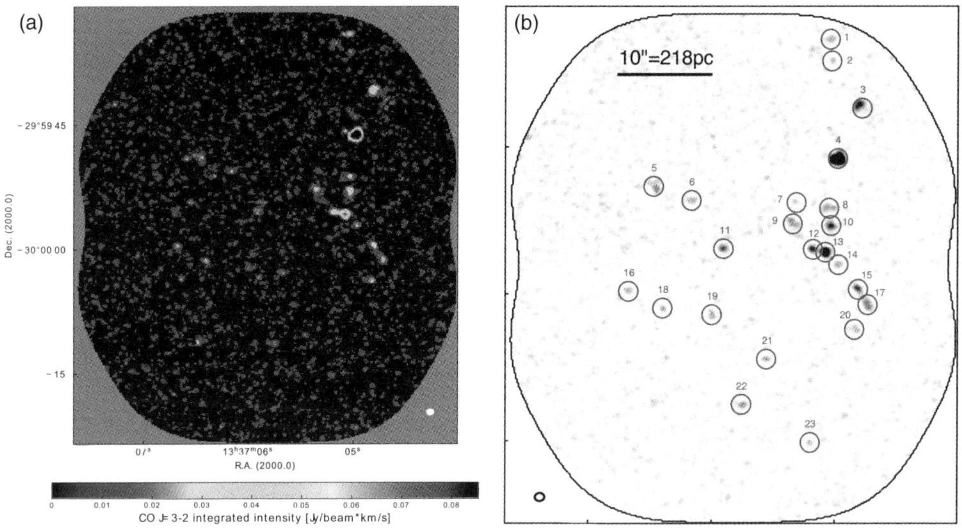

Figure 2. (a) CO(3-2) integrated intensity map. (b) Molecular clouds on the CO(3-2) peak intensity map. Each circle has a diameter of 2". The beam size of 0.96" × 0.82" (PA = −84.7 deg) is shown on the bottom-left corner, corresponding to 21 × 18 pc^2 at 4.5 Mpc.

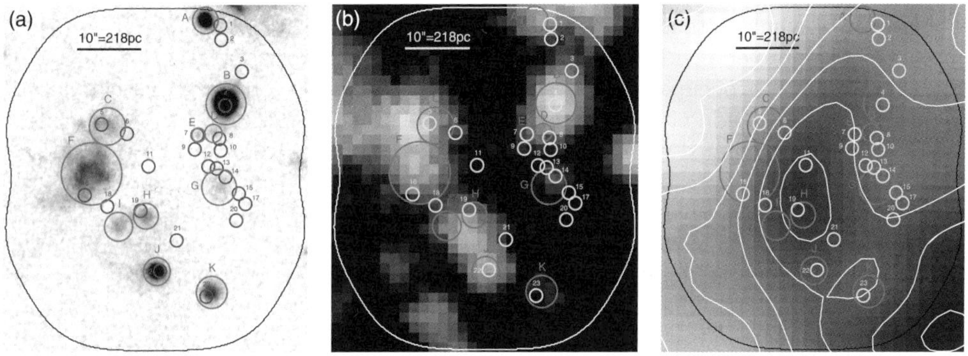

Figure 3. (a) Subaru H$_\alpha$ image at a resolution of about 1". (b) GALEX FUV & NUV-band color composite image at a resolution of about 5" (Gil de Paz et al. 2007b). (c) VLA+GBT HI 21cm image (grayscale) with contours – from the brightest – at 0.30, 0.25, 0.20, 0.15, 0.10, and 0.05 Jy/beam km s^{-1} with the 15" beam. The brightest contour corresponds to the HI surface density of $\sim 12 M_\odot$ pc^{-2} (and $\sim 16 M_\odot$ pc^{-2} for HI+He). Magenta circles denote the HII regions. Blue or yellow circles are molecular clouds and are the same as the ones in Figure 2.

($L_{\mathrm{H}\alpha}$) ranges from 1×10^{35} to 3×10^{37} erg s^{-1}. As a reference, in the Milky Way disk, the two HII regions in the Orion Nebula, M42 and M43, are located in the OMC-1 dense gas clump in the Orion A molecular cloud. They show $L_{\mathrm{H}\alpha} = 7.1 \times 10^{36}$ erg s^{-1} and 3.5×10^{35} erg s^{-1}, respectively (Scoville et al. 2001). Hence, our region has SF activity similar to, or slightly less than, that of the Orion Nebula, a relatively minor OB star forming region in the Milky Way.

3. Properties of the Detected Clouds

All clouds, except Cloud 4 and 8, are spatially unresolved; and Cloud 4 and 8 are only very marginally resolved at the resolution of 0.96" (21 pc). As a reference, the Orion

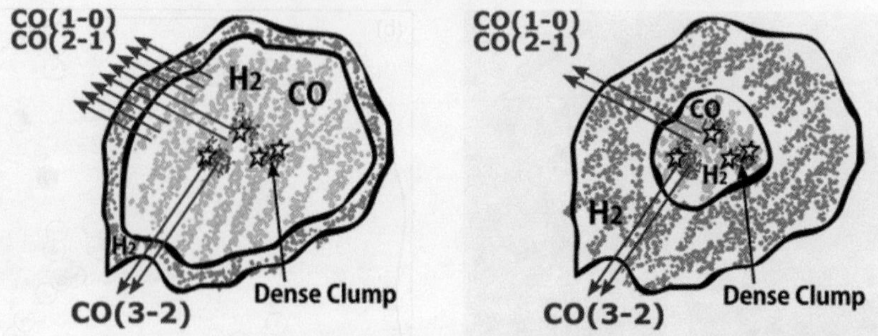

Figure 4. Schematic illustrations of the physical and chemical structures of a molecular cloud (a) in high-metallicity environment, and (b) in low-metallicity environment. The physical structures (i.e., mass distribution and SF activity) are common between (a) and (b), but their chemical structures (i.e., CO abundance) are different. The ambient radiation penetrates into the cloud, dissociates CO molecules in the cloud outer layers, and makes them CO-deficient and CO-dark, while more abundant H_2 molecules can self-shield themselves and remain (gray). These CO-dark layers thicken at low metallicity, since a lower dust abundance results in less efficient extinction and allows the radiation to penetrate deeper into the cloud (e.g., Maloney & Black 1988; van Dishoeck & Black 1988). Still, the CO molecules can remain near the cloud center (orange). We hypothesize that the dense clumps (brown), often associated with SF (yellow), are located at the heart of the cloud and likely remain intact with abundant CO molecules. The CO(3-2) emission originates from the dense clump region.

A cloud has a whole extent of ~20 pc when it is observed in the CO(1-0) or CO(2-1) emission (Sakamoto et al. 1994; Nakamura et al. 2019; Kong et al. 2018). Its star-forming dense clump OMC-1 occupies only a small area (~2pc) at the heart of the cloud and is bright in CO(3-2) (Ikeda et al. 1999). It is possible that our detected CO(3-2) peaks are mainly from dense clumps within molecular clouds, and hence, are unresolved.

The velocity dispersions of most clouds ($\sigma_{\rm v}$) are resolved. Cloud 4 has a wide width of $\sigma_{\rm v} = 6.1\,{\rm km\,s^{-1}}$ with a caveat that it could be a blend of two clouds. Again using Orion A as a reference, the velocity width of the bulk gas, measured in CO(1-0) and averaged over the whole cloud, is about $5\,{\rm km\,s^{-1}}$ at 50% of the peak brightness (from Figure 5 of Nakamura et al. 2019). The full CO(3-2) velocity width over the OMC-1 clump (size of about 2 pc) is not found in the literature; toward Orion KL (the brightest region), it is about $7\,{\rm km\,s^{-1}}$ at 50% of the peak brightness in a 134" (~0.3pc) beam (from Figure 15 of Masui et al. 2021). Hence, the dense part has a broader linewidth than the entire cloud. These are similar to the velocity widths of Cloud 4 measured in CO(3-2).

For their measured $\sigma_{\rm v}$, the XUV clouds generally have smaller (or similar) sizes compared to the clouds in the other galaxies. This makes sense if the CO(3-2) emission is primarily tracing dense clumps within the clouds, as seen in Orion A.

4. Hypothesis: Cloud Structure with Dense Clump and CO-dark Envelope

We are using the Orion A molecular cloud as a fiducial reference cloud given the similarity of its SF activity to the (brightest) clouds in the XUV disk. Orion A has an internal mass distribution common among other large and small Galactic molecular clouds, having small star-forming dense clump(s) embedded deeply in thick layers of bulk molecular gas (Nakamura et al. 1984; Enoch et al. 2008; Watanabe et al. 2017; Barnes et al. 2020). Figure 4a schematically illustrates this mass distribution. In terms of CO emissions, CO(1-0) and CO(2-1) can be excited easily at an average density and temperature in molecular clouds ($\sim 10^2\,{\rm cm^{-3}}$, $\sim 10\,{\rm K}$), and hence, the bulk gas contributes significantly to their total CO(1-0) and CO(2-1) luminosities (Kong et al. 2018;

Nakamura et al. 2019). By having the J = 3 level temperature of $E_J/k \sim 33.2\,\mathrm{K}$ and critical density of $> 10^3\,\mathrm{cm}^{-3}$, the CO(3-2) emission requires higher density and temperature than the averages for excitation and is radiated predominantly from the dense clumps (Ikeda et al. 1999). Once it is excited in thermalized gas, CO(3-2) can become brighter than CO(1-0) and CO(2-1) in total flux. We hypothesize that the clouds in the XUV disk share the common clump-envelope mass distribution of the Galactic molecular clouds. The common mass structure is expected for gravitationally-bound clouds as it is determined predominantly by the internal physics rather than the environment.

Even when the mass distribution is the same, the chemical structure and appearance of the clouds in the CO emissions depend on the metal abundance (e.g., Maloney & Black 1988; van Dishoeck & Black 1988; Wolfire et al. 2010). In high metallicity ($\sim 1Z_\odot$; Figure 4a), the bulk gas in clouds can be detected in the low excitation transitions (J =1-0 or 2-1) of CO. In low metallicity ($< 1Z_\odot$), the dust extinction is low, and the ambient stellar radiation can penetrate deeper into the clouds. This radiation dissociates more CO in their outer layers through CO line absorptions, while the abundant H2 molecules can be self-shielded. This differential photo-dissociation makes the cloud outer layers CO-deficient and CO-dark while still H2-abundant (Figure 4b).

In our hypothesis, the dense clumps are protected from photo-dissociation and would radiate a similar amount of $L_{\mathrm{CO}(3-2)}$ in high and low metallicities. Hence, the calibration of cloud mass (M_{cloud})-to-$L_{\mathrm{CO}(3-2)}$ ratio in high metallicity should be applicable in low metallicity. The conversion equation from $L_{\mathrm{CO}(3-2)}$ to M_{cloud} derived in high metallicity (Wilson et al. 2009) can be used to estimate the cloud masses. This calculation provides a cloud mass range from $M_{\mathrm{cloud}} = 8.2 \times 10^2 \mathrm{M}_\odot$ to $2.3 \times 10^4 \mathrm{M}_\odot$. The maximum mass in this range is similar to the recently-estimated mass of the Orion A cloud ($4.0 \times 10^4 \mathrm{M}_\odot$; Nakamura et al. 2019).

JK acknowledges support from NSF through 25 grants AST-1812847 and AST-2006600.

References

Gil de Paz, A., Madore, B. F., Boissier, S., et al. 2005, ApJL, 627, L29
Thilker, D. A., Bianchi, L., Meurer, G., et al. 2007, ApJS, 173, 538
Watson, L. C., & Koda, J. 2017, in Outskirts of Galaxies, ed. J. H. Knapen, J. C. Lee, & A. Gil de Paz, Vol. 434, 175
Braine, J., & Herpin, F. 2004, Nature, 432, 369
Braine, J., Ferguson, A. M. N., Bertoldi, F., & Wilson, C. D. 2007, ApJL, 669, L73
Braine, J., Gratier, P., Kramer, C., et al. 2010, A&A, 520, A107
Dessauges-Zavadsky, M., Verdugo, C., Combes, F., & Pfenniger, D. 2014, A&A, 566, A147
Braine, J., Gratier, P., Contreras, Y., Schuster, K. F., & Brouillet, N. 2012, A&A, 548, A52
Bicalho, I. C., Combes, F., Rubio, M., Verdugo, C., & Salome, P. 2019, A&A, 623, A66
Williams, J. P., de Geus, E. J., & Blitz, L. 1994, ApJ, 428, 693
Scoville, N. Z., Polletta, M., Ewald, S., et al. 2001, AJ, 122, 3017
Sakamoto, S., Hayashi, M., Hasegawa, T., Handa, T., & Oka, T. 1994, ApJ, 425, 641
Nakamura, F., Ishii, S., Dobashi, K., et al. 2019, PASJ, 71, S3
Kong, S., Arce, H. G., Feddersen, J. R., et al. 2018, ApJS, 236, 25
Ikeda, M., Maezawa, H., Ito, T., et al. 1999, ApJL, 527, L59
Masui, S., Yamasaki, Y., Ogawa, H., et al. 2021, PASJ, 73, 1100
Nakamura, T., Kodaira, S., Ishii, K., Inatani, J., & Ohishi, M. 1984, PASJ, 36, 123
Enoch, M. L., Evans, II, N. J., Sargent, A. I., et al. 2008, ApJ, 684, 1240
Watanabe, Y., Nishimura, Y., Harada, N., et al. 2017, ApJ, 845, 116
Barnes, A. T., Kauffmann, J., Bigiel, F., et al. 2020, MNRAS, 497, 1972

Maloney, P., & Black, J. H. 1988, ApJ, 325, 389
van Dishoeck, E. F., & Black, J. H. 1988, ApJ, 334, 771
Wolfire, M. G., Hollenbach, D., & McKee, C. F. 2010, ApJ, 716, 1191
Wilson, C. D., Warren, B. E., Israel, F. P., et al. 2009, ApJ, 693, 1736
Gil de Paz, A., Boissier, S., Madore, B. F., et al. 2007b, ApJS, 173, 185

Resolving the Rise and Fall of Star Formation in Galaxies
Proceedings IAU Symposium No. 373, 2023
T. Wong & W.-T. Kim, eds.
doi:10.1017/S1743921322004252

The first detection of deuterated water toward extragalactic hot cores with ALMA

Marta Sewiło[1,2], Agata Karska[3], Lars E. Kristensen[4], Steven B. Charnley[5], C.-H. Rosie Chen[6], Joana M. Oliveira[7], Martin Cordiner[5,8], Jennifer Wiseman[5], Álvaro Sánchez-Monge[9], Jacco Th. van Loon[7], Remy Indebetouw[10,11], Peter Schilke[9] and Emmanuel Garcia-Berrios[5,8,12]

[1]Exoplanets and Stellar Astrophysics Laboratory and CRESST II, NASA Goddard Space Flight Center, Greenbelt, MD 20771, USA; email: marta.m.sewilo@nasa.gov

[2]Department of Astronomy, University of Maryland, College Park, MD 20742, USA

[3]Nicolaus Copernicus University, ul. Grudzikadzka 5, 87-100 Toruń, Poland

[4]Niels Bohr Institute, University of Copenhagen, Øster Voldgade 5-7, 1350 Copenhagen K, Denmark

[5]NASA Goddard Space Flight Center, Greenbelt, MD 20771, USA

[6]Max-Planck-Institut für Radioastronomie, Auf dem Hügel 69 D-53121 Bonn, Germany

[7]Lennard-Jones Laboratories, Keele University, ST5 5BG, UK

[8]The Catholic University of America, Washington, DC 20064, USA

[9]I. Physikalisches Institut der Universität zu Köln, Zülpicher Str. 77, 50937, Köln, Germany

[10]Department of Astronomy, University of Virginia, PO Box 400325, Charlottesville, VA 22904, USA

[11]National Radio Astronomy Observatory, 520 Edgemont Rd, Charlottesville, VA 22903, USA

[12]Department of Astronomy, University of Illinois, 1002 W. Green St., Urbana, IL 61801, USA

Abstract. We discuss the first detection of deuterated water (HDO) in extragalactic hot cores. The HDO 2_{11}–2_{12} line has been detected with the Atacama Large Millimeter/submillimeter Array (ALMA) toward hot cores N 105–2 A and 2 B in the N 105 star-forming region in the low-metallicity Large Magellanic Cloud (LMC), the nearest star-forming galaxy. We compared the HDO line luminosity ($L_{\rm HDO}$) measured toward two hot cores in N 105 to those observed toward a sample of 17 Galactic hot cores and found that the observed values of $L_{\rm HDO}$ for the LMC hot cores fit very well into the $L_{\rm HDO}$ trends with $L_{\rm bol}$ and metallicity observed toward the Galactic hot cores. Our results indicate that $L_{\rm HDO}$ seems to be largely dependent on the source luminosity, but metallicity also plays a role. We provide a rough estimate of the H_2O column density and abundance ranges toward N 105–2 A and 2 B by assuming that HDO/H_2O toward the LMC hot cores is the same as that observed in the Milky Way; the obtained values are systematically lower than those measured in the Galactic hot cores. The spatial distribution and velocity structure of the HDO emission in N 105–2 A is consistent with HDO being the product of the low-temperature dust grain chemistry.

Keywords. Astrochemistry, Magellanic Clouds, Star Formation, Protostars

---

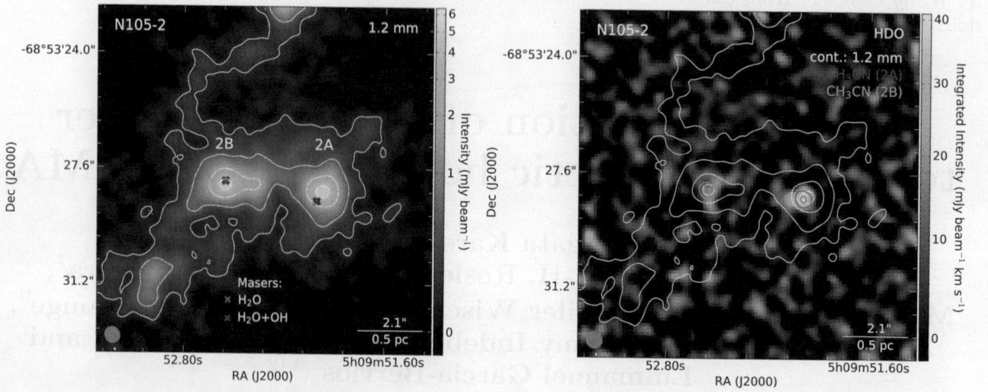

Figure 1. *Left*: The 1.2 mm continuum image of N 105–2 with the positions of hot cores 2 A and 2 B, and H_2O and OH masers indicated. *Right*: The HDO 2_{11}–2_{12} integrated intensity image. Red/green contours correspond to (50, 90)% of the CH_3CN integrated intensity peak of 0.38/0.19 Jy beam^{-1} km s^{-1} for 2 A/2 B. White contours in both panels represent the 1.2 mm continuum emission with contour levels of (3, 10, 40, 100) × the image rms noise level (σ) of 5.1×10^{-5} Jy beam^{-1}.

1. Background

Water (H_2O) is a key molecule tracing the chemical and physical processes associated with the formation of stars and planets. Deuterated water (HDO) has recently been detected for the first time toward an extragalactic star-forming region, providing us an excellent opportunity to study water chemistry in a significantly different environment than in today's Galaxy. The HDO 2_{11}–2_{12} transition at 241,561.550 MHz has been detected with ALMA toward hot cores N 105–2 A and 2 B in the star-forming region N 105 in the low-metallicity LMC (see Fig. 1; Sewiło et al. 2022a,b). Hot cores are compact ($\lesssim$0.1 pc), warm ($\gtrsim$100 K), and dense ($\gtrsim 10^{6-7}$ cm^{-3}) regions surrounding high-mass protostars very early in their evolution; a typical Galactic hot core is chemically rich, containing the products of the interstellar grain-surface chemistry (including complex organics and water). Hot cores N 105–2 A and 2 B are two of only a handful of extragalactic sources with hot core complex organic chemistry (Sewiło et al. 2022a and references therein). The metallicity of the LMC ($Z_{\rm LMC}$~0.3–0.5 $Z_{\odot}$; e.g., Russell & Dopita 1992) is similar to galaxies around the peak of star formation in the Universe (redshift~1.5), making it an ideal template for studying star formation and water chemistry in low-metallicity systems at earlier cosmological epochs where direct observations are impossible.

There are several factors that can directly impact the formation and destruction of H_2O and HDO molecules in a low-metallicity environment. The abundance of atomic O in the LMC is more than a factor of 2 lower when compared with the Galaxy (i.e., fewer O atoms are available for water chemistry). The dust-to-gas ratio in the LMC is lower, resulting in fewer dust grains for surface chemistry and less shielding than in the Galaxy. The deficiency of dust combined with the harsher UV radiation field in the LMC leads to warmer dust temperatures and consequently, less efficient grain-surface reactions. The LMC's cosmic-ray density is about 25% of that measured in the solar neighborhood, resulting in less effective cosmic-ray-induced UV radiation.

2. Results and Conclusions

The HDO data indicate that there are some minor differences (in comparison to expected) in the HDO chemistry between the LMC and the Galaxy. The HDO line luminosities ($L_{\rm HDO}$) measured toward the LMC hot cores (directly from the observed

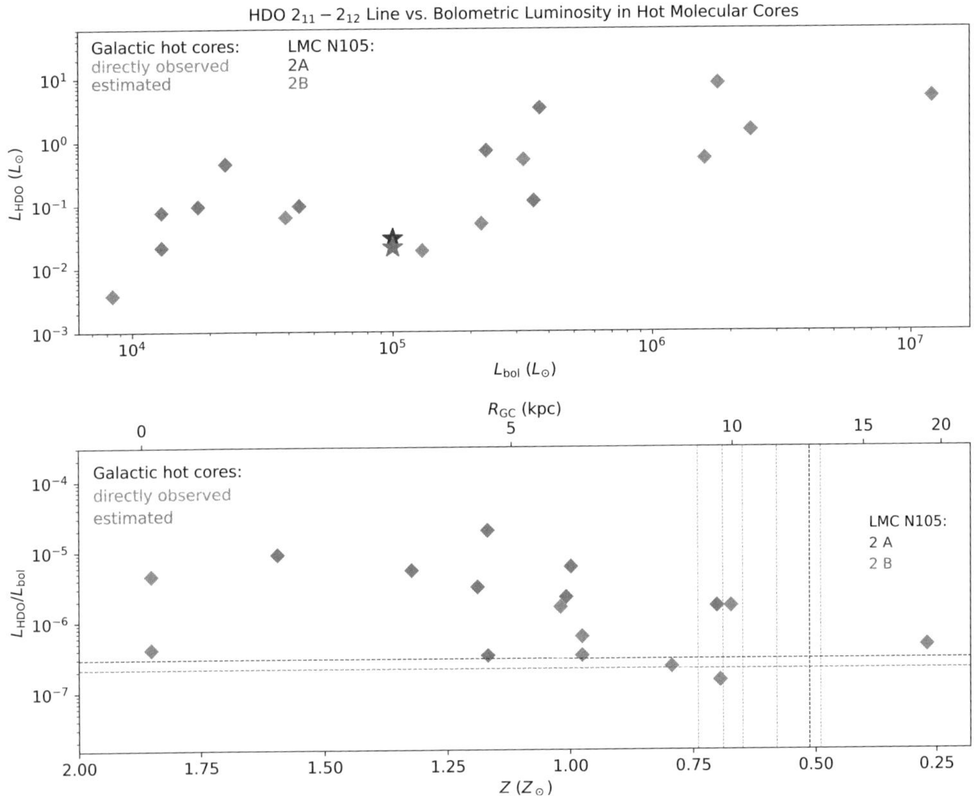

Figure 2. *Top:* The HDO 241.6 GHz line luminosities ($L_{\rm HDO}$) measured toward the LMC hot cores N 105–2 A and 2 B and those observed toward a sample of 17 Galactic hot cores as a function of the bolometric luminosity ($L_{\rm bol}$); Fig. 3 from Sewiło et al. (2022b). The values of $L_{\rm HDO}$ indicated with orange diamonds are based on the observations of the HDO $2_{11} - 2_{12}$ transition, while those indicated with blue diamonds were estimated using the observations of other HDO transitions as described in Sewiło et al. (2022b). *Bottom:* The $L_{\rm HDO}/L_{\rm bol}$ ratio as a function of the Galactocentric distance ($R_{\rm GC}$) and metallicity (Z); adapted from Fig. 4 in Sewiło et al. (2022b). Z at a given $R_{\rm GC}$ was calculated using the Balser et al. (2011)'s O/H radial gradient: $12 + \log({\rm O/H}) = -0.0446\, R_{\rm GC} + 8.962$; the vertical black line indicates $R_{\rm GC}$ where $[12 + \log({\rm O/H})]$ equals to the LMC value of 8.4 (e.g., Russell & Dopita 1992). The vertical gray lines indicate $R_{\rm GC}$ where $[12 + \log({\rm O/H})] = 8.4$ based on the O/H gradients found in other H II region studies.

integrated HDO line intensities) have been compared to those observed toward a sample of 17 Galactic hot cores covering three orders of magnitude in $L_{\rm HDO}$, four orders of magnitude in bolometric luminosity ($L_{\rm bol}$), and a wide range of Galactocentric distances (from ~0.1 to 19 kpc), and thus metallicities (from ~1.8 $Z_\odot$ toward the Galactic Center to ~0.25 $Z_\odot$ in the outer Galaxy). The results are shown in Fig. 2. The Galactic hot cores shown in the plots are (in order of increasing $L_{\rm bol}$): WB 89–789 SMM1 (an extreme outer Galaxy source), NGC 7538 S, IRAS 18089−1732, G9.62+0.19, W43 MM1, W3(H_2O), W33A, NGC 7538 IRS1, AFGL 2591, G31.41+0.31, G34.26+0.15, G29.96−0.02, G10.47+0.03A, W51 e1/e2, Sgr B2(N), W51d, and Sgr B2(M).

Our results are providing first insights into the HDO chemistry in the LMC: (1) $L_{\rm HDO}$ for the LMC hot cores fit well into the $L_{\rm HDO}$ trends with $L_{\rm bol}$ and metallicity observed toward the Galactic hot cores; (2) $L_{\rm HDO}$ seems to be largely dependent on the source luminosity, but metallicity also plays a role. Moreover, the observations show that if

HDO/H_2O in the LMC hot cores N 105–2 A and 2 B is within the range observed in the Galactic hot cores (typically $(2-8)\times10^{-4}$, van Dishoeck et al. 2021), then for the measured HDO abundances, the resulting H_2O abundance ranges toward N 105–2 A and 2 B would be systematically lower than Galactic values. The spatial distribution and velocity structure of the HDO emission in N 105–2 A is consistent with HDO being the product of the low-temperature dust grain chemistry.

Our results are in agreement with the astrochemical model predictions that HDO is abundant regardless of the extragalactic environment and should be detectable with ALMA in other external galaxies.

Acknowledgments: The material is based upon work supported by NASA under award number 80GSFC21M0002. A.K. acknowledges support from the First TEAM grant of the Foundation for Polish Science No. POIR.04.04.00-00-5D21/18-00 and from the Polish National Agency for Academic Exchange Grant No. PPI/APM/2018/1/00036/U/001.

References

Balser, D. S., Rood, R. T., Bania, T. M., & Anderson, L. D. 2011, ApJ, 738, 27
Russell, S. C., & Dopita, M. A. 1992, ApJ, 384, 508
Sewiło, M., Cordiner, M., Charnley, S. B., et al. 2022a, ApJ, 931, 102
Sewiło, M., Karska, A., Kristensen, L. E., et al. 2022b, ApJ, 933, 64
van Dishoeck, E. F., Kristensen, L. E., Mottram, J. C., et al. 2021, A&A, 648, A24

Resolving the Rise and Fall of Star Formation in Galaxies
Proceedings IAU Symposium No. 373, 2023
T. Wong & W.-T. Kim, eds.
doi:10.1017/S1743921322003994

Radiatively Cooling Superwinds in Ultracompact H II Regions

A. Danehkar

Eureka Scientific, 2452 Delmer Street, Suite 100, Oakland, CA 94602-3017, USA
email: danehkar@eurekasci.com

Abstract. Ultracompact H II regions (UC-HII) are the young, very dense cores of massive star-forming regions in dwarf galaxies, where newly formed massive OB stars are surrounded by natal molecular clouds. Thermal energy deposited by mechanical feedback from a cluster of massive OB stars can form a superwind, which may lead to a wind-blown bubble as well as radiative cooling. We investigate the formation of radiatively cooling superwinds in UC-HII using a radiative cooling module in the hydrodynamics program FLASH. We built a grid of hydrodynamic simulations to determine the dependence of radiative cooling on the cluster radius, mass-deposition rate, wind velocity, and ambient medium in UC-HII. Our findings could help to better understand star formation in massive star-forming regions, where cool superwinds could trigger the formation of molecular clumpy regions.

Keywords. Stars: winds – ISM: bubbles – radiation mechanisms – galaxies: starburst

---

The deposition of energy and momentum by young, massive OB stars into the interstellar medium (ISM) produces superwinds on parsecs to kpc scales (Heckman et al. 1990; Lehnert & Heckman 1995; Veilleux et al. 2005), whose cooling outflow feedback can promote star formation in giant molecular clouds (Fabian et al. 1984; Krause et al. 2016). The shock-ionization by superwinds often generates X-ray bubbles (Strickland et al. 1997; Martin 1999; Ott et al. 2005). Contemporary surveys of some compact star-forming regions such as those in NGC 5253 (Turner et al. 2017) and Mrk 71 (Oey et al. 2017) suggested the presence of excessive cooling effects beyond the adiabatic predictions. This is generally attributed to radiatively cooling superwinds, as shown by semi-analytical numerical results (Silich et al. 2004; Tenorio-Tagle et al. 2005) and adaptive mesh hydrodynamic simulations (Gray et al. 2019a; Danehkar et al. 2021).

One of the earliest stages in the evolution of an H II region is the so-called ultracompact H II (UC-HII) region (Wood & Churchwell 1989), which is very dense ($\gtrsim 10^4$ cm^{-3}) surrounding a compact (< 0.1 pc) stellar cluster having a lifetime of $\lesssim 0.1$ Myr (see the review by Churchwell 2002). In such compact regions, the youngest OB stars are present, which ionize the surrounding ISM and produce superwinds on a small scale ($\lesssim 10$ pc) (Hoare et al. 2007; Olivier et al. 2021), leading to hot, wind-blown bubbles (see e.g. Tsujimoto et al. 2006) and shells (De Pree et al. 2000). The cool gas formed by a superwind in UC-HII could also stimulate the formation of new stars in giant molecular clouds.

To learn more about radiatively cooling superwinds in UC-HII, we used the atomic chemistry and radiative cooling module MAIHEM (Gray et al. 2019b) in the hydrodynamics program FLASH (Fryxell et al. 2000) to run a set of hydrodynamic simulations of a spherical stellar wind model. We consider a stellar wind driven by thermal feedback from a stellar cluster described by the cluster radius $R_{\rm sc}$ being smaller than 1 pc in typical

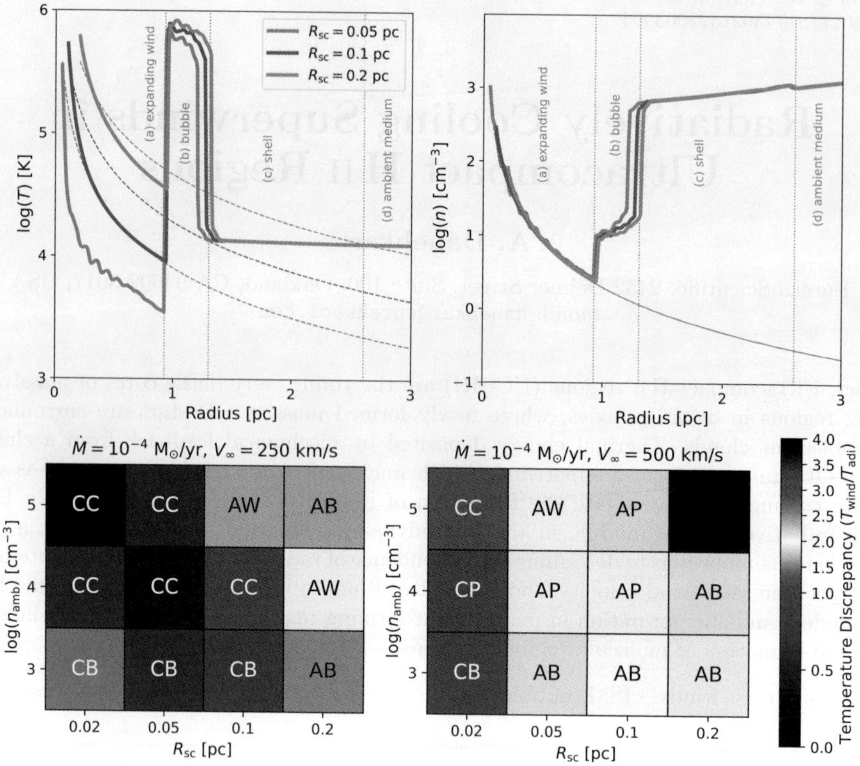

Figure 1. *Top Panels*: The temperature (T) and density (n) profiles (solid lines) along with the adiabatic predictions (dashed lines) for the models with $V_\infty = 250\,\mathrm{km\,s^{-1}}$, $\dot{M} = 10^{-4}\,\mathrm{M_\odot\,yr^{-1}}$, $t = 1\,\mathrm{Myr}$, $R_{sc} = 0.05$, 0.1, and 0.2 pc, $M_\star = 2\times10^4\,\mathrm{M_\odot}$, $n_{amb} = 10^3\,\mathrm{cm^{-3}}$, $Z/Z_\odot = 0.5$, and $t =$ 0.1 Myr. The expanding wind region, bubble, shell, and the ambient medium for $R_{sc} = 0.2$ pc are separated by dotted lines. *Bottom Panel*: The average wind temperatures T_{wind} produced by our hydrodynamic simulations with respect to the average adiabatic predictions T_{adi} for the wind velocities $V_\infty(t) = 250$ and 500 $\mathrm{km\,s^{-1}}$, mass-deposition rate $\dot{M}(t) = 10^{-4}\,\mathrm{M_\odot\,yr^{-1}}$, metallicity $Z/Z_\odot = 0.5$, ambient densities $\log n_{amb} = 3$, 4, and $5\,\mathrm{cm^{-3}}$, and stellar radii $R_{sc} = 0.02$, 0.05, 01, and 0.2 pc, total stellar mass $M_\star = 2\times10^4\,\mathrm{M_\odot}$, and age $t = 0.1$ Myr. The wind modes (AW, AB, AP, CC, CB, and NW) are classified according to the criteria by Danehkar et al. (2021).

H II regions, mass-deposition rate $\dot{M}$, wind velocity V_∞, and the radiation field made by Starburst99 (Leitherer et al. 1999) for a total stellar mass of $M_\star = 2\times10^4\,\mathrm{M_\odot}$, and ambient density n_{amb} higher than typical evolved H II regions considered in Danehkar et al. (2021). Figure 1 shows the temperature and density profiles (top panels) of three wind models simulated by MAIHEM for different cluster radii $R_{sc} = 0.05$, 01, and 0.2 pc. For comparison, the adiabatic temperature and density profiles were also plotted using dashed lines. Each model has four regions (as defined by Weaver et al. 1977), namely expanding wind, hot bubble, shell, and ambient medium, shown for $R_{sc} = 0.2$ pc in Fig. 1. With fixed wind parameters, it can be seen that models with smaller cluster radii have stronger radiative cooling.

Figure 1 presents various wind modes (bottom panel) for UC-HII regions with dense ambient densities ($n_{amb} = 10^3$, 10^4, and $10^5\,\mathrm{cm^{-3}}$), compact cluster radii ($R_{sc} = 0.02$, 0.05, 01, 0.2 pc), produced for tow different wind velocities $V_\infty = 250$ and 500 $\mathrm{km\,s^{-1}}$ and mass-deposition rate $\dot{M}(t) = 10^{-4}\,\mathrm{M_\odot\,yr^{-1}}$. We employ the superwind classification by Danehkar et al. (2021) according to energy- or momentum-conserving, the presence or absence of a bubble, and an adiabatically or radiatively cooled thermal curve. The adiabatic (AW) and catastrophic cooling (CC) modes are those without any wind-blown bubble, whereas adiabatic bubble (AB) and catastrophic cooling bubble (CB) modes have a thermal bubble. Moreover, the adiabatic, pressure-confined (AP) mode has a hot bubble being stalled by the surroundings thermal pressure. The no wind (NW) mode is assigned to a superwind suppressed by a high ambient pressure. It can be seen that reducing the cluster radii increases radiative cooling in superwinds. Moreover, we again spot the dependency of cooling on the wind velocity and mass-deposition rate similar to what Danehkar et al. (2021) obtained for more evolved H II regions with ambient densities less than $10^3\,\mathrm{cm}^{-3}$ and the typical cluster radius of 1 pc.

Recently, Danehkar et al. (2021, 2022) utilized the radial profiles of the density, temperature, and non-equilibrium ionization (NEI) states made by MAIHEM for evolved H II regions to produce the line emissivities using the photoionization program CLOUDY (Ferland et al. 2017) for combined collisional ionization and photoionization, as well as non-equilibrium photoionization conditions. Danehkar et al. (2022) found that in metal-poor regions, the O VI lines were stronger when photoionization is not in equilibrium. O VI lines were previously proposed for tracing radiative cooling in H II regions (Gray et al. 2019a,b). In particular, the nearest Lyman-break analog, Haro 11, was found to emit O VI, which could originate from radiative cooling (Grimes et al. 2007). Further numerical modeling of superwinds driven by stellar clusters with the physical properties seen in UC-HII will be great for identifying compact coolants formed by radiatively cooling in massive star-forming regions.

References

Churchwell E., 2002, ARA&A, 40, 27
Danehkar A., Oey M. S., Gray W. J., 2021, ApJ, 921, 91
Danehkar A., Oey M. S., Gray W. J., 2022, ApJ, 937, 68
De Pree C. G., Wilner D. J., Goss W. M., Welch W. J., McGrath E., 2000, ApJ, 540, 308
Fabian A. C., Nulsen P. E. J., Canizares C. R., 1984, Nature, 310, 733
Ferland G. J. et al., 2017, RMxAA, 53, 385
Fryxell B. et al., 2000, ApJS, 131, 273
Gray W. J., Oey M. S., Silich S., Scannapieco E., 2019a, ApJ, 887, 161
Gray W. J., Scannapieco E., Lehnert M. D., 2019b, ApJ, 875, 110
Grimes J. P. et al., 2007, ApJ, 668, 891
Heckman T. M., Armus L., Miley G. K., 1990, ApJS, 74, 833
Hoare M. G., Kurtz S. E., Lizano S., Keto E., Hofner P., 2007, in Protostars and Planets V, Reipurth B., Jewitt D., Keil K., eds., Tucson: University of Arizona Press, p. 181
Krause M. G. H., Charbonnel C., Bastian N., Diehl R., 2016, A&A, 587, A53
Lehnert M. D., Heckman T. M., 1995, ApJS, 97, 89
Leitherer C. et al., 1999, ApJS, 123, 3
Martin C. L., 1999, ApJ, 513, 156
Oey M. S., Herrera C. N., Silich S., Reiter M., James B. L., et al., 2017, ApJ, 849, L1
Olivier G. M., Lopez L. A., Rosen A. L., Nayak O., Reiter M., et al., 2021, ApJ, 908, 68
Ott J., Walter F., Brinks E., 2005, MNRAS, 358, 1453
Silich S., Tenorio-Tagle G., Rodríguez-González A., 2004, ApJ, 610, 226
Strickland D. K., Ponman T. J., Stevens I. R., 1997, A&A, 320, 378
Tenorio-Tagle G., Silich S., Rodríguez-González A., Muñoz-Tuñón C., 2005, ApJ, 620, 217
Tsujimoto M., Hosokawa T., Feigelson E. D., Getman K. V., Broos P. S., 2006, ApJ, 653, 409
Turner J. L., Consiglio S. M., Beck S. C., Goss W. M., et al., 2017, ApJ, 846, 73
Veilleux S., Cecil G., Bland-Hawthorn J., 2005, ARA&A, 43, 769
Weaver R., McCray R., Castor J., Shapiro P., Moore R., 1977, ApJ, 218, 377
Wood D. O. S., Churchwell E., 1989, ApJ, 340, 265

Resolving the Rise and Fall of Star Formation in Galaxies
Proceedings IAU Symposium No. 373, 2023
T. Wong & W.-T. Kim, eds.
doi:10.1017/S1743921322004902

Milky Way: structure via live potentials

Eva Durán-Camacho and Ana Duarte-Cabral
School of Physics and Astronomy, Cardiff University, Queens Buildings, The Parade, Cardiff CF24 3AA, UK
email: durancamachoe@cardiff.ac.uk

Abstract. We use the AREPO numerical code to model the structure of a Milky Way like galaxy (MW) via a suite of simulations composed of a stellar disc and bulge, a dark matter halo, and a gaseous disc under isothermal conditions. For each model, we produce longitude velocity (l-v) maps of the gas surface densities to extract the skeletons of the main features (arms, bar), and the contours defining the terminal velocities of the gas. We compare these with observations via a number of diagnostic tools, and select the model that best reproduces the main observed features of the Milky Way.

Keywords. Galaxy: structure, Galaxy: kinematics and dynamics, galaxies: spiral

1. Introduction

In order to understand the specific role that the large-scale galactic features (spiral arms/bar) have on star formation within the Milky Way, we first need to pin down its anatomy. However, the structure of the Milky Way has been a subject of study for decades and there is still no real consensus on its exact morphology, mostly due to observational limitations (e.g. Elmegreen 1985). Nevertheless, our Galaxy is unanimously believed to be spiral (e.g. Xu et al. 2016) and host a Galactic bar (e.g. Weinberg 1992).

Aiming to model the spiral/bar pattern of the Milky Way using numerical simulations, the most common approach is to set up an analytical gravitational potential that mimics the external potential from the stars and dark matter halo (e.g. Dobbs et al. 2013; Smith et al. 2020). Another approach is to use stellar and dark matter particles to recreate the galactic structure self-consistently through the N-body dynamical evolution of the system (i.e. the so-called "live potentials"). However, most models that focus on the Milky Way typically set out to produce a MW-type galaxy, rather than effectively attempting to fit the observed structure. One exception are the works of Pettit et al. (2014, 2015) who produce l-v maps from their models and compare them to observations via χ^2 statistics. Their models with live potentials are able reproduce the observed spiral arm features, but fail to generate an inner bar. The ability to reproduce the anatomy of the MW in simulations is essential to study the role of the Galactic structure on star formation, but the choice of live-versus-fixed potentials can have a strong impact on the conclusions, as it changes the gas response to the potential. In this context, our aim is to reproduce the Mily Way structure using the self-consistent evolution of live-potentials.

2. Overview

Our simulations are performed with the moving-mesh code AREPO (Springel 2010). Our primary goal is to re-create the overall structure of the MW (i.e. spiral pattern and central bar) through the N-body dynamical interaction of the stars over several Gyrs.

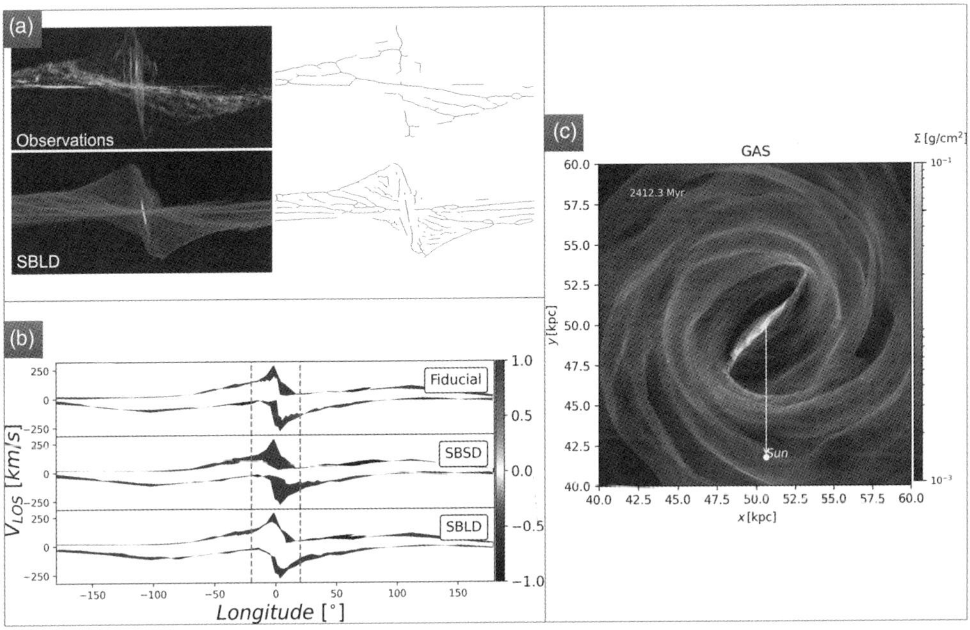

Figure 1. A) Left: *l*-*v* maps of the observations (top) and the best model at its optimal time/viewing angle. Right: the respective extracted skeletons. The ranges of these *l*-*v* are: $|V_{LOS}| \pm 280$ km/s and $|l| \pm 60°$. B) Terminal velocity comparison of the Dame et al. (2001) CO observations with our three different models (as labelled). Vertical dashed red lines delimit the inner and outer Galaxy regions ($| l |< 20°$). Red regions denote areas covered by the observations and not the models, and blue regions the reverse. C) Top-down view of the surface density of the gas of our best fit model (SBLD at a time ~ 2.4 Gyrs) with the Sun at 270°.

Therefore, we minimise the computational time spent on the hydrodynamical part of the code by having a relatively low resolution for the gas and isothermal equations of state.

The initial conditions for our MW models were setup using the moment-based code MAKENEWDISK, described in (Springel 2005). They consist of a dark matter halo, a rotationally supported disc of gas and stars, and a central stellar bulge. The dark matter halo and the stellar bulge follow an Hernquist density profile (Hernquist 1990), and the stellar disc follows an exponential surface density profile. We produce three different set ups for the stellar distribution: a Fiducial model based on the best fit model by Pettit et al. (2015); a model with a smaller bulge, and small disc (SBSD); and a model with a small bulge, and larger disc (SBLD). We also explore two different profiles for the gas distribution: same exponential surface density profile as the stars (i.e. the default of MAKENEWDISK); and a uniform surface density profile, as suggested by some observations (e.g. Ferriere et al. 2007). Full details on the parameters of these models can be found in Durán-Camacho et al. (in prep.).

To compare the structures produced by our six models to the observations, we produce *lv*-maps for each model, by looking at 20 snapshots from $1.4 - 2.9$ Gyrs through 24 different viewing angles each. We then find the best model with various different tests.

3. Results and discussion

To evaluate how our models perform at reproducing the morphology of the MW, we extract the spines that trace the main features in *lv*-space (i.e. the "skeletons") of the ^{12}CO emission from Dame et al. (2001), and of our simulations (Fig. 1, panel A). We then use a statistical tool, the Symmetrized Modified Hausdorff Distance (SMHD,

(Sormani et al. 2015), to compare the observed and modelled skeletons. We start by investigating the time evolution of the SMHD metric and select the snapshot at which the global SMHD metric was lowest for each model. We then investigate how the SMHD varies within that snapshot as a function of the observer's viewing angle, to select the optimal reproduction of the MW for each model. The best model overall is selected by comparing the SMHD metric values between models, as well as the terminal velocities reproduced by the models versus the observed ones (Fig. 1, panel B).

This analysis indicates that all models are similar at mimicking the outer Galaxy, but there is a clear difference between models in the inner Galaxy, showing SBSD as the least favourable model, and SBLD as the better one. This is mostly due to the larger velocity ranges it attains in the inner galaxy, and the better match of the skeleton structures towards the central part, with the formation of a bar that resembles that of the MW (with a bar pattern speed of $22\,\mathrm{km\,s^{-1}\,kpc^{-1}}$, and a length of $\sim 6\,\mathrm{kpc}$).

Finally, we analyse which gas profile best reproduces our MW, by creating 1D plots of the column density versus longitude. The equivalent distribution for the observations was obtained from the SED fitting of the dust continuum emission from $60 - 850\mu$m from IRAS and Planck (Wheelock et al. 1994, Collaboration 2016). We find that a uniform gas surface density distribution is a better (albeit not perfect) match to observations.

Overall, we find that the SBLD model with a uniform gas profile, at a time 2.4 Gyrs is our best fit (Fig. 1, panel C). We will use this model as the base configuration for more sophisticated modelling of the ISM in the MW.

Supplementary material

To view supplementary material for this article, please visit https://doi.org/10.1017/S1743921322004902.

References

Elmegreen, D. M. 1985, In Symposium-International Astronomical Union (Vol. 106, pp. 255–272). Cambridge University Press.

Xu, Y., Reid, M., Dame, T., Menten, K., Sakai, N., Li, J., Zheng, X. 2016, *Science Advances* 2(9), e1600878

Weinberg, M. D. 1992, *ApJ* 384, 81–94

Dobbs, C. L., Pringle, J. E. 2013, *MNRAS* 432(1), 653–667

Smith, R. J., Treß, R. G., Sormani, M. C., Glover, S. C., Klessen, R. S., Clark, P. C., Zucker, C. 2020, *MNRAS* 492(2), 1594–1613

Pettitt, A. R., Dobbs, C. L., Acreman, D. M., Price, D. J. 2014, *MNRAS* 444(1), 919–941

Pettitt, A. R., Dobbs, C. L., Acreman, D. M., Bate, M. R. 2015, *MNRAS* 449(4), 3911–3926

Springel, V. 2010, *MNRAS* 401(2), 791–851

Springel, V., Di Matteo, T., Hernquist, L. 2005, *MNRAS* 361(3), 776–794

Hernquist, L. 1990, *ApJ* 356, 359–364

Ferriere, K., Gillard, W., Jean, P. 2007, *ARAA* 467(2), 611–627

Dame, T. M., Hartmann, D., Thaddeus, P. 2001, *ApJ* 547(2), 792

Sormani, M. C., Magorrian, J. 2015, *MNRAS* 446(4), 4186–4204

Wheelock, S. L., Gautier, T. N., Chillemi, J., Kester, D., McCallon, H., Oken, C., ... Good, J. 1994, (No. NAS 1.26: 197335)

Collaboration, P., Ade, P. A. R., Aghanim, N., Akrami, Y., Aluri, P. K., Arnaud, M., ... Mennella, A. 2016

Resolving the Rise and Fall of Star Formation in Galaxies
Proceedings IAU Symposium No. 373, 2023
T. Wong & W.-T. Kim, eds.
doi:10.1017/S1743921322004926

Parsec scale CO depletion in KAGONMA 71, or a star-forming filament in CMa OB1

T. Handa[1,2], Y. Hirata[2], T. Murase[2], J. Nishi[2], Y. Shimajiri[3], T. Omodaka[2], M. Nakano[4], K. Sunada[5], T. Ito[1,6] and J. O. Chibueze[7,8]

[1]Amanogawa Galaxy Astronomy Research Center, Kagoshima Univ., 1-21-35 Korimoto, Kagoshima 890-0065, Japan
email: handa@sci.kagoshima-u.ac.jp

[2]Faculty of Science, Kagoshima Univ., 1-21-35 Korimoto, Kagoshima 890-0065, Japan

[3]ALMA project, National Astron. Obs. Japan, 1-21-3 Osawa, Mitaka 181-8588, Japan

[4]Faculty of Education, Oita University, 700 Dannoharu, Oita 870-1192, Japan

[5]Mizusawa VLBI Obs., National Astron. Obs. Japan, Hoshigamine, Oshu-shi 2531, Japan

[6]Kuamoto Univ. Kumamoto, Japan

[7]Centre for Space Research, North-West Univ., Potchefstroom 2531, South Africa

[8] Department of Phys. and Astron., Faculty of Physical Sci., Univ. of Nigeria, Carver Building, University road, Nsukka 410001, Nigeria

Abstract. The depletion of CO molecules is observed in infrared dark clouds. However, only few exsamples are found in pc-scale. An NH_3 emission is one of good counter parts of $C^{18}O$ because of similar effective critical density. Our NH_3 observations of a molecular filament associated with CMa OB1 or KAG 71, which is a target of Kagoshima Galactic Object survey with Nobeyama 45-m telescope by Mapping in Ammonia lines (KAGONMA) project. Although NH_3 data shows similarity in morphology with infrared data suggesting no depletion, $C^{18}O$ in the clumps 4 and 6 are weaker than expected based on NH_3 data. After examining the dissipation of the high-density gas, photodissociation, and depletion, we concluded that CO is depleted at least in the clump 4. It is a new example of depletion in pc-scale.

Keywords. ISM: abundances, ISM: clouds, ISM: molecules

1. Introduction

Although $C^{18}O$ is often used to estimate the column density of dense molecular gas, previous works reported depletion of CO molecules (e.g. Feng et al. 2020; Sabatini et al. 2019). To investigate such depletion in pc scale, a dense gas tracer less depleted, such as NH_3 is helpful. And NH_3 is one of the best counter part of $C^{18}O$ because of the similar effective critical density; both of which are about 2×10^3 cm^{-3}. From 72 objects of KAGONMA (Kagoshima Galactic Object survey with Nobeyama 45-m telescope by Mapping in Ammonia lines) project, we investigate the depletion in KAG 71, or a molecular filament associated with CMa OB1.

2. Observations & data

We made NH_3 observations in $(J, K) = (1,1), (2,2), (3,3)$ transitions as a part of KAGONMA project, simultaneously. The telescope beam size is $75''$ and the mapping

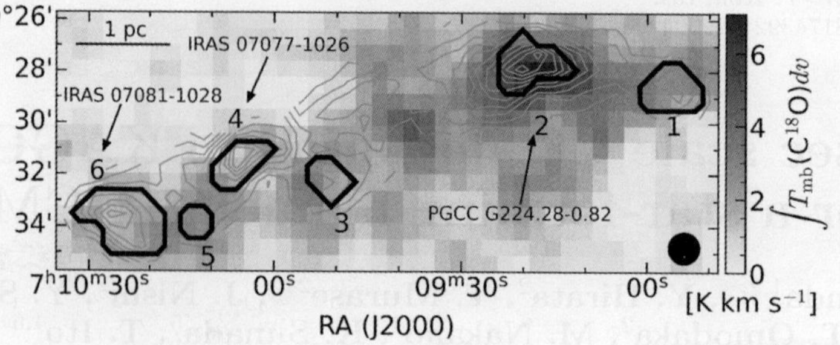

Figure 1. $NH_3(1,1)$ intensity map over FUGIN $C^{18}O$ intensity image smoothed to NH_3 beamsize. Black enclosures are the clumps 1 to 6 from east to west.

area covers about 26′× 10′ for the filament. To estimate the intensity distributions of CO, we used ^{12}CO and $C^{18}O$ data from FUGIN (Umemoto et al. 2017) and the map observed with Mopra 22-m telescope (Olmi et al. 2016). The spatial and velocity resolutions of FUGIN are 20″ and 1.3 km s^{-1}, respectively. Those of Mopra observations are 38″ and 0.09 km s^{-1}. In many investigations dust continuum flux is used to estimate the gas column density $N(H_2)$. We, therefore, used Hi-GAL data observed with *Herschel* (Molinari et al. 2010). The spatial resolutions are between 12.6″ and 36.7″ at 160 and 500 μm.

3. Results

We made the intensity distributions of NH_3 and $C^{18}O$ smoothed to be the same beamsize (fig. 1). Based on DENDROGRAM analysis (Rosolowsky et al. 2008), we identified 6 clumps in NH_3 map. We call the clumps 1 to 6 from east to west. We also made maps of the integrated intensity of $NH_3(1, 1)$ and the H_2 column density, $N(H_2)$, estimated from *Herschel* data (fig. 2(a)). The similarity of NH_3 intensity and $N(H_2)$ distributions suggests both the relative abundance of NH_3, $X(NH_3)$ and gas-to-dust ratio are almost constant in this filament. We made the map of gas kinetic temperature, T_k, from our NH_3 data (fig. 2(b)). It shows $T_k \sim 12$ K and ~ 17 K in the clumps 2 and 4, respectively. We made the distribution of $X(NH_3)$ and it is almost uniform (fig. 2(c)). However, the abundance of $C^{18}O$, $X(C^{18}O)$ estimated from Mopra data is not uniform (fig. 2(d)).

To confirm it we made the correlation plot between the intensities in $NH_3(1,1)$ and $C^{18}O$ lines from FUGIN (fig. 3). Most plots are along the strait line through the origin. It means $X(C^{18}O)$ is almost constant there. However, in the clumps 4 and 6, $C^{18}O$ intensity does not increase but slightly decreases against NH_3 intensity. In the clump 2, $C^{18}O$ intensity is higher than in the clumps 4 and 6, but smaller than the proportional line expected by overall trend.

4. Discussion

We interpret the weakness of the emission is due to depletion of $C^{18}O$ molecule on dust grains. We considered two other mechanisms shown below, but they are not the case.

(a) low volume density; dense enough to emit the N_2H^+ line (Tatematsu et al. 2017).
(b) destruction of CO molecules by UV radiation; it also destructs NH_3.

Therefore, we conclude that CO molecule is actually depleted on dust grains. The size of our clumps where we found the depletion is in pc scale.

To evaluate the degree of depletion, we define the relative depletion factor, f^{R}_{dep}, by the reciprocal value of the relative abundance of each pixel normalized by the maximum over

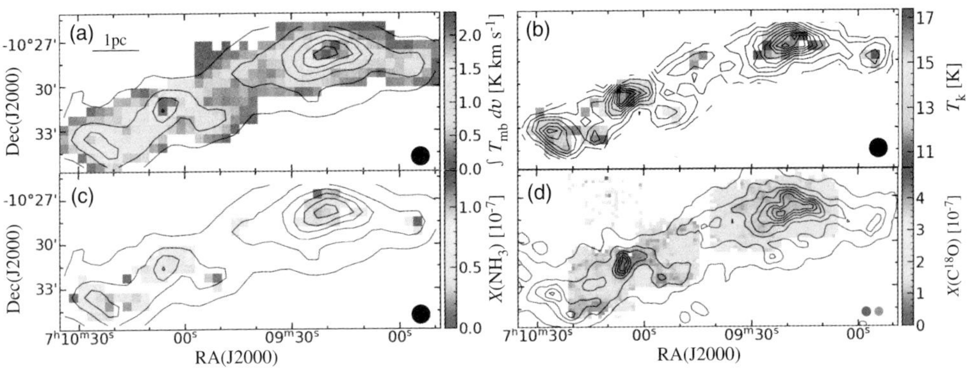

Figure 2. (a) The contour map of the column density, $N(H_2)$, estimated from *Herschel* data smoothed to NH_3 beamsize overlaid on an image of the integrated intensity of $NH_3(1,1)$ line. (b) Gas kinetic temperature image and a contour map of the integrated intensity of $NH_3(1,1)$ line. Clumps 2 is cold & 4 is warm. (c) an relative abundance image of $NH_3(1,1)$. The contour map is the contours in panel (a). (d) relative abundance image of $C^{18}O$ using dust $N(H_2)$ with the contours of $N(H_2)$.

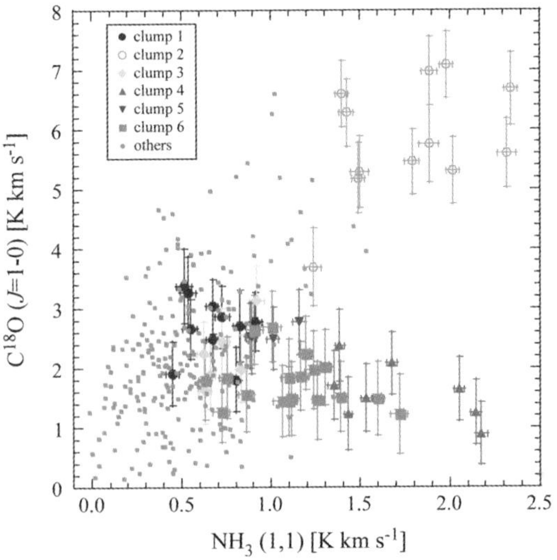

Figure 3. $NH_3(1,1)$ – $C^{18}O$ (J=1–0) integrated intensity correlation plot of the each clump and observed positions.

the filament. In the clump 2, $f^{\rm R}_{\rm dep} \simeq 2$ and in the clump 4, $f^{\rm R}_{\rm dep} \simeq 6$. They are different by a factor of 3. What makes this difference? Both temperature and density are similar; the dust temperatures in the clumps 2 and 4 are about 12 and 16 K, respectively, and the volume densities of H_2 of them using a spherically symmetric clump model are 4×10^4 and 6×10^4 cm^{-3}, respectively. To the contrary, the clump 2 have 3 YSO candidates over 100 $L_\odot$, although the clump 4 has none. Therefore, the photodesorption may prevent to CO depletion on dust.

References

Feng S., Li, D., Caselli P., *et al.* 2020, ApJ 901, 145.
Molinari S., Swinyard B., Bally J., *et al.* 2010, A&A 518, L100.

Olmi L., Cunningham M., Elia D. Jones P. 2016, A&A 594, A58.
Rosolowsky E.W., Pineda J. E., Kauffmann J., Goodman A.A. 2008, ApJ 679, 1338.
Sabatini G., Giannetti A., Bovino S., *et al.* 2019, MNRAS 490, 4489.
Tatematsu K., Liu T.; Ohashi S., *et al.* 2017, ApJS 228, 12.
Umemoto T., Minamidani T., Kuno N., *et al.* 2017, PASJ 69, 78.

Resolving the Rise and Fall of Star Formation in Galaxies
Proceedings IAU Symposium No. 373, 2023
T. Wong & W.-T. Kim, eds.
doi:10.1017/S1743921323000121

ALMA observations of the environments of G333.0162+00.7615

Toktarkhan Komesh[1,2], **Aruzhan Omar**[2], **Guido Garay**[3], **Zhandos Assembay**[2], **Nazgul Alimgazinova**[2], **Nurman Zhumabay**[4] and **Meiramgul Kyzgarina**[2]

[1]Energetic Cosmos Laboratory, Nazarbayev University, Astana 010000, Kazakhstan
email: toktarkhan.komesh@nu.edu.kz

[2]Al-Farabi Kazakh National University, Almaty 050040, Kazakhstan

[3]Departamento de Astronomía, Universidad de Chile, Las Condes, Santiago 7550000, Chile

[4]Abay Kazakh National Pedagogical University, Almaty 050010, Kazakhstan

Abstract. We have carried out ALMA observations toward the environments of G333.0162+00.7615 which was considered as a candidate of high-mass young stellar object (HMYSO) in previous studies. Our dust continuum, molecular line emission and radio recombination line emission observations show that this source is not HMYSO associated with hypercompact (HC) HII regions. Instead, we discovered two new hot cores associate with earliest stages of high mass star formation region. We estimated the rotational temperatures of these cores about 270 K from J=14→13 rotational transition of CH_3CN ladder. The moment maps show velocity gradients confirming that this cores are rotating.

Keywords. ISM: molecules — ISM: clouds — ISM: cores — stars: formation — stars: massive — ISM: kinematics and dynamics

1. Is G333.0162+00.7615 a high-mass young stellar object?

The RMS (Mottram et al. 2011) source G333.0162+00.7615 was catalogued as high-mass young stellar object (HMYSO) candidate by Urquhart et al. (2007). Due to its rising spectral index between 5.0 and 8.4 GHz, Guzmán et al. (2012) considered this source as a jet candidate.

To determine whether hot molecular cores are common around HMYSOs we observed this source using ALMA in Band 6 (256.3-259.6 GHz), dust continuum and molecular line emission arising from two molecules, SO_2 and CH_3CN (256.3-259.6 GHz). We did not detected emission in the molecular lines nor in the H29α radio recombination line (RRL), indicating that this HMYSO candidate is not associated with a a hot core nor a compact HII region. Instead, in the environments we discovered two hot cores,likely to be associated with earliest stages of high mass star formation. They are labeled as cores A and B in Figure 2 and their peak positions are at (RA, Dec) (J2000) = ($16^h15^m18.44^s$, -49°48′44.04″) and (RA, Dec) (J2000) = ($16^h15^m17.67^s$;-49°48′49.13″), respectively. These cores, located at the distance of $\sim$ 9 pc from G333.0162+00.7615, are associated with the ATLASGAL source AGAL333.018+00.766 (Contreras et al. 2013). In the following sections, we show the moment maps and rotational temperature analysis of these two cores.

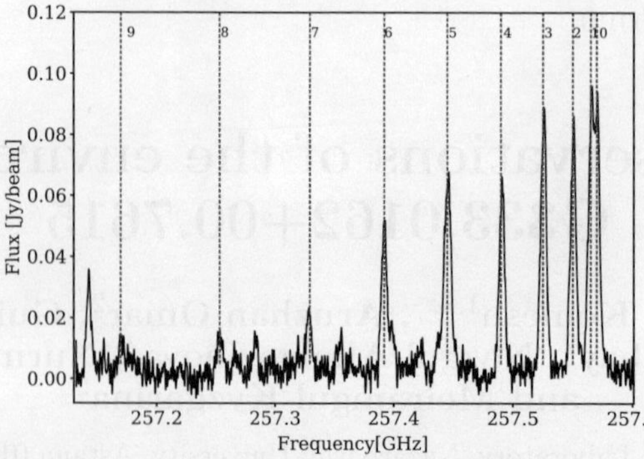

Figure 1. Methyl cyanide spectra obtained toward core A, (RA, Dec) (J2000) = ($16^h15^m18.44^s$, $-49°48'44.04''$). K components of the CH_3CN 14-13 transition are marked with dashed lines ($V_{LSRK} = -47.90$ kms^{-1}).

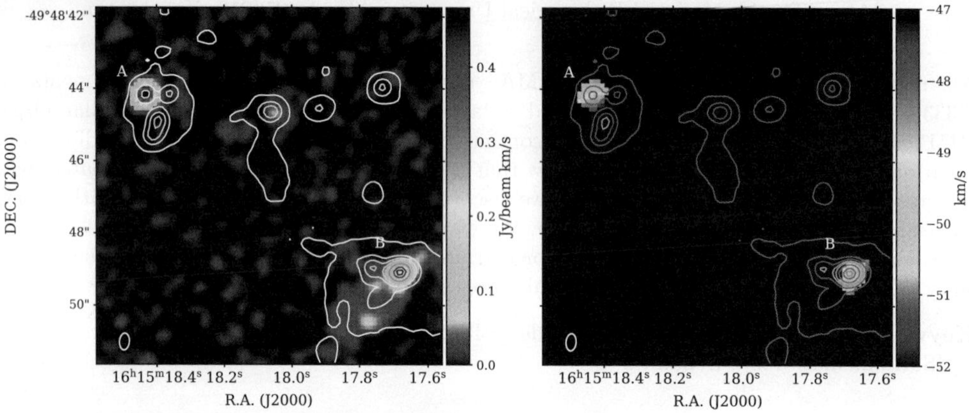

Figure 2. Image of the velocity integrated line emissions (left panel) and moment one image (right panel) of the K=3 of 14-13 ladder of CH_3CN toward the environments of G333.0162+00.7615. Superimposed are contours of the continuum emission. Contour levels are 0.001, 0.006, 0.011, 0.021, 0.031,0.05 and 0.06 Jy/beam. The white ellipse shown at the bottom left corner indicates the beam size.

2. Molecular gas

Figure 1 presents the spectrum of the $J = 14 \rightarrow 13$ rotational transition of CH_3CN integrated over a region of in $0.23''$ in size, centered on core A. Nine K components of this rotational transition were observed in this source.

Figure 2 shows the velocity integrated line emissions map (left panel) and moment 1 map (right panel) of the K=3 of 14-13 ladder of CH_3CN. Superimposed are contours of the continuum emission. Contour levels are 0.001, 0.006, 0.011, 0.021, 0.031,0.05 and 0.06 Jy/beam. Clearly seen in these maps are two compact molecular cores (left panel, A and B). The continuum emission shows several features, two of them associated with cores A and B. Interestingly, the continuum emission near core A shows three compact sources with the brightest one, located in the south, not being associated with molecular emission. The moment 1 map (right panel) clearly shows velocity gradients within the cores, from north-east (NE) to south-west (SW) with an average velocity of ~ -48 km s^{-1} in core A, and velocity gradients from SW to NE with an average velocity of ~ -50 km s^{-1} in core B. SO_2 30-30 moment maps of SO_2 (30-30) shows the same features as CH_3CN confirming that these cores are rotating.

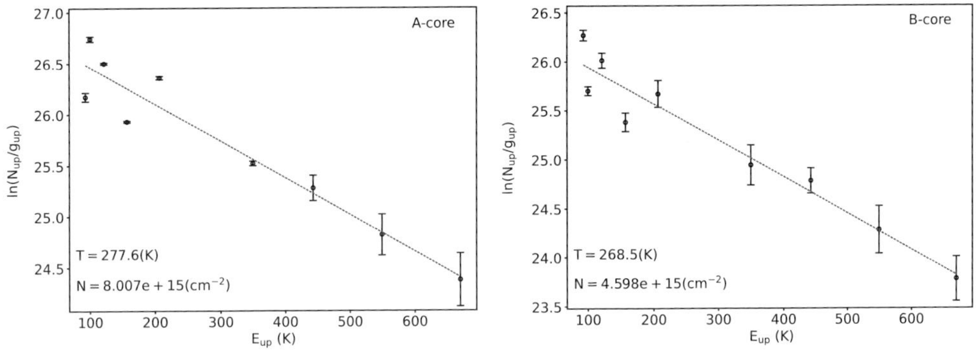

Figure 3. Rotational diagrams at the peak positions of the cores A (left panel) and B (right panel).

3. Rotation diagram analysis: estimation of gas temperature

We estimated the CH_3CN rotation temperature (T_{rot}) and column density using the population-diagram method (see Araya et al., 2005) assuming LTE and low optical depths. The column density in the (J, K) state, N_{JK}, is given by

$$\left(\frac{N_{JK}}{cm^2}\right) = 1.67 \times 10^{14} \frac{g_{JK}}{S(I,K)} \frac{J}{(J^2 - K^2)} \left(\frac{\nu_0}{GHz}\right)^{-1} \times \left(\frac{\mu}{debye}\right)^{-2} \left(\frac{\int T_B dv}{Kkms^{-1}}\right) \quad (1)$$

where g_{JK} is the statistical weight for the state in question, S(I, K) is the spin weight degeneracy factor given by equation (A4) (Boucher et al. 1980), ν_0 is the frequency of the (J, K) $\rightarrow$ (J–1, K) transition, $\mu = 3.91$ debye, and T_B is the brightness temperature.

Figure 3 displays the rotational diagrams at the peak positions of the cores. The rotational temperatures of A and B cores are 277.6 K and 268.5 K, and the column densities are 8.01×10^{15} cm^{-2} and 4.6×10^{15} cm^{-2}, respectively.

Acknowledgments

This research was funded by the Science Committee of the Ministry of Science and Higher Education of the Republic of Kazakhstan (Grant Nos. AP13067768 and AP14870504) and sponsored (in part) by the Chinese Academy of Sciences (CAS), through a grant to the CAS South America Center for Astronomy (CASSACA) in Santiago, Chile. GG acknowledges support from ANID BASAL project FB210003.

Supplementary material

To view supplementary material for this article, please visit https://doi.org/10.1017/S1743921323000121.

References

Araya, E., Hofner, P., Kurtz, S., Bronfman, L., & DeDeo, S. 2005, CH_3CN Observations toward Southern Massive Star-forming Regions. ApJS, 157(2), 279–301.

Boucher, D., Burie, J., Bauer, A., Dubrulle, A., & Demaison, J. 1980, Microwave spectra of molecules of astrophysical interest. XIX. methyl cyanide. *Journal of Physical and Chemical Reference Data*, 9(3), 659–720.

Contreras, Y., Schuller, F., Urquhart, J. S., Csengeri, T., Wyrowski, F., Beuther, H., Bontemps, S., Bronfman, L., Henning, T., Menten, K. M., Schilke, P., Walmsley, C. M., Wienen, M., Tackenberg, J., & Linz, H. 2013, ATLASGAL - compact source catalogue: 330° ¡ ℓ ¡ 21°. A&A, 549, A45.

Guzmán, A. E., Garay, G., Brooks, K. J., & Voronkov, M. A. 2012, Search for Ionized Jets toward High-mass Young Stellar Objects. ApJ, 753(1), 51.

Mottram, J. C., Hoare, M. G., Davies, B., Lumsden, S. L., Oudmaijer, R. D., Urquhart, J. S., Moore, T. J. T., Cooper, H. D. B., & Stead, J. J. 2011, The RMS Survey: The Luminosity Functions and Timescales of Massive Young Stellar Objects and Compact H II Regions. ApJ, 730(2), L33.

Urquhart, J. S., Busfield, A. L., Hoare, M. G., Lumsden, S. L., Clarke, A. J., Moore, T. J. T., Mottram, J. C., & Oudmaijer, R. D. 2007, The RMS survey. Radio observations of candidate massive YSOs in the southern hemisphere. A&A, 461(1), 11–23.

Resolving the Rise and Fall of Star Formation in Galaxies
Proceedings IAU Symposium No. 373, 2023
T. Wong & W.-T. Kim, eds.
doi:10.1017/S1743921322004306

Galaxy Zoo 3D: Identifying Bars, Spirals and Foreground Stars in MaNGA Galaxy Data

Karen L. Masters

Departments of Physics and Astronomy, Haverford College, 370 Lancaster Avenue, Haverford, Pennsylvania 19041, USA
email: klmasters@haverford.edu

Abstract. Galaxies, particularly disc galaxies, show a wide variety of internal structures (e.g. spirals, bars, and bulges). Mapping Nearby Galaxies at Apache Point Observatory (MaNGA, part of the fourth incarnation of the Sloan Digital Sky Surveys), obtained spatially resolved spectral maps for 10,010 nearby galaxies. Many results from MaNGA have collapsed this structure into azimuthally averaged radial gradients, or symmetric 2D shapes, but there is significantly more information about the effect internal structures have on the evolution of galaxies available if we can identify different internal structures. One of the simplest ways to identify irregular internal structures in galaxies is by visual inspection. By employing a citizen science technique to ask this question of N independent volunteers we have obtained quantitatively robust masks (and errors) for spirals and bars in MaNGA target galaxies. In addition to internal features the interface asked users to identify foreground stars and foreground/background galaxies.

Keywords. galaxies: structure, galaxies: spiral, techniques: spectroscopic, methods: data analysis, surveys

1. Introduction

The MaNGA survey (Bundy et al. 2015), part of the fourth phase of the Sloan Digital Sky Surveys (SDSS-IV, Blanton et al. 2017) has obtained spatially resolved spectral maps for 10,010 nearby galaxies selected from the SDSS Main Galaxy Sample (Strauss et al. 2002) using an integral field spectrometer (IFS) system on the 2.5m Sloan Telescope at Apache Point Observatory (see Gunn et al. (2006); Smee et al. (2013); Drory et al. (2015) for technical details of the telescope and instrument). These data are now all calibrated and the final public data release happened in SDSS Data Release 17 (Abdurro'uf et al. 2022). MaNGA data can be explored with the *Marvin* software (Cherinka et al. 2019).

Many published results using large samples from MaNGA collapsed this structure into azimuthally averaged radial gradients, or symmetric 2D shapes (for a list of examples see Masters et al. (2021); Section 1), but the MaNGA data contain significantly more information about the effect internal structures have on the evolution of galaxies – if we can identify different internal structures.

The Galaxy Zoo: 3D project (Masters et al. 2021) was conceived with the idea of identifying selected internal structures, galaxy centres (whether the main target or additional galaxies), and foreground stars in MaNGA target objects through a citizen science technique. The project was inspired by Galaxy Zoo (which was first described in Lintott et al. (2008)) and ran on the *Zooniverse* citizen science platform†.

In this Proceedings based on an eTalk for IAU Symposium 373 "Resolving the Rise and Fall of Star Formation in Galaxies", I describe this project, provide a guide on how

† www.zooniverse.org

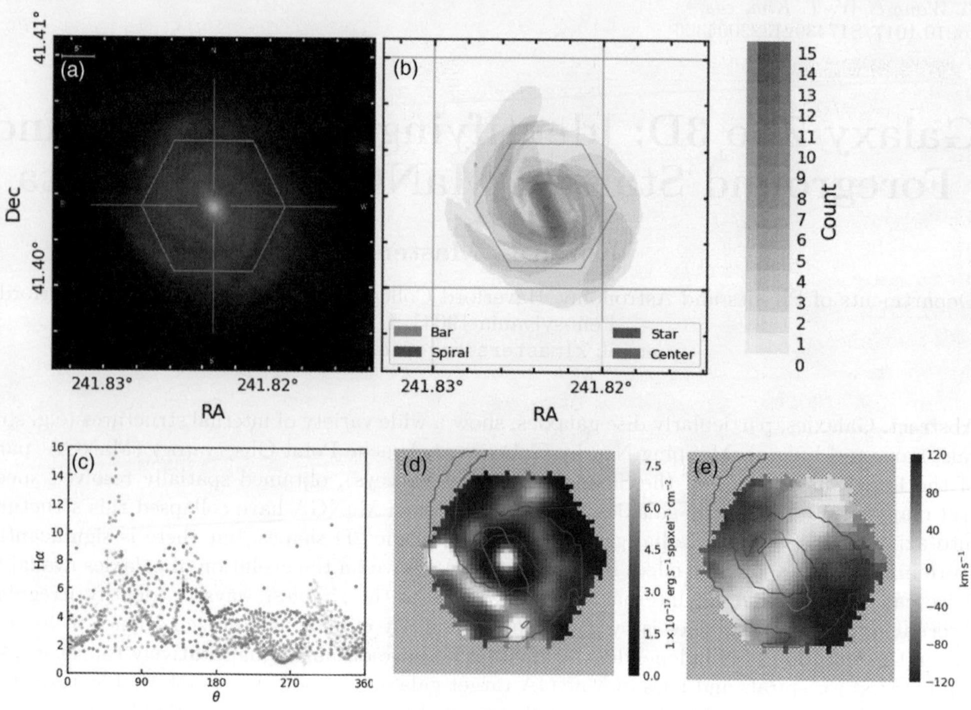

Figure 1. MaNGA galaxy (MaNGA-ID -248420) shown with a GZ:3D bar mask and spiral mask. Panel (a): image shown to *Galaxy Zoo: 3D* volunteers. Panel (b) pixel mask output. Panel (c): scatter plot of the Hα flux in each spaxel, as a function of azimuthal angle olour coded by the GZ:3D identification (e.g. orange for spaxels in the bar, blue for spiral arm spaxels). Panels (d) and (e): outlines of the bar and spiral masks overlayed on the MaNGA Hα emission map and stellar velocity map respectively. The bar contour is at the 7/15, or 47% threshold level, while the spiral is at the 3.5/15, or 23% threshold. Figure published previously as Figure 6 of Masters et al. (2021).

to make use of the results, and hi-light some already published, or in progress projects making use of them.

2. Overview

Galaxy Zoo: 3D collected crowd-sourced information about bars, spirals, galaxy centers and foreground stars for all MaNGA target galaxies. For bars and spirals, a pre-filtering was done using Galaxy Zoo 2 classifications (Willett et al. 2013) to identify MaNGA targets with such features. The images which were shown were based on the SDSS Legacy Imaging Survey (York et al. 2000), overlaid with a hexagon which showed the region covered by the MaNGA IFS. An illustration showing an example image, resulting pixel masks and some examples of how they can be used in combination with MaNGA data is shown in Figure 1. The Galaxy Zoo:3D site collected classifications in the period December 2017-October 2019 (following a beta trial phase). In total each task collected 15 independent classifications per galaxy ($N = 29,831$ galaxies in the galaxy centers/foreground stars task; $N = 5456$ in the bar drawing task, and $N = 7418$ in the spiral drawing task, which were aggregated following the methods described in Masters et al. (2021). Approximately 30% of each of these samples were ultimately observed by the MaNGA survey and therefore have accompanying IFS data.

The pixel masks and clustered galaxy centers and foreground stars were released in Masters et al. (2021) as well as through a Value Added Catalogue (VAC) in SDSS DR17 (Abdurro'uf et al. 2022). The masks were also integrated into the *Marvin* interface.†

To date works which make use of these pixel masks in combination with MaNGA data include use of the foreground star masks to flag regions were the Data Analysis Pipeline (Westfall et al. 2019) will likely fail in MaNGA galaxies; various analysis of bar properties using the bar masks (Fraser-McKelvie et al. 2019, 2020; Krishnarao et al. 2020, 2022) use of the spiral masks to indicate position of spirals (Peterken et al. 2019a,b), and to investigate dust content and star formation rates in and out of spiral arms (Greener et al. 2020; Masters et al. 2021).

The Galaxy Zoo: 3D method could easily be expanded to use samples or data beyond MaNGA, for example applied to other large scale IFS surveys, or for other applications where image segmentation masks of galaxies are desired (e.g. it's use as a training sets for machine learning methods is being explored). Interested researchers with a sample they would like to run through the Galaxy Zoo: 3D method should contact the author.

References

Abdurro'uf, Accetta, K., Aerts, C., et al. 2022, ApJS, 259, 35.
Blanton, M. R., Bershady, M. A., Abolfathi, B., et al. 2017, AJ 154, 28.
Bundy, K., Bershady, M. A., Law, D. R., et al. 2015, ApJ, 798, 7.
Cherinka, B., Andrews, B. H., Sánchez-Gallego, J., et al. 2019, AJ, 158, 74.
Drory, N., MacDonald, N., Bershady, M. A., et al. 2015, AJ, 149, 77.
Fraser-McKelvie, A., Merrifield, M., Aragón-Salamanca, A., et al. 2019, MNRAS, 488, L6.
Fraser-McKelvie, A., Aragón-Salamanca, A., Merrifield, M., et al. 2020, MNRAS, 495, 4158.
Greener, M. J., Aragón-Salamanca, A., Merrifield, M. R., et al. 2020, MNRAS, 495, 2305.
Gunn, J. E., Siegmund, W. A., Mannery, E. J., et al. 2006, AJ, 131, 2332.
Krishnarao, D., Tremonti, C., Fraser-McKelvie, A., et al. 2020, ApJ, 898, 116.
Krishnarao, D., Pace, Z. J., D'Onghia, E., et al. 2022, ApJ, 929, 112.
Lintott, C. J., Schawinski, K., Slosar, A., et al. 2008, MNRAS, 389, 1179.
Masters, K. L., Krawczyk, C., Shamsi, S., et al. 2021, MNRAS, 507, 3923.
Peterken, T., Fraser-McKelvie, A., Aragón-Salamanca, A., et al. 2019, MNRAS, 489, 1338.
Peterken, T. G., Merrifield, M. R., Aragón-Salamanca, A., et al. 2019, Nature Astronomy, 3, 178.
Smee, S. A., Gunn, J. E., Uomoto, A., et al. 2013, AJ, 146, 32.
Strauss, M. A., Weinberg, D. H., Lupton, R. H., et al. 2002, AJ, 124, 1810.
Westfall, K. B., Cappellari, M., Bershady, M. A., et al. 2019, AJ, 158, 231.
Willett, K. W., Lintott, C. J., Bamford, S. P., et al. 2013, MNRAS, 435, 2835.
York, D. G., Adelman, J., Anderson, J. E., et al. 2000, AJ, 120, 1579.

† A tutorial showing some functionality is available here: `https://sdss-marvin.readthedocs.io/en/latest/tutorials/notebooks/VAC_GZ3D_tutorial.html`

Resolving the Rise and Fall of Star Formation in Galaxies
Proceedings IAU Symposium No. 373, 2023
T. Wong & W.-T. Kim, eds.
doi:10.1017/S1743921322003799

Sub kpc-scale gas density histogram: a new statistical method to characterize galactic-scale gas structures

Ren Matsusaka[1], Toshihiro Handa[1], Yusuke Fujimoto[2,3], Takeru Murase[1], Yushi Hirata[1], Junya Nishi[1], Takumi Ito[3], Megumi Sasaki[1] and Tomoki Mizoguchi[1]

[1]Department of Physics and Astronomy, Graduate School of Science and Engineering, Kagoshima University, 1-21-35 Kôrimoto, Kagoshima, Kagoshima 890-0065, Japan

[2]Department of Computer Science and Engineering, University of Aizu Tsuruga, Ikki-machi, Aizu-Wakamatsu City, Fukushima JAPAN 965-8580

[3]Kumamoto University, Graduate School of Science and Technology, Kumamoto, 860-8555, Japan

Abstract. To understand the physical properties of the interstellar medium (ISM) in various scales, we should investigate it with pc-scale resolution over kpc scale coverage. Here, we report the sub-kpc scale Gas Density Histogram (GDH) of the Milky Way. GDH is a histogram of averaged density and corresponds to the probability density distribution (PDF) of gas volume density. We use galactic plain survey data ($l = 10° - 50°$) at ^{12}CO and ^{13}CO ($J = 1 - 0$) obtained as a part of the FOREST Unbiased Galactic plane Imaging survey with the Nobeyama 45m telescope (FUGIN). With this method and data, we are free from spatial structure and molecular cloud identification. GDH can be well fitted with single or double log-normal distribution; which we call as the low-density log-normal (L-LN) and high-density log-normal (H-LN) components. We found both the H-LN fraction (f_{H}) and L-LN width (σ_{L}) along the gas density axis show a coherent structure on the longitude-velocity diagram. It suggests that there is a relationship between the ISM property and kpc scale structure in the Milky Way.

Keywords. ISM: clouds - ISM: structure - ISM: molecules - Galaxy: structure

---

1. Introduction

Star formation is not a single stage process, although stars are formed from the interstellar medium (ISM) by gravitational collapse. Many investigations have been made on the last two stages in star formation process in these decades; they are the sages from a dense core to a star and from an inhomogeneous molecular cloud to dense cores. This progress is mainly due to improvement of high-resolution millimetre observations and computer simulations since 1980s. Now we, therefore, should address stages before them; what makes density fluctuation in a molecular cloud and how molecular clouds are made from diffuse gas. It is closely related to the star formation scenario in a galaxy.

To address the density structure of ISM, many previous investigations are focused on its geometry and statistics based on geometry; such as dendrogram analysis and core mass function. They are focused on dense ISM. So, another approach is required to address the diffuse ISM. The characteristics of the density structure of diffuse ISM should be appeared in its statistics. It is primarily independent of morphology. So, we made a histogram of diffuse ISM for an assigned volume and call it as 'gas density histogram

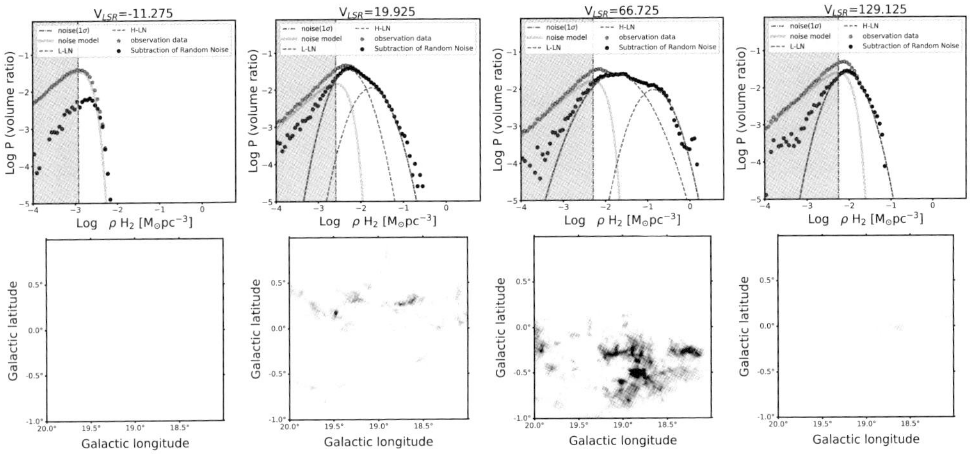

Figure 1. The GDHs and the intensity distribution on the sky of our unit area. Each area shown here is centered $(l, b) = (19^\circ, 0^\circ)$ in different velocities. The vertical dashed lines correspond to 1σ. The thick green solid line is the random noise model, and the black dots are the GDH after subtraction of the noise component. The red and blue dashed lines show the LN fitting results, respectively.

(GDH)'. GDHs from molecular lines can be made for each velocity channel on the same line of sight. For the velocity deconvolution we cannot use dust continuum data. In the case of ISM in the galactic disk of the Milky Way, we can estimate histograms not of the column density but of the volume density using the difference of kinetic distance (see Figure 1).

The star formation activity in kpc-scale concentrates to the spiral arms by many observational investigations in galactic astronomy. Theoretical studies suggest some relation between kpc-scale ISM structure and the pc-scale star formation. At the classical galactic shock scenario, Fujimoto (1968) says ISM compression due to spiral arm potential makes triggered star formation. It may be valid but proposed mechanisms consistent with detail observations, such as the internal structure of a spiral arm, has not been successfully developed. The variety of GDH at each position in the Milky Way should give a breakthrough to address the problem.

2. Data

We used the ^{12}CO and ^{13}CO $(J = 1 - 0)$ datasets from the FUGIN project (Umemoto et al. 2017). In this study, we focus on the region of $10^\circ \leq l \leq 50^\circ$, $|b| \leq 1.0^\circ$. FUGIN data is released by Japanese Virtual Observatory (JVO). Its angular and velocity resolutions are $20''$ and 0.65 km s^{-1}), respectively. To improve the sensitivity, we smoothed the $l - b - v$ data cube with a Gaussian Kernel that has a width of 5.2 km s^{-1}), or 8 channels of the spectrometer and two-dimensional spatial smoothing with a Gaussian function to be the resultant angular resolution of $1'$.

3. Results & Discussion

In this section, we present two major results of GDH from FUGIN data. One is the shape of the GDH in different regions and the other is a relationship between the GDH property and kpc scale structure in the Milky Way.

Figure 1 suggests a correlation sample between the morphology of intense emission, shape of GDH and results of the multi log normal (LN) fitting. GDH of a channel map only with diffuse less-bright gas, such as at $v = 129.125$ shows the sharp peak close to

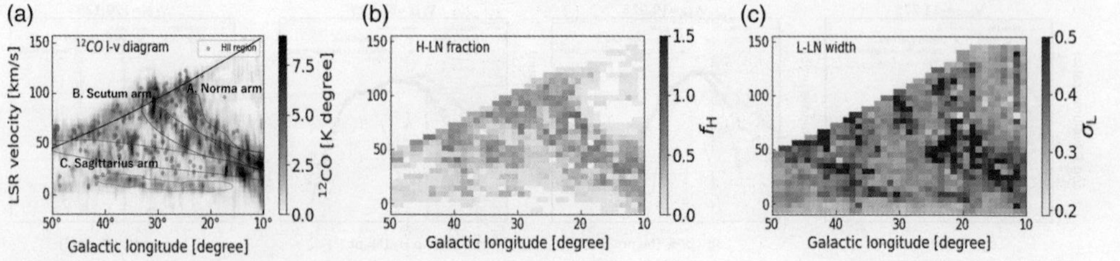

Figure 2. (a) The *l*-*v* diagram of the FUGIN ^{12}CO ($J = 1 - 0$) data. The integration range is from $-1°$ to $+1°$. The thick black line shows the curve of the terminal velocities. The coloured lines show the loci of the Galactic arms constructed by Reid M. J. et al. (2016) and the HII regions (red points: Anderson et al. 2009. (b) The *l*-*v* distribution of $f_{\rm H}$ in each $\Delta l \times \Delta b = 2° \times 2°$. The region where the H-LN cannot be defined (i.e., only diffuse gas, see section 3 is shown as $f_{\rm H} = 0$). (c) Same as (b) but the distribution of $\sigma_{\rm L}$.

the sensitivity limit and steep declining at the dense side. In contrast, GDH of a map with compact bright features, such as 66.725 km s^{-1}, GDH has a flat top structure and most likely has another peak or excess component in the dense side. At -11.275 km s^{-1}, correspond to the far outside of the Milky Way. GDH shows close shape of gaussian noise on the blank sky (see Figure 1, left). At 19.925 km s^{-1}, GDH have real features at the densest tail, which corresponds to features seen in the channel map.

Each GDH shows a straight line under the sensitivity limit. This is quantitatively consistent to the noise estimation from the data. We can easily estimate its contribution of the Gaussian noise estimated using the observed rms noise level (Figure 1, noise model). This may allows us to approach the actual GDH in the lower density side even beyond the "*completeness limit*" defined by previous N-PDF works on a molecular cloud (e.g. Alves et al. 2017).

We found every GDH of our data is well described by one or two log-normal components (Figure 1). Based on a simulation of turbulent ISM, interprets the LN GDH will be made through many random walk processes modifying ISM density by a factor at each step. It means the peak and width of GDH are the initial value and the number of steps or the modification factor by each step, respectively.our results suggest any process of modification of gas density in the sub-kpc scale, such as streaming motion, spiral arm shocks, star-forming feedback, etc., can make a LN GDH.

Figure 2 show the distribution of ^{12}CO integrated intensity, H-LN fraction ($f_{\rm H}$) and L-LN width ($\sigma_{\rm L}$) on the $l - v$ plane. In both parameters we found three ridges of high $f_{\rm L}$ and $\sigma_{\rm H}$; ridges A, B, and C are connecting from $(l, v) \simeq (15°, 25$ km s$^{-1})$ to $(27°, 100$ km s$^{-1})$, from $(l, v) \simeq (15°, 25$ km s$^{-1})$ to $(32°, 80$ km s$^{-1})$, and from $(20°, 40$ km s$^{-1})$ to $(40°, 80$ km s$^{-1})$, respectively. These ridges correspond to the spiral arms identified in the CO intensity distribution (Figure 2(a)). Ridges A, B, and C are Norma, Scutum, and Sagittarius arms, respectively (Dame et al. 2001). This indicates that molecular gas changes its density structure as it flows in the spiral arms. It suggests that uniform, diffuse gas should have a structured at encounter with the spiral arms.

References

Alves J. et al. 2017, *A&A*, 606, L2
Anderson L. et al. 2001, *ApJS*, 181, 255
Dame T. et al. 2001, *ApJ*, 547, 792
Fujimoto M. 1968, *Non-stable Phenomena in Galaxies*, p. 453
Reid et al. 2016, *ApJ*, 823, 77
Umemoto T. et al 2017, *PASJ*, 69, 78

Resolving the Rise and Fall of Star Formation in Galaxies
Proceedings IAU Symposium No. 373, 2023
T. Wong & W.-T. Kim, eds.
doi:10.1017/S1743921322004379

The density structure of molecular cloud scales: A fitting for N-PDF with multi log-normal functions

Takeru Murase[1], Toshihiro Handa[1,2], Ren Matsusaka[1], Yoshito Shimajiri[3,4,5], Masato I.N. Kobayashi[3], Mikito Kohno[6,7], Junya Nishi[1], Norimi Takeba[1] and Yosuke Shibata[1]

[1]Department of Physics and Astronomy, Graduate School of Science and Engineering, Kagoshima Univ., 1-21-35 Kôrimoto, Kagoshima, Kagoshima 890-0065, Japan

[2]Amanogawa Galaxy Astronomy Research center, Kagoshima Univ., 1-21-35 Kôrimoto, Kagoshima, Kagoshima, 890-0065, Japan

[3]National Astronomical Observatory of Japan, Osawa 2-21-1, Mitaka, Tokyo, 181-8588, Japan

[4]Laboratoire d'Astrophysique (AIM), CEA/DRF, CNRS, Université Paris-Saclay, Université Paris Diderot, Sorbonne Paris Cité, 91191 Gif-sur-Yvette, France

[5]Kyushu Kyoritsu Univ., Jiyugaoka 1-8, Yahatanishi-ku, Kitakyushu, Fukuoka, 807-8585, Japan

[6]Astronomy Section, Nagoya City Science Museum, 2-17-1 Sakae, Naka-ku, Nagoya, Aichi, 460-0008, Japan

[7]Department of Physics, Graduate School of Science, Nagoya Univ., Furo-cho, Chikusa-ku, Nagoya, Aichi, 464-8602, Japan

Abstract. We studied the probability distribution function of the column density (N-PDF) of molecular clouds based on a fit with a multi-log-normal function using the Nobeyama 45-m Cygnus X CO survey data. We identified 124 molecular clouds in ^{13}CO data using the DENDROGRAM and SCIMES algorithms. The N-PDF was constructed for 11 extended (≥ 0.4 deg^2) molecular clouds of these identified clouds. We found that every N-PDF is well-fitted with one or two log-normal (LN) distributions. We investigated the distributions of the column density, C^{18}O dense cores, and radio continuum source in each cloud and found that the N-PDF was less correlated with the star-forming activity. The LN N-PDF parameters showed two impressive features. First, the LN distribution at the low-density part had the same mean column density ($\sim 10^{21.5}$ cm^{-2}) for almost all the molecular clouds. Second, the wider LN distribution tended to show the lower mean density of the structures.

Keywords. ISM: clouds – ISM: structure – ISM: molecules – methods: analytical – stars: formation

1. Introduction

The PDF of volume density (ρ-PDF) has been used as a powerful tool for investigating the density structure of molecular clouds affected by various physical processes (e.g. Vázquez-Semadeni 1994; Ballesteros-Paredes *et al.* 2011). It is determined by the dominant physical processes governing the density structure of molecular clouds; for

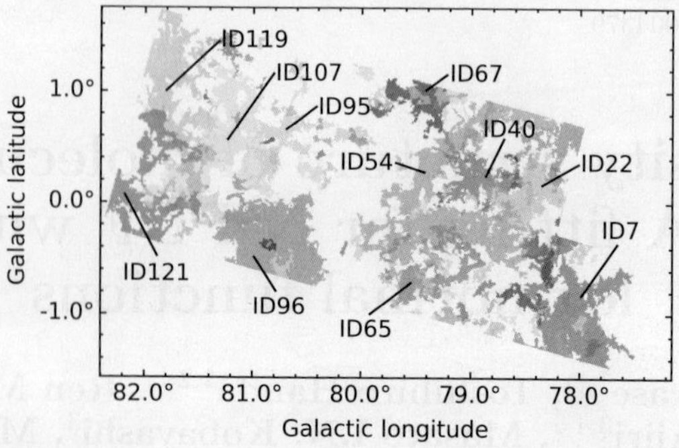

Figure 1. ^{13}CO emission segmentation performed using DENDROGRAM and SCIMES algorithms. Each color shows an individual identified cloud. ID indicates clouds with an extract projected area over $\sim$4000 pixels.

example turbulence and self-gravity should give the LN and power-law (PL), respectively. However, it isn't easy to obtain the ρ-PDF based on observations owing to the projection effect. Therefore, the N-PDF has mainly been used in observational studies.

Previous studies have shown that N-PDFs of many quiescent clouds are LN, whereas those of active star-forming clouds consist of two components: an LN and a pronounced PL tail in the high-density range. However, careful consideration is required for what functional form to fit the excess components in the high-density range. In some N-PDF studies, single or multiple power laws are applied even though the excess components of the high-density region are continuously curved on the log-log plane. Therefore, we investigated the N-PDF of molecular clouds based on a fit with a multi-LN function.

2. Data & Analysis

We used ^{12}CO ($J=1-0$) and ^{13}CO ($J=1-0$) line data from the Nobeyama 45-m Cygnus X CO Survey (Yamagishi *et al.* 2018). The effective angular resolution of the data cubes was 46″, which corresponded to 0.31 pc at a distance of 1.4 kpc to Cygnus X complex. The velocity resolution was 0.25 km s^{-1}. The typical noise levels of ^{12}CO and ^{13}CO data were 0.88 K and 0.36 K in $T_{\rm mb}$ scale, respectively.

We decomposed the intensity map into individual molecular clouds from ^{13}CO data cube using DENDROGRAM and SCIMES algorithms. Figure 1 indicates the resultant molecular clouds. We set the dendrogram parameters, `min_value` $= 3\sigma$, `min_delta` $= 2\sigma$ and `min_npix` $= 64$, where σ is the RMS noise level of the ^{13}CO data.

The column density of ^{13}CO is calculated assuming that the molecular clouds are under local thermodynamic equilibrium (LTE) with ^{12}CO and the beam filling factors of both lines are unity. For the [H_2/^{13}CO] abundance, we applied a value of 7.7×10^5 (Wilson & Rood 1994; Pineda *et al.* 2010). To obtain an accurate N-PDF, we chose 11 clouds with an extracted projected area of over $\sim$4000 pixels, i.e., 0.4 deg^2.

3. Results & Discussion

We found that some of the N-PDFs of 11 clouds have excess from a single LN distribution. These excess components of the N-PDFs showed a curve on the log-log plane. Thus, we fitted the N-PDFs with two LN distributions. The residuals after one or two LN components of the obtained N-PDFs showed no significant deviations in all column density ranges. Low-density and high-density LN distributions were referred to as LN1 and LN2, respectively.

We investigated the relationship between the number of LN distributions and star-forming activity by comparing the column density and spatial distributions of C^{18}O

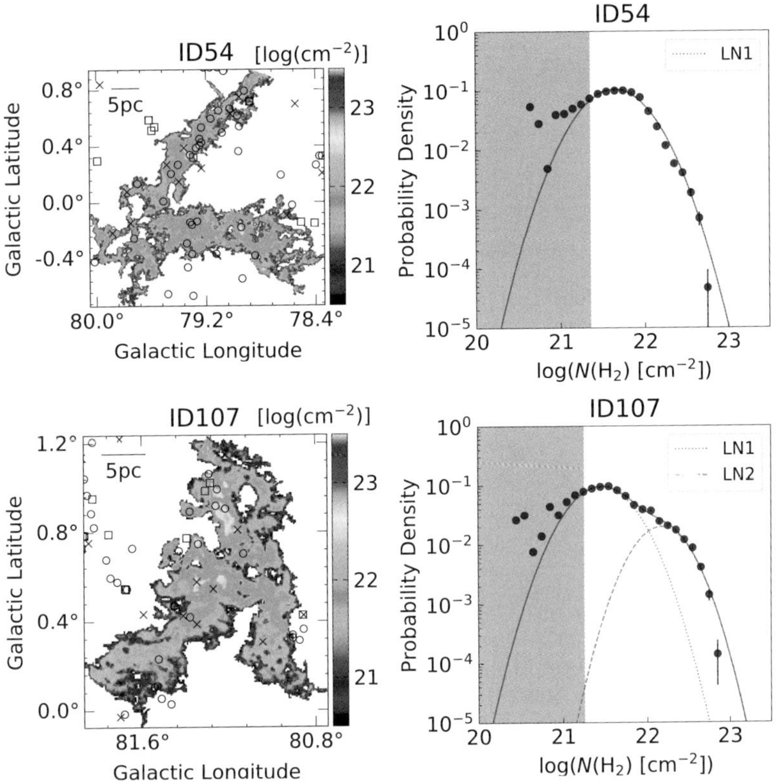

Figure 2. H_2 column density maps and their N-PDFs of typical clouds with a single LN (ID54) and double LN (ID107). The red contour indicates the completeness limit. The circles and crosses indicate the positions of the prestellar and starless cores identified by Takekoshi *et al.* (2019), respectively. The square symbol shows the center positions of the compact HII regions obtained by Urquhart *et al.* (2009).

cores and compact HII regions in each cloud (Figure 2). Consequently, we found no simple correlation between the number of $C^{18}O$ cores associated with molecular clouds and the N-PDF. Furthermore, the positional correlation between the compact HII regions and the high-density regions that compose LN2 ($\sim 10^{22.5}$ cm^{-2}) is also poor. Therefore, we considered there is little relationship between the star formation activity and N-PDF.

We performed a statistical investigation of the N-PDF using the fitting parameters of the N-PDF. In this study, we focused on the correlation between the surface fraction and width of the LN distribution with the mean column density. Panels (c) and (d) of Figure 3 indicates the correlation plots of the N-PDF parameters. We found two impressive features. First, the LN distribution at the low-density part had the same mean column density ($\sim 10^{21.5}$ cm^{-2}) for almost all the molecular clouds. Second, the width of the LN distribution tended to decrease with an increasing mean density of the structures. The two correlations suggest that a higher-density component has a smaller volume fraction and weaker turbulence within each structure.

It is unclear why the density structures of molecular clouds can be fitted only by two LN distributions. Observational studies with higher spatial resolution using various ISM probes and numerical simulations of the multi-phase ISM at later evolutional stages are required.

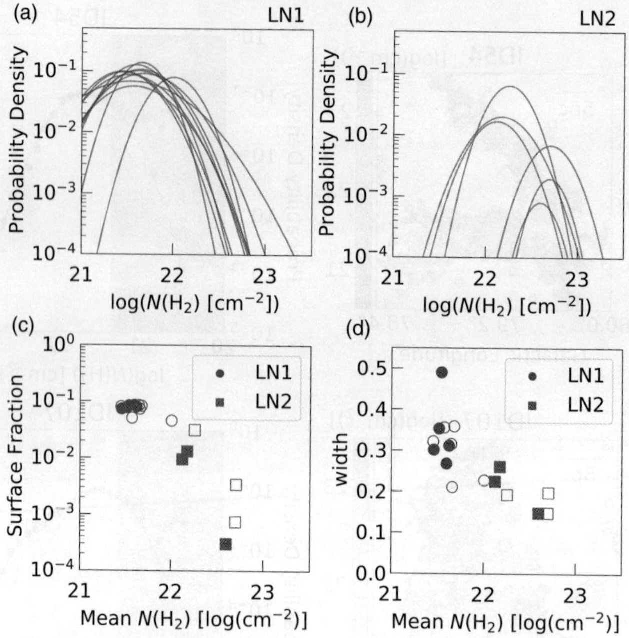

Figure 3. (a) and (b): the best-fit curves of 11 clouds for the LN1 and LN2, respectively. (c) and (d): correlation diagrams between the mean $N(H_2)$ and surface fraction, and (d) between the mean $N(H_2)$ and half-width of LN components, respectively. The circles and squares indicate LN1 and LN2, respectively. The filled symbols indicate the clouds of which the completeness limit contours were closed, and the open symbols indicate the other clouds.

References

Ballesteros-Paredes, J., Vázquez-Semadeni, E., Gazol, A., *et al.* 2011, *MNRAS*, 416, 1436

Pineda, J. L., Goldsmith, P. F., Chapman, N., *et al.* 2010, *ApJ*, 721, 686

Takekoshi, T., Fujita, S., Nishimura, A., *et al.* 2019, *ApJ*, 883, 156

Urquhart, J. S., Hoare, M. G., Purcell, C. R., *et al.* 2009, *A&A*, 501, 539

Vázquez-Semadeni, E. 1994, *ApJ*, 423, 681

Wilson, T. L., & Rood, R. 1994, *ARA&A*, 32, 191

Yamagishi, M., Nishimura, A., Fujita, S., *et al.* 2018, *ApJS*, 235, 9

Resolving the Rise and Fall of Star Formation in Galaxies
Proceedings IAU Symposium No. 373, 2023
T. Wong & W.-T. Kim, eds.
doi:10.1017/S1743921323000194

The Anomalous Ammonia Spectrum of Arp 220

Jürgen Ott[1] and **David S. Meier**[2,1]

[1]National Radio Astronomy Observatory
1003 Lopezville Road, Socorro, NM 87801, USA
email: jott@nrao.edu

[2]New Mexico Institute of Mining and Technology
801 Leroy Place, Socorro, NM 87801, USA
email: David.Meier@nmt.edu

Abstract. Ultra-luminous infrared galaxies (ULIRGs) are extreme in many ways. The major mergers trigger star formation at very high rates that cause the ISM to be dominated by infrared (IR) photons. We show the ammonia spectra toward the two cores of Arp 220, the nearest ULIRG, in three Very Large Array (VLA) bands (Ku, K, Ka). Typical decay times of the non-metastable transitions ~ 100 s and are therefore usually difficult to observe. The FIR excitation of Arp 220, however, shows that non-metastable states are widely populated up to a limiting energy of ~ 1500 K. We assume that this atypical ammonia spectrum is due to the strong FIR field that re-excites the ammonia molecule on timescales much shorter than the already short decay times. The resulting level population causes a break-down of the typical assumptions made for the use of ammonia as a molecular thermometer.

Keywords. ISM: molecules, galaxies: starburst, galaxies: ISM

1. Arp 220

Arp 220 is the prototypical, and most nearby ULIRG at a distance of only ~ 75 Mpc. This major merger has a star formation rate of hundreds of $M_\odot$/yr, which create a unique, dense, opaque, and energetic environment (Spoon et al. 2004). The large amount of dust largely retains optical and UV emission and even is opaque to the shorter wavelength infrared spectrum. The two starburst cores of Arp 220 merge and will likely create a QSO in the future.

2. The Arp 220 Ammonia Anomaly

The radio continuum reveals the two cores, 'east' and 'west' in Arp 220 (e.g. Barcos-Muñoz et al. 2015). Typically, molecular cm lines are dominated by absorption against the two cores. In most sources, the strengths of metastable $J = K$ ammonia lines and their inversion transitions follow a Boltzmann law in the rotational diagram. The slope of the line represents a temperature. For Arp 220, however, a straight line is not a good fit to the points in the Boltzmann diagram, particular for the western core (see Zschaechner et al. 2016). A 'normal' temperature determination is not possible.

3. Arp 220 K-band Spectrum

Observing the spectra in radio K-band at $\sim 18 - 26$ GHz shows that in contrast to almost any other source, Arp 220 exhibits deep absorption spectra of non-metastable levels, very similar in depth to the metastable lines (Fig. 1). The $J \neq K$ levels are usually

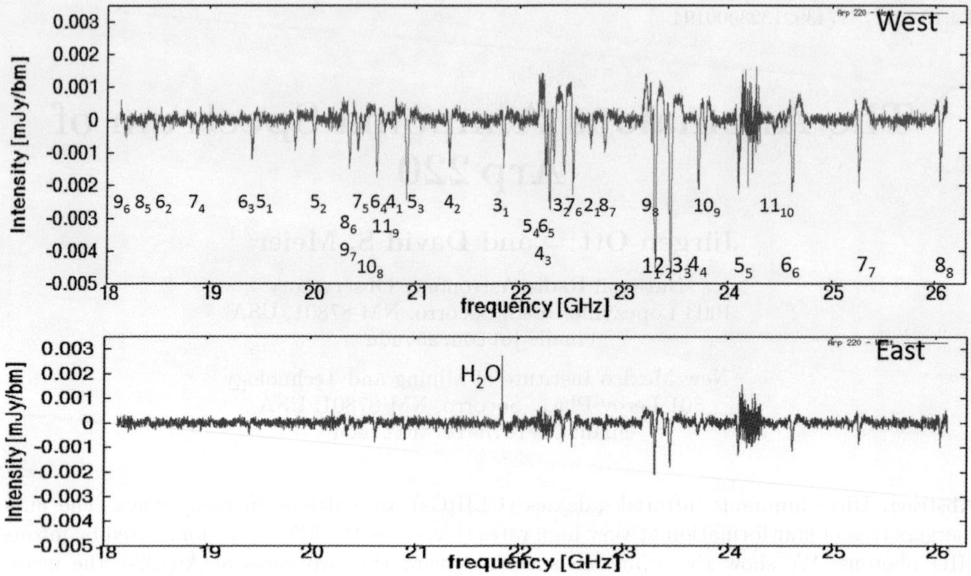

Figure 1. K-band spectra for the two cores in Arp 220.

excited by either collisions or by IR emission, from where they cascade down constant K-ladders. Without constant supply of pumping energy, ammonia would end up in the metastable $J = K$ states, which follow a Boltzmann distribution that is defined by a temperature. The constant pumping inside the strong IR field of Arp 220, however, provides a significant fraction of the gas in non-metastable states. This is particularly true for the western nucleus (see Fig. 1).

4. Ammonia level diagram

If we plot the level diagram of ammonia with K on the x-axis and energy above ground on the y-axis, it turns out that we detect almost all non-metastable lines in K-band up to an energy of about 1500 K (Fig. 2; labeled levels in black). Additional Ka (red) and Ku (blue) data are being observed to fill in essentially the entire level diagram up to that energy and will allow us to derive the most detailed radiative transfer for ammonia in a strong IR field. This will also characterize the use of ammonia as a thermometer under such conditions. The purple lines indicate transitions associated with the first vibrational level.

5. Hidden Water Maser

Our VLA spectrum of Arp 220 (Fig. 1) also reveals a single emission line, which we identify as a 22 GHz water maser line toward the eastern core. Note that the velocities of the cores conspire such that the non-metastable ammonia (3,1) absorption feature of the western core is at the exact same observing frequency. Any unresolved observation would therefore not be able to observe the water maser feature as the two lines largely cancel out (cf. Zschaechner et al. 2016).

6. Additional Lines

Together, the K, Ku, and Ka band observations also provide observations of other molecular lines, including c-C_3H_2, excited OH, the low E_u states of HC_3N, vibrationally

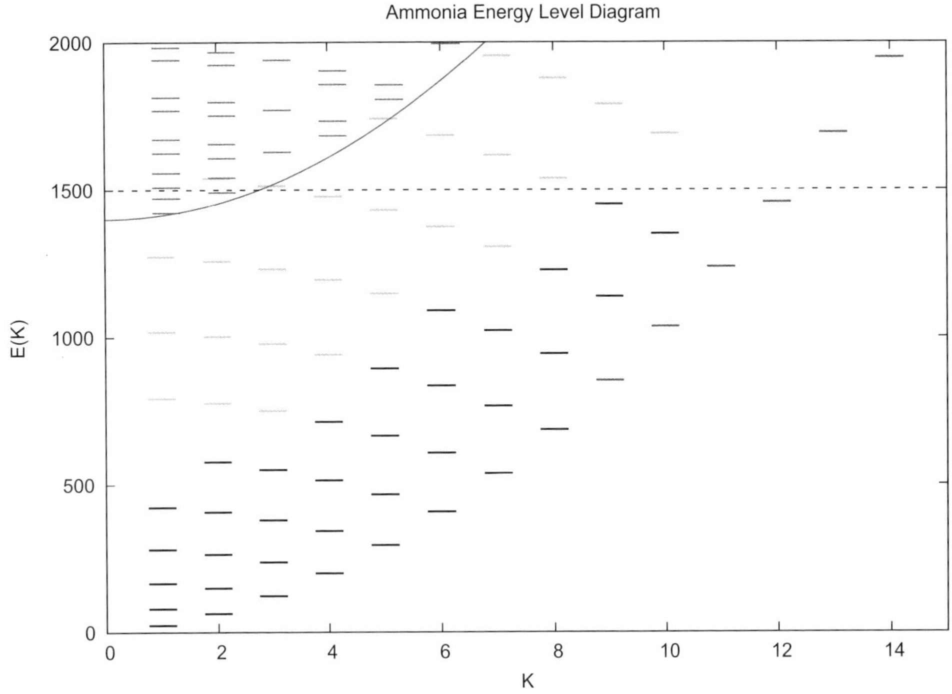

Figure 2. Level diagram for the western nucleus in ammonia in Arp 220. *Black:* Detected transitions, *Blue:* forthcoming observations of non-metastable levels in Ku band to be analyzed, *Red:* forthcoming observation of meta-stable levels observed in Ka band to be analyzed, *Purple outline and levels:* first vibrational excitation states. The dashed horizontal line show the highest excitation energies observed so far.

excited HCN, CH_3OH masers, and H_2CO. These transitions further constrain the excitation of the hot component [OH and HCN (2-1)] as well as any cool component (c-C_3H_2 and HC_3N).

References

Barcos-Muñoz, L. et al. 2015, *ApJ*, 799, 10
Spoon, H. W. W., Moorwood, A. F. M., Lutz, D., et al. 2004, *A&A*, 414, 873
Zschaechner, L. K. et al. 2016, *ApJ*, 833, 41

Resolving the Rise and Fall of Star Formation in Galaxies
Proceedings IAU Symposium No. 373, 2023
T. Wong & W.-T. Kim, eds.
doi:10.1017/S1743921323000182

SWAG: The Maps

Jürgen Ott[1], David S. Meier[2,1], Tierra Candelaria[2,1] and Dylan Ward[2,1]

[1]National Radio Astronomy Observatory
1003 Lopezville Road, Socorro, NM 87801, USA
email: jott@nrao.edu

[2]New Mexico Institute of Mining and Technology
801 Leroy Place, Socorro, NM 87801, USA
email: David.Meier@nmt.edu, tierra.candelaria@student.nmt.edu, dylan.ward@student.nmt.edu

Abstract. We present maps of the "Survey of Water and Ammonia toward the Galactic center" (SWAG). SWAG was observed over three years (∼ 550h) with the Australia Telescope Compact Array (ATCA) and covers the entire Central Molecular Zone (CMZ) at about 26" or ∼ 1 pc resolution. The observed 21.2–25.6 GHz range contains tens of spectral lines and 4 GHz of continuum. Here, we present some final maps. These include multiple NH_3 lines, radio recombination lines, shock tracers like HNCO and methanol (CH_3OH), high resolution 22 GHz water masers, and a continuum map. The maps are the foundation for ongoing comprehensive temperature mapping of the CMZ, including the identification of heating mechanisms, the characterization of water maser sources as young stellar objects or AGB stars, as well as chemistry, dynamics, and star formation studies of the ISM in this unique environment.

Keywords. Galaxy: center, radio lines: ISM, ISM: molecules, masers

1. SWAG

SWAG, the "Survey of Water and Ammonia in the Galactic center" is a comprehensive survey of molecular lines in the 21.2–25.6 GHz range. The lines include species that trace temperature, shocks, ionization, and other physical parameters. SWAG covers the entire Central Molecular Zone (CMZ), the inner ∼ 500 pc of our Galaxy, a region that is dominated by the gas accretion from disk material. The history of the gas as well as strong feedback from energetic sources in the CMZ alter the state of the gas to be warmer, more turbulent and denser than disk gas. These conditions may resemble those in early galaxy formation and studying the CMZ gas properties at about ∼ 1 pc resolution allows to draw comparisons to gas at larger lookback times. Here we present maps from SWAG.

2. The Maps

Ammonia: A key molecule of SWAG is ammonia (NH_3). NH_3 is a symmetric top, that can be used as a thermometer of the gas when observing its inversion transitions. In Fig. 1 we show the NH_3 (1,1), (3,3), and (6,6) maps as an example. SWAG also covers (2,2), (4,4), (5,5), and some non-metastable transitions like (2,1), (3,2), and (4,1). Initial temperature maps are presented in Krieger et al. (2017) with a more detailed analysis, including a combination with single dish data, forthcoming in Candelaria et al. in prep.

Water: The 22 GHz maser transition of water (H_2O) is another key tracer within SWAG. This transition is a class II maser and pumped by collisions. The underlying

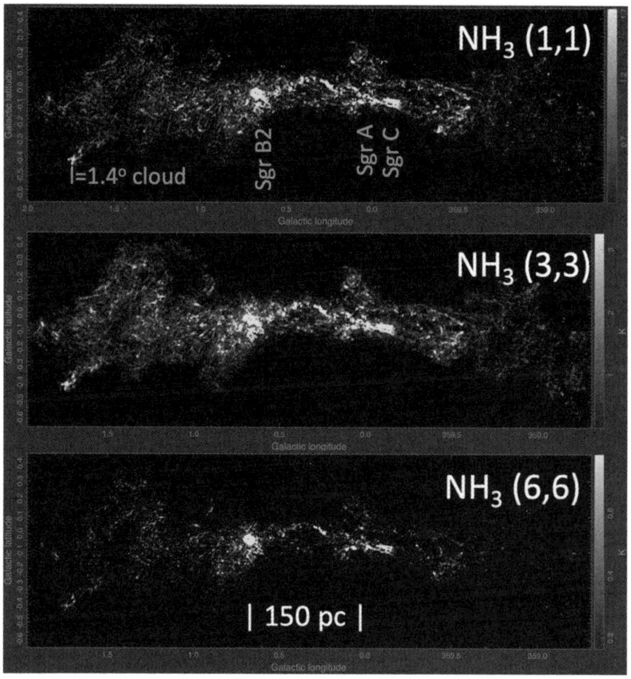

Figure 1. Ammonia (1,1), (2,2), and (3,3) maps. Some regions within the CMZ are marked in the first panel. The area between Sgr B2 and Sgr C marks the 100 pc ring, which is prominent in the dense gas signatures. The $l = 1.4°$ cloud is an accretion zone of disk material onto the 100 pc ring.

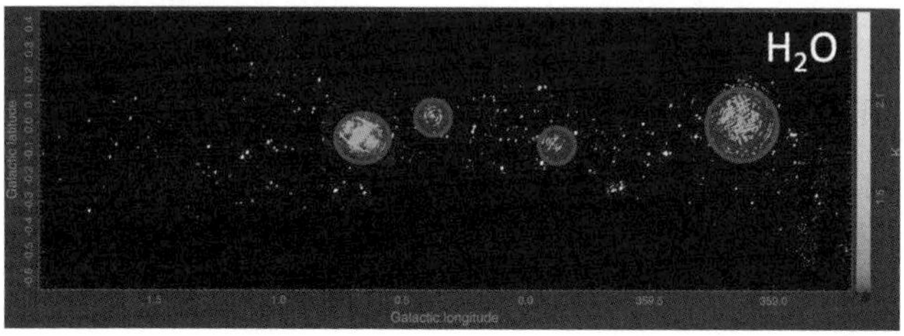

Figure 2. 22 GHz water masers. The circled, shaded regions are very strong masers that show sidelobe artifacts.

sources are typically young stellar objects (YSOs), likely associated with the CMZ molecular gas, or AGB stars across the Galaxy that shed their outer envelopes. Both sources invoke the shocks needed to turn on the H_2O masers. The SWAG map in Fig. 2 shows at least a ten-fold increase in water maser sources compared to previous survey (Ott et al. 2019; Ward et al. in prep.).

Shock tracers: Some molecules are typically formed on dust grains and only liberated by shocks of different strengths. It turns out that HNCO is mostly detected in the $l = 1.4$ deg cloud (Fig. 3). This cloud is likely the point where the gas that was transferred from the Milky Way disk to the CMZ eventually accretes on the inner 100 pc molecular ring. It appears that the gas arrives in a diversity of shock conditions. In contrast, methanol (CH_3OH), which can also be considered a shock tracer, has a very different

Figure 3. HNCO($1_{01} - 0_{00}$).

Figure 4. $CH_3OH(2_{02} - 1_{01})$.

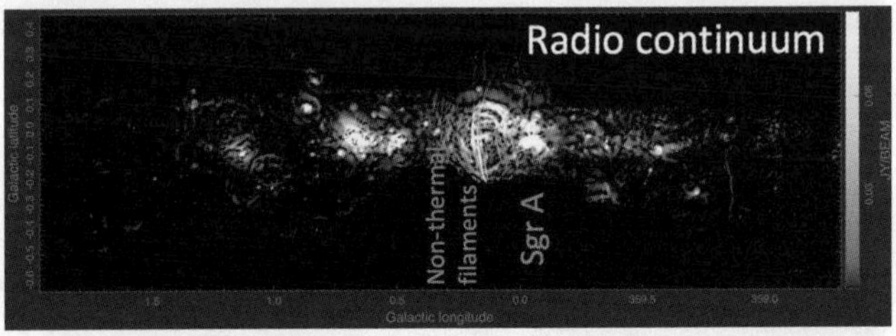

Figure 5. 1.3 cm Radio Continuum.

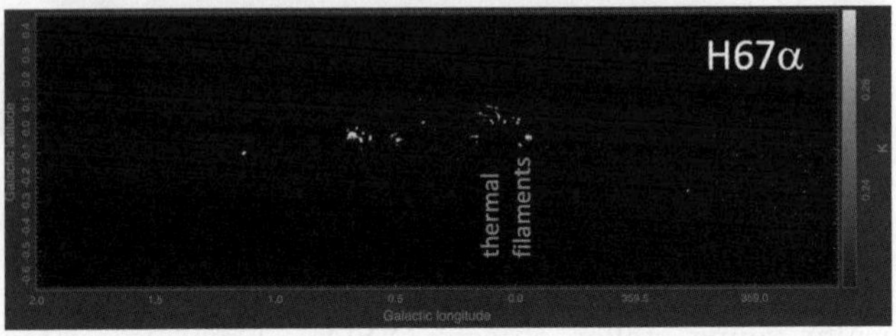

Figure 6. H67α radio recombination line.

distribution (Fig. 4). It is mostly found in the 100 pc ring itself. This may indicate different shock strengths between the gas accretion zone and the 100 pc ring and may show the evaporation of molecules from dust grains.

Radio Continuum: SWAG also observes K-band ($\sim$23 GHz) 4 GHz-wide radio continuum(Fig. 5). The resulting map shows the individual radiation processes from point and extended regions, like Sgr A, which contains the supermassive black home Sgr A*, the star forming regions Sgr B and C, thermal and non-thermal filaments, as well as individual supernovae can be detected.

Radio Recombination Lines: Radio recombination lines, such as H67a, trace ionized gas. In the case of the CMZ, this gas is mostly correlated with the thermal radio continuum, as well as individual star forming regions (Fig. 6).

References

Krieger, N. et al. 2017, *ApJ*, 850, 77
Ott, J., Meier, D. S., Ginsburg, A., et al. 2019, IAU Symposium, 343, 485

Resolving the Rise and Fall of Star Formation in Galaxies
Proceedings IAU Symposium No. 373, 2023
T. Wong & W.-T. Kim, eds.
doi:10.1017/S1743921322003891

Intrinsic dust and star-formation scaling relations in nearby galaxies

Bogdan A. Pastrav

Institute of Space Science,
Atomistilor 409, 077125 Magurele, Ilfov, ROMANIA
email: bapastrav@spacescience.ro

Abstract. Following from our recent work, we present results of a detailed analysis of a representative sample of nearby galaxies. The photometric parameters of the morphological components are obtained from bulge-disk decompositions, using GALFIT software. The previously obtained method and library of numerical corrections for dust, decomposition and projection effects, are used to correct the measured (observed) parameters to intrinsic values. Observed and intrinsic galaxy dust and star-formation related scaling relations are presented, to emphasize the scale of the biases introduced by these effects. To understand the extent to which star-formation is distributed in the young stellar disks of galaxies, star-formation connected relations which rely on measurements of scale-lengths and fluxes / luminosities of Hα images, are shown. The mean dust opacity, dust-to-stellar mass and dust-to-gas ratios of the sample, together with the main characteristics of the intrinsic relations are found to be consistent with values found in the literature.

Keywords. Galaxy: disk, Galaxy: fundamental parameters, galaxies: ISM, galaxies: spiral, galaxies: structure, (ISM:) dust, extinction, radiative transfer

1. Introduction

Accurate dust and star-formation scaling relations are essential in studies of interstellar medium (ISM) evolution, star-formation and galaxy evolution studies, or related to the duty cycle of dust and gas in galaxies. These relations can be affected by numerous biases. Among these, it is known that dust introduces the most significant effects and degeneracies in the observed (measured) photometric and structural parameters of galaxies and their main constituents - discs and bulges, with inclination (projection) and decomposition effects having a non-negligible contribution. These effects are stronger at shorter wavelengths and higher disc inclinations (Tuffs et al. 2004, Pastrav et al. 2013a,b). Obtaining intrinsic relations and show the extent of biases introduced in their characteristic parameters (zero-point, slope, correlation coefficient), is thus very important for the previously mentioned studies.

2. Method

A more detailed and complete description of the method is given in Pastrav (2020). Here we only briefly summarise the main steps. For this study we considered the images of all the unbarred late-type galaxies and lenticulars included in the SINGS (*Spitzer* Infrared Nearby Galaxies Survey; Kennicutt et al. 2003) and KINGFISH (Key Insights on Nearby Galaxies: a Far-Infrared Survey with *Herschel*; Kennicutt et al. 2011) surveys, in B band and for Hα line - in total of 24 galaxies. The B band images were already analysed in Pastrav (2020) and Pastrav (2021). All the images were extracted from the

NASA/IPAC Infrared Science Archive (IRSA) and NASA IPAC Extragalactic Database (NED). We then used GALFIT data analysis algorithm (Peng et al. 2010), for the structural analysis of this sample (bulge-disk decomposition). To fit the disc and bulge surface brightness profiles, we used the exponential and Sérsic functions available in GALFIT, and the "sky" function for an initial estimation of the image background. A complex star-masking routine was used and the integrated fluxes of the galaxies and their main constituents were also calculated, together with all the photometric parameters of interest (see Pastrav 2020). In the next step, central face-on dust opacities (τ^f_B, $\tau^f_{H\alpha}$) for B band and Hα line were derived using the empirical correlation found by Grootes et al. (2013), between the face-on dust opacity and stellar mass surface density (see Eq. (5)). This parameter is essential when determining and applying the corrections for dust effects. Further on, the dust masses were calculated with Eqs. (2), (3) in Grootes et al. (2013) (see also Eqs. A1–A5), which rely on Popescu et al. (2011) model for the dust geometry, model which takes into account Draine (2003) dust model. Next, the chain of numerical corrections for projection, dust & decomposition effects and the method described in Pastrav et al. (2013a,b) were applied to the measured parameters to derive all the intrinsic parameters involved in the analysed dust/ISM and star-formation scaling relations. Subsequently, the dust-to-stellar mass ratios and dust-to-gas ratios (observed and corrected) were calculated, using stellar and gas masses found in the literature (e.g. Rémy-Ruyer et al. 2014 & Grossi et al. 2015).

3. Results

We show in Fig. 1 some of the scaling relations derived and analysed in this study. A more complete discussion will be presented in a forthcoming paper. On the upper row, in the lefthand plot one can see an increase in τ^f_B (similar trend for Hα line) with stellar mass $M_\star$, as also recently found by van der Giessen et al. (2022) analysing a much larger sample of galaxies from SDSS & GAMA surveys. The derived average τ^f_B of the sample, 3.18 ± 0.44, is consistent with studies done on much larger samples (e.g Driver et al. 2007 found 3.8 ± 0.7; van der Giessen et al. 2022 found 4.1), while our dust mass values are within the error limits of those derived by Rémy-Ruyer et al. (2014) and Aniano et al. (2020) using completely different methods. The increasing trends in the M_{dust}/M_{HI} vs $M_\star$ (upper right plot, with M_{HI} being the neutral hydrogen gas mass, taken from the literature) and $M_{dust}/M_\star$ vs $M_{HI}/M_\star$ (middle right) is recovered after applying the corrections, with the average value for dust-to-gas ratios of -2.19 (or 0.63%) consistent with the one found by Cortese et al. (2012) of -2.1 for HI normal galaxies; the rms for the former is low (0.15 - obs. & 0.03 - intrin.) while the corresponding correlation coefficient r = 0.55 (obs)/0.36 (intrin) is consistent with $r = 0.31$ found in Cortese et al. (2012). Practically no correlation is observed for the corrected M_{HI}/M_{dust} vs stellar mass surface density $\mu_\star$ (middle left plot), in line with what was found by Cortese et al. (2012) $-$r = 0.07 vs 0.05. A weaker correlation is found for the $M_{dust}/M_\star$ vs $M_{HI}/M_\star$ relation, with $r < 0.5$. Another scaling relation is shown in the lower right panel, the ratio between the integrated $H\alpha$ flux (or luminosity) and M_{dust} vs. $M_\star$, a consequence of $SFR - M_\star$ and $M_{dust} - M_\star$ relations, with a derived slope of -1.09 ± 0.16 (obs.) / -0.95 ± 0.20 (intrin). In the lower left plot, the ratio of the intrinsic disc scale-lengths seen in optical B band and in $H\alpha$ line as a function of stellar mass is shown, with values generally lower than 1.0 as expected, only after applying the numerical corrections. A mildly increasing trend can be observed. This shows the extent of optical emission in the disc compared with the extent to which star-formation is distributed in the young stellar discs of galaxies.

Our study underlines the importance of having accurate, unbiased scaling relations in models and studies of ISM evolution and star-formation.

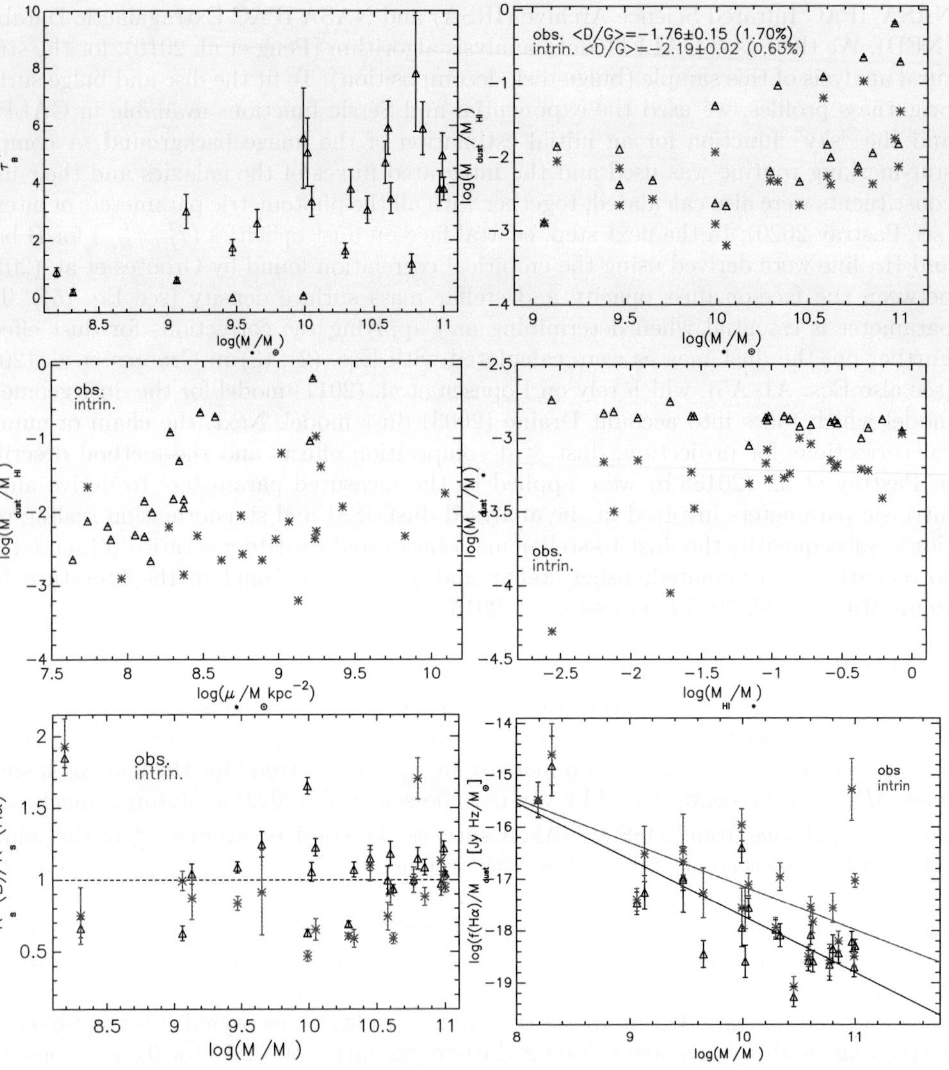

Figure 1. Dust/ISM and star-formation related scaling relations. *Upper row* - B band central face-on dust optical depth vs. galaxy stellar mass ($M_\star$) & dust-to-HI ratio (M_{dust}/M_{HI}) vs. $M_\star$; *Middle row* - M_{dust}/M_{HI} vs. stellar mass surface density ($\mu_\star$) & dust-to-stellar mass ratio $M_{dust}/M_\star$ vs. neutral gas to stellar mass ratio $M_{HI}/M_\star$; *Lower row* - B band and Hα line emission disc scale-lengths ratio vs. $M_\star$ & the ratio of Hα flux to M_{dust} vs. $M_\star$. Observed quantities are plotted with black triangles and the intrinsic ones with red stars. The error bars show the standard deviations, which for certain values are of the size of data points.

Supplementary material

To view supplementary material for this article, please visit https://doi.org/10.1017/S1743921322003891.

References

Aniano, G., Draine, B. T., Hunt, L. K., Sandstrom, K., Calzetti, D. et al. 2020, ApJ 889, 150
Cortese, L., Ciesla, L., Boselli, A. et al. 2012, *A&A*, 540, A52
Draine, B.T. 2003, *ARAA*, 41, 241
Grootes, M., Tuffs, R.J., Popescu, C.C. et al. 2013, *ApJ*, 766, 59

Grossi, M., Hunt, L. K., Madden, S. C. et al. 2015, *A&A*, 574, A126
Kennicutt, R. C., Armus, L., Bendo, G. et al. 2003, *PASP*, 115, 928
Kennicutt, R. C., Calzetti, D., Aniano, G. et al. 2011, *PASP*, 123, 1347
Pastrav, B. A., Popescu, C. C., Tuffs, R. J., Sansom, A. E., 2013a, *A&A*, 553, A80
Pastrav, B. A., Popescu, C. C., Tuffs, R. J., Sansom, A. E. 2013b, *A&A*, 557, A137
Pastrav, B. A. 2020, *MNRAS*, 493, 3580
Pastrav, B. A. 2021, *MNRAS*, 506, 452
Peng, C. Y., Ho, L. C., Impey, C. D., Rix, H.-W. 2010, *AJ*, 139, 2097
Popescu, C. C., Tuffs, R. J., Dopita, M. A. et al. 2011, *A&A*, 527, A109
Rémy-Ruyer, A., Madden, S.C., Galliano, F., Galametz, M., Takeuchi, T. T. et al. 2014, *A&A* 563, A31
Rémy-Ruyer, A., Madden, S. C., Galliano, F. et al. 2015, *A&A*, 582, A121
Tuffs, R. J., Popescu, C. C., Völk, H. J., Kylafis, N. D., Dopita, M. A. 2004, A&A, 419, 821
van der Giessen, S. A., Leslie, S. K., Groves, B., Hodge, J. A., Popescu, C. C. et al. 2022, *A&A* 662, A26

Resolving the Rise and Fall of Star Formation in Galaxies
Proceedings IAU Symposium No. 373, 2023
T. Wong & W.-T. Kim, eds.
doi:10.1017/S174392132200374X

The JWST/NIRSpec GTO programme "The Physics of Galaxy Assembly: IFS observations of high-z galaxies"

Michele Perna, on behalf of the NIRSpec Galaxy Assemby IFS GTO team

Centro de Astrobiología, (CAB, CSIC–INTA), Departamento de Astrofísica, Cra. de Ajalvir Km. 4, 28850 – Torrejón de Ardoz, Madrid, Spain
email: mperna@cab.inta-csic.es

Abstract. We present an overview of the project "The Physics of Galaxy Assembly: IFS observations of high-z galaxies", a Guaranteed Time Observations (GTO) programme of the James Webb Space Telescope (JWST). It an ambitious project aimed at investigating the internal structure of distant galaxies with the NIRSpec integral field spectrograph (IFS), having allocated 273 hours of JWST prime time. The NIRSpec capability will provide us with spatially resolved spectroscopy in the 1-5 μm range of a sample of over forty galaxies and Active Galactic Nuclei in the redshift range $3 < z < 9$. IFS observations of individual galaxies will enable us to investigate in detail the most important physical processes driving galaxy evolution across the cosmic epoch. More in detail, the main specific objectives are: to trace the distribution of star formation, to map the resolved properties of the stellar populations, to trace the gas kinematics (i.e. velocity fields, velocity dispersion) and, hence, determine dynamical masses and also identify non-virial motions (outflow and inflows), and to map metallicity gradients and dust attenuation.

Keywords. James Webb Space Telescope - galaxies: formation - galaxies: evolution - galaxies: high-redshift - galaxies: star formation

1. Introduction

The James Webb Space Telescope (JWST) promises to reveal a new view of galaxy formation in the early Universe. Thanks to its unprecedented sensitivity and spectroscopic capability in the near/mid-infrared (NIR/MIR) wavelengths, for the very first time, the rest-frame optical nebular emission lines (e.g., Hβ, [O III]$\lambda\lambda$4959,5007, Hα and [N II]$\lambda\lambda$6549,6585) of star-forming galaxies (SFGs) and active Galactic Nuclei (AGN) can be directly detected and resolved across the cosmic epoch, from the cosmic noon ($z \sim 2 - 3$; e.g. Förster Schreiber & Wuyts 2020) to the Epoch of Reionization (EoR, $z \lesssim 15$; e.g. Robertson 2022). Early Release Observations (EROs; Pontoppidan et al. 2022) results have clearly demonstrated the power of JWSTs spectroscopic observations (e.g. Brinchmann 2022; Curti et al. 2022; Tacchella et al. 2022), promising many exciting discoveries to be made over the coming years.

The focus of this paper is to present an overview of the project "The Physics of Galaxy Assembly: IFS observations of high-z galaxies" (Leads: Santiago Arribas and Roberto Maiolino), a Guaranteed Time Observations (GTO) programme for the JWST NIRSpec instrument (Jakobsen et al. 2022; Ferruit et al. 2022). We will describe the sample, the observational design, and the general goals of the project.

2. Programme overview

The project "The Physics of Galaxy Assembly: IFS observations of high-z galaxies" (GA-IFS hereinafter) is based on the use of the NIRSpec's Integral Field Spectroscopy (IFS) mode (Böker et al. 2022). It allows to cover the wavelength range $0.6 - 5.3\mu$m at spectral resolutions of R$\approx$ 100, 1000, and 2700; more in detail, low resolution (R100) observations cover the full wavelength range, while for R1000 and R2700 multiple exposures with different bands are required to cover the entire range (see Table 3 in Böker et al. 2022). The field of view (FOV) of the IFS aperture is a contiguous $3.1'' \times 3.2''$ sky area, with a sampling of $0.1''$/spaxel and a spatial resolution even lower than $0.1''$ at shortest wavelengths (Böker et al. 2022).

The GA-IFS sample consists of 42 individual systems to be observed in Cycle 1 (230 hours of JWST time), plus a number of additional targets to be selected for Cycles 2 and 3 (43 hours). The sample contains a good representation of the most massive and extended SFGs in the redshift range $3 < z < 9$, selected to increase the leverage of IFS capabilities. For instance, at $z \sim 6$, the NIRSpec IFS FOV corresponds to a physical area of $\sim 20 \times 20$ kpc, hence large enough to cover the entire galaxy emission with a sub-kpc sampling. The GA-IFS sample also comprises AGN and quasars (QSOs) in the redshift range $3 < z < 7.5$, to test the IFS capabilities in detecting extended emission associated to their host galaxy component, and map virial and non-virial motions (e.g. kpc-scale outflows and inflows). A plethora of selection criteria have been used to identify these systems: for instance, most SFGs and AGN have been selected from the COSMOS and GOODS-S fields, on the basis of their available multi-wavelength information; the sample also contains Ly-α and [O III]88μm emitters, submillimeter-bright galaxies, and the brightest and most distant QSOs (up to z $\sim$ 7.5).

For the SFGs sub-samples, we request observations at low (R100) and high (R2700) spectral resolution, to better study the continuum and line emission, respectively. For AGN and QSOs, only R2700 observations are requested, as we are interested in studying only their emission lines properties (but see Sect. 3). The specific band for the high-resolution observations mainly depends on the redshift of the individual galaxies, and it is aimed at including in a single band the most important rest-frame emission lines (e.g. covering the Hβ – Hα range, redshifted in the NIR). The exposure time for each target is settled to map all prominent rest-frame optical lines, according to flux predictions obtained from different diagnostics (taking advantage of available multi-wavelength information). No extra background exposures are requested: for R100 observations, we expect to have a relatively large number of spaxels free from galaxy emission, useful to estimate the background level. High resolution observations are instead dedicated to the study of emission line features, less affected by background emission.

In Table 1 we summarise the general properties of the selected targets, as well as the observational design for each class of sources in our sample. For more details, we refer to the APT of each program of the GTO: PID 1215, 1216, 1217, 1218, 1219, 1220, 1222, 1262, 1263, 1264. The GTO project has allocated 273 hours of JWST prime time; most of the program will be executed during the first three JWST cycles (2022–2024).

3. Addressing the study of galaxy formation and assembly with NIRSpec IFS

GA-IFS high resolution (R2700) observations, with a spectral range encompassing the main rest-frame optical emission lines, allows a plethora of diagnostic tools to study the structural, chemical, and dynamical properties of the ionised gas with a sub-kpc sampling. More in detail, we will:

Table 1. Observations and targets.

#	Class and type of selection criteria	redshift range	Average exp. time and configuration(s) per target
14	SFGs: mainly optical and UV-selected; a few objects are IR-selected	$3 < z < 6.3$	$\sim 4-5$ hr (R2700) ~ 1 hr (R100)
9	SFGs at the EoR: Ly-α, [O III]88μm emitters	$6.5 < z < 9.1$	$\sim 3-5$ hr (R2700) ~ 1 hr (R100)
13	AGN: X-ray, optical and/or MIR selected	$3 < z < 5$	~ 1 hr (R2700)
6	type 1 AGN/QSOs	$6.3 < z < 7.5$	~ 3 hr (R2700, all targets) ~ 1 hr (R100, two targets*)

Notes:
Column (1): Number of targets; (2): Class of targets and selection criteria; (3): redshift range; (4): average exposure time and spectral configuration per individual target.
*dedicated PSF star observations will be acquired, and used to model and subtract the QSO component and reveal the stellar emission.

• trace the gas kinematics (i.e. velocity fields, velocity dispersion) and, hence, determine dynamical masses and also identify non-virial motions (outflow and inflows);
• map metallicity gradients (e.g. using strong-line diagnostics) and dust attenuation (using the Balmer decrement);
• map the distribution of star formation traced by the Balmer emission.

Moreover, the R100 continuum emission (complemented with available R2700 emission line fluxes) will allow us a full characterisation of the SFGs in our sample, for instance tracing the distribution of dust, and mapping the resolved properties of the stellar populations (e.g. star formation history, stellar mass, age, and metallicity). R100 observations will be also obtained for two QSOs at $z > 6$: dedicated PSF star observations will be used to model and subtract the nuclear component and reveal the stellar emission.

To summarise, the GA-IFS sample has been chosen to maximize the novelty and the impact of the very first NIRSpec IFS outcomes; our project will also pave the way for more efficient observational strategies in future NIRSpec IFS programs.

References

Brinchmann, J. 2022, *MNRAS*, arXiv:2208.07467
Böker, T., Arribas, S., Lützgendorf, N., & Zeidler P. 2022, *A&A*, 661, 82
Curti, M., D'Eugenio, F., Carniani, S., & Wallace, I., E., B. 2022, *MNRAS*, arXiv:2207.12375
Ferruit, P., Jakobsen, P., Giardino, G., & Zeidler, P. 2022, *A&A*, 661, 81
Förster Schreiber, N. M. & Wuyts, Stijn, 2020 *ARA&A*, 58, 661
Jakobsen, P., Ferruit, P., Alves de Oliveira, C., & Zincke, C. 2022, *A&A*, 661, 80
Pontoppidan, K. M., Barrientes, J., Blome, C., & Nota, A. 2022, *ApJ*, 936, 14
Robertson, Brant E. 2022, *ARA&A*, 20, 121
Tacchella, S., Johnson, B. D., Robertson, B. E., & Witstok, J. 2022, *MNRAS*, arXiv:2208.03281

Resolving the Rise and Fall of Star Formation in Galaxies
Proceedings IAU Symposium No. 373, 2023
T. Wong & W.-T. Kim, eds.
doi:10.1017/S1743921323000017

The evolution of carbon-chain chemistry from prestellar to protostellar cores in Taurus Molecular Cloud

Jenny M. Ramos[1,2] **and Yao-Lun Yang**[3,4]

[1]National University of San Marcos, Lima, 15081, Peru

[2]Chalmers Astrophysics & Space Sciences Summer (CASSUM) Research Program 2021
email: `jenny.ramos@unmsm.edu.pe`

[3]RIKEN Cluster for Pioneering Research, Wako-shi, Saitama, 351-0106, Japan

[4]Department of Astronomy, University of Virginia, Charlottesville, USA 22904
email: `yaolunyang.astro@gmail.com`

Abstract. The discovery of abundant carbon-chain molecules in protostellar cores motivates the development of the warm carbon-chain chemistry. To understand the role of warm carbon-chain chemistry in star-forming regions, we studied C_2H and c-C_3H_2 in 15 embedded protostars in the Taurus molecular cloud, whose evolutionary stages range from prestellar to Class I/II, using data from the Submillimeter Telescope (SMT). We calculated the excitation temperature, column density, and abundance of C_2H and c-C_3H_2 in each source. We compared those properties with evolutionary indicators of the protostars. We also estimated the kinetic temperature using RADEX. Finally, we compared the abundance of C_2H and c-C_3H_2 in our survey with that in the survey of protostellar cores in the Perseus molecular cloud. While we are unable to identify new WCCCs, our results suggest that the abundances of C_2H and c-C_3H_2 could be an indicator to find WCCC candidates.

Keywords. carbon-chain molecules, protostars, Taurus Molecular Cloud, astrochemistry.

---

1. Introduction

The study of carbon-chain molecules in low-mass protostars helps us understand the past and evolution of carbon-chain chemistry in our solar system. Carbon-chain molecules are usually abundant in prestellar cores. As the core density increase, C^+ is converted to CO in a timescale of 3×10^5 years, reducing the production of carbon-chain molecules. Additionally the abundance of carbon-chains molecules decrease due to ionic destruction of H^+ and He^+ as well as reactions with oxygen (Sakai & Yamamoto 2013). Hence, observations often only detected carbon-chains molecules in dense cores and prestellar cores. Interestingly, Sakai et al. (2008) discovered abundant carbon-chains in a Class 0 protostar IRAS 04368+2557 in L1527. The carbon-chain molecules locate in a lukewarm region (Sakai et al. 2008) close to the embedded protostar (a few thousand AU; Sakai & Yamamoto 2013). To explain this type of chemistry, Sakai et al. (2008) proposed the warm carbon-chain chemistry (WCCC). In the WCCC scenario, the carbon-chain molecules form from gaseous CH_4, which has a sublimation temperature of 25 K. This sublimation temperature is higher than that of CO (20 K; Law et al. 2018) lower than that of H_2O (100 K; Sakai et al. 2008; Sakai & Yamamoto 2013). As the temperature increases due to accretion heating, CH_4, if presents on ices, can sublimate and react with other molecules to form carbon-chains molecules. The carbon-chain chemistry in

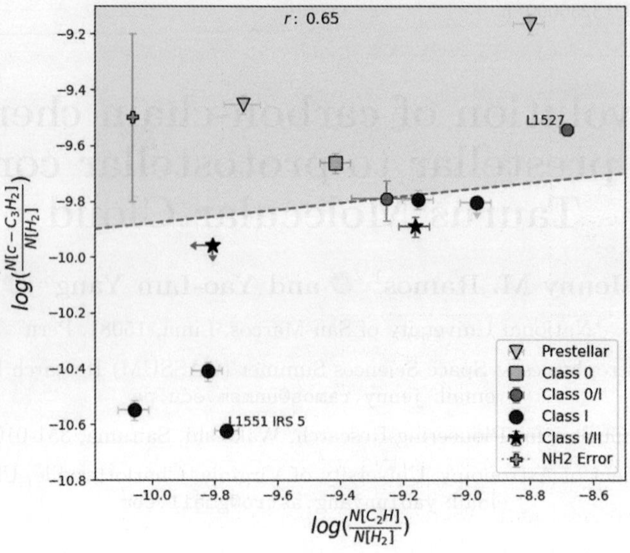

Figure 1. Abundance of carbon-chain molecules in 12 prestellar and protostellar sources in Taurus molecular cloud.

L1527 has been extensively studied (e.g., Sakai et al. 2008, 2010a, 2010c). Surveys of the carbon-chain chemistry in protostellar cores are available for 36 Class 0/I protostars in the Perseus molecular cloud and a few sources in the Taurus molecular cloud (Higuchi et al. 2018; Law et al. 2018), providing statistical characteristics of carbon-chain chemistry in the early stage of star formation. Increasing the sample size in the Taurus molecular cloud offers a great opportunity to study the carbon-chain chemistry in different environments. Thus, we conducted a survey of 15 protostars to measure the emission of C_2H and c-C_3H_2, characterizing the carbon-chain chemistry in these sources.

2. Results

For C_2H, we measured transitions of $N=3\rightarrow2$, $J=7/2\rightarrow5/2$, $F=4\rightarrow3$, and $N=3\rightarrow2$, $J=7/2\rightarrow5/2$, $F=3\rightarrow2$; and for c-C_3H_2, we measured transitions of $6(1, 5)\rightarrow5(2, 4)$, $6(2, 5)\rightarrow5(1, 4)$, and $3(2, 1)\rightarrow2(1, 2)$. We detect C_2H in 14 sources and c-C_3H_2 in 13 sources. We calculated the excitation temperature and the column density, assuming local thermodynamic equilibrium (LTE). The excitation temperatures of c-C_3H_2 range from 5 to 11 K, where L1551 IRS 5 and L1527 have the highest excitation temperatures. C_2H and c-C_3H_2 has column density ranging from 2.6×10^{12} cm^{-2} to 1.5×10^{14} cm^{-2} and 2.6×10^{12} cm^{-2} to 2.0×10^{13} cm^{-2}, respectively. The column densities and abundances between the c-C_3H_2 and C_2H are tightly correlated (Figure 1), confirming that both species are related in the formation pathways of carbon-chain molecules, where c-C_3H_2 is a secondary product of C_2H.

The abundances of C_2H and c-C_3H_2 are generally higher in prestellar sources and lower in more evolved sources (Class 0/I-I/II); (Figure 1), except for L1527, which is a known WCCC source. Therefore, the abundance of C_2H and c-C_3H_2 could serve as an indicator for the search of WCCC candidates. Moreover, there are three pairs of sources, where each pair consists of a protostar and its adjacent cold envelope. In each pair, both sources share similar environments except for the warmer temperature in the protostar. Thus, the chemical variation within each pair provides a unique opportunity to characterize the chemical change from cold to warm environments. Figure 2 shows that the protostars have notably different chemistry of C_2H and c-C_3H_2. To further constrain

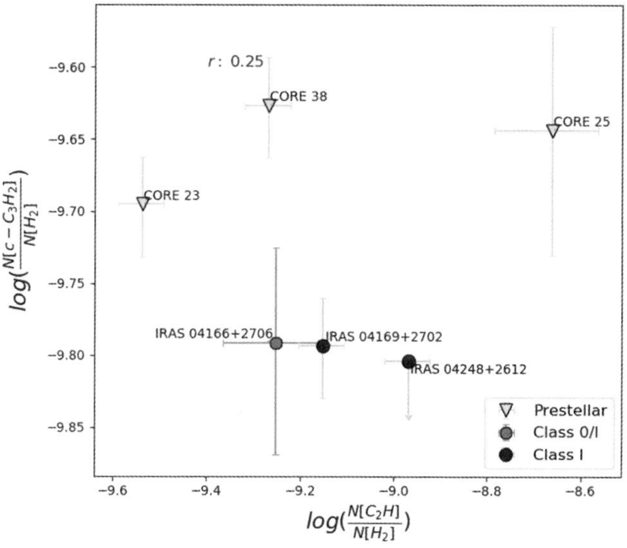

Figure 2. Abundance of carbon-chain molecules in the three pairs of protostar and cold envelope: CORE 23-IRAS 04166+2706, CORE 25-IRAS 04169+2702, and CORE 38-IRAS 04248+2612.

the chemical evolution with respect to the star formation, we compared the bolometric luminosity and bolometric temperature with the abundances of C_2H and c-C_3H_2. The bolometric luminosity and bolometric temperature are weakly negatively correlated with the abundance of C_2H and c-C_3H_2, suggesting that the protostellar evolutionary stage has no direct impact on the production of carbon-chain molecules.

We also derived the kinetic temperature of c-C_3H_2 in 4 of 15 sources using RADEX (van der Tak et al. 2007), a non-LTE radiative transfer code. The kinetic temperatures range from 10 to 28 K. L1551 IRS 5 has the highest kinetic temperature followed by L1527 with 22 K. Interestingly, the abundance of C_2H and c-C_3H_2 in the Taurus protostars is consistent with that in the Perseus protostars, both of which show correlation between these two species (Higuchi et al. 2018).

3. Conclusions

We studied the C_2H and c-C_3H_2 to characterize the carbon-chain chemistry in 15 protostars of Taurus molecular cloud. We estimated column densities and abundances of C_2H and c-C_3H_2, finding a high correlation between them. In addition, the abundances of the C_2H and c-C_3H_2 could be an indicator for identifying WCCC candidates (Figure 1). We also found significant chemical change from cold to warm environments (Figure 2). The correlation between the abundances of C_2H and c-C_3H_2 and the protostellar parameters is not very significant. Comparing with the Class 0/I protostars in the Perseus molecular cloud, the carbon-chain chemistry in Taurus sources is similar to that in Perseus (Higuchi et al. 2018). We derived kinetic temperatures for 4 of 15 sources and detected no additional WCCC source, except for L1527.

References

Higuchi, A. E., Sakai, N., Watanabe, Y., et al. 2018, ApJS, 236, 52. doi:10.3847/1538-4365/aabfe9
Law, C. J., Öberg, K. I., Bergner, J. B., et al. 2018, ApJ, 863, 88. doi:10.3847/1538-4357/aacf9d
Sakai, N., Sakai, T., Hirota, T., et al. 2008, ApJ, 672, 371. doi:10.1086/523635

Sakai, N., Saruwatari, O., Sakai, T., et al. 2010, A&A, 512, A31. doi:10.1051/0004-6361/200913098

Sakai, N., Sakai, T., Hirota, T., et al. 2010, ApJ, 722, 1633. doi:10.1088/0004-637X/722/2/1633

Sakai, N. & Yamamoto, S. 2013, Chemical Reviews, 113, 8981. doi:10.1021/cr4001308

van der Tak, F. F. S., Black, J. H., Schöier, F. L., et al. 2007, A&A, 468, 627. doi:10.1051/0004-6361:20066820

Resolving the Rise and Fall of Star Formation in Galaxies
Proceedings IAU Symposium No. 373, 2023
T. Wong & W.-T. Kim, eds.
doi:10.1017/S1743921322003726

The physical properties of IR-bright Dust-Obscured Galaxies

Nofoz Suleiman[1], Yoshiki Toba[2], Sándor Frey[3] and L. Viktor Tóth[1,4]

[1]Astronomy Department, Eötvös Loránd University, 1117 Budapest, Hungary
email: n.suleiman@astro.elte.hu

[2]National Astronomical Observatory of Japan, Japan

[3]Konkoly Observatory, ELKH CSFK, Hungary

[4]University of Debrecen, Institute of Physics, Hungary

Abstract. This work aims to represent the physical properties of a sample of infrared-bright dust-obscured galaxies (DOGs) studied by Suleiman et al. (2022) by fitting the spectral energy distributions (SEDs). Twenty-eight DOGs were examined at redshifts $0.47 \leq z \leq 1.63$ discovered by combining images of the Subaru Hyper Suprime-Cam (HSC) survey, VISTA Kilo-degree Infrared Galaxy (VIKING) survey, and the *Wide-field Infrared Survey Explorer* (*WISE*) all-sky survey, and detected at *Herschel* Spectral and Photometric Imaging Receiver (SPIRE) bands. The results show a correlation between the star formation rate (SFR) and the dust luminosity of Suleiman et al. (2022) DOG sample, the SFR ranges of the sample according to different redshifts, and a comparison between Suleiman et al. (2022) sample and other samples of DOGs.

Keywords. catalogs — galaxies: active — infrared: galaxies — surveys

1. Introduction

Extreme star formation generates enormous amounts of dust, which plays a crucial role in shaping the observed SED of a massive evolving galaxy in different phases. Dust absorbs most of the ultraviolet (UV) and optical photons and re-emits them in the far-IR (FIR) and sub-millimeter (sub-mm) bands. As a result, starburst-dominated and active galactic nuclei (AGN)–starburst composite systems will appear to be IR luminous, leading to the observed populations: ultraluminous IR galaxies (ULIRGs), sub-mm galaxies (SMGs) and dust-obscured galaxies (DOGs; Dey et al. 2008). The study of IR luminous galaxies at high redshift will aid in understanding the extreme scenarios in the early stages of massive galaxy evolution.

Color-based selection criteria were proposed for the efficient interpretation of a statistically significant sample of dusty ULIRGs at $z \sim 1.5 - 3$ (Dey et al. 2008). Taking advantage of the unprecedented sensitivity and angular resolution at IR wavelengths of the *Spitzer Space Telescope*, a population of optically faint ($22 < R_{\rm Vega} < 27$) and mid-IR bright ($F_\nu(24\ \mu{\rm m}) > 0.3$ mJy) DOGs were selected, defined as those sources having $F_\nu(24\ \mu{\rm m})/F_\nu(R) > 1000$. The comoving number density of DOGs shows its peak at $z \sim 1 - 2$ (e.g., Dey et al. 2008, Toba et al. (2017), Suleiman et al. 2022) that corresponds to the peak of star-formation rate (SFR) density and the growth rate of SMBHs (e.g., Richards et al. 2006). This strongly suggests that DOGs are linked to the most active objects in terms of the co-evolution between galaxies and SMBHs. In this context, DOGs with a high IR luminosity conceivably shelter a rapidly growing SMBH, and therefore they are important for understanding the evolution of galaxies and SMBHs.

2. Sample selection and used data

The IR-bright DOGs sample was discovered by Noboriguchi et al. (2019). They selected IR-bright DOGs by combining the *WISE* all-sky data and the deep optical imaging data obtained with the Subaru HSC. Furthermore, they provided NIR imaging data to preselect relatively red objects among the objects observed by HSC, in the same way as Toba et al. (2015).

In Suleiman et al. (2022), we used the *Herschel* point source catalog (HPSC; Marton et al. 2015) beside catalogs which were already used in Noboriguchi et al. (2019): the HSC (optical), VIKING (NIR), and *WISE* catalog data (MIR).

3. Methodology

The main physical properties of IR-bright galaxies were determined in Suleiman et al. (2022) by fitting their SEDs. The SEDs covered the rest-frame wavelength ranges from FUV to FIR (from 0.15 to 106 μm). They have been modeled using the CIGALE† software package (Burgarella et al. 2005, Noll et al. 2009). When constructing the SEDs, CIGALE manages a balance between the energy that is absorbed by dust and the energy that is re-emitted in the IR.

We used CIGALE to determine the physical characteristics of our DOGs, including SFRs, IR luminosities, stellar masses, dust attenuation, and others. Furthermore, CIGALE contains a useful methodology for evaluating the validity of the parameters and any potential degeneracies by building a mock catalog that is examined using techniques similar to those used to analyze the observed data.

4. Results and discussion

The dust luminosity is a good tracer of SFR, the reason is that the bulk of star formation is obscured, and dust emission is generated by the absorption of photons produced by massive stars. It was interesting to check the relation of the SFR with dust luminosity produced by CIGALE findings because CIGALE seeks to account for both obscured (radiated in the FIR) and unobscured (radiated in the UV) star formation. Additionally, CIGALE accounts for that portion of dust luminosity which is heated by earlier stellar populations. Our result in Suleiman et al. (2022; Eq.(1); $SFR[M_{\odot}\,\mathrm{yr}^{-1}] = (3.6 \pm 0.5) \times 10^{-44} L_{\mathrm{dust}}[\mathrm{erg\,s^{-1}}]$, where L_{dust} refers to the infrared luminosity integrated over the full MIR and FIR spectrum ($8 - 1000$ μm)) is marginally consistent with Kennicutt et al. (1998) in (Eq.(2); $SFR[M_{\odot}\,\mathrm{yr}^{-1}] = 4.5 \times 10^{-44} L_{\mathrm{dust}}[\mathrm{erg\,s^{-1}}]$) with 2σ confidence.

We examined the distribution of SFR values in the sample. The SFR varies from 26 to almost 1900 $M_{\odot}\,\mathrm{yr}^{-1}$ at different z (see Fig. 7 in Suleiman et al. 2022). The median and mean values are 600 and 755 $M_{\odot}\,\mathrm{yr}^{-1}$, respectively (Suleiman et al. 2022).

In addition, we compared the SFR range for five DOG samples at different redshifts with our sample (see Fig. 8 in Suleiman et al. 2022). The luminosities seen in Hot DOGs would require extreme SFRs of the order of 10^4 $M_{\odot}\,\mathrm{yr}^{-1}$ (Eisenhardt et al. 2012, Assef et al. 2015).

References

Assef, R., Eisenhardt, P., *et al.* 2015, *ApJ*, 804
Burgarella, D., Buat, V., *et al.* 2005, *MNRAS*, 360

† `https://cigale.lam.fr/`

Dey, A., Soifer, B. T., Desai, V., *et al.* 2008, *ApJ*, 667
Eisenhardt, P., Wu, J., Tsai, C., *et al.* 2012, *ApJ*, 755
Kennicutt, Jr., Robert, C. 1998, *ARA&A*,
Marton, G., Schulz, B., Altieri, B., *et al.* 2015, *arXiv preprint arXiv:1510.08325*
Noboriguchi, A., Nagao, T., *et al.* 2019, *ApJ*, 876
Noll, S., Burgarella, D., *et al.* 2009, *A&A*, 507
Richards, G., Strauss, M., *et al.* 2006, *AJ*, 131
Suleiman, N., Noboriguchi, A., *et al.* 2022, *PASJ*, 74
Toba, Y., Nagao, T., *et al.* 2015, *PASJ*, 67, 86
Toba, Y., Nagao, T., *et al.* 2017, *ApJ*, 835

Resolving the Rise and Fall of Star Formation in Galaxies
Proceedings IAU Symposium No. 373, 2023
T. Wong & W.-T. Kim, eds.
doi:10.1017/S1743921322004495

ALMA resolved views of molecular filaments/clumps in the Large Magellanic Cloud: A possible gas flow penetrating one of the most massive protocluster systems in the Local Group

Kazuki Tokuda

Department of Earth and Planetary Sciences, Faculty of Sciences, Kyushu University, Nishi-ku, Fukuoka 819-0395, Japan
National Astronomical Observatory of Japan, National Institutes of Natural Sciences, 2-21-1 Osawa, Mitaka, Tokyo 181-8588, Japan

Abstract. We present spatially resolved molecular filaments and clumps in the high-mass star-forming regions N159E-Papillon, W-South, and W-North in the Large Magellanic Cloud (LMC). Our ALMA observations in CO isotopes and millimeter continuum revealed remarkable hub-filament systems with a typical width of 0.1 pc. The most massive clump in the observed regions, N159W-North MMS-2, shows an especially massive/dense nature whose total H_2 mass and peak column density are $\sim 10^4\,M_\odot$ and $\sim 10^{24}\,\mathrm{cm}^{-2}$, respectively, and harbors massive ($\sim 100\,M_\odot$) starless core candidates. The hub-filamentary clouds in the three regions share a common orientation and have 10–30 pc scale head-tail structures with active star formation at the tips. Their striking similarity proposes a "teardrops-inflow" model, i.e., substructured conversing H I flow, that explains the synchronized, extreme star formation across ∼50 pc, including one of the most massive protocluster clumps in the Local Group.

Keywords. Interstellar filaments, Large Magellanic Cloud, Protostars, Local Group

1. Introduction

The Large Magellanic Cloud (LMC) is an ideal laboratory to study high-mass star formation thanks to its uncontaminated face-on view at the closest distance (∼50 kpc) among the external galaxies. ALMA offers a fantastic opportunity to resolve the star formation mechanism at a $\sim 10^4$ au (0.06 pc) scale. N159E/W is an active high-mass star-forming region younger than $\sim 10^5$ years with the most intense CO emission among the molecular clouds discovered by the single-dish NANTEN survey (Fukui et al. 2008), where UV ionization by the stars has not yet dispersed the parental cloud as in R136. The tidal interaction between the LMC and the Small Magellanic Cloud (SMC) drove a kpc H I flow, which originated in the past close encounter between the two galaxies (Fukui et al. 2017; Tsuge et al. 2019). The N159 GMC may be the end product of a cloud strongly compressed by the H I flow and enriched to the large mass we observe currently.

2. ALMA observations toward N159 molecular clouds in the LMC

In ALMA Cycle 4, we carried out Band 6 (211–275 GHz) observations toward three massive star-forming regions, N159E-Papillon, N159W-South, and N159W-North. We

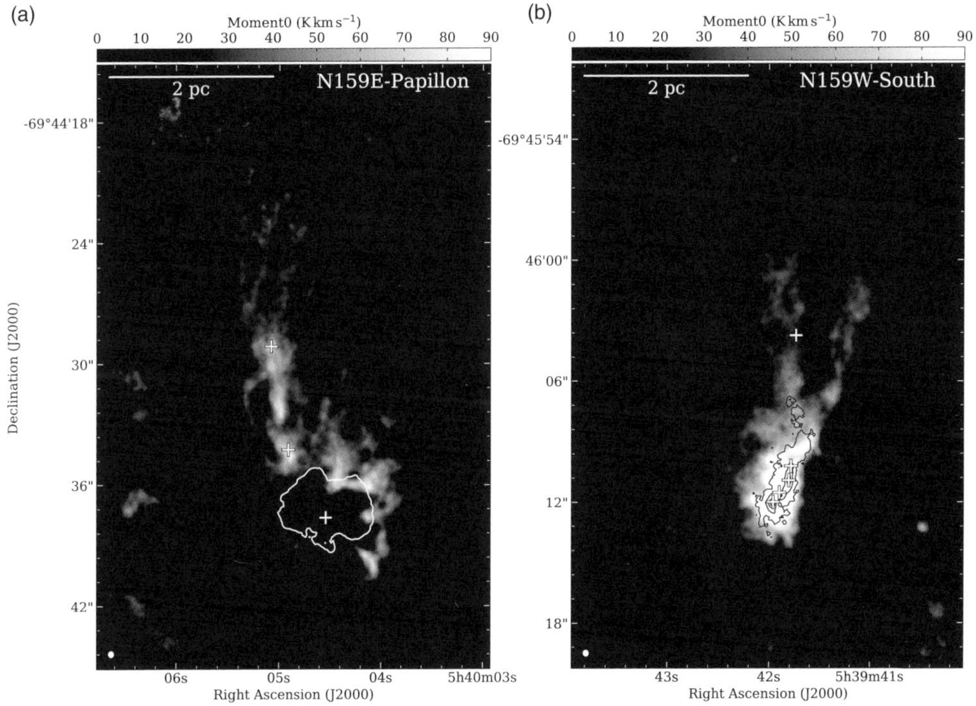

Figure 1. ALMA molecular cloud images toward the N159E-Papillon and N159W-South regions. (a) The grayscale image shows the ^{13}CO integrated intensity image. White contours show the Papillon Nebula H II region. White crosses represent high-mass protostellar sources. The white ellipse in the lower left corner shows the beam size. (b) Same as (b) but for the N159W-South region. Black contours show the 1.3 mm continuum emission.

selected the target regions based on our previous molecular cloud observations by Fukui et al. (2015) and Saigo et al. (2017). The frequency setting mainly targeted the molecular lines of ^{12}CO(2–1), ^{13}CO(2–1), C^{18}O(2–1), and the 1.3 mm continuum. The synthesized beam sizes of the molecular line and continuum images are $0.''27 \times 0.''23$ and $0.''26 \times 0.''23$, respectively.

2.1. *N159E-Papillon and W-South regions*

ALMA observations revealed remarkable hub-filament systems with a typical width of 0.1 pc in ^{13}CO, where a young H II region, the Papillon Nebula (see white contour in Figure 1a), and embedded high-mass protostars along with six protostellar outflows have been newly discovered/resolved (Figure 2). All these young protostellar objects have an age of 10^4 to 10^5 years, whereas they are scattered over a distance spanning $\sim$50 pc.

For N159E-Papillon (Figures 1a and 2a), the filamentary distributions are seen as "pillars" similar to those in the Galactic famous region, such as M16, extending into the Papillon Nebula. Although part of the filaments is already ionized in the H II region, the column density of the pillars is an order of magnitude higher than that of the pillars in M16.

Along the 1.3 mm continuum filament in N159W-South (Figure 2b), we performed the Gaussian fitting to the cross-section taken along the filament crest. The average width (FWHM) of the filament is $\sim$0.14 pc. The resultant line mass (=total mass/length) of the filament is $\sim 2 \times 10^3\, M_\odot\, \mathrm{pc}^{-1}$. These properties are consistent with those of "ridges" in Galactic high-mass star-forming clouds (e.g., NGC 6334, André et al. 2016). Note that

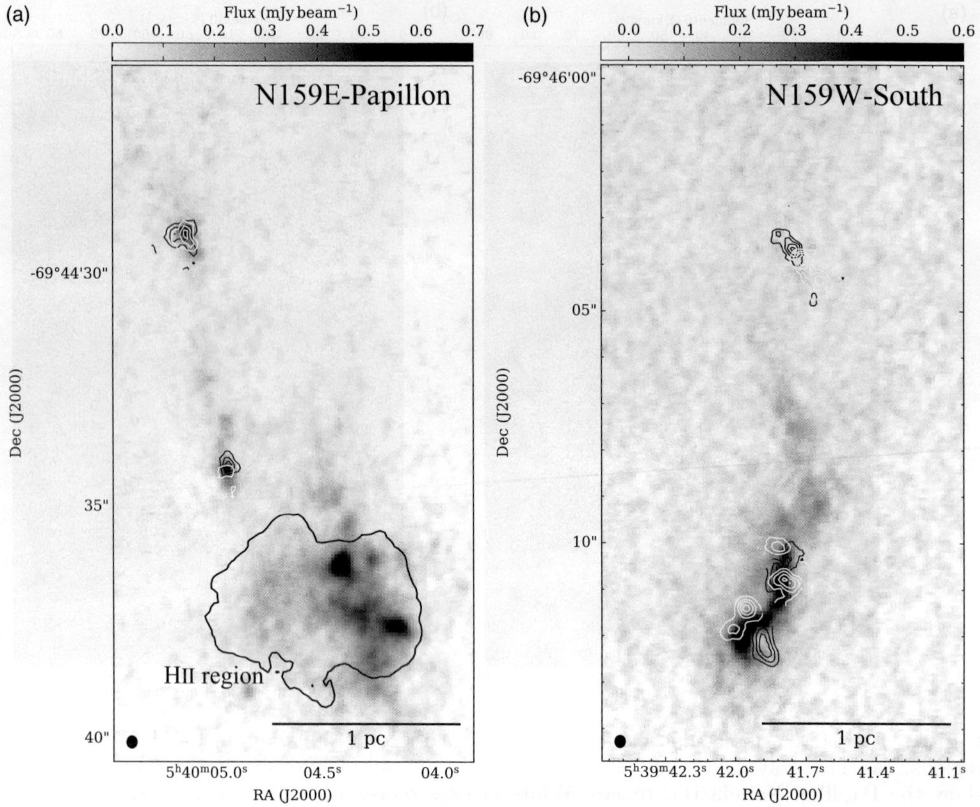

Figure 2. Millimeter continuum emission and ^{12}CO outflow distributions toward the N159E-Papillon and W-South regions. (a) The grayscale image shows the 1.3 mm continuum image. White contours show the Papillon Nebula H II region. Redshifted and blueshifted outflow lobes in ^{12}CO(2–1) are shown in white and black contours, respectively. The black ellipse in the lower left corner shows the beam size. (b) Same as (a) but for the N159W-South region.

the directions of the outflows are roughly perpendicular to the orientation of the ridge filament (Figure 2b).

2.2. *N159W-North region: One of the most massive prorocluster system in the Local Group*

The ^{12}CO and ^{13}CO high-resolution data depicted highly complex spatial and velocity structures of the N159W-North molecular cloud (Figure 3a). The ^{12}CO brightness temperature is higher than $\sim$30 K throughout the observed region and even higher than $\sim$50 K, especially on the southern side of the cloud (see "Warm side" in Figure 3a). One of the outstanding features is that several filaments, whose lengths and line masses are a few parsecs and $\sim$100 $M_\odot$ pc^{-1}, extend from high-mass star-forming dense cores toward the southern direction (Figure 3a).

Based on ^{12}CO high-velocity emission, which appears to be of outflow origin, we found five new bright star-forming cores in 1.3 mm continuum emission at the infrared quiescent spots. We also identified $\sim$0.1 pc scale starless cores with a mass of $\sim$100 $M_\odot$ that were not discovered in the previous extragalactic studies (Figure 3b). The dense molecular region (main ridge) traced by the thermal dust continuum and C^{18}O harbors the abovementioned proto- and prestellar cores. The total mass of the cluster-forming clump

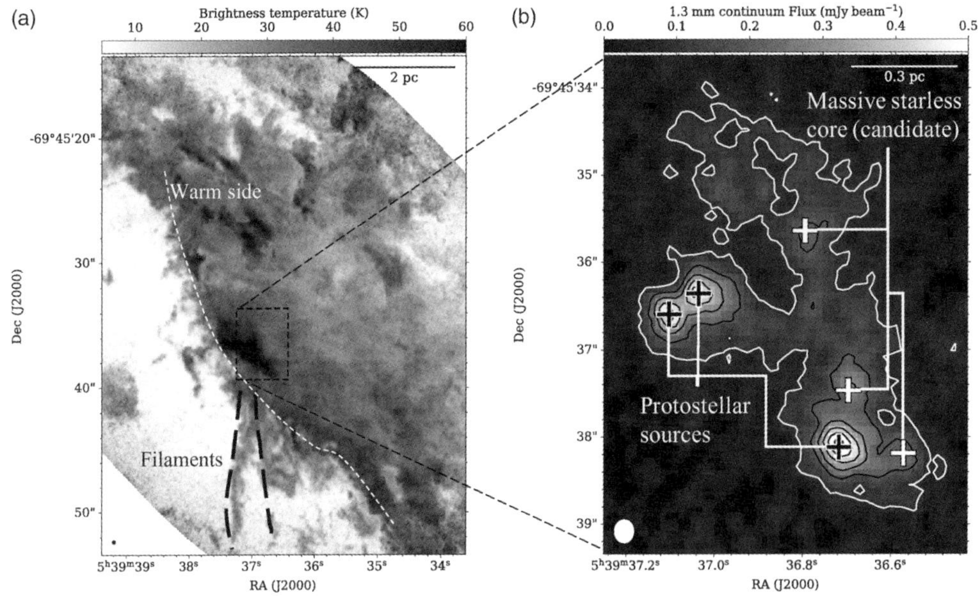

Figure 3. (a) Explanatory map of distinct features (see the text) in the N159W-North region on the high-resolution ^{12}CO(2–1) peak brightness temperature image. The black ellipse in the lower left corner gives the beam size. (b) The grayscale image and contours show an enlarged view of the 1.3 mm continuum emission toward MMS-2. The white ellipse in the lower left corner shows the beam size.

exceeds $10^4\,M_\odot$ despite its compact size of a few parsecs, making it one of the most massive, dense categories in the Local Group of galaxies (Tokuda et al. 2022 and references therein).

We investigated an archival CO(1–0) data set with a larger field of view and found a conical-shaped, ~30 pc long complex extending toward the northern direction. These characteristics qualitatively resemble those of the V-shaped or head-tailed filaments in the N159E-Papillon and N159W-South regions.

3. Discussion

The line mass of the filaments in the three regions is as large as $\sim$ a few $\times\,10^2$–$10^3\,M_\odot\,\mathrm{pc}^{-1}$. Such supermassive filaments should be in a "supercritical" and are considered to fragment and radially collapse within the freefall time. We speculate that the massive filaments that we see in the present observations formed quite recently and may be gravitationally unstable objects. Large-scale conversing flow is a promising trigger in the rapid formation of multiple filamentary complexes across more than 50 pc. A recent H I study by Fukui et al. (2017) suggested that kiloparsec-scale triggering of star formation occurred about 1 Myr ago through the compression in the converging H I flows in the R136-N159 region (Section 1). We suggest that the H I converging flows possibly triggered the coeval formation of three systems in N159. The formation of the filaments driven by the collision between two clouds is consistent with the magnetohydrodynamic simulations by Inoue et al. (2018), which show a hub-filamentary distribution with a pivot at the O-star formation site. The present picture explains the similar filamentary morphology of N159E and N159W, both of which show unevenly extended filaments to the north despite the large separation. The separation does not allow communication between the two nearly coeval star-forming spots within the O-star formation timescale of less than ~1 Myr.

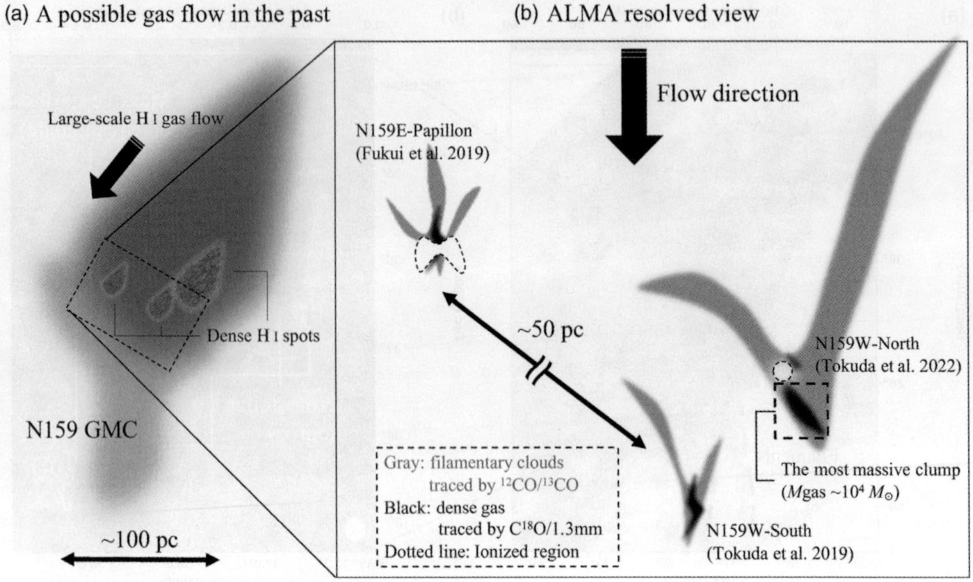

Figure 4. Schematic view of the star formation process in N159E/W motivated by the ALMA studies (see also Tokuda et al. 2022).

Combining our current observational understanding of N159E/W, three systems across more than 50 pc show active star formation simultaneously. Massive clumps, especially N159W-North-like objects (see Section 2.2), are highly rare in the Local Group, indicating that a typical galactic environment cannot easily produce YMC precursors. We hypothesize a "teardrops inflow model" (Figure 4) to explain the synchronized extreme cluster formation possibly driven by a quite dynamic, substructured flow induced by a galactic-scale phenomenon.

3.1. *Bibliography*

References

Fukui, Y., Kawamura, A., Minamidani, T., et al. 2008, *ApJS*, 178, 56
Fukui, Y., Tsuge, K., Sano, H., et al. 2017, *PASJ*, 69, L5
Tsuge, K., Sano, H., Tachihara, K., et al. 2019, *ApJ*, 871, 44
Fukui, Y., Harada, R., Tokuda, K., et al. 2015, *ApJL*, 807, L4
Saigo, K., Onishi, T., Nayak, O., et al. 2017, *ApJ*, 835, 108
Fukui, Y., Tokuda, K., Saigo, K., et al. 2019, *ApJ* 886, 14
Tokuda, K., Fukui, Y., Harada, R., et al. 2019, *ApJ*, 886, 15
André, P., Revéret, V., Könyves, V., et al. 2016, *A&A*, 592, A54
Tokuda, K., Minami, T., Fukui, Y., et al. 2022, *ApJ*, 933, 20
Inoue, T., Hennebelle, P., Fukui, Y., et al. 2018, *PASJ*, 70, S53

Resolving the Rise and Fall of Star Formation in Galaxies
Proceedings IAU Symposium No. 373, 2023
T. Wong & W.-T. Kim, eds.
doi:10.1017/S1743921322004173

Extended neutral hydrogen filamentary network in NGC 2403

Simone Veronese[1,2] **and W. J. G. de Blok**[1,2,3]

[1]Netherlands Institute for Radio Astronomy (ASTRON)

[2]Kapteyn Astronomical Institute, University of Groningen

[3]Department of Astronomy, University of Cape Town

Abstract. We present new neutral hydrogen (H I) observations of the nearby galaxy NGC 2403 to determine the nature of a low-column density cloud that was detected earlier by the Green Bank Telescope.

We find that this cloud is the tip of a complex of filaments of extraplanar gas that is coincident with the main disk. The total H I mass of the complex is 2×10^7 $M_\odot$ or 0.6% of the total H I mass of the galaxy. The main structure, previously referred to as the 8-kpc filament, is now seen to be even more extended, along a 20 kpc stream.

Keywords. Galaxies: evolution, Galaxies: interactions, Radio lines: galaxies, Techniques: interferometric

1. Introduction

Neutral hydrogen (H I) is the fuel of the star formation processes, but there is not enough of it in galaxies to sustain star formation over the age of the Universe (Sancisi et al. 2008; Madau & Dickinson 2014). Hence, there must be accretion of H I onto galaxies and we believe most of it comes from the Inter Galactic Medium (IGM) (Somerville & Davé 2015; Danovich et al. 2015); yet no unambiguous evidence has been found (Barnes et al. 2001; de Blok et al. 2002; Pisano et al. 2004; Giovanelli et al. 2007; Walter et al. 2008; Heald et al. 2011). We combined new and archival (Fraternali et al. 2002; Walter et al. 2008) 21-cm observations of the nearby galaxy NGC 2403 taken by the Very Large Array to better study an accreting cloud candidate observed with the Green Bank Telescope (GBT) (de Blok et al. 2014) and explored its connection with the previously reported 8 kpc anomalous velocity H I filament found in the galaxy (Fraternali et al. 2002).

2. Results

The combination of the new and archival data allowed us to reach a column density sensitivity of 1.8×10^{18} cm^{-2} with a spatial resolution of 52×49 arcsec (1.04×0.98 kpc). The GBT cloud is detected as being the tip of the 8 kpc filament, which is now seen as a 20 kpc long stream. Two 10 kpc long H I filamentary structures have also been detected close to the 20 kpc filament (see right panel of Fig. 1). The filaments have a collective mass of 2×10^7 $M_\odot$.

3. Discussion

One of the questions we tried to answer is whether the GBT cloud and the filaments are formed due to a gas accretion event. However, other explanations for these features are possible. For example, the structures could be the result of a galactic fountain that

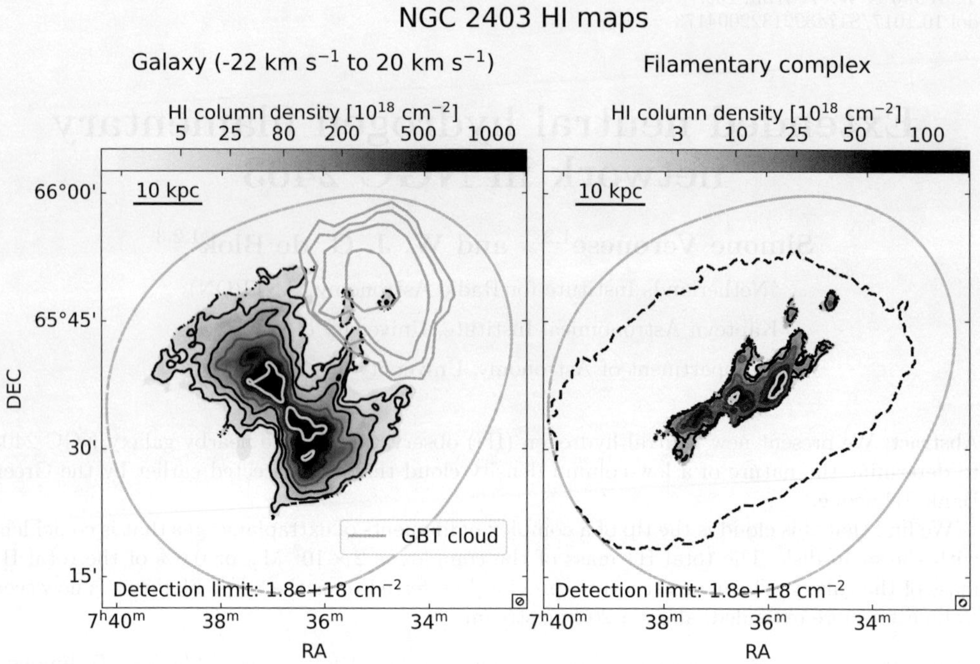

Figure 1. *Left panel*: primary beam corrected H I column density map of NGC 2403 (in reversed grey-scale colormap) overlaid with the the GBT cloud candidate (green) from de Blok et al. (2014). The map was produced by integrating over the channels from -22 km s^{-1} to 20 km s^{-1} w.r.t. the systemic velocity. This range corresponds to the one used by de Blok et al. (2014) to computing the candidate GBT cloud moment 0 map. Grey-scale contours levels are from 5×10^{18} cm^{-2} to 1×10^{21} cm^{-2}. Intermediate levels are given above the colorbar. The 3σ 1-channel column density limit is reported at the bottom. The beam of the interferometric data is shown in the bottom right, while in the top-left corner we show a 10-kpc scale for reference. The green contours, instead, define the (6.25, 12.5, 25, 62.5) $\times 10^{17}$ cm^{-2} column density in the GBT data. The light-gray line denote the 20% cut-off threshold of the primary beam response. *Right panel*: primary beam corrected H I column density map of the filamentary network. Grey-scale contours levels are (3, 10, 25, 50, 100) $\times 10^{18}$ cm^{-2} column density. The spatial scale is the same as in the left panel, as well as the definition of the light gray line. We indicate the galaxy edge with the dashed-black contour.

ejects gas from the disk. Yet, it is difficult to form a straight 20 kpc long filament in a pure galactic fountain model, as the gas ejecta reach distances of at most 10 kpc even with extremely high kick velocities (Fraternali & Binney 2006, 2008; Marasco et al. 2019).

The origin of the filamentary complex might be also a galactic interaction. In the event of a merging or disrupting dwarf, one might expect to see a disturbed stellar population from the dwarf, most likely coinciding with or near the filaments. Barker et al. (2012) performed an in-depth study of the stellar properties in NGC 2403 but did not identify any peculiarities in the distribution of the stars and concluded that the galaxy has evolved in isolation with no significant recent merging event. Those information, available at the time of the IAU General Assembly 2022, brought us to the conclusion that the filaments are likely the result of accretion from the IGM.

However, recent re-reduction of deep optical wide-field data presented in Carlin et al. (2019) reveal a ~ 50 kpc long stellar stream originating from the NGC 2403 dwarf satellite DDO 44 and connecting with the NW part of the stellar disk of NGC 2403 at the location of the northern tip of the 20 kpc filament. Thus, the most plausible explanation for the origin of the H I filaments in NGC 2403 is now a previous interaction with the dwarf satellite DDO 44.

Acknowledgements

This work has received funding from the European Research Council (ERC) under the European Union's Horizon 2020 research and innovation programme (grant agreement No 882793 'MeerGas').

References

Barker, M. K., Ferguson, A. M. N., Irwin, M. J., Arimoto, N., & Jablonka, P. 2012, MNRAS, 419, 1489
Barnes, D. G., Staveley-Smith, L., de Blok, W. J. G., et al. 2001, MNRAS, 322, 486
Carlin, J. L., Garling, C. T., Peter, A. H. G., et al. 2019, ApJ, 886, 109
Danovich, M., Dekel, A., Hahn, O., Ceverino, D., & Primack, J. 2015, MNRAS, 449, 2087
de Blok, W. J. G., Keating, K. M., Pisano, D. J., et al. 2014, A&A, 569, A68
de Blok, W. J. G., Zwaan, M. A., Dijkstra, M., Briggs, F. H., & Freeman, K. C. 2002, A&A, 382, 43
Fraternali, F. & Binney, J. J. 2006, MNRAS, 366, 449
Fraternali, F. & Binney, J. J. 2008, MNRAS, 386, 935
Fraternali, F., van Moorsel, G., Sancisi, R., & Oosterloo, T. 2002, AJ, 123, 3124
Giovanelli, R., Haynes, M. P., Kent, B. R., et al. 2007, AJ, 133, 2569
Heald, G., Józsa, G., Serra, P., et al. 2011, A&A, 526, A118
Madau, P. & Dickinson, M. 2014, Annual Review of Astronomy and Astrophysics, 52, 415
Marasco, A., Fraternali, F., Heald, G., et al. 2019, A&A, 631, A50
Pisano, D. J., Barnes, D. G., Gibson, B. K., et al. 2004, ApJ, 610, L17
Sancisi, R., Fraternali, F., Oosterloo, T., & van der Hulst, T. 2008, A&A Rev., 15, 189
Somerville, R. S. & Davé, R. 2015, ARA&A, 53, 51
Walter, F., Brinks, E., de Blok, W. J. G., et al. 2008, AJ, 136, 2563

Chapter 2:
Sustaining Star Formation: Gas Conditions & Environment

Resolving the Rise and Fall of Star Formation in Galaxies
Proceedings IAU Symposium No. 373, 2023
T. Wong & W.-T. Kim, eds.
doi:10.1017/S1743921322004896

Probing the Conditions for the Atomic-to-Molecular Transition in the Interstellar Medium

Gyueun Park[1,2], Min-Young Lee[1,2], Shmuel Bialy[3], Blakesley Burkhart[4,5], Joanne Dawson[6], Carl Heiles[7], Di Li[8,9,10], Claire Murray[11,12], Hiep Nguyen[13], Anita Petzler[6] and Snežana Stanimirović[14]

[1]Korea Astronomy & Space Science Institute, 776 Daedeok-daero, Yuseong-gu, Daejeon 34055, Republic of Korea
email: gpark@kasi.re.kr

[2]Department of Astronomy & Space Science, University of Science and Technology, 217 Gajeong-ro, Yuseong-gu, Daejeon 34113, Republic of Korea
email: mlee@kasi.re.kr

[3]Department of Astronomy, University of Maryland, College Park, MD 20742, USA

[4]Department of Physics and Astronomy, Rutgers University, Piscataway, NJ 08854, USA

[5]Center for Computational Astrophysics, Flatiron Institute, 162 Fifth Avenue, New York, NY 10010, USA

[6]Department of Physics and Astronomy, Macquarie University, NSW 2109, Australia

[7]Department of Astronomy, University of California, Berkeley, CA 94720, USA

[8]National Astronomical Observatories, Chinese Academy of Sciences, Beijing 100101, China

[9]NAOC-UKZN Computational Astrophysics Centre, University of KwaZulu-Natal, Durban 4000, South Africa

[10]Research Center for Intelligent Computing, Zhejiang Laboratory, Hangzhou 311100, China

[11]Space Telescope Science Institute, 3700 San Martin Drive, Baltimore, MD 21218, USA

[12]Department of Physics & Astronomy, Johns Hopkins University, MD 21218, USA

[13]Research School of Astronomy and Astrophysics, The Australian National University, Canberra, ACT 2611, Australia

[14]Department of Astronomy, University of Wisconsin, Madison, WI 53706-15821, USA

Abstract. We examine the physical conditions required for the formation of H_2 in the solar neighborhood by comparing H I emission and absorption spectra toward 58 lines of sight at $b < -5^\circ$ to CO(1–0) and dust data. Our analysis of CO-associated cold and warm neutral medium (CNM and WNM) shows that the formation of CO-traced molecular gas is favored in regions with high column densities where the CNM becomes colder and more abundant. In addition, our comparison to the one-dimensional steady-state H I-to-H_2 transition model of Bialy et al. (2016) suggests that only a small fraction of the clumpy CNM participates in the formation of CO-traced molecular gas. Another possible interpretation would be that missing physical and chemical processes in the model could play an important role in H_2 formation.

Keywords. ISM: atoms – ISM: clouds – dust, extinction – ISM: molecules – ISM: structure – radio lines: ISM

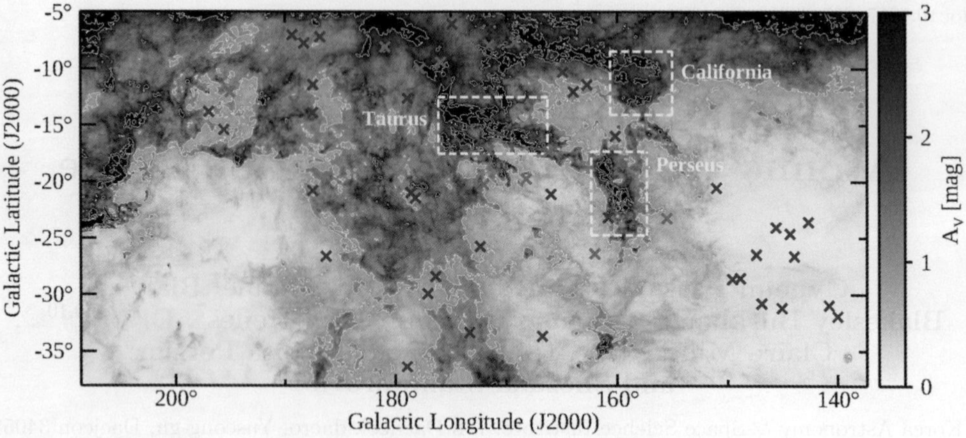

Figure 1. Our 58 LOSs are marked on the *Planck* A_V map at $b < -5°$ (Planck Collaboration et al. 2016) with contours 1, 3, and 5 mag in white. The 19 LOSs marked in red crosses show clear CO(1–0) detection, while the remaining 39 LOSs marked in blue crosses show no CO(1–0) detection. The yellow boxes represent the approximate locations of local molecular clouds (Perseus, Taurus, and California).

1. Introduction

Molecular clouds, the principal stellar nurseries, form out of the surrounding diffuse gas through the conversion of atomic (H I) to molecular hydrogen (H_2). Among the two main flavors of H I (cold and warm neutral medium; CNM and WNM), the colder and denser CNM (spin temperature $T_s \simeq 60$–260 K; gas density $n \simeq 7$–$70\,\mathrm{cm^{-3}}$) is expected to be more critical for the conversion, and yet its exact roles remain largely unknown due to a difficulty in direct measurements.

To overcome observational challenges and provide unique insights into the roles of the CNM in the formation of molecular gas, we obtained Arecibo H I emission and absorption spectra (angular and velocity resolution of 3.5′ and 0.16 km s^{-1}, respectively) toward 58 lines of sight (LOSs) at $b < -5°$ (Figure 1; Stanimirović et al. 2014; Nguyen et al. 2019) and analyzed them with TRAO/PMO CO(1–0) emission spectra (angular and velocity resolution of 48″ and 0.32 km s^{-1}, respectively). For 19 CO-detected LOSs where we have optical depth (τ_{CNM}), spin temperature (T_s), and column density (N_{CNM} and N_{WNM}) estimates, the CO spectra were fitted and decomposed into individual Gaussians to derive line parameters such as the central velocity (ν_{CO}) and line width (FWHM; $\Delta\nu_{\mathrm{CO}}$).

In this article, we present our study on the H I-to-H_2 transition from two perspectives: observational (comparison between H I and CO) and theoretical (comparison between our observations and the analytic model of Bialy & Sternberg 2016 (B16 hereafter)) approaches.

2. Conditions for the Formation of Molecular Gas: Observational Perspective

2.1. *Kinematic Association between H I and CO*

To investigate the H I conditions that are favorable to form molecular gas, we first compared the central velocities of H I and CO. In Figure 2, we included H I components that are closest to CO emission in velocity and displayed the absolute velocity differences between H I and CO in cumulative distribution function (CDF).

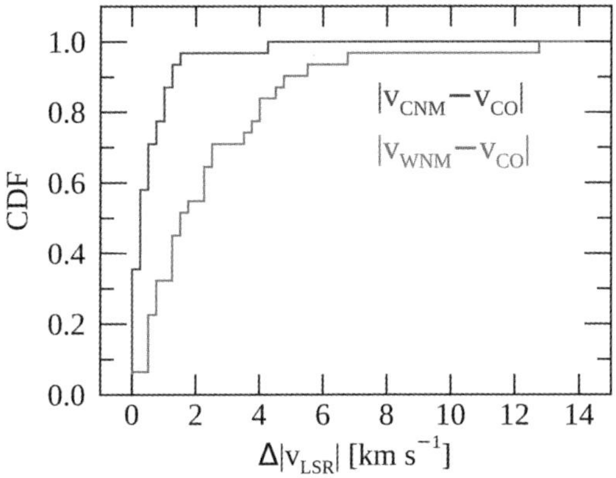

Figure 2. CDF of the absolute velocity difference between H I and CO (blue: $|\nu_{\rm CNM} - \nu_{\rm CO}|$; red: $|\nu_{\rm WNM} - \nu_{\rm CO}|$). For this plot, only the CNM and WNM components that are closest to the observed CO emission in velocity are considered.

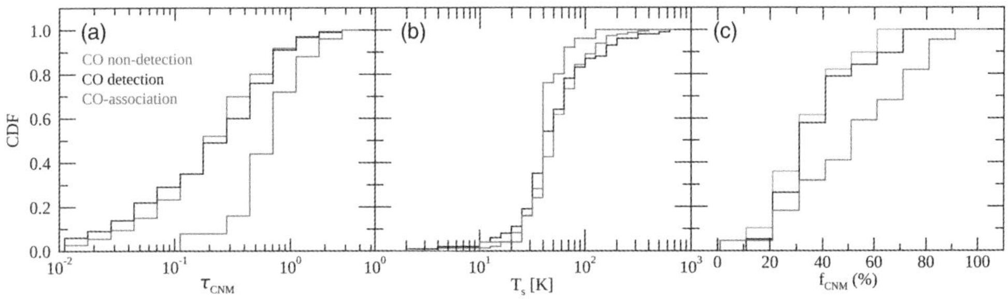

Figure 3. CDFs of (a) optical depth, (b) spin temperature, and (c) CNM fraction. In each panel, the three groups are shown in different colors (gray: CO non-detection; black: CO detection; red: CO-association).

As shown in Figure 2, the velocity difference between CNM-CO has a range of 0.01–4.3 km s^{-1} with a median of 0.4 km s^{-1}, while that of WNM-CO ranges from 0.03 km s^{-1} to 12.8 km s^{-1} with a median of 1.9 km s^{-1}. The systematically smaller velocity difference between the CNM and CO suggests that the CNM is kinematically more closely associated with CO emission. If velocity is taken as a proxy for distance, our result implies that the CO-traced molecular gas forms in CNM environments.

2.2. *H* I *Properties*

Next, we investigated the properties of CO-associated CNM components by classifying the observed CNM into three groups: (1) CO non-detection (all components toward the 39 CO not-detected LOSs); (2) CO detection (all components toward the 19 CO-detected LOSs); (3) CO-association (components whose central velocities are within $\nu_{\rm CO} \pm \Delta\nu_{\rm CO}$). We note that the CO-association is a subset of the CO detection.

For each group, we show the distributions of the individual optical depth ($\tau_{\rm CNM}$) and spin temperature ($T_{\rm s}$), as well as the integrated CNM fraction (CNM-to-total H I column density ratio; $f_{\rm CNM}$), in Figure 3. Generally, the CO non-detection and detection groups are comparable in terms of the CNM properties. On the contrary, the CO-association has several distinctive features compared to the CO non-detection and detection groups. For example, its minimum $\tau_{\rm CNM} \sim 0.1$ and maximum $T_{\rm s} \sim 130$ K are larger and smaller by

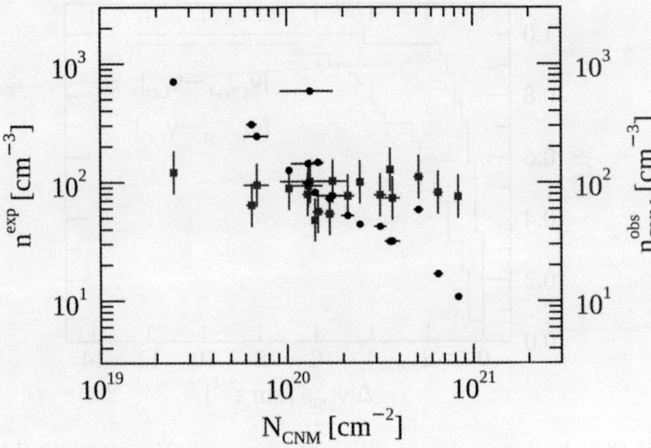

Figure 4. Comparison between the B16-based total densities (circles in black) and the observationally inferred CNM densities (squares in blue) as a function of the CNM column density. For this plot, the 19 CO-detected LOSs are included. As B16 predicts total densities expected for H_2 formation, $n_{\rm CNM}^{\rm obs}$ should be lower than $n^{\rm exp}$. The variation in the assumed thermal pressure ($\log_{10}(P/k_{\rm B}) = 3.58 \pm 0.18$ K cm^{-3}) is indicated as the 1σ error bars for $n_{\rm CNM}^{\rm obs}$.

a factor of ten and five, respectively, than those for the other two groups. Moreover, its median $f_{\rm CNM} \sim 0.6$ is 1.5 times larger than that of the CO non-detection and detection groups. Overall, our result implies that the formation of CO-traced molecular gas is favored in environments where the CNM becomes colder and more abundant.

3. Conditions for the Formation of Molecular Gas: Theoretical Perspective

3.1. *Density versus CNM column density*

With the aim of probing the underlying principles of the H I-to-H_2 transition, we compared the observed CNM properties to the B16 model. B16 predicted the total H I column density in a one-dimensional slab of gas and dust irradiated by isotropic UV radiation as a function of UV radiation field strength ($I_{\rm UV}$), dust absorption cross-section ($\tilde{\sigma_{\rm g}}$), and total hydrogen density ($n = n(\text{H I}) + n(\text{H}_2)$) by assuming H_2 formation in steady-state and chemical equilibrium. Given that our LOSs probe solar neighborhood environments ($\tilde{\sigma_{\rm g}} \sim 1$), the total H I column density can simply be expressed as a function of $I_{\rm UV}$ and n:

$$N(\text{H I}) = 8.4 \times 10^{20} \ \ln\left(\frac{18.4 I_{\rm UV}}{n} + 1\right) \ (\text{cm}^{-2}). \tag{1}$$

Using our observed CNM column density ($N_{\rm CNM}$) and spin temperature ($T_{\rm s}$) and assuming the measured thermal pressure of $\log_{10}(P/k_{\rm B}) = 3.58 \pm 0.18$ from Jenkins & Tripp (2011), we can make a comparison between the observationally inferred CNM density:

$$n_{\rm CNM}^{\rm obs} = \left(\frac{10^{3.58 \pm 0.18}}{T_{\rm s}}\right) \ (\text{cm}^{-3}), \tag{2}$$

and the theoretically expected total density:

$$n^{\rm exp} = \frac{18.4 \ I_{\rm UV}}{\exp\left(N_{\rm CNM}/8.4 \times 10^{20}\right) - 1} \ (\text{cm}^{-3}), \tag{3}$$

where *Planck*-based $I_{\rm UV}$ estimates in units of the Draine field (Planck Collaboration et al. 2016) are:

$$I_{\rm UV} = \left(\frac{T_{\rm dust}}{17.5\ {\rm K}}\right)^{\beta+4}. \tag{4}$$

For Equation 1, we substituted N(H I) into $N_{\rm CNM}$. This is based on the fact that the observed uniform $I_{\rm UV} \sim 1$ likely represents isotropic UV radiation attenuated by the widespread WNM: the impact of the WNM on the H I-to-H_2 transition is hence already considered. As B16 predicts total densities expected for H_2 formation, the CNM density ($n^{\rm obs}_{\rm CNM}$) should be a lower limit on the total gas density ($n^{\rm exp}$).

Since the real structure of gas is unknown, we assumed that only the CNM close to CO (within $\nu_{\rm CO} \pm \Delta\nu_{\rm CO}$; CO-association) is highly effective in shielding molecular gas from dissociating UV photons. This assumption of conservative geometry would be the case of a small volume filling factor, making the penetration of UV radiation into the CNM easier.

Figure 4 shows that the predicted total densities are consistent with the lower limits provided by our measurements at low column densities ($N_{\rm CNM} \lesssim 10^{20}$ cm^{-2}). On the other hand, the observationally inferred CNM densities are higher than the B16-based total densities at high column densities ($N_{\rm CNM} \gtrsim 10^{21}$ cm^{-2}), resulting in by up to an order of magnitude discrepancy.

3.2. *Limitations and Implications of B16*

The discrepancy between our observations and the B16 model at high CNM column densities indicates that only a small fraction of the CNM along a LOS ($\lesssim 30\%$ on average; estimated from the LOSs at $N_{\rm CNM} > 10^{20}$ cm^{-2} for the CO-association) contribute to H_2 formation. In other words, the distribution of the CNM should be clumpy with a small volume filling factor.

While the above interpretation is certainly viable, our analysis may have some limitations. For example, one possible source of the discrepancy is the assumed thermal pressure of 2500–5800 K cm^{-3} for the CNM. Recently, Goldsmith et al. (2018) found a larger variation in $P/k_{\rm B}$ ($\sim 10^3$–10^4 K cm^{-3}) in the ISM. If $P/k_{\rm B}$ of the CNM varies by an order of magnitude as Goldsmith et al. (2018) suggested, the discrepancy between our observations and the B16 model would decrease. Yet, the discrepancy would still exist at $N_{\rm CNM} \gtrsim 10^{21}$ cm^{-2}.

Another possible source of the discrepancy is physical and chemical processes that are not considered in the simple steady-state H_2 formation model of B16, e.g., cosmic-rays and turbulence. These missing ingredients could play an important role in the H I-to-H_2 transition and need further studies.

4. Summary

We performed TRAO/PMO CO(1–0) observations toward 58 LOSs at $b < -5^\circ$ where we have Arecibo H I absorption and emission spectra. For 19 out of the 58 LOSs with clear CO detection, we compared the H I and CO spectra and analyzed the CNM properties to constrain the physical conditions for the H I-to-H_2 transition in the solar neighborhood. Our primary results are as follows:

(1) The CNM is kinematically (spatially if velocity is considered as a proxy for distance) more closely associated with CO than the WNM.
(2) The formation of CO-traced molecular gas is favored in environments where the CNM becomes colder and more abundant.

(3) The comparison with the B16 model suggests that the distribution of the CNM should be clumpy with a small volume filling factor, and only a small fraction of the total CNM along a LOS contributes to H_2 formation. Another possibility is that missing processes in the model, such as cosmic-rays and turbulence, play an important role in the H I-to-H_2 transition.

References

Bialy, S. & Sternberg, A. 2016, *ApJ*, 822, 83
Goldsmith, P. F., Pineda, J. L., Neufeld, D. A., *et al.*, 2018, *ApJ*, 856, 96
Jenkins, E. B. & Tripp, T. M. 2011, *ApJ*, 734, 65
Nguyen, H., Dawson, J. R., Lee, M.-Y., *et al.*, 2019, *ApJ*, 880, 141
Planck Collaboration, Aghanim, N., Ashdown, M., *et al.*, 2016, *A&A*, 596, A109
Stanimirović, S., Murray, C. E., Lee, M.-Y., *et al.*, 2014, *ApJ*, 793, 132

Resolving the Rise and Fall of Star Formation in Galaxies
Proceedings IAU Symposium No. 373, 2023
T. Wong & W.-T. Kim, eds.
doi:10.1017/S1743921322004653

The SUNBIRD Survey: Insights into small-scale star formation mechanisms through near-infrared study of Young Massive Clusters

Zara Randriamanakoto[1], **P. Väisänen**[1,2] **and P. Ranaivomanana**[3]

[1]South African Astronomical Observatory
P.O Box 9, Observatory 7935, South Africa
email: zara@saao.ac.za

[2]Southern African Large Telescope
P.O Box 9, Observatory 7935, South Africa

[3]Department of Astrophysics/IMAPP, Radbourd Observatory
P.O Box 9010, 6500 GL, Nijmegen, The Netherlands

Abstract. We draw the K-band luminosity functions (CLFs) of young massive clusters (YMCs) hosted by 34 SUNBIRD targets to evaluate the impact of the host galaxy environment on their YMC properties. The depth and high resolution of the NIR images (PSF ~ 0.1") allow us to test whether CLF power-law slopes (α) of high star-forming galaxies are similar to those of gas-poor low star formation rate (SFR) galaxies. We found that α ranges between 1.53 and 2.41 with a median value of 1.87 ± 0.23. We also performed correlation searches between α and the host global properties and noticed that α decreases with an increasing SFR and SFR density. On sub-galactic scales, CLF slopes of cluster-rich galaxies differ by ~ 0.5. Our NIR CLF analyses suggest that the extreme environment of high SFR galaxies such as the SUNBIRD sample is likely to affect the formation mechanisms of YMCs and hence to govern the ongoing small-scale SF processes of the host galaxy.

Keywords. galaxies: interactions, galaxies: star clusters, infrared: galaxies

1. Introduction

Young massive star clusters (YMCs) are usually found in large quantity in high star formation rate (SFR) galaxies such as interacting starburst galaxies ($10.6 \lesssim \log(L_{IR}/L_{\odot}) < 11$) and luminous infrared galaxies (LIRGs, $11 \lesssim \log(L_{IR}/L_{\odot}) < 12$). Low SFR galaxies such as normal spiral galaxies, nearby dwarf galaxies and circumnuclear rings of otherwise unremarkable galaxies are also known to host these young and compact objects. With their masses as massive as $10^4 - 10^8\,M_{\odot}$, they represent the most massive and extreme form of star formation (SF) activity in galaxies (e.g. Efremov 1995; Randriamanakoto et al. 2019). These make them good tracers of small-scale SF mechanisms. And since their properties are similar to those of global clusters (GCs) in terms of mass and density, they are also thought to be young GCs that are still forming today (see e.g. Bastian 2016 for a review). Still embedded in their dust cocoons, YMC formation mechanisms, however, are not fully understood and hence deserve a more comprehensive investigation. One hot topic in this research area is whether the host environments influence the formation and evolution mechanisms of their cluster population. It is therefore relevant to use near-infrared (NIR) observations which can probe deeper into the galaxy dusty nuclear regions, believed to be the nurseries of the most massive YMC candidates (Pflamm-Altenburg et al. 2013; Adamo et al. 2017; Randriamanakoto et al. 2019).

A pilot study by Randriamanakoto et al. (2013b) considered a representative sample of 8 local LIRGs with high SFRs observed with NIR adaptive optics (AO) instruments to construct the K-band cluster luminosity functions (CLFs, $dN/dL \propto L^{-\alpha}$) of the targets. The intrinsic shape and power-law slope α of the CLF are usually interpreted, not only because the CLF could be a mere reflection of the cluster initial mass function but also to help define a first order approximation of how the host environment affects the YMC physical and photometric properties. Randriamanakoto et al. (2013b) found that LIRGs have shallower slopes ($\alpha \approx 1.9$) compared to those of gas-poor galaxies. This discrepancy is probably due to resolution bias and simple statistics (e.g. Whitmore et al. 2014; Mulia et al. 2016) and/or external factors such as the extreme environments of interacting LIRGs (e.g. Cook et al. 2016; Larson et al. 2020). We note however that no firm conclusions could not be drawn from this pilot study given the small sample size used in the work.

2. The SUNBIRD sample and the cluster catalogues

The local LIRGs studied by Randriamanakoto et al. (2013b) are part of the SUNBIRD survey (SUperNovae and starBursts in the InfraReD or Supernovae UNmasked By InfraReD Detection, Väisänen et al. 2014; Kool et al. 2018). This NIR survey is an ongoing project mainly aimed at searching for optically hidden core collapse supernovae in the dusty regions of 42 starburst and luminous IR galaxies (Mattila et al. 2007; Kool et al. 2018). High spatial resolution instruments (PSF $\sim$ 0.1") mounted on telescopes that are equipped with AO imaging (Gemini-N/ALTAIR/NIRI, Gemini-S/GeMS/GSAOI, VLT/NACO, and Keck II/NIRC2) were used to observe the K-band images of the targets. Other research projects focusing on the stellar population properties (Ramphul 2018) and the YMC populations (Randriamanakoto et al. 2013a,b) of the galaxy sample also used the SUNBIRD sample to probe further the physical details of SF activity in strongly star-forming galaxies.

By considering a larger sample of 34 targets from the SUNBIRD survey, the current work is a follow-up study of the pilot project done by Randriamanakoto et al. (2013b). Its goal is to deliver more robust K-band CLF analyses and thus contribute to addressing the differing views on the correlation between YMCs and their galaxy environments. The work is unique in a way that previous CLF works in the literature were mainly based either on optical observations (and hence miss the detection of highly embedded young YMCs) and/or on targets located less than ~ 20 Mpc away to avoid dealing with resolution bias (but then exclude cluster-rich galaxies at larger distances). With a wide range of morphologies and interaction stages, the SUNBIRD targets studied in this work have an average SFR of $\sim 30\,M_{\odot}\,yr^{-1}$ and luminosity distances D_L below 135 Mpc, except for IRAS 19115−2124 or the Bird where $D_L \sim 206$ Mpc. We note that 8 of the 34 galaxies had their CLFs already published in the pilot study.

The same catalogues used by Randriamanakoto et al. (2013a) to establish the first ever relation between the NIR brightest cluster magnitude and the host's SFR are considered to build the CLFs in this work. A comprehensive description of the aperture photometry (with apertue radii of $\sim$ 0.1") as well as the selection steps carried out to generate the final YMC catalogues can be found in Randriamanakoto et al. (2013b). The number of YMC candidates, N, in each catalogue ranges between 25 and 414. Targets with low number statistics of YMCs are not necessarily cluster-poor but their high inclination and/or a degraded AO correction are likely to challenge the extraction of these compact sources.

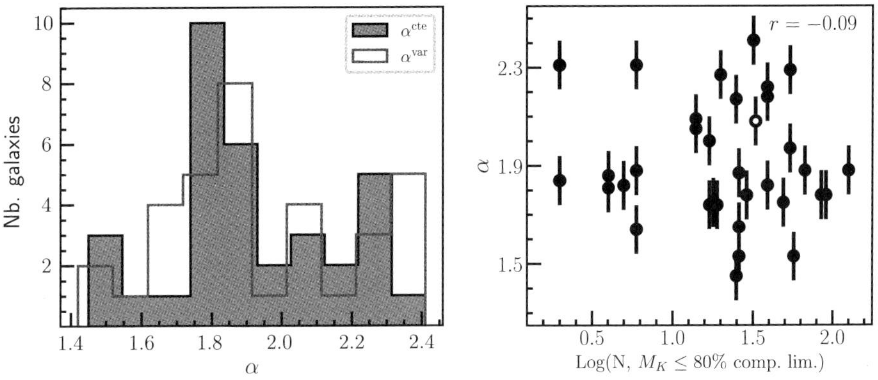

Figure 1. *Left*: Distributions of the power-law α for the whole sample using a constant (cte, grey) and a variable binning (var, blue). The median values of α are 1.86 ± 0.24 and 1.88 ± 0.28, respectively. *Right*: α^{cte} plotted against the number of YMCs with K-band magnitudes brighter than the 80 percent completeness level for each galaxy. The open circle denotes data points of the Bird.

3. The power-law slopes of the individual CLFs

Individual K-band CLFs of the 26 SUNBIRD targets (i.e. part of the sample not included in the pilot study) before and after completeness corrections were constructed using a constant and then a variable binning. We then fit a single power-law function to the corrected data until the 80 percent completeness level estimated from a Monte-Carlo (MC) completeness simulation. A constant binning of the CLFs results in $1.53 < \alpha^{cte} < 2.41$ with a median and average values of $\alpha^{cte}_{med} = 1.87 \pm 0.23$ and $\alpha^{cte}_{aver} = 1.93 \pm 0.23$, respectively. This range becomes $1.42 < \alpha^{var} < 2.37$ in case of a variable binning, and with $\alpha^{var}_{med} = 1.88 \pm 0.27$ and $\alpha^{var}_{aver} = 1.94 \pm 0.27$. The left panel of Fig. 1 compares the distributions of the power-law slopes derived from the two binning methods. The values of α from the two methods are consistent within their uncertainties and they are also in agreement with the computed slopes in Randriamanakoto et al. (2013b) where $\alpha \sim 1.9$. Therefore, any observed flattening in the CLF could not be mainly due to the choice the binning. In the remainder of this work, we consider α^{cte} from the constant binning, hereafter also referred to as the slope α.

We also conducted other analyses to identify any possible uncertainties and biases and to test the robustness of our results. These include varying the PSF size of the MC completeness simulations, assessing the effects of resolution bias and extinction. In particular, we verified whether statistical bias and stochastic sampling strongly affect the CLFs of galaxies with low-number statistics of YMCs. The right panel of Fig. 1 plots the slope α against log N, which is the number of YMCs with magnitudes brighter than the 80 percent completeness limit in this case. With a correlation coefficient $r = -0.09$, there is no clear trend between the two entities. Overall, none of the various uncertainties and biases investigated in this work did not seem to have significantly flatten the CLFs of the YMC catalogues. What could then be the reason(s) behind a CLF slope smaller than the canonical value of $\alpha \approx 2$?

4. Composite CLFs of supergalaxies split as a function of the host global properties

We also built CLFs from composite "supergalaxies" using a constant binning and corrected for observational completeness. The data are created by combining the individual catalogues in different ways with the aim to increase the number statistics of YMCs for more robust results. Data points from the Bird are not included in this analysis to avoid

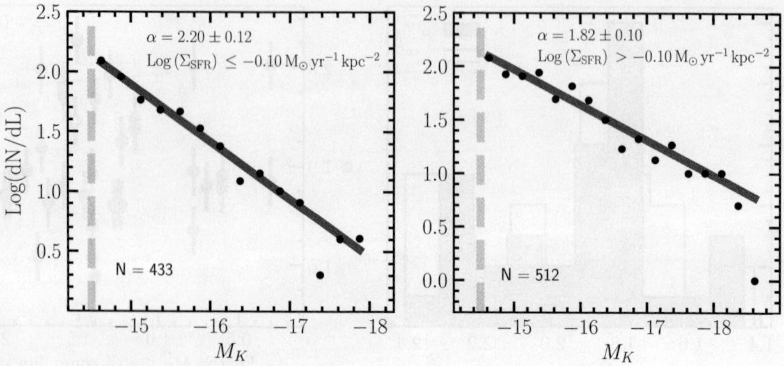

Figure 2. Completeness-corrected composite CLFs of the SUNBIRD targets with the individual star cluster catalogues (excluding the Bird) divided as a function of the host galaxy's SFR density, below (*left*) and above (*right*) $\log(\Sigma_{\rm SFR}) = -0.10\,{\rm M_\odot\,yr^{-1}}$. The vertical line denotes the 80 percent completeness level until which a single power-law function (blue line) is fitted to the corrected data. The number of YMCs considered in the fitting process along with the computed slopes are added as insets in the CLF plots.

bias. By combining the YMCs of 33 galaxies, the composite CLF which is fitted down to a magnitude limit $M_K = -14.5$ mag, i.e. where 80 percent of the SUNBIRD data should be complete, has a bright end slope of $\alpha = 1.98 \pm 0.10$. This value is consistent with the average slope of the individual CLFs in Section 3.

4.1. *Sample split as a function of the SFR density*

Among the composite CLFs of supergalaxies created in this work (e.g. based on the galaxy distances, SFR, specific SFR = SFR/Mass), we show in Fig. 2 the cases where the main sample is split with respect to the SFR density of the host galaxy: $\log(\Sigma_{\rm SFR}) \leq -0.10\,{\rm M_\odot\,yr^{-1}\,kpc^{-2}}$ (left panel) and $\log(\Sigma_{\rm SFR}) > -0.10\,{\rm M_\odot\,yr^{-1}\,kpc^{-2}}$ (right). This cutoff was chosen given that it is the average of the SFR density. The computed power-law slopes are $\alpha = 2.20 \pm 0.12$ and $\alpha = 1.82 \pm 0.10$, respectively. The value of α increases by $\Delta\alpha = 0.38$ with a decreasing SFR density. Such trends were also observed while dividing the sample as a function of the SFR ($\Delta\alpha \sim 0.44$ for a cutoff of ${\rm SFR} = 30\,{\rm M_\odot\,yr^{-1}}$) and the specific SFR ($\Delta\alpha = 0.40$ for a cutoff of $\log({\rm sSFR}) = 9.68\,{\rm yr^{-1}}$). We note that host global properties used here and in Section 5 were mostly taken from Ramphul (2018).

4.2. *Composite CLFs of cluster-rich galaxies on sub-galactic scales*

We also constructed composite CLFs of cluster-rich galaxies with N $\gtrsim$ 300 (IC 2522, ESO 221−IG008, NGC 3110, and NGC 6000). While the power-law CLF slope of such supergalaxy is $\alpha = 2.18 \pm 0.09$, the composite CLFs of two distinct subsamples segregated by the cluster physical locations in the field of the galaxy have slopes $\alpha = 1.72 \pm 0.11$ for the nuclear regions and $\alpha = 2.51 \pm 0.09$ for the arms or other starbust regions of the field.

This analysis on sub-galactic scales was also performed using the catalogues of each individual cluster-rich galaxy. Table 1 summarizes the computed values of α for a nuclear (Col 8) vs outer (Col 10) regions assuming their respective completeness limits (Cols 7 & 9). CLF slopes of the nuclear regions are shallower compared to those of the outer regions for NGC 3110 and NGC 6000. Similar values are found for both regions of the other two galaxies. Table 1 lists as well the computed CLF slopes of the whole catalogue for each target while estimating the completeness limits based on the background contour levels (Col 3) and then the galaxy morphology (Col 5). The values of the two parameters remain

Table 1. The values of α and the 80 % completeness limits in various regions of cluster-rich galaxies.

Galaxy name	N	Contour levels		Galaxy morphology		Nuclear regions		Outer regions	
		CL	α_1	CL	α'_1	CL	α	CL	α
(1)	(2)	(3)	(4)	(5)	(6)	(7)	(8)	(9)	(10)
IC 2522	302	−12.2	2.41 ± 0.08	−12.2	2.40 ± 0.08	−12.4	2.55 ± 0.19	−12.2	2.31 ± 0.10
ESO 221−IG008	414	−11.9	2.00 ± 0.11	−11.7	1.98 ± 0.09	−11.8	2.06 ± 0.11	−12.1	2.11 ± 0.16
NGC 3110	279	−13.8	2.31 ± 0.08	−13.9	2.25 ± 0.07	−14.5	1.92 ± 0.08	−13.7	2.43 ± 0.09
NGC 6000	285	−12.5	1.82 ± 0.19	−12.3	1.81 ± 0.19	−12.9	1.44 ± 0.07	−12.3	2.24 ± 0.10

Notes: The different regions and the corresponding 80 % completeness limits (CLs) were defined based on the background contour levels and then considering the nuclear and outer regions of the galaxy.

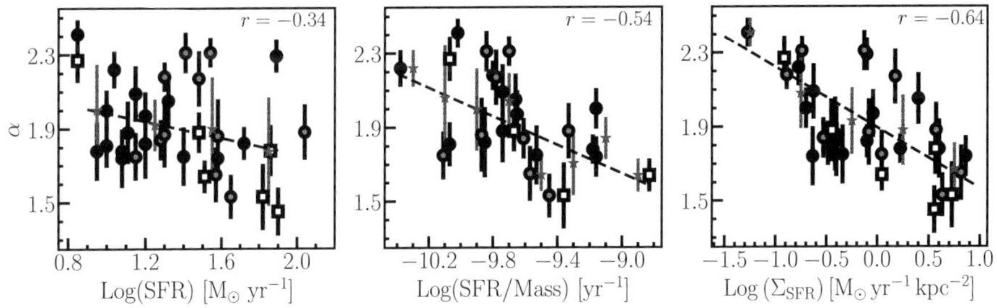

Figure 3. The power-law slope plotted against the host galaxy's SFR (*left*), specific SFR (*middle*), and SFR density (*right*). The data points are labelled as a function of the host galaxy distance: black circles for $D_L \leq 60$ Mpc, grey circles for $60 < D_L \leq 100$ Mpc, and open squares for $D_L > 100$ Mpc. Red stars represent derived data points from a constant binning of α and the dashed line a linear fit to these data.

consistent within the error estimates which imply that using equally-spaced background levels to output the 80 percent completeness limits applied to the individual CLFs in Section 3 are reasonable.

5. Correlation searches between the host global properties

Correlation searches between α and the host global properties were also conducted to check whether the trends seen in the composite CLFs on both global and sub-galactic scales are real. Fig. 3 shows in particular the trends with the galaxy's SFR (left), specific SFR (middle), and SFR density (right). The values of the correlation coefficient are $r = -0.34, -0.54, -0.64$, respectively. These coefficients become $r = -0.97, -0.86, -0.97$ once we bin the data points (red stars). These clearly indicate the existence of a (mild) correlation between α and the three SFR indicators, with the strongest correlation associated with α vs. $\Sigma_{\rm SFR}$ plot. Correlation searches with other global properties such as the stellar mass, the light-weighted metallicity, or the stellar visual extinction were also done but only very weak or no trend could only be observed.

6. Conclusions and way forward

This work summarizes major results from drawing the constructed K-band cluster luminosity functions of 34 SUNBIRD targets imaged with NIR AO telescopes (a comprehensive report on this research project is available in Randriamanakoto et al. 2022). These main findings are as follows: smaller values of the (median and average) slopes compared to the canonical index of $\alpha \sim 2$, a flattening CLF slope with an increasing SFR indicator (especially in the case of the SFR density), and the difference by > 0.4 in the CLF slopes for individual cluster-rich catalogues and supergalaxies split as a function of the cluster physical locations or the host SFR, specific SFR, and SFR density.

We demonstrated that internal factors such as size-of-sample effects and other systematic biases cannot only explain the observed trends in this work. Previous works by e.g. Whitmore et al. (2014), Mulia et al. (2016), and Cook et al. (2019) merely associate similar trends to size-of-sample effects which they believe are the main responsible for any artificially flattened CLF. This also means that the cluster formation mechanisms generally remain the same irrespective of the host SFR level. However, other works that also recorded unusual YMC properties of high-star forming galaxies explain the observed trends as a consequence of some underlying process involving the host galaxy environment (e.g. Bastian et al. 2012; Cook et al. 2016; Randriamanakoto et al. 2019). For instance, environments with high gas surface density are generally of higher insterstellar medium pressure and are thus ideal birthsites of strongly bound clusters such as YMCs (e.g. Kruijssen 2012; Espada et al. 2018).

Given that rigourous assessment of any possible biases have been conducted before interpreting our results, we are thus likely in favour of the latter explanation, i.e. the CLF and hence the cluster SF processes being influenced by its host environment. However, we are aware that CLFs can only provide first order approximation of the effects of internal and/or external factors on the cluster population and that this work has as well its own limitations (e.g. drawing binned CLFs, using a single filter or comparing our NIR-based results with optical CLF studies of low SFR galaxies). Therefore, further analyses need to be implemented such as the use of other fitting techniques that involve maximum-likelihood estimation of cumulative CLFs. Moreover, a comprehensive multi-wavelength study of the cluster population of the SUNBIRD targets will allow the modelling of the cluster ages and masses to provide better constraints of the cluster SF formation processes and ultimately the small-scale SF mechanisms of the host galaxy.

Acknowledgements

ZR acknowledges funding from the South African Astronomical Observatory, which is a facilitiy of the National Research Foundation, and financial support from the L'Oréal - UNESCO For Women In Science sub-Saharan Africa regional Progamme.

References

Adamo, A., Ryon, J. E., Messa, M., et al. 2017, *ApJ*, 841, 131

Bastian, N. 2016, *EAS Publications Series*, 80, 5

Bastian, N., Adamo, A., Gieles, M., et al. 2012, *MNRAS*, 419, 2606

Cook, D. O., Dale, D. A., Lee, J. C., et al. 2016, *MNRAS*, 462, 3766

Cook, D. O., Lee, J. C., Adamo, A., et al. 2019, *MNRAS*, 484, 4897

Efremov, Y. N. 1995, *AJ*, 110, 2757

Espada, D., Martin, S., Verley, S., et al. 2018, *ApJ*, 866, 77

Kool, E. C., Ryder, S., Kankare, E., et al. 2018, *MNRAS*, 473, 5641

Kruijssen, J. M. D. 2012, *MNRAS*, 426, 3008

Larson, K. L., Díaz-Santos, T., Armus, L., et al. 2020, *ApJ*, 888, 92

Mattila, S., Väisänen, P., Farrah, D., et al. 2007, *ApJL*, 659, L9

Mulia, A. J., Chandar, R., & Whitmore, B. C. 2016, *ApJ*, 826, 32

Pflamm-Altenburg, J., González-Lópezlira, R. A., & Kroupa, P. 2013, *MNRAS*, 435, 2604

Ramphul, R. 2018, PhD thesis, Department of Astronomy, University of Cape Town, South Africa

Randriamanakoto, Z., Escala, A., Väisänen, P., et al. 2013a, *ApJL*, 775, L38

Randriamanakoto, Z., Väisänen, P., Ranaivomanana, P., et al. 2022, *MNRAS*, 513, 4232

Randriamanakoto, Z., Väisänen, P., Ryder, S.D., et al. 2013b, *MNRAS*, 431, 554

Randriamanakoto, Z., Väisänen, P., Ryder, S. D., et al. 2019, *MNRAS*, 482, 2530

Väisänen, P., Randriamanakoto, Z., Escala, A., et al. 2014, *Massive Young Star Clusters Near and Far: From the Milky Way to Reionization*, p. 185

Whitmore, B. C., Chandar, R., Bowers, A. S., et al. 2014, *AJ*, 147, 78

Resolving the Rise and Fall of Star Formation in Galaxies
Proceedings IAU Symposium No. 373, 2023
T. Wong & W.-T. Kim, eds.
doi:10.1017/S1743921322003957

A search for correlations between turbulence and star formation in LITTLE THINGS and THINGS galaxies

Bruce G. Elmegreen[1], Deidre Hunter[2], Zorayda Martinez[2,3], Haylee Archer[2], Caroline E. Simpson[4] and Phil Cigan[5]

[1]IBM Research, T.J. Watson Research Center, 1101 Kitchawan Road, Yorktown Heights, NY 10598

[2]Lowell Observatory, 1400 West Mars Hill Road, Flagstaff, Arizona 86001, USA

[3]Department of Astronomy and Planetary Science, Northern Arizona University, 527 S. Beaver Street, Flastaff, Arizona 86011

[4]Department of Physics, Florida International University, CP 204, 11200 SW 8th Street, Miami, FL 33199 USA

[5]George Mason University, 4400 University Drive, Fairfax, VA 22030-4444, USA

Abstract. Azimuthal variations in HI velocity dispersion do not correlate with variations in the star formation rate per unit area, SFR/A, suggesting that local star formation does not increase HI turbulence significantly. These variations are determined for each pixel in HI and FUV maps of THINGS and LITTLE THINGS galaxies by subtracting the average radial profiles from the measured quantities. The kinetic energy density and HI surface density increase slightly with SFR/A, suggesting that feedback goes into pushing the local dense gas around without increasing the velocity dispersion. We suggest that star formation feedback does not promote large-scale stability against gravitational forces through turbulence regulation, and that gravitational energy from recurrent instabilities drives turbulence on galactic scales.

Keywords. star formation, gravitational instability, turbulence, feedback

1. Introduction

Young stars power their local gas with radiation, winds, and supernovae, but where and how far this energy goes is not known. Some clusters clear their residual gas on small scales quickly (Chevance *et al.* 2002), while OB associations clear away gas on larger scales, making expanding HII regions and supernova bubbles. The fraction of all disk gas that is directly energized by this feedback is difficult to measure because there are other sources of gas motion on larger scales that could overwhelm feedback. Examples are gaseous gravity on the scale of the interstellar Jeans length, which is typically several hundred parsecs and where ISM turbulent pressures balance gaseous self-gravity, and spiral density waves on the scale of the Toomre length, which is typically larger than the Jeans length, and where Coriolis forces balance disk gravity.

We are interested in whether feedback from star formation can power turbulence in the atomic phase of interstellar gas, which determines the warm-phase scale height and an important component of pressure resistance to large-scale gravity. This is the resistance, when combined with the Coriolis force, that determines the Toomre Q parameter. One possibility is that Q may self-regulate by spiral wave generation to a value close to marginal stability, $Q \sim 1.5$. Such regulation is suggested by simulations (e.g., Bertin *et al.* 2001; Kim & Ostriker 2007; Booth & Clarke 2019; Bethune *et al.* 2021) and is in agreement with Q ∼constant observations at low redshift in spiral galaxies (Leroy *et al.* 2008) and at high redshift in clumpy galaxies (Tacconi et al. 2020). With Q regulation, gravity is an important source of turbulence (e.g., Elmegreen *et al.* 2003; Genzel *et al.* 2011; Goldbaum *et al.* 2016; Orr *et al.* 2020). That is, when turbulence decays in a vertical disk crossing time, two-dimensional disk instabilities become stronger and the resulting local collapse pumps energy from disk rotation into the gas to recover the marginally stable state. If Q is regulated in this way, independent of star formation, then large-scale turbulence could be independent of star formation in some regions too. We expect that generally both gaseous self-gravity and star formation feedback pump interstellar turbulence, as in models by Grisdale *et al.* (2017), Nusser & Silk (2021) and others.

Here we want to determine how the local HI linewidth scales with the local star formation rate density. Unlike previous studies that correlate the average galactic gas velocity dispersion with the total star formation rate (e.g., Green *et al.* 2014), or the local velocity dispersion inside a galaxy with the local star formation rate (e.g., Stilp *et al.* 2013), we determine only the locally excess turbulent speed as a function of locally excess star formation rate density to remove any galactic mass dependence or radial dependence. The results come from two publications, Hunter *et al.* (2021) and Elmegreen *et al.* (2022).

The HI observations are from the LITTLE THINGS (Hunter *et al.* 2012) and THINGS (Walter *et al.* 2008) surveys. The HI surface density is corrected for inclination but not for He and heavy elements; the HI velocity dispersion is from the second moment, MOM2, and is not adjusted for inclination, assuming the motions are isotropic. The star formation rate per unit area, corrected for inclination, comes from FUV observations by GALEX† (Martin et al. 2005). There are 36 LITTLE THINGS dwarfs and 10 spirals plus a dwarf in THINGS. GALEX observations of FUV are converted to SFR/Area; they measure the average star formation rate in the last ~ 100 Myr. The resolution of the HI is $6''$ from the JVLA‡, and for GALEX it is $4''$ convolved to the same $6''$, The pixel size for the resulting maps is $1.5''$ (but $3.5''$ for DDO 216 and Sag DIG and $1.0''$ for NGC 2403). Note that $6'' = 110$ pc for the average dwarf at 3.8 Mpc and 240 pc for the average spiral at 8.3 Mpc.

The method is to first determine the average radial profiles of the Kinetic Energy Density, $KED = 0.5 \times 1.36 \times \Sigma_{HI} V_{\rm disp}^2$ (corrected for He and heavy elements by the factor 1.36), HI velocity dispersion, $V_{\rm disp}$, inclination-corrected HI surface density, $\Sigma_{\rm HI}$, and inclination-corrected SFR/Area, in annuli of width equal to 1 HI beam for dwarfs and 10 HI beams for spirals. Then, for each pixel, we subtract the average radial profile values from the observed values for KED, $V_{\rm disp}$, Σ_{HI} and $SFR/Area$. Finally, we plot the excess values of KED, $V_{\rm disp}$ and Σ_{HI} versus the log of the excess SFR/Area (positive values only), pixel by pixel, and measure the trends.

2. Results

Figure 1 shows the distributions of excess KED, $V_{\rm disp}$ and $\Sigma_{\rm HI}$ versus excess SFR/A on a pixel-by-pixel basis for all positions that have HI and GALEX observations in the spiral galaxy NGC 5055. Generally the inner parts of the disk correspond to higher excess SFR/A, and at these values there is a wider range of excess KED and $\Sigma_{\rm HI}$ than at lower excess SFR/A. The most common values of excess KED, $V_{\rm disp}$ and $\Sigma_{\rm HI}$, where

† GALEX was operated for NASA by the California Institute of Technology under NASA contract NAS5-98034

‡ The JVLA is a facility of the National Radio Astronomy Observatory. The National Radio Astronomy Observatory is a facility of the National Science Foundation operated under cooperative agreement by Associated Universities, Inc.

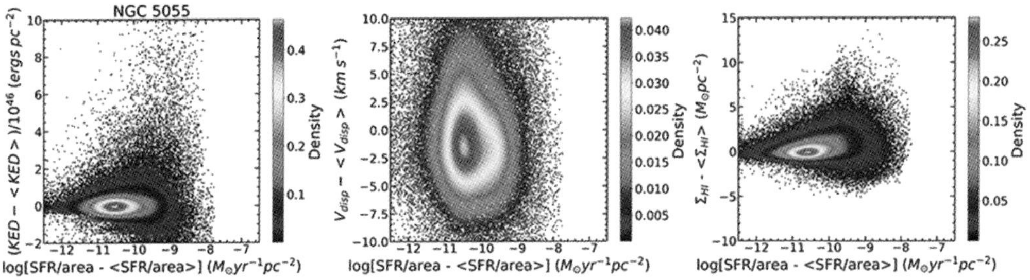

Figure 1. Pixel-pixel plots for NGC 5055 showing the excess KED, $V_{\rm disp}$, and $\Sigma_{\rm HI}$ against the log of the excess star formation rate per unit area, SFR/A. The density of pixels is indicated by color. The excess is defined to be the difference between the local value and the value from the average radial profile.

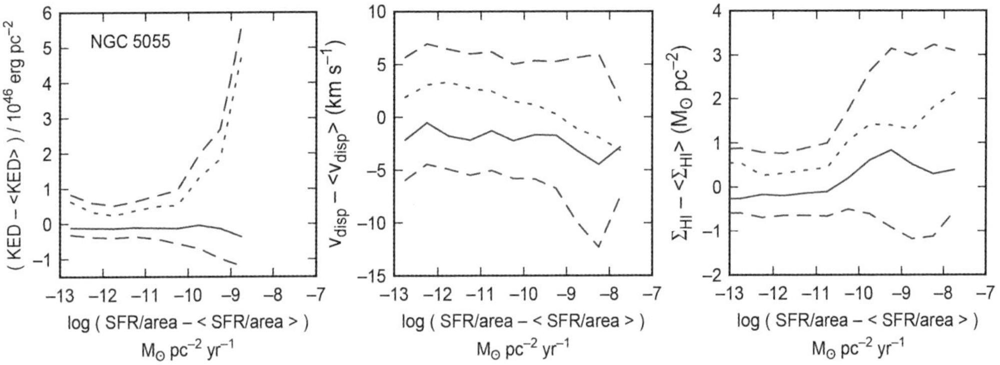

Figure 2. Representation of the pixel distributions in Figure 1. The solid lines are values at the peak densities, the dashed lines are rms deviations, and the dotted line is the difference between the positive and negative rms values added to the values at the peak.

the pixel densities are the greatest, show slight increases in excess KED and $\Sigma_{\rm HI}$ with excess SFR/A but no obvious increase in excess $V_{\rm disp}$.

We represent these pixel distributions by four curves versus excess SFR/A: one that is the value of the quantity at the peak pixel density, two that are the rms dispersions of values above and below this value at peak pixel density, and a fourth that is the value at the peak density plus the difference between the upper and lower rms values. This latter curve is considered to be a statistical upper limit to the trend of the parameter value versus excess SFR/A.

Figure 2 shows these four curves for NGC 5055 (solid, dashed, and dotted, respectively). The most probable value of excess $\Sigma_{\rm HI}$ increases slightly with excess SFR/A, as does the upper limit for that and for the KED, but excess $V_{\rm disp}$ decreases with excess SFR/A. Figure 3 shows the upper limit curves for all of the THINGS galaxies (NGC 5055 included). The trends are the same: excess KED and $\Sigma_{\rm HI}$ increase slightly with excess SFR/A and excess $V_{\rm disp}$ decreases.

Figure 4 shows the results for LITTLE THINGS dwarfs in a slightly different way. Each blue point is the average of the peak pixel-density values at low excess SFR/A and each red cross is the average of the peak pixel-density values at high excess SFR/A, with one or two points per galaxy. The curves for the KED are fits for values of the feedback efficiency, which is the fraction of the supernova energy at the given SFR/A that is needed to produce the KED. These fits are made at two representative disk thicknesses, H, which are fairly large for dIrr galaxies. The conversion from KED and SFR/A to efficiency η comes from Bacchini *et al.* (2020). The curves on the right-hand panel for

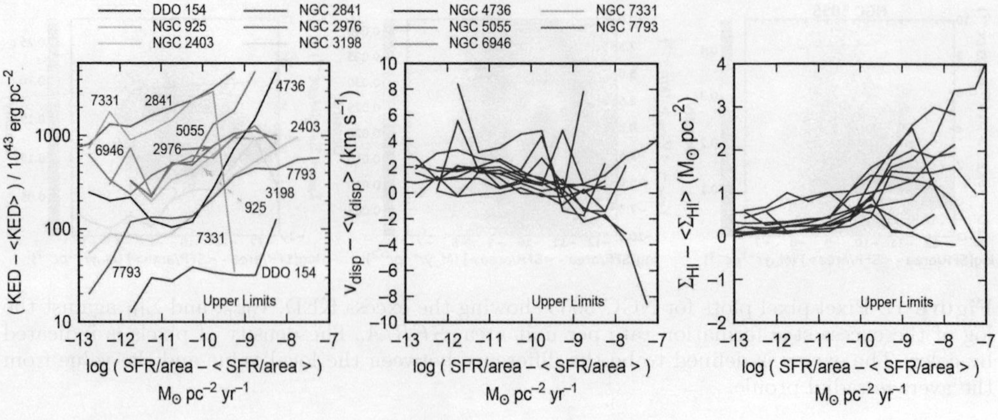

Figure 3. Statistical upper limits to the values of the three quantities from the pixel density plots. The upper limits for the excess KEDs are plotted on a log scale with galaxy names indicated. The upper limit to excess $V_{\rm disp}$ decreases with increasing SFR/A. while excess KED and $\Sigma_{\rm HI}$ increase slightly. There is no evidence that feedback from star formation generates turbulence in the local HI gas.

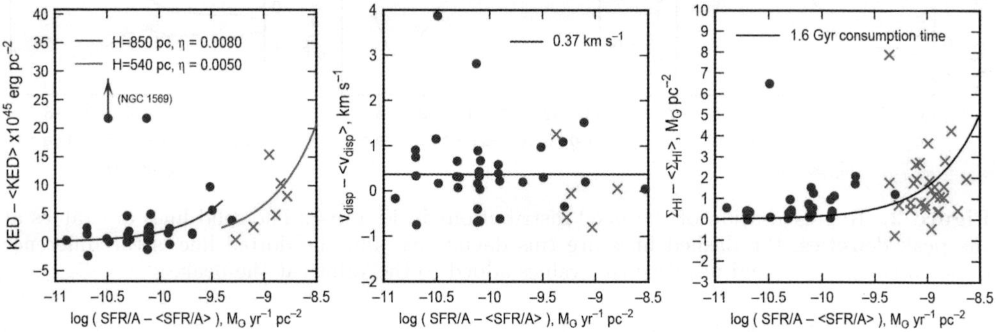

Figure 4. Mean excess KED, $V_{\rm disp}$, and $\Sigma_{\rm HI}$ versus the mean of the log of the (positive) excess SFR/A for LITTLE THINGS galaxies. Dots correspond to the low SFR/A concentrations in the outer disks and crosses correspond to high SFR/A in the inner disks.

excess $\Sigma_{\rm HI}$ are from the best-fit ratio of the two plotted quantities, giving the average consumption time for excess $\Sigma_{\rm HI}$ at the plotted excess SFR/A. As for the THINGS galaxies, the excess $V_{\rm disp}$ at the peak of the pixel densities is about constant with excess SFR/A.

3. Conclusions

The velocity dispersion of HI in nearby spiral and dwarf irregular galaxies shows no local increase in regions of star formation. The kinetic energy density increases a small amount in these regions, but only because the HI surface density increases at constant velocity dispersion. This result implies that star formation feedback may be deposited mostly on small scales where it breaks apart molecular clouds and dissipates quickly at high density. Feedback does not appear to energize turbulence in the atomic phase, which limits its ability to control the dynamical state of the interstellar medium on large scales. Because gravitational processes like the response of gas to spiral density waves and gas collapse on the scale of the turbulent Jeans length can cause large-scale motions, these more pervasive sources of energy seem favored in determining galaxy-scale gas dynamics.

References

Bacchini, C., Fraternali, F., Iorio, G., Pezzulli, G. Marasco, A., Nipoti, C. 2020, *A&A*, 641, A70
Bertin G., Lodato G., 2001, Â&A, 370, 342
Bethune W., Latter H., Kley W., 2021, *A&A*, 650, 49
Booth R. A., Clarke C. J., 2019, MNRAS, 483, 3718
Chevance, M., Kruijssen, J.M.D., Krumholz, M.R., *et al.* 2022, *MNRAS*, 509, 272
Elmegreen, B.G., Elmegreen, D.M., Leitner, S.L. 2003, *Ap.J.*, 590, 271
Elmegreen, B.G., Martinez, Z., Hunter, D.A. (2022) *Ap.J.* 928, 143
Genzel, R., Newman, S., Jones, T. *et al.*, 2011, *Ap.J.*, 733, 101
Goldbaum, N.J., Krumholz, M.R., Forbes, J.C. 2016, *Ap.J.*, 827, 28
Green, A.W., Glazebrook, K., McGregor, P.J. 2014, *MNRAS*, 437, 1070
Grisdale, K., Agertz, O., Romeo, A.B., Renaud, F., Read, J.I. 2017, *MNRAS*, 466, 1093
Hunter, D. A., Ficut-Vicas, D., Ashley, T., Brinks, E., Cigan, P., *et al.* 2012, *A.J.*, 144, 134
Hunter, D.A., Elmegreen, B.G., Archer, H., Simpson, C.E., Cigan, P. (2021), *AJ*, 161, 175
Kim, W.-T., Ostriker, E.C. 2007, *Ap.J.*, 660, 1232
Leroy, A.K., Walter, F., Brinks, E., et al. 2008, *A.J.*, 136, 2782
Nusser, A., Silk, J. 2021, *MNRAS*, 509, 2979
Orr, M. E., Hayward, C. C., Medling, A. M., et al. 2020, MNRAS, 496, 1620
Stilp, A. M., Dalcanton, J. J., Skillman, E., *et al.* 2013, *Ap.J.*, 733, 88
Tacconi, L.J., Genzel, R., Sternberg, A. 2020, *ARA&A*, 58, 157
Walter, F., Brinks, E., de Blok, W. J. G., *et al.* 2008, *A.J.*, 136, 2563

Resolving the Rise and Fall of Star Formation in Galaxies
Proceedings IAU Symposium No. 373, 2023
T. Wong & W.-T. Kim, eds.
doi:10.1017/S1743921322004380

Exploring the evolution of giant molecular clouds in one of the nearest spiral galaxies M33

Ayu Konishi[1], Kazuyuki Muraoka[1], Kazuki Tokuda[2,3], Shinji Fujita[4], Rin I. Yamada[5], Fumika Demachi[5], Kengo Tachihara[5], Yasuo Fukui[5], Masato I. N. Kobayashi[3], Tsuge Kisetsu[6], Akiko Kawamura[3] and Toshikazu Onishi[1]

[1]Department of Physics, Graduate School of Science, Osaka Metropolitan University, 1-1 Gakuen-cho, Naka-ku, Sakai, Osaka 599-8531, Japan

[2]Department of Earth and Planetary Sciences, Faculty of Sciences, Kyushu University, Nishi-ku, Fukuoka 819-0395, Japan

[3]National Astronomical Observatory of Japan, National Institutes of Natural Sciences, 2-21-1 Osawa, Mitaka, Tokyo 181-8588, Japan

[4]Institute of Astronomy, the University of Tokyo, 2-21-1 Osawa, Mitaka, Tokyo 181-0015, Japan

[5]Department of Physics, Nagoya University, Chikusa-ku, Nagoya 464-8602, Japan

[6]Dr. Karl Remeis Observatory and ECAP, Universität Erlangen-Nürnberg, Sternwartstraße 7, 96049, Bamberg, Germany

Abstract. The evolution of giant molecular clouds (GMCs), which are the main sites of star formation, is essential for unraveling how stars form and how galaxies evolve. We analyzed the M33 CO(J = 2–1) data with spatial resolution of 39 pc obtained by ALMA-ACA 7 m array combined with IRAM 30 m. We identified 736 GMCs and classified them into three types; Type I: associated with no H II regions, Type II: associated with H II regions with the Hα luminosity $L(\mathrm{H}\alpha) < 10^{37.5}$erg s^{-1}, Type III: associated with H II regions with $L(\mathrm{H}\alpha) > 10^{37.5}$erg s^{-1}. We found that mass, size, and velocity dispersion of GMCs slightly increase in the order of Type I, II, and III GMCs. Type III GMCs mainly exist in the spiral arm, while many of Type I and Type II GMCs are distributed in the inter-arm. Assuming that the star formation proceeds steadily, we roughly estimated the total GMC lifetime of 30 Myr.

Keywords. ISM: molecules, stars: formation, galaxies: spiral, galaxies: Local Group

1. Introduction

Giant molecular clouds (GMCs) are the primary reservoirs of cold, dense molecular gas and the birthplace of stars in galaxies. Massive stars can cause supernova explosions, release and supply heavy elements to interstellar space, and have a profound effect on the evolution of galaxies. Large-scale molecular gas survey toward the Large Magellanic Cloud (LMC) by the NANTEN 4 m telescope found that the GMCs are classified into three types according to activities of massive star formation: Type I shows no signature of massive star formation; Type II is associated with only H II region(s); and Type III with both H II region(s) and young stellar cluster(s) (YSCs) (Fukui et al. 1999; Kawamura et al. 2009). They argued that these types indicate the evolutionary sequence of GMCs. Such studies of the GMC type classification were extended to the nearest spiral galaxy M33, which is one of the best targets to study the GMC evolution because of its proximity ($D\sim$840 kpc, Freedman et al. 2001) and the moderate inclination angle ($i\sim55°$, Koch et al. 2018). The CO all-disk survey of M33 by IRAM 30 m telescope detected over five hundred GMCs (Gratier et al. 2012; Druard et al. 2014). Corbelli et al. (2017) compared the GMCs with multi-wavelength data set and classified the GMCs into different evolutionary stages. More recently, Similar studies on star-formation and the GMCs evolution have been advanced toward the nearby spiral galaxies thanks to the Physics at High Angular Resolution in Nearby GalaxieS (PHANGS) ALMA survey (e.g., Schinnerer et al. 2019; Pan et al. 2022). In this study, we reexamine the GMC evolution in M33 based on the higher spatial (~39 pc) and higher velocity (~0.7 km s^{-1}) resolution CO(J = 2–1) data by the 7 m antennae of the Atacama Compact Array (ACA).

2. Large-scale CO(J = 2–1) survey in M33 by the ACA 7 m array

The ^{12}CO, ^{13}CO, $C^{18}O$ (J = 2–1) observation toward M33 was performed with the ACA stand-alone mode (project code 2018.A.00058.S; PI: K. Muraoka, 2017.1.00901.S and 2019.1.01182.S; PI: E. Rosolowsky). ALMA data reduction was carried out using the Common Astronomy Software Application (CASA) package (McMullin et al. 2007) version 5.4.0 and the `tclean` task implemented in the CASA package (Cornwell 2008) with input parameters; weighting = natural, deconvolver = `multi-scale`, usemask = `auto-multithresh`. The final synthesized beam is 7."31 × 6."50 (30 pc × 26 pc), and its position angle is −28.92 deg for ^{12}CO data with a grid size of 1."5. The ACA survey setting and data analyses in detail were described in (Muraoka et al. 2022).

We combined the ACA 7 m array and IRAM 30 m ^{12}CO data with the `feathering` task. We spatially smoothed the ^{12}CO data to be ~39 pc in order to compare our data with the GMC-type classification in LMC reported by Kawamura et al. (2009) at the same resolution. Our ^{12}CO data sensitivity is ~0.02–0.04 K at a velocity resolution of 0.7 km s^{-1}.

3. Results & Discussion

3.1. *Identification of GMCs*

Figure 1 (a) shows the ^{12}CO(J = 2–1) emission from M33 obtained by the ACA 7 m array + IRAM 30 m with a spatial resolution of ~39 pc. We applied cloud identification algorithm `PYCPROPS` to the ^{12}CO cube data. We briefly describe this algorithm and identification procedure because the details are described in Rosolowsky et al. (2021). `PYCPROPS` first extracts all "local maxima" of input cube data and then removes some of them which are considered noise with threshold parameters. We set the local maxima to be more than 3 σ (`delta` parameter) and a number of cube pixels of them (`min_pix` parameter) corresponds to twice the beam size of the ^{12}CO(J = 2–1) data. `PYCPROPS` extrapolated parameters of identified structures such as second moment of the major ($\sigma_{\rm maj}$) and minor ($\sigma_{\rm min}$) axes, velocity dispersion, and integrated intensity in order to remove sensitivity bias when we measure physical properties of clouds. We deconvolved a radius R of each cloud by the beam size $\sigma_{\rm beam}$ of ~39 pc as follows:

$$R = 1.91\sqrt{(\sigma_{\rm maj}^2 - \sigma_{\rm beam}^2)^{0.5}(\sigma_{\rm min}^2 - \sigma_{\rm beam}^2)^{0.5}}. \quad (3.1)$$

Finally, we obtained 736 GMCs with masses between $10^{3.6}\,M_\odot$ and $10^{6.5}\,M_\odot$ using a CO-to-H_2 conversion factor $X_{\rm CO}$ = 4.0 × 10^{20} cm^{-2} / (K km s^{-1}) and CO(J = 2–1) / CO(J = 1–0) ratio R_{21} = 0.6 (Muraoka et al. 2022).

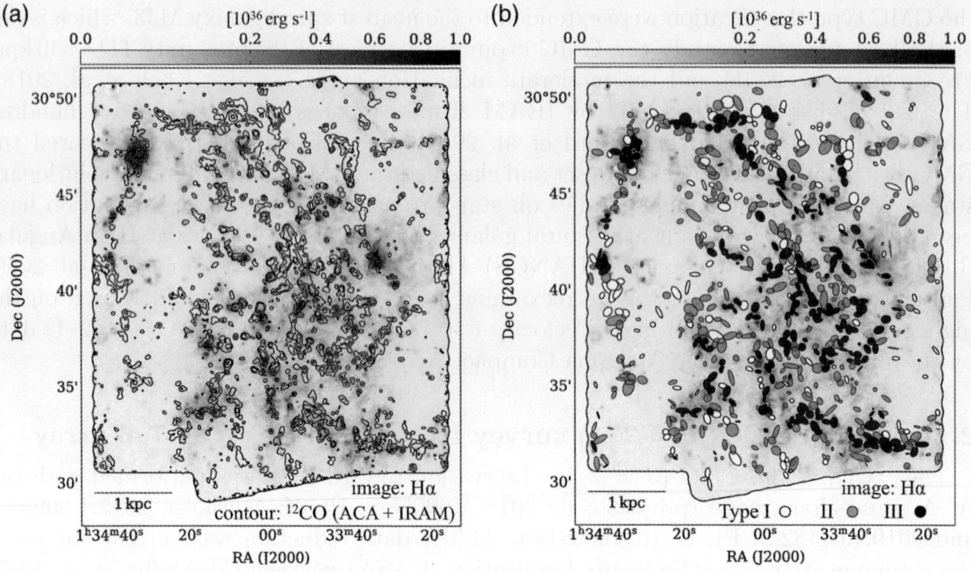

Figure 1. (a) ^{12}CO(J = 2–1) emission from M33 obtained by the ACA 7 m array + IRAM 30 m with a spatial resolution of ∼39 pc. Gray scale shows the Hα image by KPNO 2.1 m (Hoopes & Walterbos 2000; Hoopes et al. 2001) which cut to fit the ACA observation field. ^{12}CO contour levels are 0.15, 0.5, 2 K, respectively. (b) Distribution of GMCs in each type on Hα image. White, gray, and black ellipses represent Type I, II, and III GMCs, respectively.

3.2. *Classification of GMCs based on the Hα luminosity*

There are large variations in star formation activity between GMCs; e.g., the association with star clusters and/or H II regions. In LMC, Yamaguchi et al. (2001) found that Hα luminosity, L(Hα), of the H II regions associated with GMCs increases as GMCs evolve. The frequency distributions of Hα luminosity are offset between Type II GMCs which is associated with only H II regions and Type III GMCs which is associated with both H II regions and YSCs, and overlapping at L(Hα) of $\sim 10^{37.5}$erg s^{-1}.

In this study, we try classifying GMCs into following the three types based on the association with H II regions and Hα luminosities;

Type I : GMCs associated with no H II regions (inactive massive star-forming region)
Type II : GMCs associated with H II regions with $L(\mathrm{H}\alpha) < 10^{37.5}$erg s^{-1}
Type III : GMCs associated with H II regions with $L(\mathrm{H}\alpha) > 10^{37.5}$erg s^{-1}

We defined the association of GMCs and H II regions if the spatial extent of a GMC is overlapped with the Hα emitting region identified by `astrodendro` (Rosolowsky et al. 2008) which is suitable for finding hierarchical structures such as H II regions and diffuse ionized gas. We set three `astrodendro` parameters; `min_value`: the minimum intensity value is 2.0×10^{34} erg s^{-1}, `min_delta`: the minimum height to be "leaf" is 1.0×10^{34} erg s^{-1} which corresponds to about 5 sigma, `min_npix`: the minimum number of two-dimensional pixels in the RA, Dec axes is 25 (1 pix = 1.5"). `astrodendro` eventually identified 769 H II regions as leaves. This type classification yields 182 Type I GMCs, 316 Type II GMCs, and 238 Type III GMCs. Table 1 summarize the classification results and Figure 1 (b) shows the distribution of GMCs in each type on Hα image.

Table 1. GMC type and their physical properties.

GMC Type	**Type I**	**Type II**	**Type III**
Number	182 (25 %)	316 (43 %)	238 (32 %)
Mass ($M_\odot$)	$\sim$7.1× 10^4	$\sim$1.6× 10^5	$\sim$3.2× 10^5
Radius (pc)	$\sim$34	$\sim$44	$\sim$50
Velocity dispersion (km s^{-1})	$\sim$2.43	$\sim$3.03	$\sim$3.37
Timescale (Myr)	7	13	10

Notes: The median of mass, radius, and velocity dispersion of GMCs.

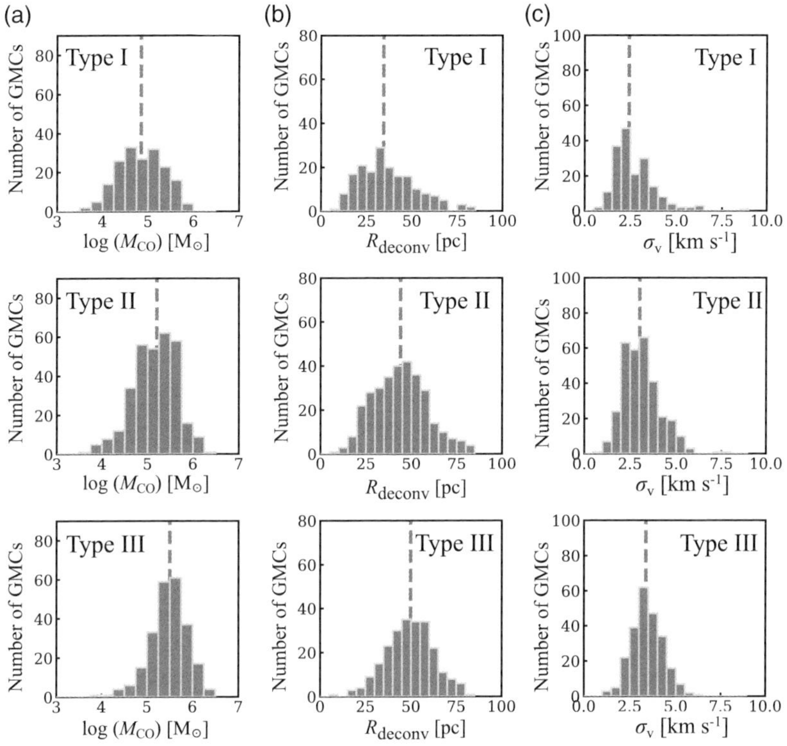

Figure 2. Frequency distributions of GMC properties, mass (a), radius (b), and velocity dispersion (c). The upper, middle, and lower panels indicate those of Type I, II, and III GMCs, respectively. Vertical dashed lines indicate the median for histograms.

3.3. *Basic properties of GMCs*

Figure 2 indicates histograms of mass, radius, and velocity dispersion for the three GMC types. The properties of Type I, II, and III GMCs are represented in the upper, middle, and lower panels, respectively. By comparing properties of GMCs, we found that the mass, size, and velocity dispersion slightly increase in the order of Type I, II, and III GMCs. Typical mass of Type II and III GMCs is larger than $10^5\ M_\odot$ while that of Type I GMCs is smaller than $10^5\ M_\odot$. This suggests that massive star-forming GMCs likely have their mass of $> 10^5\ M_\odot$. The GMCs are considered to form and evolve by continuous mass-accretion from local surrounding H I gas (e.g., Hennebelle & Pérault 1999; Koyama & Inutsuka 2000, 2002; Vázquez-Semadeni et al. 2007; Fukui et al. 2009; Inoue & Inutsuka 2012). In addition, cloud-cloud collisions can make a major contribution the mass growth ($\sim 10^6\ M_\odot$) and star formation in GMCs (e.g., Kondo et al. 2021; Tokuda et al. 2020; Muraoka et al. 2020).

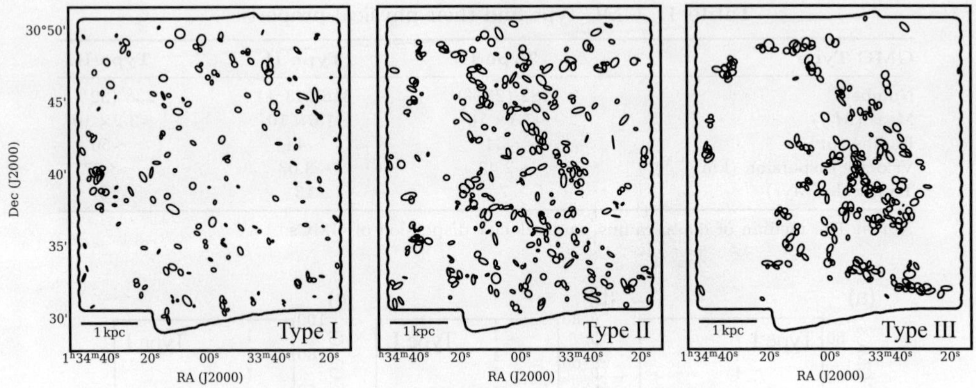

Figure 3. Distribution of GMCs in each type.

3.4. *Lifetime of GMCs in M33 based on the previous studies in LMC*

We estimate the GMC lifetime in M33 with reference to the previous studies of GMC evolution in LMC (Kawamura et al. 2009). First, the timescale for Type II GMC of 13 Myr was originally determined in LMC, and we assume that Type II GMC in M33 has the same timescale as the LMC. This is because Type II GMCs are considered to be less sensitive to the detection limit than Type I GMCs whose typical mass ($<\sim 10^5\, M_\odot$) is lowest in three types, and also less affected by gas dissipation due to stellar feedback than Type III GMCs which shows most active star formation. If we assume time-steady star formation rate in M33, the timescale of each type GMC is proportional to the number of them. From the number of Type I, II, III GMCs are 182 ($\sim$25 %), 316 ($\sim$43 %), and 238 ($\sim$32 %), we estimated the timescale of Type I, II, and III GMCs to be 7 Myr, 13 Myr, and 10 Myr, respectively. The total lifetime of a GMC is estimated to 30 Myr in M33.

3.5. *The comparison of spatial distribution between GMCs and spiral arm*

Recent observations of GMCs in Local Group and nearby galaxies suggested that galactic scale events such as spiral shock and tidal interaction of galaxies play an important role in the massive star formation (e.g., Fukui et al. 2017; Tsuge et al. 2019; Tokuda et al. 2020; Muraoka et al. 2020). We investigated the spatial distribution of GMCs in each type and the relationships between the evolution of GMCs and galactic dynamics. Figure 3 shows the significant difference in the randomness of GMC distributions according to their types. Type I GMCs randomly distributed while Type III GMCs have a distribution like spiral arm. Then, we compared the distribution between GMCs and stellar spiral arms. M33 is a flocculent spiral galaxy, thus its spiral arms are less prominent. In order to clarify nonaxisymmetric structures such as the spiral arm, we created an axisymmetric model disk of M33 from the IRAC 3.6 μm data (Dale et al. 2009), and subtracted it from the original 3.6 μm image to form the residual image (Regan & Vogel 1994, Figure 4 a). The spiral arm structures are clearly seen. Figure 4 (b) shows the comparison between the 3.6 μm residual image and the distribution of type III GMCs. Most of Type III are distributed along the arms. We also found Type I GMCs are mainly distributed in the inter-arm and Type II GMCs are distributed not only in the arm but also in the inter-arm. These results suggest that GMCs tend to gradually evolve and move from both sides (leading and trailing sides) towards the arms, and the spiral arm plays an important role in massive star formation even in flocculent spiral galaxy M33.

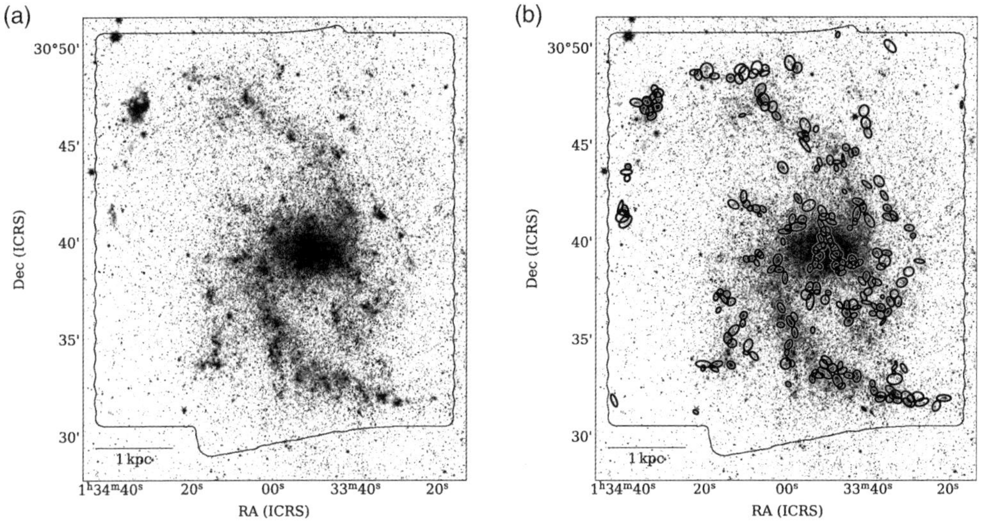

Figure 4. (a) Model-subtracted IRAC 3.6 μm image of M33. The spiral arm structures are clearly seen. The solid line shows the ACA observed area. (b) Distribution of Type III GMCs on (a). The ellipses represent Type III GMCs.

4. Summary

We conducted CO(J = 2–1) survey in the nearest spiral galaxy M33 by ACA 7 m array and analyzed the combined ACA 7 m and IRAM 30 m ^{12}CO data with spatial resolution of ~39 pc. We detected 736 GMCs and classified them into three types based on the association with H II regions and Hα luminosities. Properties of GMCs, the mass, size, and velocity dispersion slightly increase in the order of Type I, II, and III GMCs. Type I GMCs are mainly distributed in the inter-arm and Type II GMCs are distributed not only in the arm but also in the inter-arm. Type III GMCs are mostly distributed along the arms. This suggests that the spiral arm plays an important role in massive star formation. If we assume that the timescale of Type II GMCs in M33 is 13 Myr as in LMC, the timescales of Type I, II, III GMCs in M33 are estimated to be 7 My, 13 Myr, and 10 Myr, respectively. These timescales yield a GMC lifetime of 30 Myr.

References

Cornwell, T.,J. 2018, IEEE Journal of Selected Topics in Signal Processing, 2, 793
Corbelli, E., Braine, J., Bandiera, R., et al. 2017, *A&A*, 601, A146
Dale D. A., Cohen, S. A., Johnson, L. C. et al., 2009, *ApJ*, 703, 1
Druard, C., Braine, J., Schuster, K. F. et al., 2009, *A&A*, 567, A118,
Freedman, W. L., Madore, B. F., Gibson, B. K., et al. 2001, *ApJ*, 553, 47
Fukui, Y., Mizuno, N., Yamaguchi, R., et al. 1999, *PASJ*, 51, 745
Fukui, Y., Kawamura, A., Wong, T., et al. 2009, *ApJ*, 705, 1
Fukui, Y., Tsuge, K., Sano, H., et al. 2017, *PASJ*, 69, 3
Gratier, P., Braine, J., Rodriguez-Fernandez, N. J., et al. 2012, *A&A*, 542, A108.
Hennebelle, P. & Pérault, M. 1999, *A&A*, 351, 309
Hoopes, C. G. & Walterbos, R. A. M. 2000, *ApJ*, 541, 597
Hoopes, C. G., Walterbos, R. A. M., & Bothun, G. D. 2001, *ApJ*, 559, 878
Inoue, T. & Inutsuka, S.-. ichiro. 2012, *ApJ*, 759, 35
Kawamura, A., Mizuno, Y., Minamidani, T., et al. 2009, *ApJS*, 184, 1
Koch, E. W., Rosolowsky, E. W., Lockman, F. J, et al. 2018, *MNRS*, 479, 2505
Kondo, H., Tokuda, K., Muraoka, K., et al. 2021, *ApJ*, 912, 66
Koyama, H. & Inutsuka, S.-I. 2000, *ApJ*, 532, 980

Koyama, H. & Inutsuka, S.-. ichiro. 2002, *ApJL*, 564, L97
McMullin, J. P., Waters, B., Schiebel, D., Young, W., & Golap, K. 2007, Astronomical Data Analysis Software and Systems XVI, 376, 127
Muraoka, K., Kondo, H., Tokuda, K., et al. 2020, *ApJ*, 903, 94
Muraoka, K., Konishi, A., Tokuda, K., et al. 2022, *ApJ*, submitted
Pan, H. A., Schinnerer, E., Hughes, A., et al. 2022, *ApJ*, 927, 1
Regan M. W. & Vogel S. N. 1994, *ApJ*, 434, 536
Rosolowsky, E. W., Pineda, J. E., Kauffmann, J., et al. 2008, *ApJ*, 679, 1338
Rosolowsky, E. W., Hughes, A., Leroy, A., et al. 2021, *MNRS*, 502, 1
Tokuda, K., Muraoka, K., Kondo, H., et al. 2020, *ApJ*, 896, 36
Tsuge, K., Sano, H., Tachihara, K., et al. 2019, *ApJ*, 871, 1
Schinnerer, E., Hughes, A., Leroy, A., et al. 2019, *ApJ*, 887, 49
Vázquez-Semadeni, E., Gómez, G. C., Jappsen, A. K., et al. 2007, *ApJ*, 657, 870
Yamaguchi, R., Mizuno, N., Mizuno, A., et al. 2001, *PASJ*, 53, 985

Resolving the Rise and Fall of Star Formation in Galaxies
Proceedings IAU Symposium No. 373, 2023
T. Wong & W.-T. Kim, eds.
doi:10.1017/S1743921322003763

Gas-Star Formation Cycle in Nearby Galaxies

Hsi-An Pan[1], Eva Schinnerer[2], Annie Hughes[3], Adam Leroy[4], Brent Groves[5] and The PHANGS Team

[1]Department of Physics, Tamkang University, No.151, Yingzhuan Road, Tamsui District, New Taipei City 251301, Taiwan
email: hapan@gms.tku.edu.tw

[2]Max-Planck-Institut für Astronomie, Königstuhl 17, D-69117 Heidelberg, Germany

[3]CNRS, IRAP, Av. du Colonel Roche BP 44346, F-31028 Toulouse cedex 4, France

[4]Department of Astronomy, The Ohio State University, 140 West 18th Ave, Columbus, OH 43210, USA

[5]International Centre for Radio Astronomy Research, The University of Western Australia, Crawley, WA 6009, Australia

Abstract. Star formation, from cold giant molecular clouds to diverse population of stars, is a complex process involving a wide variety of physical processes. In this work, we constrain the link between the gas-star formation cycle and several secular and environmental probe of galaxies. Specifically, we quantify the spatial correlation between molecular gas and star-forming regions for 49 nearby galaxies using the ALMA and narrowband-Hα imaging from the PHANGS survey. At the resolution (150 pc) at which the individual molecular clouds and star-forming regions can be identified, we find that molecular clouds and star-forming regions do not necessarily coexist. The decoupled molecular clouds and star-forming regions are a signature of evolutionary cycling and feedback of the star formation process. Therefore, the impact of galactic-scale conditions and environments must be considered for a complete understanding of how stars form in galaxies and how this process influences the evolution of the host galaxies.

Keywords. ISM: clouds, stars: formation, galaxies: evolution

1. Introduction

Star formation is a complex process that requires tracing cloud formation, fragmentation, and evolution within the local parsec to 100 pc scale and galactic to large-scale environment at kilo-parsec to mega-parsec scales. Schmidt (1959) observed a tight correlation between the star formation rate (SFR) and the volume density of gas in the Milky Way. Later on, Kennicutt (1998) found that the galaxy-integrated SFR surface density (Σ_{SFR}) and gas surface density (Σ_{gas}) also follow a similar power-law relation for nearby galaxies. The relation is now known as the Schmidt-Kennicutt relation. Recently the improvement in spatial resolution brought by instruments such as GALEX, VLA, IRAM 30-m telescope, and BIMA have allowed the Schmidt-Kennicutt relation to be probed on sub-kpc scales (e.g., Bigiel et al. 2008). They show that, on sub-kpc scales, Σ_{SFR} is directly related to molecular gas (Σ_{H_2}) with a power-law index around unity, while the relationship between Σ_{SFR} and total gas (Σ_{gas}) is rather loose. The linear relationship between Σ_{SFR} and Σ_{H_2} implies a roughly constant H_2 consumption time; namely, star formation proceeds in a uniform way in molecular gas in nearby spiral galaxies.

However, cloud-scale ($\sim$ 100 pc) observations in the Local Group and a few nearby star-forming galaxies suggest a more complex picture. The scatter in the Schmidt-Kennicutt relation increase with improving spatial resolution (e.g., Onodera et al. 2010, Kreckel et al. 2018). This is because when the resolution approaches the scale of individual giant molecular clouds (GMCs) and star-forming regions (e.g., HII regions), tracers of massive stars and cold molecular gas become spatially distinct. On the small scale, the scatter in the Schmidt-Kennicutt relation has been interpreted as a results of the evolution of GMCs and the impact of destructive stellar feedback on the structure of interstellar medium (ISM) (e.g., photoionization, stellar winds, and supernova explosions). On the other hand, on the large scale, various environmental mechanisms, such as galactic shear, noncircular motions, gas flows along and through stellar dynamical structures (e.g., bars and spiral arms), and accretion flows also play a role to determine when and which pockets of the GMCs collapse (e.g., Jeffreson et al. 2020), and therefore the relative distribution of GMCs and star-forming regions.

In this work (Pan et al. 2022), we develop a simple and robust method that quantifies the relative spatial distributions of molecular gas and recent star formation, as well as the spatial scale dependence of the relative distributions. The method considers the presence or absence of molecular gas traced by CO emission and star formation traced by Hα emission in a given region (i.e., sight-line or pixel) at a given observed resolution. The method, in principle, is able to provide more information on the gas-to-star formation cycle than the Schmidt-Kennicutt relation as the Schmidt-Kennicutt relation, by definition, only consider the regions where both molecular gas and recent star formation coexist, while at cloud-scale ($\sim$ 100 pc) resolution, such region only accounts for a small fraction of regions with emission.

2. Data and Analysis

The data of this work is taken from the PHANGS (Physics at High Angular resolution in Nearby GalaxieS)† survey. PHANGS is a multi-wavelength survey to observe the tracers of the full star formation process (from gas to star clusters) in a diverse sample of nearby galaxies. The typical spatial resolution achieved with the multi-wavelength observations is about 100 pc. The galaxies are lying on or near the star-forming main sequence. More details on the survey design and scientific motivation are presented in Leroy et al. (2021). In this work, we focus on the molecular gas and ionized (Hα) gas observed by the PHANGS-ALMA and PHANGS-Hα (narrowband) surveys, respectively. 49 galaxies are used in this work.

Our analysis uses Hα as a tracer for the location of recent high-mass star formation. A two-step unsharp masking technique is used to remove the diffuse ionized gas (DIG) from the Hα images. A step-by-step, quantitative description of the method is described in Schinnerer et al. (2019) and Pan et al. (2022). On average, about 65% of the Hα emission across the sample is removed by the DIG removal process.

The CO images are treated using a similar scheme. We clip the CO images at our best-matching resolution of 150 pc using a $\Sigma_{\mathrm{H_2}}$ threshold of 10 $M_\odot$ pc^{-2} accounting for galaxy inclination. This corresponds to a 3σ $\Sigma_{\mathrm{H_2}}$ sensitivity of our CO map with the lowest sensitivity at this spatial scale.

Finally, we measure the presence or absence of the two tracers at each resolution in an field-of-view extending to 0.6 R_{25}, which is the largest radial extent probed by our data in all galaxies, corresponding to $\sim$6.4 kpc on average. We classify each sight line (pixel) within 0.6 R_{25} in the CO and Hα images into one of four categories: *CO-only* – only CO emission is present, *Hα-only* – only Hα emission is present, *overlap* – both CO and

† www.phangs.org

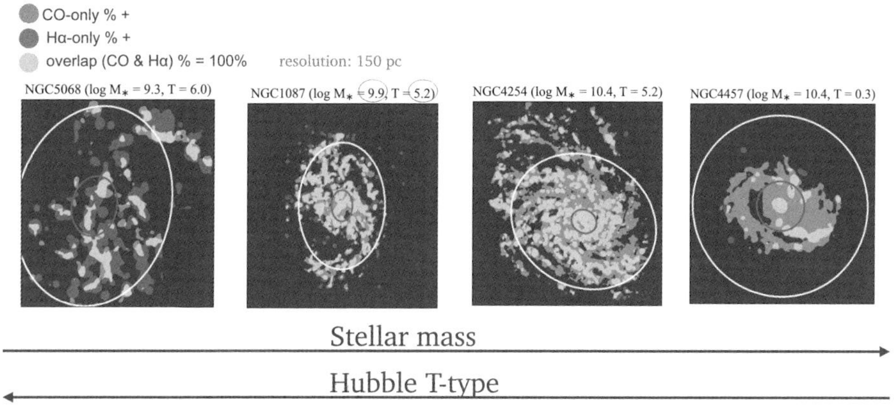

Figure 1. Examples of the spatial distribution of different sight lines. Galaxy maps show the regions of *CO-only* (blue), *Hα-only* (red), and *overlap* (yellow) sight lines at 150 pc resolution. The inner ellipses (magenta) mark the central region, defined as the central 2 kpc in deprojected diameter. The outer ellipses (white) indicate the $0.6R_{25}$ regions where we measure the global sight line fractions.

Hα emission are present, and *empty* – neither CO nor Hα emission is present. In this work, we do not consider the *empty* sight lines but measure the fraction of sight lines in *CO-only*, *Hα-only*, and *overlap* regions at each resolution.

3. Results and Discussion

3.1. *Sight-line Distribution*

Figure 1 shows the examples of the spatial distribution of different sight lines. As seen in the figure, our galaxies show diverse relative distribution of CO and Hα emission. In our sample, some galaxies, such as NGC 5068, are dominated by *Hα-only* regions, while some galaxies, e.g., NGC 4457 are dominated by *CO-only* regions.

3.2. *Sight line Fractions vs. Global Galaxy Properties*

Figure 2 shows box plots of sight line fractions as a function of stellar mass (left) and Hubble type (right). Galaxies are divided into three groups according to their stellar mass or Hubble type. The darker colors indicate increasing stellar mass or decreasing Hubble-type value. The inner horizontal belt-like symbol and line in the boxes represent the median and mean of the distribution, respectively The left panel of Figure 2 shows a tendency for more massive galaxies to have higher *CO-only* fractions. The opposite trend is exhibited by *Hα-only* fractions. The median *overlap* fractions remain at a nearly constant value, ~30%, as a function of stellar mass. In the right panel of Figure 2, the fraction of *CO-only* sight lines deceases with Hubble type, and the fraction of *Hα-only* sight lines increases with Hubble type. The trends with Hubble type and stellar mass are consistent in the sense that late-type galaxies tend to be less massive.

Moreover, Figure 2 implies that if the CO and Hα images at 150 pc resolution are used to construct the Schimdt-Kennicutt relation as a diagnostic of star formation process, only 30% of sight lines (the *overlap* regions) would show in the relation. As a result, we might be missing important variation in gas depletion time (the slope and intercept of the Schmidt-Kennicutt relation) and change in cloud life cycle/time variations (i.e., scatter of the relation).

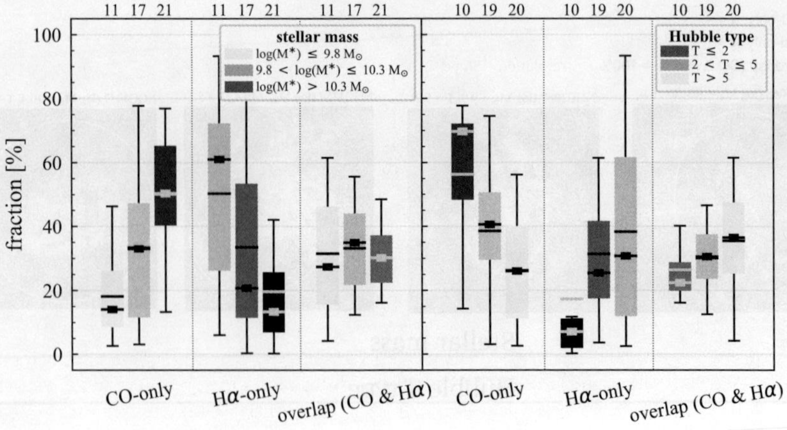

Figure 2. Global sight line fractions at 150 pc resolution as a function of stellar mass (left) and Hubble type (right). The boxes show the interquartile ranges (IQR; the Q1/25 percentile to Q3/75 percentile), and the vertical whiskers extend to Q1 - 1.5 × IQR and Q3 + 1.5 × IQR. The inner horizontal belt-like symbol and line in the boxes represent the median and mean of the distribution, respectively. The number of galaxies in each stellar mass and Hubble-type bin are shown at the top of the plots.

3.3. *Sight-line Fractions vs. Dynamical Structures*

In addition to global galaxy properties, we can also classify galaxies according to their dynamical structures, such as the presence of stellar bar and spiral arms. We classify galaxies into four groups:

- No Structure (NS): galaxies without a bar and grand-design (GD) spiral arms (e.g., galaxies with flocculent/multiple arms are in this category),
- Bar:galaxies with a bar but no GD spiral arms,
- Bar+GD: galaxies with a bar and GD spiral arms,
- GD: galaxies with GD arms but without a bar.

The corresponding box plots are shown in Figure 3.

We find a sign for galaxies with only grand-design spiral arms to show higher fraction of *CO-only* regions and lower *Hα-only* regions. Some galaxies in our sample show a pronounced offset between the different sight line types with a sequence of *CO-only* to *overlap* and to *Hα-only* when going from up- to downstream (assuming the spiral arms are trailing, e.g., NGC 4321, NGC 0628, NGC 1566, and NGC 2997). The offset of different types of sight lines is consistent with expectations for a spiral density wave. These offsets are almost exclusively found in galaxies with well-defined grand-design spiral arms and presumably lead in turn to the high *CO-only* fraction in the disk.

3.4. *Physical Interpretation of Different Types of Sight lines*

What do the three types of sight line mean physically? We find a large reservoir of molecular gas traced by CO not associated with star-forming regions in our galaxies. One of the most interesting result in this work is that the molecular gas surface density of *CO-only* regions are not necessarily low, ranging from 10 (the applied threshold) to several thousands solar mass per parsec square. These high-density *CO-only* regions could be pre-star-forming clouds, they are still undergoing gravitational collapse; or, massive stars may be already formed in part of the *CO-only* gas, but their Hα emission is obscured by dust. Some CO-only regions might be a diffuse, dynamically hot component (Pety et al. 2013) that is not prone to star formation or may be analogous to the gas in the centers of early-type (elliptical) galaxies. It is also possible that high-mass star formation is suppressed in

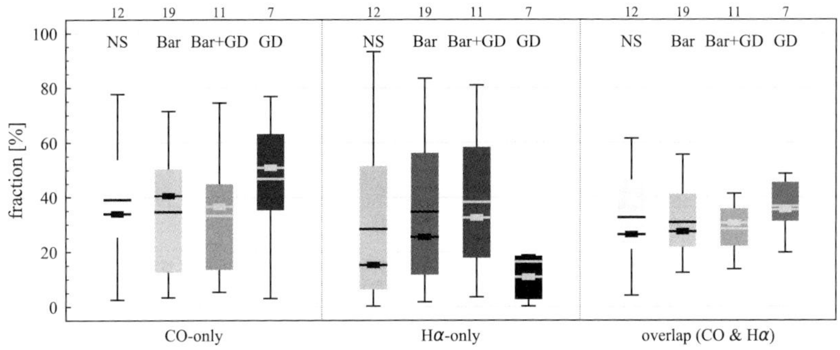

Figure 3. Sight line fractions at 150 pc resolution for galaxies without structures (bar or grand-design spiral arms; NS), galaxies with a bar but without grand-design spiral arms (Bar), galaxies with both a bar and grand-design spiral arms (Bar+GD), and galaxies with grand-design spiral arms but no bar (GD).

some *CO-only* regions, forming stars that are not massive enough to produce detectable Hα emission.

The *overlap* regions are typical star-forming regions where molecular gas and stars coexist. Finally, about one-third of sight-lines show Hα emission not associated with molecular gas. Molecular gas might be completely consumed by star formation in these regions, so star formation will halt at some point in these regions in the near future unless there is additional supply of molecular gas. It is also possible that these regions have experienced significant stellar feedback which destroys the parent molecular clouds of the star forming regions.

In conclusion, the different types of sight-lines reprcsent different phases of star formation. Therefore, different relative distribution of CO and Hα implies different gas-to-star formation cycle in galaxies.

3.5. *Impact of Spatial Scale*

With high-resolution images, we can investigate the impact of spatial scale on the distributions of CO and Hα emission. Figure 4 shows the sight line fractions for three example galaxies as a function of spatial scale from 150 pc to 1.5 kpc. The figure shows that, no matter which type of the sight-line dominate the images at 150 pc resolution, at 1.5 kpc resolution, the *overlap* region become the dominant one. This is simply because the CO and Hα images become similar when we lower the resolution. For all sight line categories, the variations with spatial scale become less evident at $>$ 500 pc resolution, indicating a critical resolution requirement to resolve the gas-to-star formation cycle in nearby galaxies.

4. Summary

In this work, we adopt a simple and reproducible method developed to quantify the relative spatial distributions of molecular gas and recent star formation. We classify each sight line (i.e., pixel) at each resolution according to the overlap between the tracers: *CO-only*, *Hα-only*, and *overlap* (CO and Hα). These three categories can be translated into the following star formation phases: *CO-only* – molecular gas currently not associated with star formation traced by Hα, *overlap* – star-forming molecular clouds, and *Hα-only* –regions of young massive stars. The best common resolution of our galaxies (150 pc) is sufficiently high to sample individual star-forming units and to separate different types of sight lines.

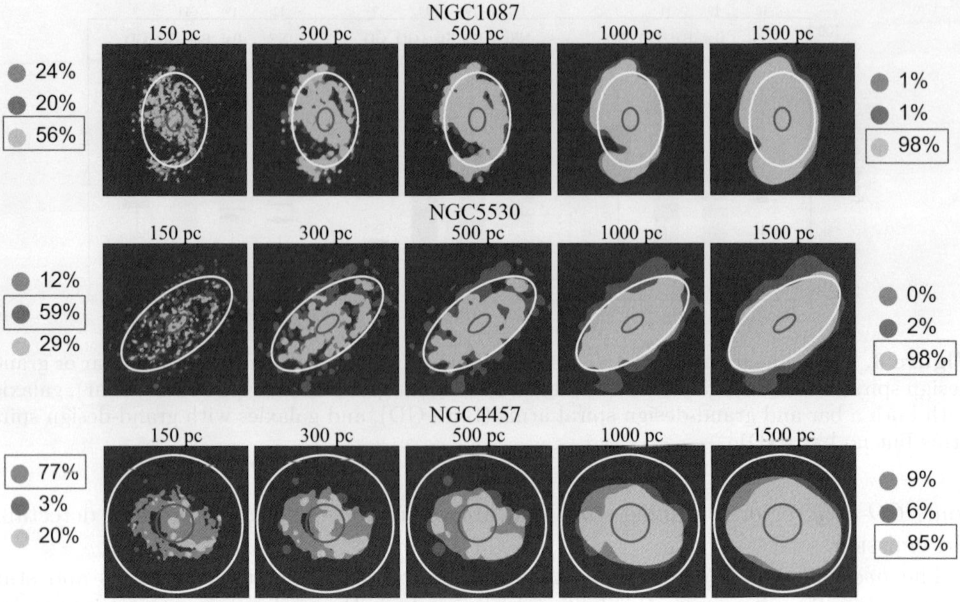

Figure 4. Sight line map as a function of spatial scale. The colors and symbols are the same as in Figure 1.

At 150 pc resolution, we find strong correlations between the sight-line fractions and global galaxy properties (stellar mass and Hubble type). In addition, galactic dynamics further contributes to organizing the spatial configuration of molecular gas and star-forming regions within galaxies. Since different types of sight lines represent different phases of star formation, the different relative amount and spatial distribution of molecular gas traced by CO and star-forming regions traced by Hα implies different gas-to-star formation cycle in nearby galaxies.

Finally the sight line fractions show a strong dependence on the spatial scale (resolution). Any relation between the sight line fractions and galaxy properties are only evident when the resolution is higher than 500 pc, suggesting the importance of cloud-scale resolution on the studies of star formation in nearby galaxies.

References

Bigiel, F., Leroy, A., Walter, F., et al. 2008, AJ, 136, 2846. doi:10.1088/0004-6256/136/6/2846

Jeffreson, S. M. R., Kruijssen, J. M. D., Keller, B. W., et al. 2020, MNRAS, 498, 385. doi:10.1093/mnras/staa2127

Kreckel, K., Faesi, C., Kruijssen, J. M. D., et al. 2018, ApJL, 863, L21. doi:10.3847/2041-8213/aad77d

Leroy, A. K., Schinnerer, E., Hughes, A., et al. 2021, ApJS, 257, 43. doi:10.3847/1538-4365/ac17f3

Onodera, S., Kuno, N., Tosaki, T., et al. 2010, ApJL, 722, L127. doi:10.1088/2041-8205/722/2/L127

Pan, H.-A., Schinnerer, E., Hughes, A., et al. 2022, ApJ, 927, 9. doi:10.3847/1538-4357/ac474f

Pety, J., Schinnerer, E., Leroy, A. K., et al. 2013, ApJ, 779, 43. doi:10.1088/0004-637X/779/1/43

Schinnerer, E., Hughes, A., Leroy, A., et al. 2019, ApJ, 887, 49. doi:10.3847/1538-4357/ab50c2

Schmidt, M. 1959, ApJ, 129, 243. doi:10.1086/146614

Resolving the Rise and Fall of Star Formation in Galaxies
Proceedings IAU Symposium No. 373, 2023
T. Wong & W.-T. Kim, eds.
doi:10.1017/S1743921322004008

Characterizing star formation in the innermost kiloparsec of the galaxy NGC 1386

Gustavo Bruzual[1], **Almudena Prieto**[2,3], **Gladis Magris C.**[4] and **Juan A. Fernández-Ontiveros**[5]

[1]Instituto de Radioastronomía y Astrofísica, UNAM, México

[2]Instituto de Astrofísica de Canarias (IAC), E-38200 La Laguna, Tenerife, Spain

[3]Dept. Astrofísica, Universidad de La Laguna, E-38206 La Laguna, Tenerife, Spain

[4]Centro de Investigaciones de Astonomía (CIDA), Mérida, Venezuela

[5] Centro de Estudios de Física del Cosmos de Aragón, Plaza San Juan 1, 44001 Teruel, Spain

Abstract. We characterize the star formation going on in the inner kiloparsec region of the galaxy NGC 1386 as derived from the analysis of a multiwavelength dataset covering the optical, near-IR and mid-IR at subarsec resolution. We detect 61 point sources, distributed in a ring of 960 pc radius around the center of the galaxy. From SED fitting we conclude that these are low mass ($440 < M/M_{\odot} < 79000$) young clusters, with age distributed from 1 to 10 Myr, with median at 3.6 Myr. Comparison of the Hα luminosity of the clusters derived from the Hα+[N II] narrow band image with that expected from the fitted ionizing continuum shows that a large fraction of the ionizing photons escapes from the clusters. Moreover, a large fraction of these photons escapes from the regions around the star forming ring.

Keywords. galaxies: individual NGC1386, galaxies: star formation, galaxies: ISM, methods: data analysis

1. Introduction

We present preliminary results of the star formation properties of the circumnuclear region of NGC 1386, a nearby (15.3 Mpc, Jensen et al. 2003) Sa galaxy. This is part of an ambitious program, "The central parsecs of the nearest galaxies"†)(PARSEC), aimed to determine the physical processes occurring in the inner regions of the nearest galaxies, with particular focus on the relation between the nuclear star formation and its parental molecular gas and dust material (e.g. Prieto et al. (2009)). We use multiwavelength, from optical to near-IR, high angular resolution data that allow us to identify clusters of size $\approx$ 8 pc, and calculate its age, mass and extinction by fitting its spectral energy distribution with stellar population models.

2. Data

We use near-IR images on *J*, *H*, *Ks* bands, obtained with the AO assisted instrument NaCo, located at the VLT facility in Paranal Observatory, Chile as part of the PARSEC project. To complement IR observations, we use optical images from the HST scientific archive acquired with the Wide-Field Planetary Camera 2 and the Near Infrared Camera

† `http://research.iac.es/proyecto/parsec/main/index.php`

and Multi-Object Spectrometer with the broad band filters: F547M, F606W, F791W, F814W, F110M, F110W, and the narrow-band filters F656N, F658N, FR656N_1 and FR680N, centred on the emission lineHα+[N II] and its adjacent continuum

3. Model

We use DynBaS (Magris *et al.* 2015) standard spectral energy distribution fitting algorithm, with the assumption that each detected cluster in NGC 1386 is well represented as a single-age and single-metallicity population of stars (SSP) (DynBaS1D), that evolves passively. The SSPs are from the Charlot & Bruzual models as described in Plat et al 2019, computed for the solar metallicity evolutionary tracks of the Padova's group (Bressan et al. 2017). We use a Chabrier (2003) IMF in the standard interval between 0.1 and 100 $M_\odot$. For each cluster, DynBas1D provides the model *age*, *mass*, and A_V that best fit the observed SED and computes the corresponding intrinsic Hα luminosity ($L(\mathrm{H}\alpha)_0^{\mathrm{mod}}$), assuming a complete conversion of the hydrogen ionizing photons.

Due to the low mass derived for the clusters (see Fig. 1), in a range where the initial mass function might not be fully populated, additionally to the standard DynBas fitting we use a stochastic implementation, described in Prieto *et al.* (2022), where we use 1000 SSPs, with a stochastic sampling of the IMF as described in Bruzual (2002).

4. Results

Fig. 1 shows the distribution of the physical quantities derived for the clusters in NGC 1386, and its mutual relations. We find the following results:

• Age: Model age distribution shows clusters in NGC 1386 were mainly formed less than 6 Myr ago, with a median at 3.6 Myr, and a low dispersion (first and third quartiles at 2.75 and 4.3 Myr). We found no evidence of older events of star formation in the innermost part of this galaxy, if ocurred, they must have been of too low mass to be detected at the present.

• Mass: Detected clusters correspond to low mass objects, between 440 and 79000 $M_\odot$ with median at 4100 $M_\odot$. The most massive clusters are found to be the youngest, in the 2 Myr range, with $A_V \approx 3-4$ mag, consistent with young ages being in denser and dustier environments.

• Av: Clusters are mildly extincted with A_V between 0 and 5 mag. A comparison of A_V with the A_V derived from the VLT/Ks-band – HST/F814W image shows a fair agreement (Prieto *et al.* 2022). We find a positive correlation between *mass* and A_V, which we interpret as lower mass clusters are detected only in low dust environment. Also consistent with the fact that only massive clusters are able to retain material from their original cloud (e.g. Tenorio-Tagle et al 1999). However this correlation is not reported in more massive clusters (e.g. Turner et al. 2021) and should be considered to further analysis.

• Hα luminosity: We find the Hα luminosity of the clusters derived from the Hα+[N II] narrow band image is significantly less that expected from the fitted ionizing continuum. Moreover, total $L(\mathrm{H}\alpha)_0^{\mathrm{mod}}$ from clusters exceeds the total measured Hα in the ring (cluster+intracluster emission). This can occur in three non-exclusive scenarios: a fraction of the ionizing photons are locally absorbed within the HII region, and/or the HII regions are matter-bounded, and/or with a covering factor < 1, and the ionizing radiation finds a way to escape from the respective HII region. The consequences of these scenarios on the star formation rate–L(Hα) calibration is presented in Prieto *et al.* (2022).

• Stochasticity: None of the above general results are affected by the fluctuations in the IMF sampling in low mass stellar populations (gray dots and histograms compared with

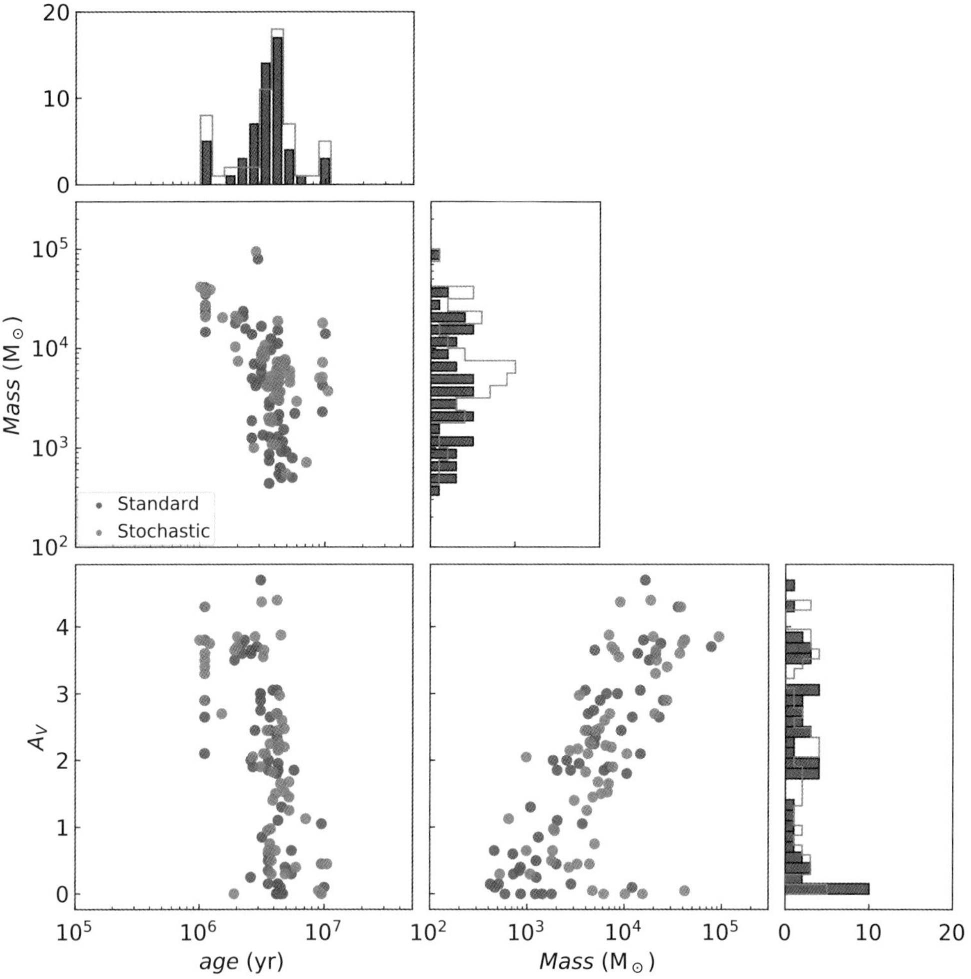

Figure 1. Distribution of derived *age*, *mass*, A_V for the clusters in NGC 1386 for both the *standard* (blue dots) and *stochastic*(gray dots) fitting. Histograms on each axis show the marginalized distributions of each parameter.

blues in Fig. 1). Even though physical parameters derived using *standard* or *stochastic* model can be different for individual clusters, the global conclusions are the same.

References

Bressan A., Marigo, P., Girardi, L., Salasnich, B., Dal Cero, C., Rubele, S., Nanni, A. 2012, *MNRAS*, 427, 127

Bruzual A. G., 2002, in Geisler D. P., Grebel E. K., Minniti D., eds, Vol. 207, Extragalactic Star Clusters. p. 616 (arXiv:astro-ph/0110245)

Charlot S., Bruzual G. 2022, *In preparation*

Chabrier G., 2003, *PASP*, 115, 763

Magris C., G, Mateu P., J., Mateu, C., Bruzual A., G., Cabrera-Ziri, I., Mejía-Narváez, A.. 2015,. *PASP*, 127, 16

Turner, J.A. *et al.* 2021, *MNRAS*, 502, 1366

Prieto, M.A., Fernandez-Ontiveros, J.A., Bruzual, G., Burkert, A., Schartmann, M., Charlot, S., 2019, *MNRAS*, 485, 3264

Prieto, M.A. et al 2022, *In preparation*

Resolving the Rise and Fall of Star Formation in Galaxies
Proceedings IAU Symposium No. 373, 2023
T. Wong & W.-T. Kim, eds.
doi:10.1017/S1743921323000376

A UVIT Look at Star Formation in Nearby Interacting/Merging Galaxies

Mousumi Das

Indian Institute of Astrophysics
2nd Block, 100 Feet Rd, Koramangala, Bengaluru,
Karnataka 560034, India
email: mousumi@iiap.res.in

Abstract. The interactions and mergers of gas rich galaxies are known to produce star formation which often leads to nuclear activity as well. The star formation is ideally mapped using FUV and NUV emission, since UV traces star formation for longer timescales compared to Hα emission. It is also emitted over a broader range of stellar masses in galaxies. In this study we present FUV and NUV observations of merging and interacting galaxies in our nearby universe conducted using the UVIT. We present the example of a merging system MRK212 that has dual AGN and the triple AGN system NGC7733-7734. The UV emission is associated with the tidal arms, individual nuclei, resonance rings, nuclear spirals as well as AGN/stellar feedback. We also find that radio emission is often closely associated with the UV emission, arising from both star formation as well as AGN activity, and perhaps kpc-scale AGN feedback. We find that a comparison of optical IFU imaging with FUV in NGC7733-7734 reveals unique properties associated with the interaction including the third AGN buried in a tidal arm.

Keywords. galaxies: spiral, galaxies: interacting, galaxies: active, galaxies: UV

1. Introduction

Galaxy interactions and mergers cause enormous changes in the gravitational potential of galaxies, even if the interactions are distant and at separations of tens of kpc. The change in potential results in cloud collisions and hence star formation resulting in the formation of very bright merger remnants. The tidal interaction also causes the formation of extended tidal arms and rings, both of which may host star forming complexes (SFCs). As the galaxies come closer, there is usually gas infall to the nuclear regions, often resulting in nuclear starburst activity and the triggering of active galactic nuclear (AGN) activity. Both star formation and AGN emit copiously at UV wavelengths. Hence, both FUV and NUV are ideal wavelengths to trace galaxy evolution via star formation and AGN activity. In this paper we present a variety of UV observations of closely interacting and merging galaxies. Our aim is to examine the star formation on kpc scales in the form of SFCs and understand how they are related to the merging process. Details of the projects are given in some recent papers (Yadav et al. 2021).

2. Observations

The observations were conducted mainly in the far-UV (FUV) band (1300–1800 A) with the Ultraviolet Imaging Telescope (UVIT), on board the AstroSat Satellite. The details of the filters and instrument are briefly described in (Das et al. 2021). The advantage of using the UVIT is that it has a wide field of view of approximately $28'$ and a spatial resolution $\sim 1.5''$, which is good enough to resolve the large star forming clusters in nearby galaxies. Our UVIT observations of the galaxies presented in this paper have observing times of 15Ks and 6Ks. The data analysis was done with the standard UVIT pipeline. The UV emission arises from massive stars that give out FUV and NUV emission as well as Hα emission. But FUV and NUV trace star formation for upto 10^8 years whereas Hα traces for it for 10^6 to 10^7 years only. So UV can detect star formation for longer duration in galaxies. Also, for good spatial resolution such as that in UVIT, the FUV and NUV emission can detect individual star forming clusters. The UV emission also arises from AGN activity i.e. mainly from the accretion disk and makes the nucleus appear as a compact, bright source in the centers of galaxies.

3. UVIT observations Dual AGN in MRK212

During mergers, as the galaxies come closer they eventually become gravitationally bound. When their bulges are at separations of few to several kpc their supermassive black holes (SMBHs) can start accreting mass. Then they often become AGN and their nuclei become very luminous. When both nuclei are AGN and are at separations of approx 1-10 kpc they form dual AGN (DAGN). DAGN are very rare and are mainly detected using radio observations and X-ray observations Rubinur et al. (2019). Alternatively the nuclei can be mixed starburst-AGN pairs or starforming nuclei pairs, both of which are bright at UV wavelengths. The luminosity can be due to star formation or it can be associated with the gas outflows due to starburst and/or AGN activity consituting stellar or AGN feedback. There will be negative feedback due to the outflows pushing out the infalling gas, but there may also be positive feedback caused by to star formation in the shocked gas falling back onto the disk. There can also be interaction between the outflows if the nuclei are close enough. Such activity will trigger star formation and lead to the growth of hot gas around galaxies.

A good example is the study of the dual nuclei in the galaxy MRK212, which we have studied using radio observations (EVLA and uGMRT), FUV (UVIT) observations and optical spectroscopy (Himalayan Chandra Telescope) (Rubinur et al. 2021). The galaxy is a major merger and nearly face-on in orientation. Both nuclei are bright in FIRST radio images. The HCT spectra of the 2 nuclei confirmed that they fall in the star forming and AGN class in the BPT plot and so this is a DAGN system. Figure 1 shows the UVIT FUV observations of MRK212. The star forming knots are clearly detected along spiral or tidal arms. They may represent super star clusters that have formed due to the merger. There must be gas collected between the two bulges which has given rise to star formation between the two nuclei. Such gas has been detected in ALMA observations of similar mergers (e.g. NGC6240) and here we are seeing the star formation associated with that gas. The S2 nucleus has 2 knots of star formation that coincide with the 2 radio lobes detected in radio EVLA 8.5 GHz observations. The outflows from the AGN in S2 could have triggered the star formation and so this could be a signature of AGN feedback on kpc scales. We may also be seeing signatures of AGN wind triggered star formation i.e. positive AGN feedback in the knots 3 and 4 which coincide with the radio lobes associated with nucleus S2.

4. The Triple AGN in NGC7733-7734

As part of a study of interacting galaxies, we observed the southern interacting galaxy pair NGC7733 and NGC7734 in the JHK band using the near infrared 1.5m telescope IRSF in the South Africam Astronomical Observatory (SAAO). The northern knot in NGC773 was previously identified as a star forming region, but is clearly a bulge in the NIR images. The nature of the nuclei can be determined using the empirical plot called the BPT plot, which is based on the ratio of the emission lines [OIII], [NII], Hα and Hβ.

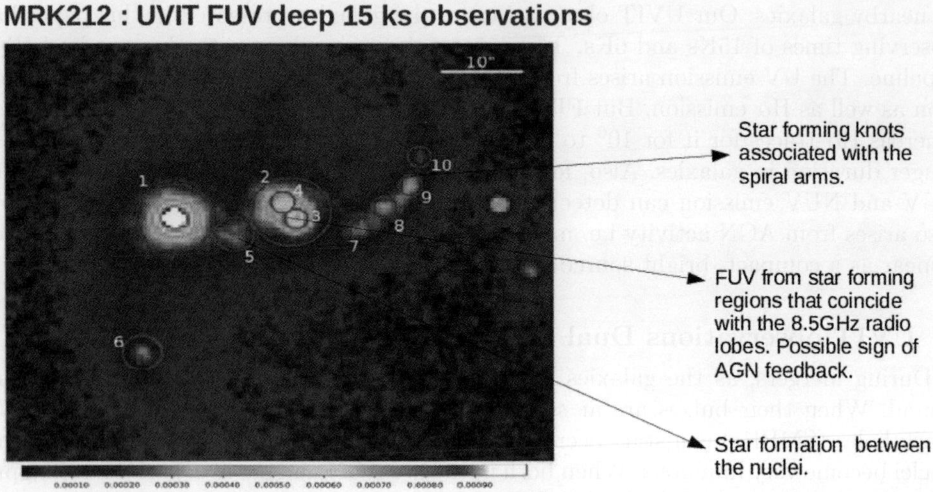

Figure 1. The FUV image of MRK212 had 15ks of observing time. The S2 nucleus has 2 knots of star formation that coincide with the 2 radio lobes detected at 8.5 GHz radio observations. The outflows from the AGN in S2 could have triggered the star formation and could be a signature of AGN feedback on kpc scales. Details are given in Rubinur et al. (2021).

We aslo used the MUSE archival data of the interacting pair. MUSE gives the spectra of the spaxels and so the BPT position of the individual pixels can be determined. It is clear that NGC7733 nucleus and NGC7733N are both Seyfert type whereas NGC7734 is a LINER type nucleus. Both NGC7733 and NGC7734 show extensive star formation in the spiral arms, but it is stronger in NGC7733. So the small group can be considered as a triple AGN system, although all 3 nuclei are not really within a common envelope. The nuclear SMBH mass in NGC733 and NGC7734 are few times 10^8 solar mass, whereas NGC7733N has the least massive SMBH mass of a few times 10^7 solar mass. The galaxy NGC7733 shows emission from an extended narrow line region (ENLR) which is due to the photionized gas around the AGN. It extends out to 18 kpc from the nucleus of NGC7733. Apart from the triple AGN, the FUV image shows many rings of star formation associated with the two galaxies, and especially with the northern galaxy NGC7734. There is also star formation along the bar of the latter galaxy as seen in the FUV image. Details are given in Yadav et al. (2021).

5. Conclusions

UV studies can offer us new insights into the star formation and nuclear activity in nearby interacting galaxies. This is mainly because UV traces star formation for longer timescales compared to Hα. So it can detect fainter star formation in tidal arms, diffuse disks and in extended tidal tails. When the UV is combined with optical integral field spectroscopy (IFS) data such as MUSE or SDSS-MANGA it can reveal hidden nuclear activity and star formation processes.

References

Das, M., Yadav, J., Patra, N., et al. 2021, ApJ, 42, 2, 85
Rubinur, K., Kharb, P., Das, M. 2019, MNRAS, 484, 4933
Rubinur, K., Kharb, P., Das, M., et al. 2021, MNRAS, 500, 3, 3908
Yadav, J., Das, M., Barway, S., and Combes, F. 2021, A&A, 651, L9
Yadav, J., Das, M., Barway, S., and Combes, F. 2021, A&A, 657L, L10

Resolving the Rise and Fall of Star Formation in Galaxies
Proceedings IAU Symposium No. 373, 2023
T. Wong & W.-T. Kim, eds.
doi:10.1017/S1743921322004914

Star formation history for the starburst dwarf galaxy in the Local Group, IC 10

Mahtab Gholami[1], Atefeh Javadi[1], Jacco Th. van Loon[2], Habib Khosroshahi[1] and Elham Saremi[1,3,4]

[1]School of Astronomy, Institute for Research in Fundamental Sciences (IPM), Tehran, 19568-36613, Iran
email: mhtab.gholami@gmail.com

[2]Lennard-Jones Laboratories, Keele University, ST5 5BG, UK

[3]Instituto de Astrofisica de Canarias, C/ Vìa Làctea s/n, 38205 La Laguna, Tenerife, Spain

[4]Departamento de Astrofisica, Universidad de La Laguna, 38205 La Laguna, Tenerife, Spain

Abstract. IC 10 as a starburst dwarf galaxy in the Local Group (LG) has a large population of newly formed stars that are massive and intrinsically very bright in comparison with other LG galaxies. Using the Isaac Newton Telescope (INT) with the Wide Field Camera (WFC) in the i-band and V-band, we performed an optical monitoring survey to identify the most evolved asymptotic giant branch stars (AGBs) and red supergiant stars (RSGs) in this star-forming galaxy, which can be used to determine the star formation history (SFH). The E(B - V) as an effective factor for obtaining the precise magnitude of stars is measured for each star using a 2D dust map (SFD98) to obtain a total extinction for each star in both the i-band and V-band. We obtained the photometric catalog for 53579 stars within the area of 0.07 deg^2 (13.5 kpc^2), of which 762 stars are classified as variable candidates after removing the foreground stars and saturated ones from our catalog. To reconstruct the SFH for IC 10, we first identified 424 long-period variable (LPV) candidates within the area of two half-light radii ($2r_h$) from the center of the galaxy. We estimated the recent star formation rate (SFR) at ~ 0.32 $M_\odot$ yr^{-1} for a constant metallicity Z = 0.0008, showing the galaxy is currently undergoing high levels of star formation. Also, a total stellar mass of 0.44×10^8 $M_\odot$ is obtained within $2r_h$ for that metallicity.

Keywords. stars: AGB – stars: RSG – stars: LPV – stars: formation – galaxies: dwarf – galaxies: irregular – galaxies: starburst – galaxies: evolution – galaxies: star formation – galaxies: individual: IC 10

---

1. Introduction

A dwarf irregular galaxy, IC 10 (UGC 192) is one of the most interesting systems and is the nearest starburst dwarf located within the LG (e.g., Gerbrandt et al. 2015). The existence of young, intermediate, and old-age stellar populations has been studied in IC 10 (e.g., Massey & Armandroff 1995, Sakai et al. 1999, Borissova et al. 2000) in which the highest density of Wolf-Rayet stars indicates the recent burst of star formation (e.g., Cosens et al. 2022). On the other hand, the light of the galaxy is affected by gas-dust clouds of the Milky Way (MW) galaxy because IC 10 is located in the Galactic plane ($l = 118^\circ.9$, $b = -3^\circ.3$). Therefore, studying this galaxy is restricted due to the strong extinction of MW (Richer et al. 2001). The greatest readily available tracers of stellar populations are evolved stars with high luminosity, such as AGBs and RSGs (Javadi et al. 2011b). For the AGBs and RSGs which are at the final stage of their evolution, their luminosity would be directly proportional to their birth mass, and they are exceptional objects for reconstructing the SFH for a galaxy and tracing stellar populations from as recently formed as 10 Myr to as ancient as 10 Gyr (Javadi et al. 2011b). In this paper, we conducted an optical survey of the majority of dwarf galaxies in the LG between June 2015 and October 2017 with the 2.5-m INT/WFC over nine epochs (Saremi et al. 2020). We identified the LPV candidates of IC 10, then, based on their luminosity distribution, we determined the SFH using the method that we successfully applied previously to a number of LG galaxies in our survey (e.g., Javadi et al. 2017, Hashemi, Javadi, & van Loon 2018, Saremi et al. 2021, Navabi et al. 2021); Here, we describe our method for reconstructing the SFH of IC 10 using our identified LPV candidates.

2. Identification of LPV candidates

A photometry catalog was obtained for 53579 stars in the entire area of CCD4 (0.07 deg^2 (13.5 kpc^2)) using the DAOPHOT/ALLSTAR software package (Stetson 1987). We estimated amount of foreground contamination for the galaxy using the stars in common between our catalog and the Gaia Early Data Release 3 catalog (Gaia EDR3) (Lindegren et al. 2021). Then, we determined the foregrounds based on the parallax and proper motion measurements of stars. We estimated the percentage of contamination of our catalog by foreground stars $\sim 7\%$. Finally, we removed them from our catalog for further analysis. As a result, 762 stars are identified as variable stars. Among them, 424 stars are identified as LPV candidates due to large amplitude variability, mostly more than 0.20 mag within the area of $2r_h$ centered on the galaxy. Moreover, we applied the extinction correction from SFD98 dust map (Schlegel, Finkbeiner & Davis 1998) in both i-band (i_0) and V-band (V_0) magnitudes for all stars in our catalog.

3. Star formation history for IC 10

We derived the SFH of IC 10 following the same method applied to other LG dwarf galaxies described in previous papers (e.g., Saremi et al. 2021, Navabi et al. 2021). The SFH has been estimated by calculation of the SFR, ξ, in $M_\odot yr^{-1}$ based on the LPV candidates. The value of SFR, ξ, shows the amount of gas mass that was converted into numerous stars per year. The number of newborn stars can be determined from their total mass, which is measured by using an initial mass function (IMF) (Kroupa 2001) in which the minimum and maximum ranges of stellar mass have been determined as 0.02 $M_\odot$ and 200 $M_\odot$, respectively. The $\xi(t)$ in each age bin (dt), is obtained using the following equation in which the dn' is the number of LPV candidates in each age bin.

$$\xi(t) = \frac{dn'(t)}{\delta t} \frac{\int_{min}^{max} f_{IMF}(m) m \, dm}{\int_{m(t)}^{m(t+dt)} f_{IMF}(m) \, dm} \tag{3.1}$$

In the equation 3.1, the range of the stellar mass for LPV candidates is defined between m(t) and m(t+dt) while the δt is the time duration of pulsation of those stars. The f_{IMF} is defined using $f_{IMF} = Am^{-\alpha}$ where A and α are recognized as the constant of normalization and the mass range dependence, respectively (Kroupa 2001). Fig. 1 shows the SFH for a constant metallicity of $Z = 0.0008$ over the time period of 5 Myr to 13.8 Gyr ago. The horizontal and vertical error bars indicate the age range and statistical errors in each bin, respectively. The statistical errors for each age bin are also calculated assuming Poisson statistics. As can be seen in the Fig. 1, the figure for SFH shows a peaking around 10 Myr ago ($\log t \sim 7$) at 0.32 ± 0.07 $M_\odot$ yr^{-1} which indicates high star-forming activity in IC 10 in the recent years. Also, there is another major peak coming up at 0.05 ± 0.07

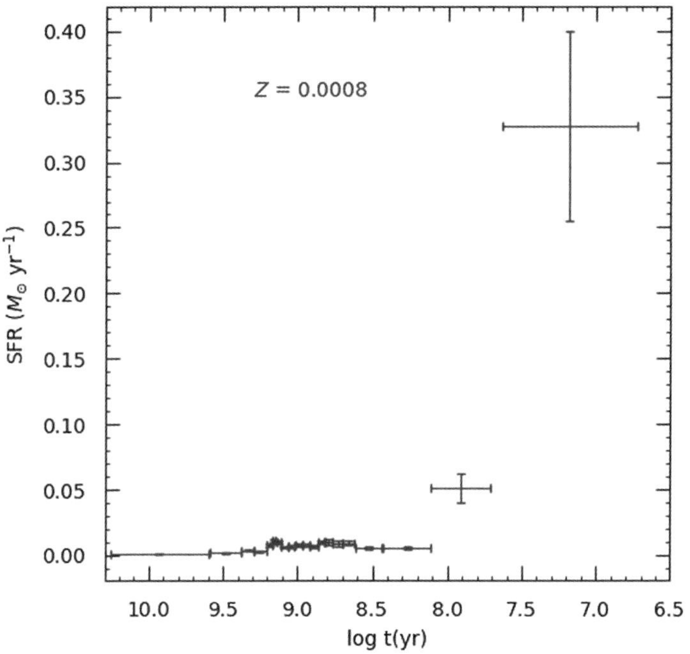

Figure 1. SFH of IC 10 for constant metallicity of Z = 0.0008. The horizontal and vertical error bars represent the age range and the statistical errors, respectively in each bin derived for LPV candidates within $2r_h$ area.

$M_\odot$ yr^{-1} at around 80 Myr ago. It explains the high star-forming activity in that bin followed by a significant increase in recent years. Also, the total stellar mass within $2r_h$ of the galaxy is obtained $(0.44 \pm 0.08) \times 10^8$ $M_\odot$.

4. Conclusion

We have conducted an optical monitoring survey of the majority of dwarf galaxies in the LG with WFC at INT. Based on our obtained 424 LPV candidates within $2r_h$, we explain our method to reconstruct the SFH of the starburst dwarf, IC 10. For each age bin, we calculated the SFR with a constant metallicity Z = 0.0008; the SFH is prominent at recent times at 0.32 ± 0.07 $M_\odot$ yr^{-1} around 10 Myr ago. It shows the current intense star formation activity of IC 10.

References

Borissova, J., Georgiev, L., Rosado, M., et al., 2000, *A&A*, 363, 130

Cosens, M., Wright, S. A., Murray, N., et al., 2022, *ApJ*, 929, 74

Gerbrandt, Stephanie A. N. and McConnachie, Alan W. and Irwin, Mike, 2015, *MNRAS*, 454, 1000

Hashemi S. A., Javadi A., van Loon J. Th., 2018, *MNRAS*, 483, 4751

Javadi A., van Loon J. Th., Mirtorabi M. T., 2011b, *MNRAS*, 414, 3394

Javadi, A., van Loon, J. T., Khosroshahi, H. G., Tabatabaei, F., Hamedani Golshan, R., & Rashidi, M., 2017, *MNRAS*, 464, 2103

Kroupa P., 2001, *MNRAS*, 322, 231

Lindegren, L., Klioner, S. A., Hernández, J., et al., 2021, *A&A*, 649, *A2*

Massey, P. & Armandroff, T. E., 1995, *AJ*, 109, 2470

Navabi, M., et al., 2021, *ApJ*, 910, 2, 127

M. G. Richer, A. Bullejos, J. Borissova, M. L. McCall, H. Lee, R. Kurtev, L. Georgiev, R. L. Kingsburgh, R. Ross, M. Rosado, 2001, *A&A*, 370, 34
Sakai, S., Madore, B. F., & Freedman, W. L. 1999, *ApJ*, 511, 671
Saremi, E., et al., 2020, *ApJ*, 894, 135
Saremi, E., et al., 2021, *ApJ*, 923, 2, 164
Schlegel, D.J., Finkbeiner, D.P., & Davis, M., 1998, *ApJ*, 500, 525
Stetson, P. B., 1987, *PASP*, 99, 191

Resolving the Rise and Fall of Star Formation in Galaxies
Proceedings IAU Symposium No. 373, 2023
T. Wong & W.-T. Kim, eds.
doi:10.1017/S1743921322004781

Effect of gas percentage during minor mergers on the star formation in galaxies

S. N. Hasan[1], Swetha Thakkalapally[2] and Priya Hasan[3]

[1]Department of Mathematics, Maulana Azad National Urdu University, Hyderabad 500032, India,
email: hasan.najam@gmail.com

[2]Department of Astronomy, Osmania University, Hyderabad 500007, India
email: swetha3737@gmail.com

[3]Department of Physics, Maulana Azad National Urdu University, Hyderabad 500032, India,
email: priya.hasan@gmail.com

Abstract. We study the effect of minor mergers on star formation using simulations. We use GADGET4 code which has both collisionless and hydrodynamical particles. Our goal is to establish a relation between gas percentage present in the galaxies and the star formation in the merged galaxy. We use 1:10 minor mergers and we run the isolated simulations with varying gas percentages in the primary galaxy. We observe that the gas particles convert into stars due to the impact of the minor merger. As the gas percentage increases in the primary disk of the galaxy, more number of stars are formed. We also observed that newly formed star particles settle down in the disk of the primary galaxy and increase the thickness of the disk. We also observe that the thickness of the stellar disk containing the old stars also increases due to the impact of the merger.

Keywords. Numerical Simulations, Galaxy Interactions, Mergers.

1. Introduction

Major and minor mergers are the main contributors for the changing morphology of galaxies. Major mergers play an important role in transforming a disk dominated spiral galaxy into a spheroid while minor mergers explain the origin of thick disks and diffuse stellar halos and also contribute to star bursts when sufficient gas is present (Moster *et al.* 2010). In general, minor mergers are more frequent than major mergers in the Universe as there are far more smaller galaxies than the massive ones.

The main goal of our study is to investigate the effect of the presence of gas and it's fraction on the stellar disc during a minor merger. The most popular tool to study galaxy interactions and evolution is numerical simulations. For this we run simulations without gas and with different percentages of gas in the disk of the primary galaxy and study the resultant galaxy after the merger. In this study, we also include star formation during the merger.

Table 1. Simulation parameters of primary and secondary galaxies.

Galaxy	M_{vir} $10^{10}M_\odot$	M_{disc} $10^{10}M_\odot$	M_{bulge} $10^{10}M_\odot$	R_{disc} kpc	R_{bulge} kpc	N_{halo}	N_{disc}	N_{bulge}	SDM kpc	SD kpc	SH kpc
Primary	100	4	1.0	3.0	0.5	10^6	5×10^5	10^5	0.1	0.1	0.05
Secondary	10	0	1.5	0	0.2	10^5	0	5×10^4	0.1	NA	NA

Table 2. Simulation parameters of the primary galaxy in each run.

Primary Galaxy	M_{disc} $10^{10}M_\odot$	$M_{Stellardisc}$ $10^{10}M_\odot$	$M_{gaseousdisc}$ $10^{10}M_\odot$	$N_{Stellardisc}$	$N_{gaseousdisc}$
0%	4×10^{10}	4×10^{10}	0	2×10^5	0
10%	4×10^{10}	3.6×10^{10}	0.4×10^{10}	2×10^5	1×10^5
20%	4×10^{10}	3.2×10^{10}	0.8×10^{10}	2×10^5	2×10^5
30%	4×10^{10}	2.8×10^{10}	1.2×10^{10}	2×10^5	3×10^5
40%	4×10^{10}	2.4×10^{10}	1.6×10^{10}	2×10^5	4×10^5

2. Numerical Simulations

2.1. *Numerical Code and Galaxy Models*

We use the parallel Tree-SPH code GADGET4 (Springel *et al.* 2021) for hydrodynamical simulations†. In this code, gravitational interactions are computed with hierarchical multipole expansion, gas dynamics and smoothed particle hydrodynamics. We include radiative cooling and star formation in our simulations.

For the purpose of simulations we use, the Center for Security, Theory & Algorithmic Research (CSTAR), International Institute of Information Technology, Hyderabad.

We create the initial conditions using DICE (Disk Initial Conditions Environment) code (Perret et al. 2014). We apply the method described in Springel *et al.* (2005) to construct both primary and secondary galaxies. The dark matter halo is modelled using NFW profile (Navarro, Frank and White 1997). The stellar and gaseous disks are modelled by exponential profiles and the bulge is modelled using the Hernquist profile (Hernquist 1990).

2.2. *Simulation Parameters*

We create initial conditions with the primary galaxy at the center and a secondary galaxy at a distance of 100 kpc in all the runs with mass ratio of 1:10 between primary and secondary galaxy. Table 1 summarizes the different parameters used for creating primary and secondary galaxies. These values are kept constant in all the runs. The primary galaxy contains a dark matter halo, stellar and/or gaseous disc and a bulge. The secondary galaxy contains a dark matter halo and a bulge. They have been softened as SDM, SD and SH respectively.

We create five sets of initial conditions with varying gas percentages of 0, 10, 20, 30 and 40 in the disk of the primary galaxy. The mass of the total disk is constant in all the runs but the mass of gaseous disc keeps increasing and accordingly the number of particles in the stellar disc also increases. Table 2 lists the parameters of the primary galaxy in each of the runs. The simulations are run from redshift 1 to redshift 0 (present time) using the GADGET4 code.

† The code discussed in this study has been made publicly available at `https://wwwmpa.mpa-garching.mpg.de/gadget4`.

Table 3. Results.

Simulation	Initial number of gas particles present	Number of gas particles turned to new stars	Percentage of the gas particles turned to stars
10%	1×10^5	37308	37.3%
20%	2×10^5	91395	45.7%
30%	3×10^5	159074	53%
40%	4×10^5	227258	56.8%

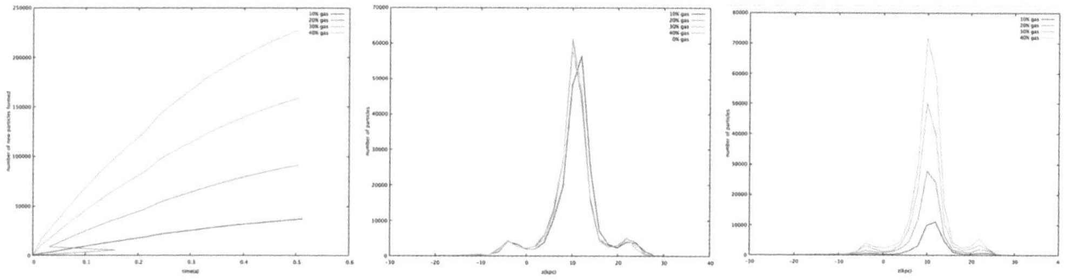

Figure 1. Number of new particles formed in each simulation with respect to time (left), Distribution of old stellar disc particles after the merger(center) and Distribution of newly formed particles in the disc(right).

3. Results and discussion

Table 3 summarizes the results of the new particles formed after the simulation. It is observed that as the percentage of the gas in the primary galaxy increases from 10% – 40%, the number of new stars formed increases from 37.3% – 56.8%.

Figure 1 describes the distribution of old and new stellar particles in the disc after the simulation. As the gas percentage increases, the fraction of gas particles turning into new stars also increase. These new stars formed settle into a thicker disc. It is also observed that as the gas percentage in the initial galaxy is increased, number of new particles formed also increases and so is the thickness of the disc.

4. Acknowledgment

The authors would like to thank Dr. Subhadip Mitra for providing the facilities at the Center for Security, Theory & Algorithmic Research (CSTAR), IIIT, Hyderabad for running our simulations.

References

Springel, Volker, et al. 2021, *MNRAS*, 506, 2, 2871–2949

Perret, V., Renaud, F., Epinat, B., et al. 2014, *A&A*, 562, A1

Navarro, Julio F., Carlos S. Frank, and Simon D.M. White 1997, *ApJ*, 490, 2, 493

Hernquist, Lars 1990, *ApJ*, 356, 359–364.

Springel, Volker, Tiziana Di Matteo, and Lars Hernquist 2005, *MNRAS*, 361, 3, 776–794

Benjamin P. Moster, Andrea V. Macciò, Rachel S. Somerville, Peter H. Johansson, Thorsten Naab 2010, *MNRAS*, 403, 2, 1009–1019

Resolving the Rise and Fall of Star Formation in Galaxies
Proceedings IAU Symposium No. 373, 2023
T. Wong & W.-T. Kim, eds.
doi:10.1017/S1743921322004264

The curious case of NGC 5068 and its neutral hydrogen fingers

J. Healy[1], W.J.G. de Blok[1,2,3] and F.M. Maccagni[1,4]

[1]Netherlands Institute for Radio Astronomy (ASTRON), Postbus 2, 7990 AA Dwingeloo, The Netherlands,

[2]Kapteyn Astronomical Institute, University of Groningen, PO Box 800, 9700 AV Groningen, The Netherlands,

[3]Department of Astronomy, University of Cape Town, Private Bag X3, 7701 Rondebosch, South Africa,

[4]INAF – Osservatorio Astronomico di Cagliari, via della Scienza 5, 09047, Selargius (CA), Italy

Abstract. How galaxies replenish their gas supply in order to sustain star formation, is a research topic of many of the new and upcoming neutral atomic hydrogen (H I) surveys on the SKA precursor instruments. This replenishment, or accretion, of gas in the form of H I is likely to occur at column densities one or two orders of magnitude below previous observational limits and it has, so far, not been unambiguously detected. We present recent deep H I observations of NGC 5068, an isolated nearby star-forming galaxy observed by MeerKAT as part of the MHONGOOSE survey. This survey is the deepest H I survey of nearby galaxies until the advent of the SKA and is reaching column densities of $N_{\rm H\,I} \sim 2 \times 10^{18}\,{\rm cm}^{-2}$ at 30″ resolution. The combination of the resolution and sensitivity of the MeerKAT HI data have revealed "fingers" of low column density gas extending out from the main H I disk of the galaxy. While the origin of these fingers remains a mystery for now, the dynamics of the main galaxy disk and the outer disk in which the fingers are located, as well as the morphology of the fingers, does not seem to suggest a previous merger event.

Keywords. ISM, galaxies, NGC 5068

1. Introduction

Studies of the gas and star formation in galaxies have shown that the observed amount of neutral gas, in particular the neutral atomic hydrogen (H I), is insufficient to sustain the observed star formation (Larson 1972; Sancisi et al. 2008). One of the goals of the MeerKAT HI Observations of Nearby Galactic Objects - Observing Southern Emitters (MHONGOOSE; de Blok et al. 2016, 2020), a large survey project using the MeerKAT Radio Telescope (Jonas & MeerKAT Team 2016), is to observe the low column density H I gas in 30 nearby galaxies to uncover how galaxies sustain their ongoing star formation. In this work, we present deep and detailed new high sensitivity H I observations of one of the MHONGOOSE galaxies, NGC 5068. This galaxy is a nearby ($d = 5.2 \pm 0.22$ Mpc, Anand et al. 2021), face-on spiral galaxy of moderate mass and star formation ($M_\star = 2.57 \times 10^9\,M_\odot$ and SFR $= 0.288\,M_\odot\,{\rm yr}^{-1}$, Leroy et al. 2019, 2021).

2. What the H I reveals about NGC 5068

The deep, high resolution H I data observed by MeerKAT reveals a complex picture of the H I reservoir of NGC 5068. Fig. 1 shows NGC 5068 in gray scale as observed

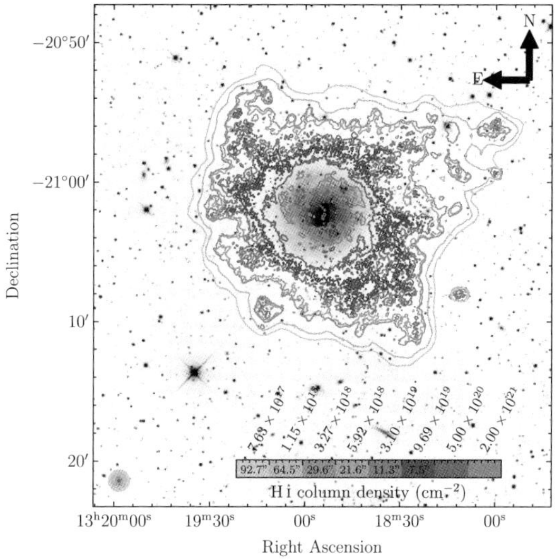

Figure 1. Optical MeerLICHT q-band image of NGC 5068 with 3σ H I column density contours from the different resolutions overlaid.

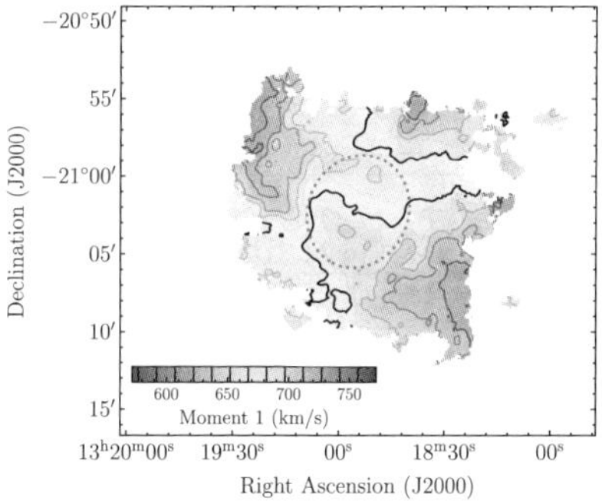

Figure 2. Velocity field (moment 1) for NGC 5068. Contours are spaced at $15\,\mathrm{km\,s^{-1}}$, with the black contour representing the systemic velocity of the galaxy.

by MeerLICHT† in the optical q-band with the overlaid contours indicating the 3σ H I column density ($N_{\mathrm{H\,I}}$) of at the six different resolutions ($N_{\mathrm{H\,I},3\sigma} = 7.76 \times 10^{17}\,\mathrm{cm^{-2}}$ at $90''$ to $N_{\mathrm{H\,I},3\sigma} = 9.6 \times 10^{19}\,\mathrm{cm^{-2}}$ at $7.5''$) of the MeerKAT data. On the edge of the star forming (SF) disk, Fig. 1 shows clumpy "fingers" of low-column density gas extending outwards, as well as low column density clumps to the north west of the galaxy.

The velocity field (Fig. 2) reveals a regular rotating inner disk coincident with the SF disk (indicated by the dotted grey ellipse), suggesting that the inner disk is largely unaffected by whatever is driving complex motion of the outer disk. The velocity dispersion map (Fig. 3) also shows a region of high dispersion gas encircling the SF disk, further suggestion activity affecting gas is in the outskirts. It is clear from the position velocity slice, shown in Fig. 4, extracted from the high dispersion regions (indicated by the black line in Fig. 3) that the high velocity dispersion is in fact driven the presence of clumpy low column density gas.

† MeerLICHT (`http://www.meerlicht.uct.ac.za`) is the 0.6 m optical telescope located in South Africa that is tied to MeerKAT and has the same field of view.

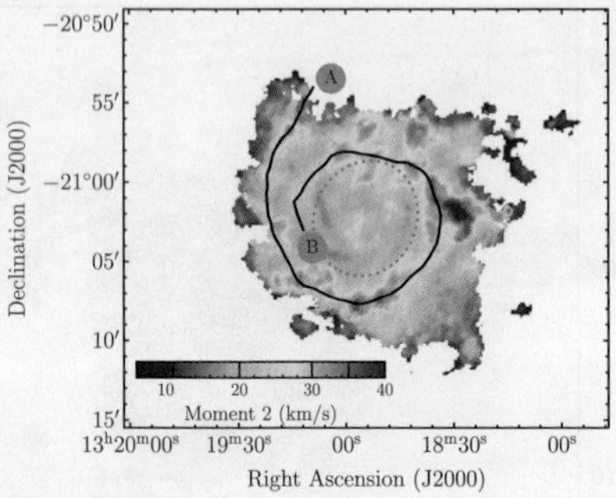

Figure 3. Moment 2 map for NGC 5068. The black line indicates the path over which the position-velocity (PV) slice shown in Fig. 4 is extracted. The "A" and "B" markers correspond to the same markers in the PV diagram.

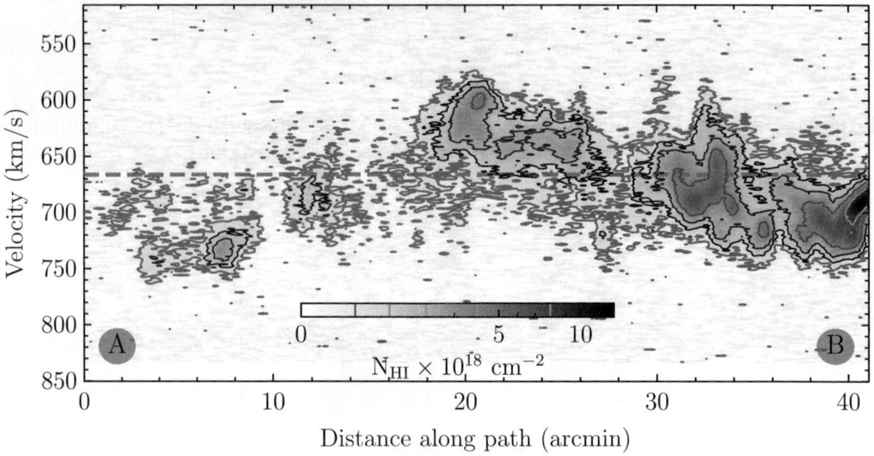

Figure 4. Position velocity slice extracted along the path indicated in Fig. 3. "A" and "B" match the corresponding markers in the moment 2 map indicating the start and end of the PV slice.

3. Conclusions

There are a number of possible origins of the "fingers" and clumpy gas on the north western side of the galaxy. Below, we discuss three possible scenarios:

Previous interaction with a neighbour: despite NGC 5068 being a member of a loose group of galaxies (Pisano et al. 2011; Kourkchi & Tully 2017), a recent ($t < 1$ Gyr) interaction is unlikely as the closest neighbouring galaxy has a projected separation of > 400 kpc. For an imprint of such an interaction to remain in the H I of NGC 5068, the neighbour would need to moving at $v \gtrsim 390\,\mathrm{km\,s^{-1}}$.

Outer gas from a minor merger: this is unlikely as we find low column density gas throughout the H I disk.

Accretion of cold gas: this is a likely scenario as the gas in the outskirts of the H I disk is very clumpy (see the clouds particularly in the north west region of the galaxy in Fig. 1). More evidence to support this scenario is that the high velocity dispersion gas (Fig. 3) is driven by low column density component outside of the SF region. The clumps of gas to the north west of the galaxy are also not rotating with the general rotation of the outer disk or the main SF disk.

Despite the last scenario being the most likely, it is not possible given the currently available data to completely rule out the first two scenarios. Moreover, accretion of gas can also come in two forms: re-accretion of previously expelled gas or accretion of fresh gas from the cosmic web. Disentangling these two requires further modelling and will be an interesting challenge.

The authors acknowledge funding from the European Research Council (ERC) under the European Union's Horizon 2020 research and innovation programme (grant agreement No 882793/MeerGas).

References

Anand, G. S., Lee, J. C., Van Dyk, S. D., et al. 2021, Monthly Notices of the Royal Astronomical Society, 501, 3621

de Blok, W. J., Adams, E. A., Amram, P., et al. 2016, in Proceedings of Science (Sissa Medialab Srl)

de Blok, W. J. G., Athanassoula, E., Bosma, A., et al. 2020

Jonas, J. L. & MeerKAT Team. 2016, in MeerKAT Science: On the Pathway to the SKA, ed. R. Taylor, F. Camilo, L. Leeuw, & K. Moodley (Stellenbosch: Proceedings of Science), 1

Kourkchi, E. & Tully, R. B. 2017, The Astrophysical Journal, 843, 16

Larson, R. B. 1972, Nature, 236, 21

Leroy, A. K., Sandstrom, K. M., Lang, D., et al. 2019, The Astrophysical Journal Supplement Series, 244, 24

Leroy, A. K., Schinnerer, E., Hughes, A., et al. 2021, The Astrophysical Journal Supplement Series, 257, 43

Pisano, D. J., Barnes, D. G., Staveley-Smith, L., et al. 2011, Astrophysical Journal, Supplement Series, 197, 28

Sancisi, R., Fraternali, F., Oosterloo, T., & Van Der Hulst, T. 2008, Astronomy and Astrophysics Review, 15, 189

Resolving the Rise and Fall of Star Formation in Galaxies
Proceedings IAU Symposium No. 373, 2023
T. Wong & W.-T. Kim, eds.
doi:10.1017/S1743921322003829

Formation of the Stellar System and the Central Gravitation-Potential Vessel of Galaxies

Keiichi Kodaira[1,2,3] and Veselina Kalinova[1]

[1]Max Planck Institute for Radioastronomy,
Auf dem Hügel 69, 53121 Bonn, Germany
email: kodaira_keiichi0815@soken.ac.jp

[2]National Astronomical Observatory of Japan,
Osawa2-21-1, Mitaka-shi, Tokyo, Japan PC 181-8588

[3]SOKENDAI, International Village
Hayama-machi, Miura-gun, Kanagawa-ken, Japan, PC 240-0193

Abstract. The formation of the global stellar system of galaxies are studied through the circular velocity curves of CALIFA nearby galaxies by sequencing the depth and size of the central gravitation-potential vessel and its dynamical mass, relative to the masses of the stellar system and of the parent halo, with the population or age parameters, to explore the dynamical characteristics of the dissipative contracting baryonic matter.

Keywords. galaxies: kinematics and dynamics, galaxies: structure, galaxies: evolution

---

1. Stellar mass, dynamical mass, and halo mass

The stellar systems of the present-day normal galaxies reside in the central part of the parent halos under dynamical influences of baryonic matter (stars, dust, gas, black hole) and dark matter, in the central gravitation-potential vessel (CGPV), which was deepened by the dissipative contraction of baryonic matter. The resolved field spectroscopic data of 231 nearby galaxies from CALIFA (Calar Alto Legacy Integral Field Area; Sánchez et al. 2012) survey were used to derive the circular velocity curve (CVC or V_c(R); Kalinova et al. 2017), proxies of the dynamical mass within CGPV (Mdyn; Kalinova et al., in prep.), and mass of the stellar system (M_*; Kalinova et al. 2021). The $M_* - M_{dyn}$ relation (Fig. 1, left) shows bimodality, similar to the $M_* - M_{halo}$ relation (Fig. 1, right) after the stellar-to-halo relation of Moster et al. (2013). We find an empirical relationship between the dynamical mass and the mass of the parent halo,

$$\log M_{dyn} = 0.68 \log M_{halo} + 3.1 \tag{1.1}$$

which is mono-modal in contrast to other relations typically bimodal, a low-mass branch for young disk-type galaxies and a high-mass branch for aged ellipsoid-type galaxies with the mode-transition at $\log M_* \sim 10.7$ ($\log M_{dyn} \sim 11.1$, $\log M_{halo} \sim 12.3$), suggesting that M_{dyn} might be one of important parameters to inter-relate the galaxy and the parent halo.

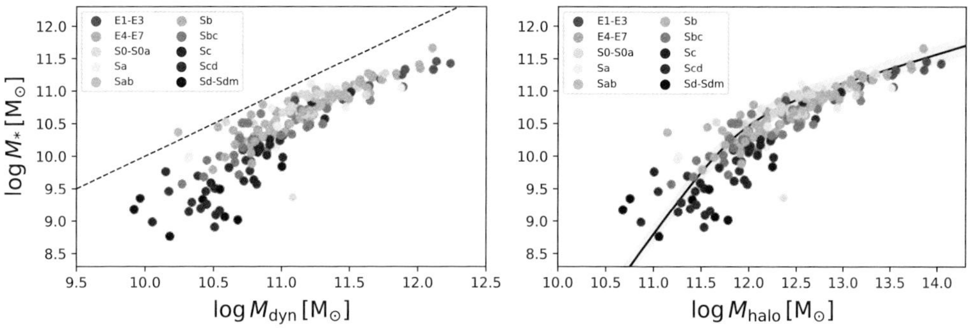

Figure 1. CALIFA-galaxy points are transferred from the $M_* - M_{dyn}$ relation (left) to the $M_* - M_{halo}$ relation (right) by applying the transformation of eq. 1.1, and fitted by the stellar-to-halo relation (black continuous line with an uncertainty grey band) of Moster et al. (2013). The dotted line in the left panel represents the one-one relation in the $M_* - M_{dyn}$ plane, while the colour of the points indicate the morphological type of the galaxies.

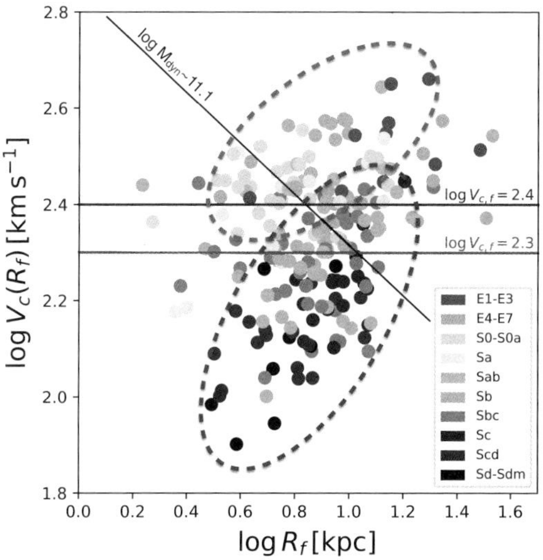

Figure 2. Amplitude versus size parameter diagram of the dynamical mass.

2. Bimodality and the dichotomy around the mode transition in the amplitude and size diagram of the CGPV

With CVC in hand, we introduce a proxy of the dynamical mass in the CGPV, $M_{dyn,f}$, with the standard analytical dependence on $\eta V^2(R/G)$ (V: velocity parameter, R: length parameter, G : gravitation constant, η : scaling factor);

$$\log M_{dyn,f} = 2 \log\, V_c(R_f) + \log\, R_f - \log G, \qquad (2.1)$$

$R_f \equiv 1.5 R_e$ (R_e–the effective radius of the galaxy), where the CVCs become flat and CALIFA velocity data are of high quality.

This proxy of the dynamical mass, $M_{dyn,f}$, is compared with the global dynamical mass M_{dyn}, to confirm the finding by Aquino-Ortiz et al. 2018 that all of the well devised dynamical-mass proxies can be regarded to be statistically the same, except for the scaling factor, η (in the present case $\eta \sim -0.15$). Taking the advantage of the analytical expression of $M_{dyn,f}$, we explore the nature of the bi-modality of the dynamical structure in terms of the potential-amplitude index, $V_c(R_f)$, and the size index, R_f (Fig. 2).

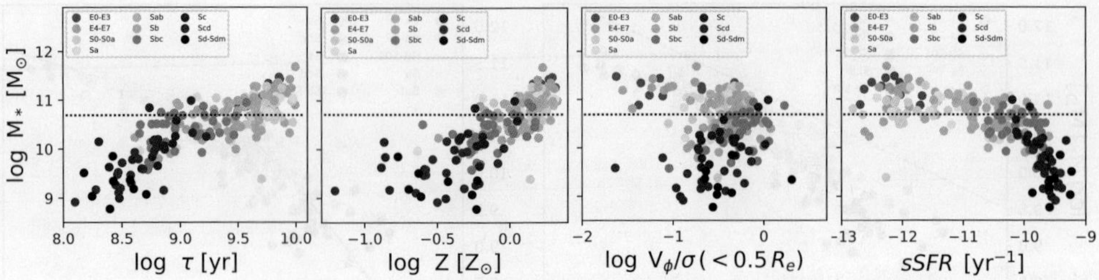

Figure 3. log M_*-sequence diagram (from left to right) of formation age (log τ), metallicity (Z), ordered-over-random motion (V_ϕ/σ) within 0.5 R_e, and specific star-formation rate (sSFR). The horizontal line indicates the changing mode at log M_* ~10.7 M$_\odot$.

The mode-transition is seen around log $V_c(R_f)$ ~2.4 (blue line), changing the dynamical structure from the disk-type to the ellipsoid-type. The transition of the dynamical structure occurs in the $M_* - M_{dyn}$ diagram (Fig. 1) around log M_{dyn} ~11.1, which is indicated in Fig. 2 (black line), traversing the low-mass part of the red sequence and the highest-mass part of the blue sequence. The transition domain between the bimodal branches shows a kind of dichotomy, depending not only on the mass-concentration degree towards the center but also on the second parameter involving rotation characteristics to form the disk-system.

3. Characteristics of bimodal stellar systems and of the mode transition

In order to find any clues about the baryon-condensation history in CGPV, we examine the M_*- dependences of the population age τ, the heavy-element index Z, the ordered-over-random motion V_ϕ/σ in $R_e < 0.5$ (where V_ϕ is the azimuthal velocity and σ is the line-of-sight velocity), and the specific star-formation rate (sSFR) of the nearby CALIFA galaxies (Fig. 3) (parameter data are taken from Kalinova et al. 2017). Our findings are:

(1) High-mass (log $M_* > \sim 10.7$) spheroidal stellar systems (e.g. E0–E7) with dominant dispersion-motion were formed in early epochs, but later with poor baryon supply leading to no evident star-formation activity. They are embedded in correspondingly high-mass CGPVs (log $M_{dyn} > \sim 11.1$) and high-mass halos (log $M_{halo} > \sim 12.3$).

(2) Low-mass (log $M_* < \sim 10.7$) disk stellar systems (Sc–Sdm) with highly ordered rotational motion were formed in late epochs with continuous baryon supply to keep further star-formation activity. They are embedded in low-mass CGPVs (log $M_{dyn} < \sim 11.1$) and in low-mass halos (log $M_{halo} < \sim 12.3$). Among them, the CALIFA galaxies with the lowest mass (log $M_* < \sim 9.7$, Scd–Sdm) are the youngest, and the star-formation activity is high in the shallow CGPVs, which might have resulted in the slight decrease of the V_ϕ/σ ratio.

(3) The transition zone between the above two dominant dynamical types of stellar systems lies around log M_* ~10.7 (log $M_{dyn} \sim$ 11.1, corresponding to log $M_{halo} \sim$ 12.3), where we notice a kind of dichotomy: On one side baryon is condensed into prominent (bulge + disk) stellar systems (e.g. Sab–Sbc) with strong ordered motion, while, on the other side at earlier epoch than the former, almost same amount of baryon condensed into the stellar systems with a dominant ellipsoidal/spheroidal component (e.g. S0–Sa), in which marginal-present gas-disks are prevented from active star formation, apparently due to the stabilizing gravitation-potential of the strong-concentrated stellar systems.

This mass-dependence study indicates that initial baryon infall in massive halos produced the central ellipsoidal stellar system whose dynamics/activity hindered succeeding in-fall of angular-momentum-rich baryonic matter to form the rotating disk, while in low-mass halos not enough low-angular-momentum baryon was available to form the dominant central ellipsoidal stellar system. These processes inevitably depend on the varying dynamical characteristics, in particular on the angular-momentum spectra, of the infalling baryonic matter of the parent halos, influenced by the merging/accretion prehistory. The present study suggests that massive halos tend to have more baryonic matter of low specific angular-momentum relative to low-mass halos.

References

Moster, B. P., Naab, T., & White, S. D. M. 2013, *MNRAS*, 428, 3121
Sánchez, S. F., Kennicutt, R. C., Gil de Paz, A., et al. 2012, *A&A*, 538, A8
Aqúino-Ortiz, E., Velenzuela, O., Sánchez, S. F., et al., 2018, *MNRAS*, 479, 2
Kalinova, V., Colombo, D., Rosolowsky, E., et al., 2017, *MNRAS*, 469, 3
Kalinova, V., Colombo, D., Sánchez, S. F., et al., 2021, *A&A*, 648, A64

Resolving the Rise and Fall of Star Formation in Galaxies
Proceedings IAU Symposium No. 373, 2023
T. Wong & W.-T. Kim, eds.
doi:10.1017/S1743921323000303

Revealing the Relation between Star Formation Activity of Jellyfish Galaxies and Ram Pressure Stripping

Jeong Hwan Lee[1,2], Myung Gyoon Lee[1], Jae Yeon Mun[3], Brian S. Cho[1] and Jisu Kang[1]

[1]Seoul National University, Seoul 08826, Republic of Korea

[2]Kyungpook National University, Daegu 41566, Republic of Korea

[3]Australian National University, Canberra, ACT 2611, Australia

Abstract. Jellyfish galaxies are starburst galaxies with ram-pressure-stripped tails and blue star-forming knots. These galaxies show a snapshot of star formation enhancement triggered by ram pressure stripping (RPS), being important targets for studying the RPS-induced star formation in gas-rich galaxies. Here we investigate the star formation activity of five jellyfish galaxies in massive clusters, using Gemini GMOS/IFU observations. From the Hα-derived star formation rates (SFRs), we find that our sample shows higher SFR excess to the star formation main sequence than the jellyfish galaxies in low-mass clusters. From the compiled sample of jellyfish galaxies in low-mass to high-mass host clusters, we suggest that the star formation activity of jellyfish galaxies has positive correlations with host cluster mass and degree of RPS. These relationships imply that higher ram pressure environments tend to trigger stronger starbursts in jellyfish galaxies in the early stage of RPS.

Keywords. Galaxy evolution; Galaxy environments; Ram pressure stripping; Starburst galaxy; Jellyfish galaxy; Integral Field Spectroscopy

1. Introduction

Ram pressure stripping (RPS; Gunn & Gott 1972) is the gas stripping mechanism related with the interaction between gas and intracluster medium (ICM). This process is one of the main drivers of galaxy evolution in dense environments. In the long-term view, RPS quenches the star formation activity by transforming gas-rich galaxies into gas-deficient galaxies. However, RPS enhances the star formation activity of galaxies within a few hundred Myrs since the first infall to the cluster center (Bekki & Couch 2003). For instance, some ram-pressure-stripped galaxies show star-forming regions in their disks and tails, exhibiting "jellyfish-like" morphology (Ebeling et al. 2014). Thus, these jellyfish galaxies are ideal laboratories for studying the relationship between star formation activity and RPS.

Integral-field spectroscopy (IFS) has been widely used to investigate the star formation activity of jellyfish galaxies. The GAs Stripping Phenomena in Galaxies with MUSE (GASP; Poggianti et al. 2017) is one of the most successful IFS survey observing jellyfish galaxies at low redshift ($\langle z\rangle \sim 0.05$), demonstrating the star formation enhancement of the GASP jellyfish galaxies compared to the non-RPS control sample (Vulcani et al. 2018). However, previous IFS studies have been limited to jellyfish galaxies in relatively low-mass clusters, which are likely to show mild RPS effects on galaxies. In contrast,

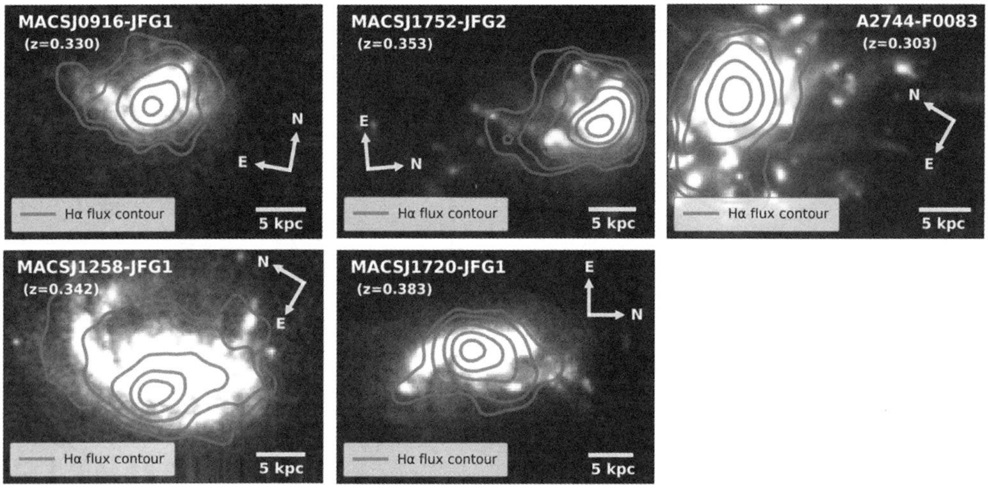

Figure 1. The Hubble Space Telescope (HST) optical images of the five jellyfish galaxies in this study. The Hα flux distributions observed by the Gemini GMOS/IFU are overlaid with magenta lines.

massive clusters at intermediate redshift are expected to exert much more extreme RPS to galaxies due to high ICM density and dynamical instability (McPartland et al. 2016). Thus, jellyfish samples in massive clusters can be helpful to reveal the relation between the RPS-driven star formation activity and their host cluster properties.

In this study, we investigate the relation of star formation rates (SFRs) of jellyfish galaxies with their host cluster velocity dispersion ($\sigma_{v,\mathrm{cl}}$) and ram pressure (P_{ram}). We compile the Hα-derived star formation rates (SFRs) of jellyfish galaxies in clusters with a wide range of cluster velocity dispersion from 500 km s^{-1} to 1500 km s^{-1}. For the jellyfish sample in massive clusters, we use the Gemini GMOS/IFU data of five jellyfish galaxies represented in Figure 1 (Lee et al. 2022a). These GMOS/IFU observations can provide a great opportunity to understand how strong RPS in massive clusters affect the star formation activity in jellyfish galaxies.

2. Data and Methods

We obtained the GMOS/IFU data of five jellyfish galaxies in the MAssive Cluster Survey (MACS; Ebeling et al. 2001) and Hubble Frontier Fields (HFF; Lotz et al. 2017) clusters at $z > 0.3$ with four Gemini science programs since 2019 March (27.4 hours in total, PI: J. H. Lee). We derived SFRs from the extinction-corrected Hα luminosity adopting the dust extinction laws in Cardelli et al. (1989) and the initial mass function in Chabrier (2003), which are consistent with the methods for SFR estimation in the GASP studies. Along with our GMOS/IFU results, we also collected the Hα-derived SFRs of jellyfish galaxies in the GASP clusters (Gullieuszik et al. 2020), A901/2 multicluster (Roman-Oliveira et al. 2019), and A1758N cluster (Ebeling & Kalita 2019). The MACS and HFF clusters show higher cluster velocity dispersions ($\sigma_{v,\mathrm{cl}} \gtrsim 1000$ km s^{-1}) than the GASP clusters ($\langle\sigma_{v,\mathrm{cl}}\rangle = 730$ km s^{-1}) and A901/2 ($\langle\sigma_{v,\mathrm{cl}}\rangle = 800$ km s^{-1}), implying that the five jellyfish galaxies in this study are subject to stronger RPS than those in low-mass clusters. The ram pressure exerted on individual galaxies ($P_{\mathrm{ram}} \sim \rho_{\mathrm{ICM}}\Delta v^2$) was calculated using the scaling relation between the cluster velocity dispersion and ICM density, as suggested in Gullieuszik et al. (2020). Lee et al. (2022a) described more detailed information about data and methods.

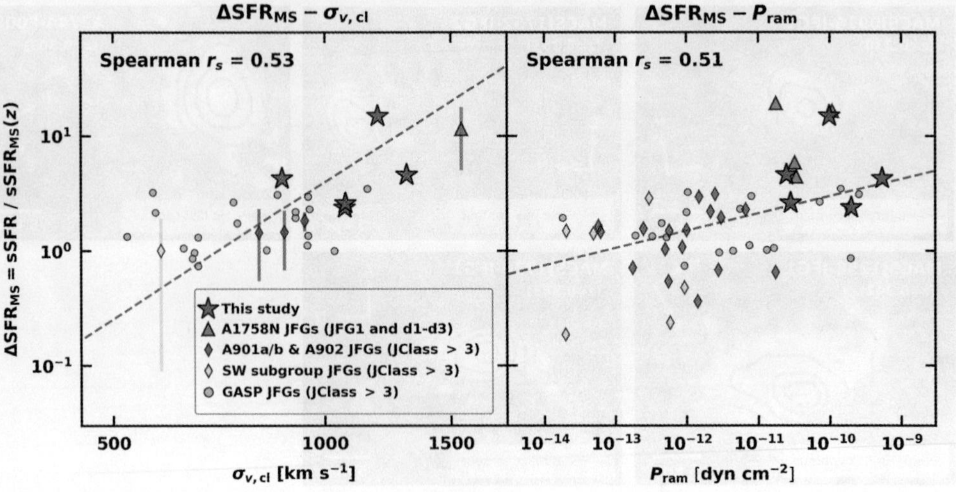

Figure 2. SFR excess to the star formation main sequence (ΔSFR_{MS}) of jellyfish galaxies as functions of host cluster velocity dispersion (left) and ram pressure (right). Gray dashed lines denote the results of linear fitting for the relations.

3. Results and Discussions

Figure 2 illustrates the main results of this study (Lee et al. 2022b). We estimate the SFR excess (ΔSFR_{MS}) relative to the star formation main sequence at the redshift of each host cluster (Speagle et al. 2014) as an indicator of RPS-induced star formation activity in jellyfish galaxies. We plot the SFR excess as functions of host cluster velocity dispersion (*left*) and ram pressure on each galaxy (*right*). We select the jellyfish galaxies with clear RPS signatures, using a visual classification index (JClass > 3) used in Poggianti et al. (2017).

This figure shows that the SFR excess of jellyfish galaxies positively correlates with cluster velocity dispersion and ram pressure. This implies that the star formation activity of jellyfish galaxies is more enhanced as they are in more massive clusters and undergo stronger RPS. In massive clusters, an increase of ram pressure due to dense ICM and cluster dynamics (e.g., cluster collisions or mergers) can lead to efficient gas compression and intense starburst in gas-rich galaxies (Owers et al. 2012; Roediger et al. 2014). Recently, Roberts et al. (2022) also found a positive correlation between host halo mass and SFR excess of ram-pressure-stripped galaxies in the Coma cluster and the SDSS groups/clusters. We suggest that these correlations show the short-term triggering effect of RPS on the star formation activity of jellyfish galaxies. In the future, more IFS studies of jellyfish samples in various environments would be helpful to strengthen the trends between the star formation activity and RPS.

Supplementary material

To view supplementary material for this article, please visit https://doi.org/10.1017/S1743921323000303.

References

Bekki, K. & Couch, W. J. 2003, *ApJL*, 596, L13
Cardelli, J. A., Clayton, G. C., & Mathis, J. S. 1989, *ApJ*, 345, 245
Chabrier, G. 2003, *PASP*, 115, 763
Ebeling, H., Edge, A. C., & Henry, J. P. 2001, *ApJ*, 553, 668
Ebeling, H., Stephenson, L. N., & Edge, A. C. 2014, *ApJL*, 781, L40

Ebeling, H. & Kalita, B. S. 2019, *ApJ*, 882, 127
Gullieuszik, M., Poggianti, B. M., McGee, S. L., et al. 2020, *ApJ*, 899, 13
Gunn, J. E. & Gott, J. R. 1972, *ApJ*, 176, 1
Lee, J. H., Lee, M. G., Mun, J. Y., et al. 2022a, *ApJ accepted*, arXiv:2209.07189
Lee, J. H., Lee, M. G., Mun, J. Y., et al. 2022b, *ApJL*, 931, L22
Lotz, J. M., Koekemoer, A., Coe, D., et al. 2017, *ApJ*, 837, 97
McPartland, C., Ebeling, H., Roediger, E., et al. 2016, *MNRAS*, 455, 2994
Owers, M. S., Couch, W. J., Nulsen, P. E. J., et al. 2012, *ApJL*, 750, L23
Poggianti, B. M., Moretti, A., Gullieuszik, M., et al. 2017, *ApJ*, 844, 48
Roberts, I. D., Parker, L. C., Gwyn, S., et al. 2022, *MNRAS*, 509, 1342
Roediger, E., Bruggen, M., Owers, M. S., et al. 2014, *MNRAS*, 443, L114
Roman-Oliveira, F. V., Chies-Santos, A. L., Rodríguez del Pino, B., et al. 2019, *MNRAS*, 484, 892
Speagle, J. S., Steinhardt, C. L., Capak, P. L., et al. 2014, *ApJS*, 214, 15
Vulcani, B., Poggianti, B. M., Gullieuszik, M., et al. 2018, *ApJL*, 866, L25

Resolving the Rise and Fall of Star Formation in Galaxies
Proceedings IAU Symposium No. 373, 2023
T. Wong & W.-T. Kim, eds.
doi:10.1017/S1743921322004392

The Impact of the Group Environment on the Molecular Gas and Star Formation Activity

Bumhyun Lee[1,2,3], Jing Wang[2], Aeree Chung[3], Luis C. Ho[2,4], Juan Molina[5,2], Yongjung Kim[6], Shun Wang[2,4], Bi-Qing For[7,8], Bärbel S. Koribalski[9,10], Kristine Spekkens[11], Albert Bosma[12], Benne W. Holwerda[13] and Lourdes Verdes-Montenegro[14]

[1]Korea Astronomy and Space Science Institute, 776 Daedeokdae-ro, Daejeon 34055, Republic of Korea
bhlee301@gmail.com

[2]Kavli Institute for Astronomy and Astrophysics, Peking University, Beijing 100871, People's Republic of China

[3]Department of Astronomy, Yonsei University, 50 Yonsei-ro, Seodaemun-gu, Seoul 03722, Republic of Korea

[4]Department of Astronomy, School of Physics, Peking University, Beijing 100871, People's Republic of China

[5]Department of Space, Earth and Environment, Chalmers University of Technology, Onsala Space Observatory, 439 92 Onsala, Sweden

[6]Department of Astronomy and Atmospheric Sciences, College of Natural Sciences, Kyungpook National University, Daegu 41566, Republic of Korea

[7]International Centre for Radio Astronomy Research (ICRAR), The University of Western Australia, 35 Stirling Highway, Crawley WA 6009, Australia

[8]ARC Centre of Excellence for All-Sky Astrophysics in 3 Dimensions (ASTRO 3D), Australia

[9]CSIRO Space and Astronomy, Australia Telescope National Facility, P.O. Box 76, NSW 1710, Australia

[10]School of Science, Western Sydney University, Locked Bag 1797, Penrith, NSW 2751, Australia

[11]Royal Military College of Canada, PO Box 17000, Station Forces, Kingston, Ontario, K7K7B4, Canada

[12]Aix Marseille Univ, CNRS, CNES, LAM, Marseille, France

[13]University of Louisville, Department of Physics and Astronomy, 102 Natural Science Building, 40292 KY Louisville, USA

[14]Instituto de Astrofísica de Andalucía (CSIC), Spain

Abstract. At least half of the local galaxies reside in galaxy groups, which indicates that the group is the common environment where galaxies evolve. Therefore, it is important to probe how significantly galaxies are affected by group environmental processes, in order to obtain a better understanding of galaxy evolution. We carried out a new CO imaging survey for 31 galaxies in the IC 1459 and NGC 4636 groups, using the Atacama Compact Array, to study the effect of the group environment on the molecular gas properties and the star formation activity. With our resolved CO data, combined with high-resolution H I images, we find asymmetric CO and H I distributions in the group galaxies. Compared to isolated galaxies, group members have relatively low molecular gas fraction and low star formation rate. These results suggest that the group environment can change the properties of cold gas components and star formation in group galaxies.

Keywords. Molecular gas, Galaxies, Galaxy groups, Galaxy evolution

1. Introduction

Galaxy groups are common places where at least 50% of a whole population of galaxies in the local universe live and evolve. In particular, many galaxies experience external perturbations for the first time in the group environment before they join larger structures such as galaxy clusters. To begin with, tidal interactions among group galaxies are expected to commonly occur due to the low velocity dispersion of the group. In addition, the detection of X-ray in some groups indicates that a hot intragroup medium (IGrM) can also affect galaxies, mostly the interstellar medium (e.g., Cortese, Catinella, & Smith 2021). Those environmental processes can change the physical properties of group galaxies, making their color redder, and stopping their star formation. The combination of these processes can also account for many quenched galaxies at the outskirts of a cluster (known as "preprocessing", Morokuma-Matsui et al. 2021). Therefore, probing the influence of the group environment on member galaxies is a key to better understanding galaxy evolution in general. Recently, we carried out a CO imaging study of group galaxies to investigate how the group environment affects the cold molecular gas content and star formation processes. In this work, we present CO images of group members showing signs of external perturbations, together with H I images. We also compare the global properties (e.g., star formation rate (SFR), H_2 gas fraction) of group galaxies with those of isolated galaxies. The details of this work are presented in Lee et al. (2022).

2. Sample and Observations

From the Group Evolution Multiwavelength Study (GEMS)-H I survey (Kilborn et al. 2009) with the Parkes radio telescope, we selected 31 target galaxies with H I detections from the IC 1459 (I1459G) and NGC 4636 (N4636G) groups. Recent H I observations with the Australian Square Kilometre Array Pathfinder (ASKAP, Serra et al. 2015; Koribalski et al. 2020) covering these two groups provide high-resolution imaging data which we use to identify the environmental effects. Our sample includes dwarf and spiral galaxies with stellar masses ranging from $1.6 \times 10^6 - 4.9 \times 10^{10}\ M_\odot$. Using the Atacama Compact Array (ACA), we observed ^{12}CO(J=1–0) of 31 group members from October 2019 to December 2019. To detect the edge of the molecular disk, we aimed to reach a 5σ gas surface density sensitivity of $4\ M_\odot\ \mathrm{pc}^{-2}$.

3. Results

3.1. *Peculiar CO and* H I *structures*

Sixteen out of the 31 galaxies in these two groups observed were detected in CO, such that imaging data could be obtained at ~1 kpc scale. Figure 1 displays both CO (color scale) and H I (contours) distributions of two interesting cases (IC 5264 and NGC 4632) that show signs of violent external perturbations (for other group members, see Lee et al. 2022). IC 5264 (1459G) has an asymmetric CO distribution with CO disk size of 6.5 kpc in the west and 7.5 kpc in the east. The CO disk is also more extended toward the south. The H I morphology is similar to the CO morphology, showing an extended H I tail-like structure in the east and a marginally truncated H I disk in the west. IC 5264 is thought to be experiencing both ram pressure stripping and tidal interaction. NGC 4632

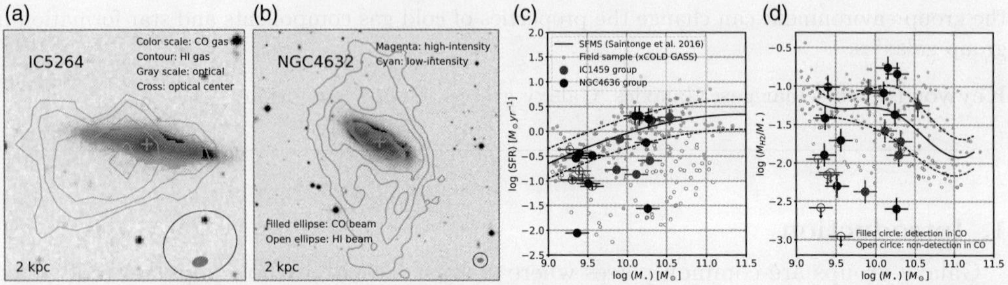

Figure 1. (a) & (b): CO (color scale) and H I (contours) distributions of two group members (left: IC 5264 (I1459G) and right: NGC 4632 (N4636G)) overlaid on the DSS2 blue images. (c) & (d): comparisons of SFR and H_2 gas fraction ($M_{H2}/M_\star$) between our group sample and the field sample. In panel (c), the star-forming main sequence (Saintonge et al. 2016) and the 0.3 dex scatter are shown in a solid line and dashed lines, respectively. In panel (d), a solid line and dashed lines indicate the median trend and the 1σ standard deviation, respectively, which are derived from the field sample. This figure is reconstructed from results presented in Lee et al. (2022).

(N4636G) has a large H I polar ring-like structure. Its CO distribution is irregular with a more extended disk in the east and a local peak. For this galaxy, tidal interaction or a minor merger may cause these complex structures of H I and CO.

3.2. *Comparisons of SFR and H_2 gas fraction between group and isolated galaxies*

We compare the SFR and H_2 gas fraction ($M_{H2}/M_\star$) of our group members with those of 240 isolated galaxies (i.e., the field sample) from the extended CO Legacy Database for GASS (xCOLD GASS; Saintonge et al. 2017). In Figure 1(c), the fraction (48%) of the star-forming galaxies in our group sample is similar to that (50%) in the field sample. However, many low-mass members (log $(M_\star/M_\odot) < 10$) in our group sample have suppressed SFRs (i.e., below the star-forming main sequence). For the H_2 gas fraction (Figure 1(d)), our group sample shows a high fraction (57%) of H_2 deficient galaxies compared to the field sample (26%). In particular, most low-mass group galaxies tend to be deficient in H_2. Interestingly, these low-mass galaxies are not deficient in H I. One possible scenario is that the conversion of H I to H_2 in the low-mass members is severely suppressed due to the environmental effects, resulting in low SFRs of the low-mass group galaxies. Further discussion of this is given in Section 5.2.1 of Lee et al. (2022).

The results in this work indicate that the group environmental effects can change not only the cold gas (CO and H I) distribution but also the global properties (e.g., SFR and H_2 gas fraction) of group members. This is particularly true for the N4636G, which is falling into the Virgo cluster. Our findings from the N4636G seem supporting evidence for the group preprocessing, but this study has been conducted with a small sample (only 31 galaxies). Follow-up studies observing various group members are required to obtain a more concrete and robust result of group environmental effects. In particular, more CO imaging observations with the ACA will be very helpful to reveal the environmental effects on the molecular gas of group galaxies.

Acknowledgements. This work was supported by the Korea Astronomy and Space Science Institute grant funded by the Korea government(MSIT) (Project No. 2022-1-840-05). Support for this work was also provided by the National Research Foundation of Korea to the grant No. 2018R1D1A1B07048314. BL acknowledges support from the National Science Foundation of China (12073002, 11721303, 11991052) and the National Key R&D Program of China (2016YFA0400702). LVM acknowledges financial support from grants SEV-2017-0709, RTI2018-096228-B-C31 (MCIU/AEI/FEDER,UE), and IAA4SKA (Ref. R18-RT-3082).

References

Cortese L., Catinella B., Smith R., 2021, *PASA*, 38, e035
Kilborn V. A., Forbes D. A., Barnes D. G., 2009, *MNRAS*, 400, 1962
Koribalski B. S., Staveley-Smith L., Westmeier T., *et al.*, 2020, *Ap&SS*, 365, 118
Lee B., Wang J., Chung A., *et al.*, 2022, *ApJS*, 262, 31
Morokuma-Matsui K., Kodama T., Morokuma T., *et al.*, 2021, *ApJ*, 914, 145
Saintonge A., Catinella B., Cortese L., *et al.*, 2016, *MNRAS*, 462, 1749
Saintonge A., Catinella B., Tacconi L. J., *et al.*, 2017, *ApJS*, 233, 22
Serra P., Koribalski B., Kilborn V., *et al.*, 2015, *MNRAS*, 452, 2680

Resolving the Rise and Fall of Star Formation in Galaxies
Proceedings IAU Symposium No. 373, 2023
T. Wong & W.-T. Kim, eds.
doi:10.1017/S1743921322003805

Colour gradients of low-redshift galaxies in the DESI Legacy Imaging Survey

Li-Wen Liao[1,2] **and Andrew P. Cooper**[1,2,3]

[1]Institute of Astronomy and Department of Physics, National Tsing Hua University, Hsinchu 30013, Taiwan

[2]Center for Informatics and Computation in Astronomy, National Tsing Hua University, Hsinchu 30013, Taiwan

[3]Physics Division, National Center for Theoretical Sciences, Taipei 10617, Taiwan

Abstract. Radial colour gradients within galaxies arise from gradients of stellar age, metallicity, and dust reddening. Large samples of colour gradients from wide-area imaging surveys can be used to constrain galaxy formation models. Here we measured colour gradients for low-redshift galaxies using photometry from the 9th DESI Legacy Imaging Survey (LS), which reaches $r \sim 24$ over ~14,000 deg^2. We investigate empirical relationships between colour gradients, $M_\star$, and sSFR. We compared our results with the prediction of the Illustris TNG-100 simulation using SDSS mock images.

Keywords. galaxies: formation, galaxies: statistics, galaxies: structure, galaxies: general

---

1. Introduction

Radial colour gradients in galaxies arise from age and metallicity gradients (Tinsley & Larson 1978). These gradients are closely related to the distribution of the stellar population, which is influenced by galaxy assembly history and formation processes. Therefore, colour gradients are expected to vary with observational parameters, such as colour, mass, and sSFR (e.g., de Jong 1996; Tortora et al. 2010). For example, late-type galaxies are expected to grow 'inside out' through gradual gas accretion, leading to the result that old stars centralise at a centre region of galaxies and producing a negative colour gradient. For early-type galaxies major mergers could redistribute the stellar component and result in relatively flat colour gradients.

In our work, we used the 9th data release from the DESI Legacy Imaging Survey (Dey et al. 2019), which has a coverage of ~14,000 deg^2 and provides catalogues containing effective radius (R_{eff}), and fluxes within circular apertures with radii [0.5, 0.75, 1.0, 1.5, 2.0, 3.5, 5.0, 7.0] arcsec around each source. We plot surface brightness profiles and colour profiles computed from these aperture fluxes. Colour gradients are defined as $\nabla_{X-Y} = \delta(X-Y)/\delta(R/R_{\text{eff}})$, the slope of $X-Y$ colour versus R/R_{eff}. We choose $R_1 = 0.5R_{\text{eff}}$ and $R_2 = R_{\text{eff}}$. Galaxies will have a negative colour gradient if their central region is more red than their outer part. Galaxies will have a negative colour gradient. The spectroscopic redshifts of galaxies are obtained from the SDSS. Details about sample selection can be found in Liao & Cooper (2022).

2. Colour gradient as a function of $M_\star$ and sSFR

Fig. 1 shows $g-r$ colour gradients (∇_{g-r}) versus stellar mass ($M_\star$) and specific star formation rate (sSFR). We divided galaxies into red and blue using a colour magnitude

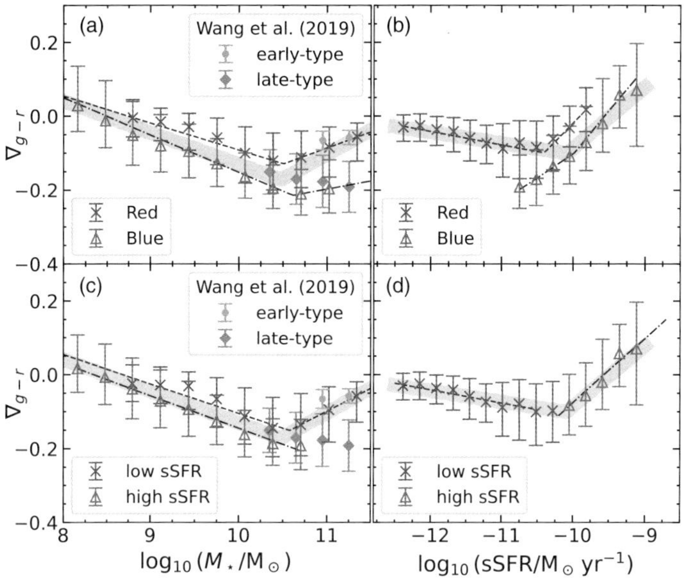

Figure 1. Colour gradients as a function of $M_\star$ and sSFR for red/blue galaxies (upper panels) and high-/low-sSFR galaxies (lower panels). The error bars indicate the 25-75 percentile range. The solid diamond and dots show colour gradients computed from the Wang et al. (2019).

diagram, and high-sSFR and low-sSFR galaxies using their sSFRs. The results show that the colour gradients of galaxies with $M_\star$ less than $10^{10.5}\,\mathrm{M}_\odot$ decrease (become steeper) as the stellar mass increases. This colour gradient trend holds for red/blue and high-sSFR/low-sSFR galaxies, suggesting that galaxies may undergo a similar formation process in this regime. One broadly accepted scenario is the "inside-out" formation scenario, in which galaxies form gradually as gas is accreted and cooled in growing dark matter halos (Larson 1974; De Lucia & Blaizot 2007). The oldest stars are located at the centres of galaxies, in the earliest-forming and deepest part of the gravitational potential. In contrast, younger stars are formed with lower binding energy, with the support of angular momentum. This age difference creates negative colour gradients.

For galaxies with $M_\star > 10^{10.5}\,M_\odot$, colour gradients are flatter with increasing mass, especially for red/low-sSFR galaxies. The inflection of colour gradient trends around $M_\star \sim 10^{10.5}\,M_\odot$ coincides with a critical host halo virial mass, above which the AGN feedback is thought to dominate for galaxies massive (Kauffmann et al. 1996; Guo & White 2008). The suppression of the halo cooling flow by AGN reduces the star-formation rate of galaxies (quenching). Therefore, for galaxies above the transition mass, major mergers drive their subsequent stellar mass growth. During major mergers, the stellar components can be redistributed through violent relaxation, resulting in colour gradients that flatten with successive generations of merging. An alternative explanation for the flattening trends of colour gradient to high mass is that, in a passively evolving galaxy, a fixed age gradient results in a flatter colour gradient for larger absolute age, because younger populations redden more rapidly.

3. Comparision to Illustris TNG100

The Next Generation Illustris Simulation, IllustrisTNG, is a suite of cosmological megnetohydrodynamic simulations (Pillepich et al. 2018; Springel et al. 2018; Nelson et al. 2018; Naiman et al. 2018; Marinacci et al. 2018). To compare our results with these simulations, we measured the colour gradients of galaxies with mock SDSS *griz* images

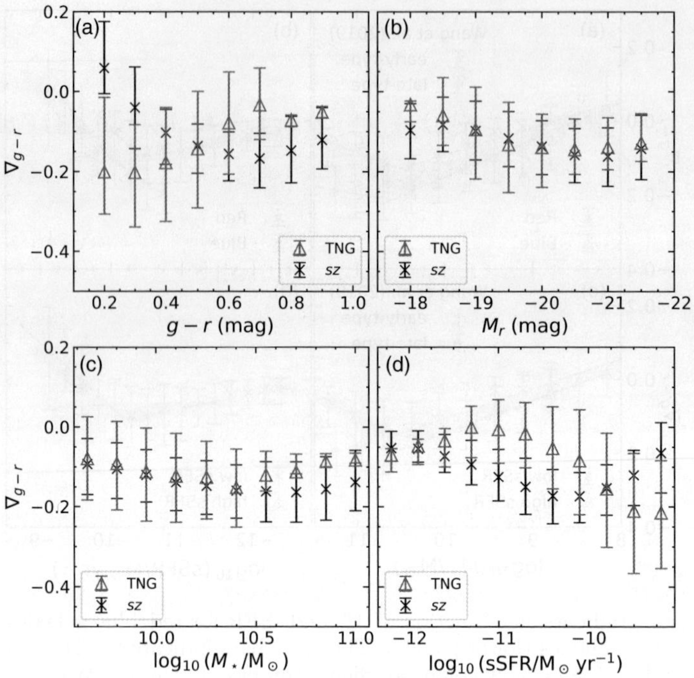

Figure 2. ∇_{g-r} for TNG100 (triangles) and out mass-limited sample (crosses) as a function of (a) $g-r$, (b) M_r, (c) $M_\star$, and (d) sSFR. The error bar label 25th and 75th percentile.

in the dataset provided by Rodriguez-Gomez et al. (2017). In Fig. 2, we plot the colour gradients of galaxies in IllustrisTNG as a function of $g-r$, M_r, $M_\star$ and sSFR. The trends of $g-r$ and sSFR are very different to the observational results. There may be many reasons for these differences; we speculate that the discrepancy could be mitigated by a decrease in the late-time sSFR of TNG galaxies. In contrast to $g-r$ and sSFR, the trends of M_r and $M_\star$ agree well with the observations, especially for low-mass galaxies. The results support the model of supernova feedback in TNG.

References

De Lucia G., Blaizot J., 2007, MNRAS, 375, 2
Dey A., et al., 2019, AJ, 157, 168
Guo Q., White S. D. M., 2008, MNRAS, 384, 2
Kauffmann G., Charlot S., White S. D. M., 1996, MNRAS, 283, L117
Larson R. B., 1974, MNRAS, 166, 585
Liao L.-W., Cooper A., 2022, arXiv e-prints, p. arXiv:2209.14166
Marinacci F., et al., 2018, MNRAS, 480, 5113
Naiman J. P., et al., 2018, MNRAS, 477, 1206
Nelson D., et al., 2018, MNRAS, 475, 624
Pillepich A., et al., 2018, MNRAS, 475, 648
Rodriguez-Gomez V., et al., 2017, MNRAS, 467, 3083
Springel V., et al., 2018, MNRAS, 475, 676
Tinsley B. M., Larson R. B., 1978, ApJ, 221, 554
Tortora C., Napolitano N. R., Cardone V. F., Capaccioli M., Jetzer P., Molinaro R., 2010, MNRAS, 407, 144
Wang W., et al., 2019, MNRAS, 487, 1580
de Jong R. S., 1996, in Minniti D., Rix H.-W., eds, Spiral Galaxies in the Near-IR. p. 43 (`arXiv:astro-ph/9509001`), doi:10.1007/978-3-540-49739-4˙6

Resolving the Rise and Fall of Star Formation in Galaxies
Proceedings IAU Symposium No. 373, 2023
T. Wong & W.-T. Kim, eds.
doi:10.1017/S1743921322004483

Bar-Driven Star and Star Cluster Formations and Gas Fueling to Galactic Center

Hidenori Matsui[1], **Toshiyasu Masakawa**[2], **Asao Habe**[3] **and Takayuki R. Saitoh**[4]

[1]National Institute of Technology, Asahikawa College, Asahikawa, Shunkodai 2-2-1-6, Asahikawa, Hokkaido, 071-8142, Japan

[2]The Open University of Japan

[3]Graduate School of Science, Hokkaido University, Kita 10 Nishi 8, Kita-ku, Sapporo, Hokkaido 060-0810, Japan

[4]Department of Planetology, Graduate School of Science, Kobe University, 1-1 Rokkodai-cho, Nada-ku, Kobe, Hyogo 657-8501, Japan

Abstract. In order to study gas evolution in the central region of a barred galaxy, we have performed numerical simulations of gas in the potential of the barred galaxy. We have found that the bar potential produces a gas ring within the central 1 kpc region. In the gas ring, active star and star cluster formations take place. Since the gas ring is dense enough to become self-gravitationally unstable, gas clouds form in the ring. These gas clouds interact gravitationally and collide with the other clouds. Such interaction and collision reduces their angular momentums effectively, and finally gas clouds fall into the galactic center. These processes triggers episodic gas fueling to the galactic center.

Keywords. galaxies: evolution, galaxies: nuclei, galaxies: star formation, method: numerical

1. Introduction

Barred galaxies are generally known to have a large amount of gas in their central kpc region. For examples, in Milky Way, the central region called central molecular zone (CMZ) has rich molecular gas and gas clouds (Morris & Serabyn 1996). Such abundant gas is expected to play important roles in the galactic central evolution. In the central region, active star formations have been observed in some barred galaxies (Benedict et al. 2002). Observations also have implied the existence of candidates of intermediate-mass black holes (IMBHs) in Milky Way (Oka et al. 2017). Gas supply from kpc scale to pc scale would induce growth of a central supermassive black hole (SMBH) or activity of active galactic nuclei (AGN). Thus, understanding of gas evolution in the central region of barred galaxies is crucial for understanding of galaxy formation. Here, we report results of our numerical simulations of barred galaxies.

2. Method

In order to study evolution of gas in the central region of a barred galaxy, we have performed numerical simulations of barred galaxies by using parallel SPH/Nbody code "ASURA" (Saitoh et al. 2008). In our simulations, a wide range of radiative cooling ($10\,\mathrm{K} < T < 10^8\,\mathrm{K}$), star formations from cool and dense gas ($T < 100\,\mathrm{K}$, $n_{\mathrm{H}} > 100\,\mathrm{cm}^{-3}$), and heating from supernovae (SNe) feedback are taken into account, which enables us

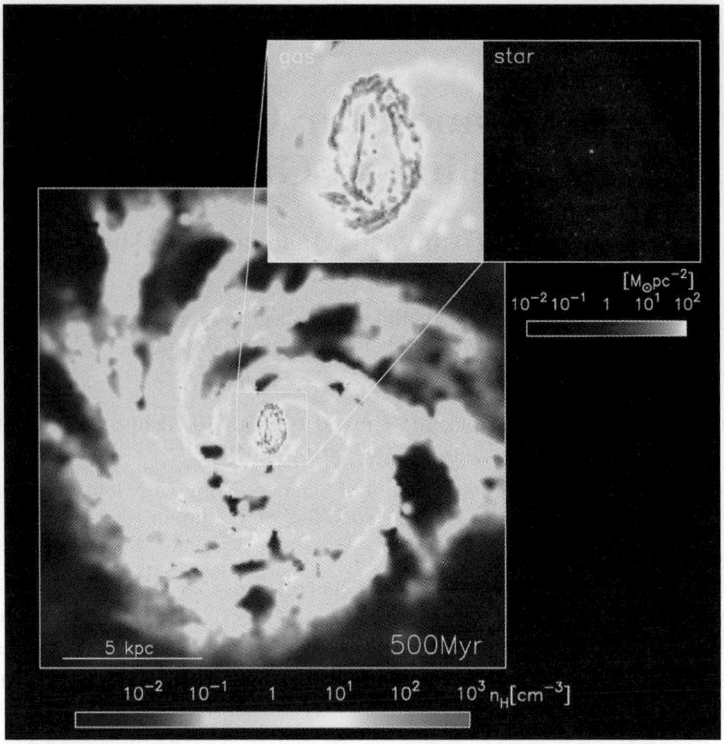

Figure 1. Gas density map at 500 Myr. The upper left and right panels show gas density and surface density of newly formed stars within the central 2 kpc. In our simulations, the coordinate system rotating with the bar potential is used, and the bar is along horizontal direction.

to understand formations of gas clouds (Saitoh et al. 2008) and star and star clusters (Saitoh et al. 2009; Matsui et al. 2012, 2019).

We have simulated gas motion in the external potential of a barred galaxy. The exponential disk with total mass of $3.5 \times 10^{10}\ M_{\odot}$ and scale length of 7 kpc is adopted for the initial gas distribution. Since the number of SPH is 10^6, the mass of a SPH particle is $3.5 \times 10^4\ M_{\odot}$. A stellar disk, a stellar bulge, a stellar bar, a dark halo, and a central SMBH are given by external potential. The models of the stellar disk, the stellar bulge, and the dark matter halo are Miyamoto-Nagai potential, Plummer model, and NFW profile, respectively. In these models, Milky Way like parameters are adopted (Saitoh et al. 2008; Sofue 2012; Gillessen et al. 2017). The stellar bar potential is given by the quadrupole model (Monari et al. 2016). The bar potential gradually grows from 200 Myr to 500 Myr in order to avoid unnatural reaction of gas.

3. Results

3.1. *Episodic gas fueling to the galactic center*

Figure 1 shows the density map of gas at 500 Myr. In the figure, the bar is along the horizontal direction. The bar potential produces a vertically elongated ring, which is generally called the x-2 ring, with respect to the bar. The size of ring is less than 1 kpc. As the bar potential grows, the gas ring becomes dense. Since the gas ring is dense enough to become self-gravitationally unstable and clumpy during bar growth phase, gas clouds form in the ring.

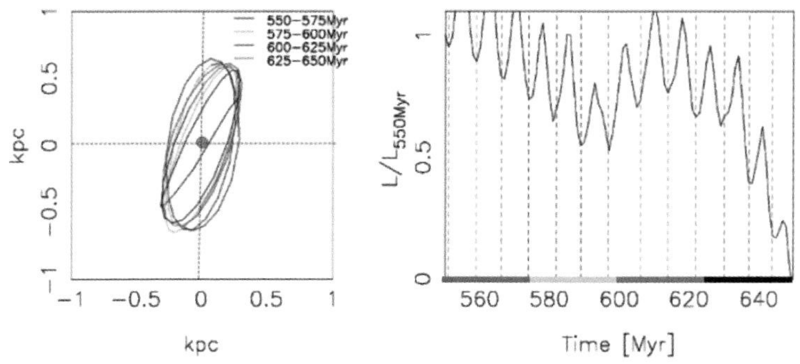

Figure 2. The orbit of one gas cloud (left) and time evolution of the angular momentum of the cloud (right). In the right panel, the dot at the center shows the galactic central SMBH. In the left panel, the vertical dashed lines show the time that the cloud reaches the apocenter.

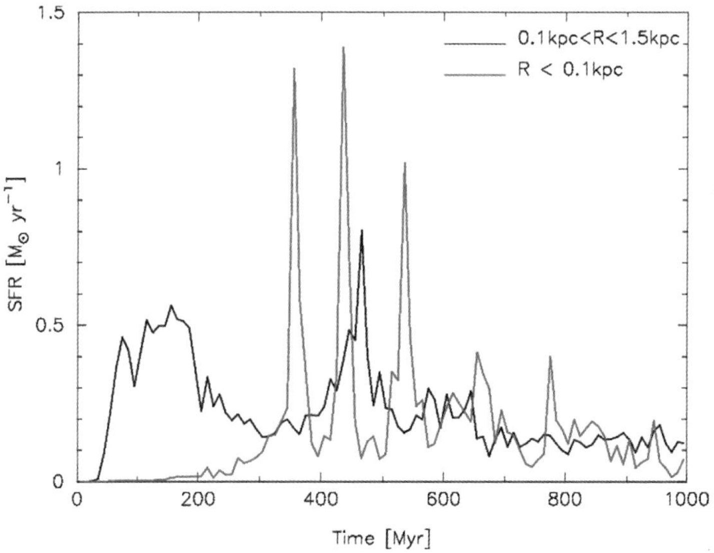

Figure 3. Time dependence of SFR. The black and red lines show SFR at 0.1 kpc $< R <$ 1.5 kpc and $R < 0.1$ kpc, respectively.

In the ring, gravitational interactions and collisions between clouds take place. Due to their interactions and collisions, some gas clouds begin to deviate from the ring. Figure 2 shows the time evolution of the distance from the galactic center (left) and the angular momentum (right) of one deviated cloud. We can see that the cloud finally falls into the galactic center due to the decrease of its angular momentum within a few hundred Myr. This is because the deviated cloud interacts gravitationally and collides with the other clouds, which still rotate on the ring, at the apocenter of the orbits. The interactions and collisions reduce the angular momentum of the cloud effectively. We can also see the short time oscillations of its angular momentum. This is caused by a bar potential.

The falls of gas clouds trigger gas fueling to the galactic center episodically. As a result, mass within the central 10 pc increases by $8 \times 10^7\ M_\odot$ for 500 Myr.

3.2. *Star formations*

Figure 3 shows the time evolution of star formation rate (SFR). Since a large number of gas is collected during the bar growth phase, gas ring becomes dense enough to induce active star formations and increase SFR. After the bar grows, SFR gradually decreases. These results agree with Baba & Kawata (2020). When gas clouds fall into the galactic center, a nuclear starburst takes place. As a result, some peaks of SFR appear.

4. Discussion

In our simulations, some star clusters form in the galactic central region. If runaway collisions of massive stars occur within the cluster, IMBH would form (Portegies Zwart et al. 2004). These clusters might play an important role in IMBH formations observed in CMZ (Oka et al. 2017).

This work was supported by JSPS KAKENHI Grant Number 21K03621. Numerical computations were carried out on Cray XC50 at Center for Computational Astrophysics (CfCA), National Astronomical Observatory of Japan.

References

Baba, J., & Kawata, D. 2020, MNRAS, 492, 4500
Benedict, G. F., Howell, D. A., Jørgensen, I., Kenney, J. D. P., & Smith, B. J. 2002, AJ, 123, 1411
Gillessen, S., Plewa, P. M., Eisenhauer, F., et al. 2017, ApJ, 837, 30
Matsui, H., Tanikawa, A., & Saitoh, T. R. 2019, PASJ, 71, 19
Matsui, H., Saitoh, T. R., Makino, J., et al. 2012, ApJ, 746, 26
Monari, G., Famaey, B., Siebert, A., et al. 2016, MNRAS, 461, 3835
Morris, M., & Serabyn, E. 1996, ARA&A, 34, 645
Oka, T., Tsujimoto, S., Iwata, Y., Nomura, M., & Takekawa, S. 2017, Nature Astronomy, 1, 709
Portegies Zwart, S. F., Baumgardt, H., Hut, P., Makino, J., & McMillan, S. L. W. 2004, Nature, 428, 724
Saitoh, T. R., Daisaka, H., Kokubo, E., et al. 2008, PASJ, 60, 667
—. 2009, PASJ, 61, 481
Sofue, Y. 2012, PASJ, 64, 75

Resolving the Rise and Fall of Star Formation in Galaxies
Proceedings IAU Symposium No. 373, 2023
T. Wong & W.-T. Kim, eds.
doi:10.1017/S1743921322004562

Collision-induced formation of dark-matter-deficient galaxies

Koki Otaki[1,2] and **Masao Mori**[3]

[1]Degree Programs in Pure and Applied Sciences, Graduate School of Science and Technology, University of Tsukuba, Tennodai 1-1-1, Tsukuba, Ibaraki 305-8577, Japan

[2]Degree Programs in Systems and Information Engineering, Graduate School of Science and Technology, University of Tsukuba, Tennodai 1-1-1, Tsukuba, Ibaraki 305-8577, Japan
email: otaki@ccs.tsukua.ac.jp

[3]Center for Computational Sciences, University of Tsukuba, Tennodai 1-1-1, Tsukuba, Ibaraki 305-8577, Japan

Abstract. The standard galaxy formation model predicts that galaxies form within a Cold Dark Matter (CDM) halo and that galaxies are dominated by dark matter. However, recent observations have discovered dark-matter-deficient galaxies with much less dark matter mass than theoretical predictions, and the process of their formation has been discussed. Here, we investigate the physical processes of galaxy formation by collisions between gas-rich dark matter subhalos within the context of the CDM paradigm. We investigate the formation process of dark-matter-deficient galaxies by running three-dimensional simulations of the collision process between dark matter subhalos (DMSHs) with the same mass of $10^9\,M_\odot$ colliding the velocity of $100\,\mathrm{km\,s^{-1}}$. We then compared the effect of different supernova feedback models, the subgrid physics of the simulation, on the collision-induced formation of galaxies. The results show that the strong feedback model ejects gas out of the system more efficiently than the weak feedback model, leading to lower star formation rates and the formation of a more extended galaxy. Finally, dark-matter-deficient galaxies with stellar masses of $\sim 10^7\,M_\odot$ and $\sim 10^8\,M_\odot$ are formed in the weak and strong feedback models, respectively.

Keywords. galaxies: formation, galaxies: evolution

1. Introduction

In the standard galaxy formation model, cold dark matter (CDM) drives the hierarchical structure formation in the universe. Based on the CDM model, the correlation between stellar components and dark matter halos has been numerous studies from both theoretical and observational viewpoints. Almost all studies about the relationship between the stellar mass and the dark matter halo mass in galaxies have shown that the mass fraction of the dark matter in galaxies is expected to be more than 90% (e.g., Behroozi *et al.* 2013).

However, van Dokkum *et al.* (2018) recently reported that the satellite galaxy NGC1052-DF2, a member of the elliptical galaxy NGC1052 group, has very little dark matter component compared to the theoretical predictions. Its stellar mass is $2 \times 10^8\,M_\odot$, whereas its dynamical mass is $< 3.4 \times 10^8\,M_\odot$ within a radius of 7.6 kpc. Additionally, NGC1052-DF4 in the NGC1052 group has also been discovered as a galaxy with similar properties (van Dokkum *et al.* 2019). These two galaxies are classified as ultra-diffuse galaxies (UDGs). UDGs are peculiar galaxies with extremely low surface brightness,

$\mu(g,0) > 24$ mag arcsec^{-2}, and large effective radii, $r_e > 1.5$ kpc (van Dokkum *et al.* 2015; Koda *et al.* 2015).

Furthermore, different observations have also reported the existence of other dark-matter-deficient galaxies. Mancera Piña *et al.* (2019) identified six H I-rich UDGs and Guo *et al.* (2020) found 19 dwarf galaxies which have high baryon fraction.

So far, only a few theoretical studies of the formation of such exotic galaxies have been discussed. Recently, it has been pointed out that high-velocity collisions between gas-rich dwarf galaxies are capable of forming dark-matter-deficient galaxies (Silk *et al.* 2019; Shin *et al.* 2020). Shin *et al.* (2020) showed that dark-matter-deficient galaxies formed when two dwarf galaxies collide with each other at a relative velocity of 300 km s^{-1} using self-gravitating hydrodynamics simulations.

Otaki & Mori (2022a) simulated head-on collisions between dark matter subhalos using a three-dimensional simulation code. They show that a dark-matter-dominated galaxy is formed when the collision velocity is 20 km s^{-1}, and a dark-matter-deficient galaxy is formed when the collision velocity is 100 km s^{-1} In this paper, we investigate the effects of subgrid physics for supernova feedback on subhalo collision and collision-induced formation of galaxies.

2. Simulations of dark matter subhalos collision

We ran simulations of head-on collisions between two dark matter subhalos as the formation process of dark-matter-deficient galaxies. The initial conditions and simulation setup are as in Otaki & Mori (2022a) thus the two colliding DMSHs each have a mass of 10^9 M$_\odot$ and the mass ratio between dark matter and gas is 5.36. The density profile of the dark matter is assumed to be the Navarro-Frenk-White (NFW) profile, and the gas is in hydrostatic equilibrium with the dark matter potential. The DMSH centres are initially at $(x, y, z) = (\pm 5, 0, 0)$ kpc, and the initial bulk-velocities of the DMSHs are $(v_x, v_y, v_z) = (\mp 100, 0, 0)$ km s^{-1}.

We use the N-body/Smoothed Particle Hydrodynamics (SPH) simulation code developed by Otaki & Mori (2022a), but modified to compare the effect of supernova feedback. Our feedback model assumes that SPH particles in the vicinity of newly born stellar particles receive the thermal energy released by the supernova explosions and evolve adiabatically without energy loss by the radiative cooling during the lifetime of a massive star with 8 M$_\odot$ (Mori *et al.* 1997, 1999). This model is referred to below as the strong feedback model. On the other hand, a weak feedback model does not turn off the radiative cooling calculation of the SPH particles when they receive energy from the supernova feedback. The cooling time is very short compared to the dynamical time of the star forming region (Otaki & Mori 2022b), therefore the energy injection by the supernova explosions does not affect the thermodynamic evolution of the gas significantly.

Figure 1 shows the time evolution of the dark matter subhalos collision with (a) the weak feedback model and (b) the strong feedback model. In each panel, the upper panels are dark matter distribution, the lower panels show baryon, in which the blue and red points represent gas and stars, respectively. At 50 Myr, the center of the subhalo collides and the gas density is enhanced at the colliding surface, which triggers star formation. The dark matter component in the dark matter subhalo is passed through without merger and a dark-matter-deficient galaxy is formed.

In the case of the weak feedback model, the energy injected into the gas by the supernova feedback is lost by radiative cooling before it is converted to kinetic energy, so the gas remains at the collision interface. Compared to the weak feedback model, the strong feedback model simulation results in a more extended stellar system because the energy injection from stellar particles to gas particles makes the strong outflow and most of the

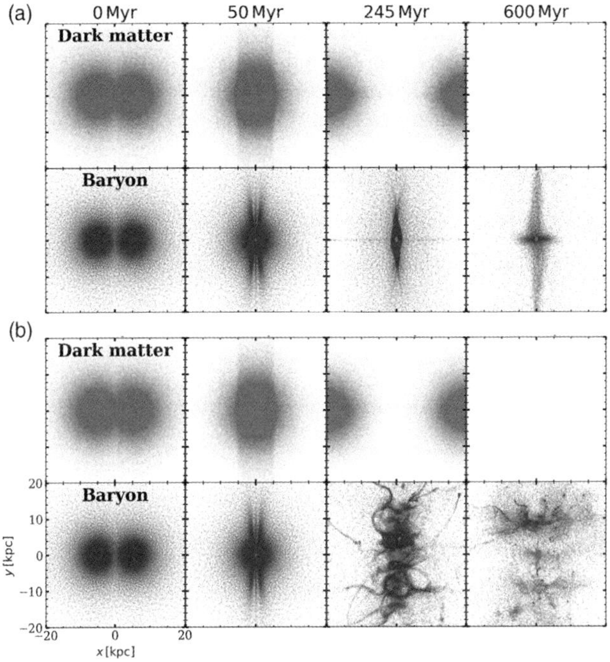

Figure 1. Snapshots of the head-on collision between dark matter subhalos with $10^9\,M_\odot$ for a collision velocity of $100\,\mathrm{km\,s^{-1}}$ in weak feedback model (a) and strong feedback model (b). The gray, blue, and red dots represent dark matter, gas, and stars, respectively. From left to right, $t = 0$, 50, 245, and 600 Myr, respectively. The gray, blue, and red points represent dark matter, gas, and stars, respectively.

gas is ejected from the system. This suggests that the strength of the feedback affects the formation process of the ultra-diffuse galaxies. Both simulation results showed star formation rates of a few $M_\odot\,\mathrm{yr}^{-1}$ at the time of the collision, but the star formation histories differ significantly thereafter. The star formation rate rapidly decreases in the strong feedback model, but not so much in the weak feedback model. Finally, galaxies with stellar masses of $\sim 10^8\,M_\odot$ and $\sim 10^7\,M_\odot$ are formed in the weak and strong feedback models, respectively.

Acknowledgements

Numerical computations were performed with computational resources provided by the Multidisciplinary Cooperative Research Program in the CCS, University of Tsukuba. This work was supported by JSPS KAKENHI Grant Numbers JP21J21888, JP20K04022.

References

Behroozi, P. S., Wechsler, R. H. and Conroy, C. 2013, The Astrophysical Journal, 770
van Dokkum, P., *et al.* 2018, Nature, 555
van Dokkum, P., *et al.* 2019, The Astrophysical Journal, 874
van Dokkum, P., *et al.* 2015, The Astrophysical Journal, 798
Koda, J., *et al.* 2015, The Astrophysical Journal, 807
Mancera Piña, P. E., *et al.* 2019, The Astrophysical Journal, 883
Guo Q., *et al.* 2020, Nature Astronomy, 4
Silk, J. 2019, Monthly Notices of the Royal Astronomical Society: Letters, 488

Shin, E. J. 2019, The Astrophysical Journal, 899
Otaki, K. and Mori, M. 2022, Journal of Physics: Conference Series, 2207
Mori, M., *et al.* 1997, The Astrophysical Journal Letters, 478
Mori, M., Yoshii, Y. and Nomoto, K. 1999, The Astrophysical Journal, 511
Otaki, K. and Mori, M. 2022, Lecture Notes in Computer Science, 13378

Resolving the Rise and Fall of Star Formation in Galaxies
Proceedings IAU Symposium No. 373, 2023
T. Wong & W.-T. Kim, eds.
doi:10.1017/S174392132200463X

The interplay of internal and external processes in the buildup of disk galaxies: thick-disk star formation histories in AURIGA simulations

Francesca Pinna[1], Daniel Walo-Martín[2,3] and Robert J.J. Grand[2,3]

[1]Max Planck Institute for Astronomy, Koenigstuhl 17, D-69117 Heidelberg, Germany
email: `pinna@mpia.de`

[2]Instituto de Astrofísica de Canarias, Calle Vía Láctea s/n, E-38205 La Laguna, Tenerife, Spain

[3]Departamento de Astrofísica, Universidad de La Laguna, Av. del Astrofísico Francisco Sánchez s/n, E-38206, La Laguna, Tenerife, Spain

Abstract. Recent integral-field spectroscopy observations have revealed that thick- and thin-disk star-formation histories are regulated by the interplay of internal and external processes. We analyze stellar-population properties of 24 spiral galaxies from the AURIGA zoom-in cosmological simulations, to offer a more in-depth interpretation of observable properties. We present edge-on maps of stellar age, metallicity and [Mg/Fe] abundance, and we extract the star-formation and chemical-evolution histories of thin and thick disks. Both show signs of the interplay between internal chemical enrichment and gas and star accretion. Thick disks show particularly complex stellar populations, including an in-situ component, formed from both slowly enriched and accreted more pristine gas, and an additional significant fraction of ex-situ stars.

Keywords. techniques: spectroscopic, galaxies: evolution, galaxies: structure

1. Introduction

Spatially resolved stellar-population analyses of massive galactic disks allow us to draw the spatial and temporal distribution of star formation during the mass assembly of disk-dominated galaxies. Properties of thick disks, old, metal poor and enhanced in α elements, trace mostly the early stages of galaxy formation, while younger metal-rich thin disks tell us about later evolutionary phases (e.g., Yoachim *et al.* 2008a, Gallart *et al.* 2019). While we know that thin disks form mostly in situ, with its intense and extended star formation being fuelled by gas accretion, there is no common agreement on how thick disks form. Three main formation scenarios were proposed to explain their properties: their stars may have formed in situ, at high redshift, from turbulent gas (Brook *et al.* 2004), in satellites that were later directly accreted (Abadi *et al.* 2003), or in a preexisting thinner disk which was later dynamically heated (e.g., Di Matteo *et al.* 2011).

Observational studies of thick disks are generally based on edge-on galaxies, where fainter thick disks can be morphologically decomposed from the bright thin disks (e.g., Comerón *et al.* 2018). Most of the spectroscopic studies, which revealed some variety in thick-disk properties, have supported one or another of the mentioned formation scenarios (e.g., Yoachim *et al.* 2008a, Kasparova *et al.* 2016, Comerón *et al.* 2019). On the other hand, resolved star-formation and chemical-evolution histories of thick disks, extracted

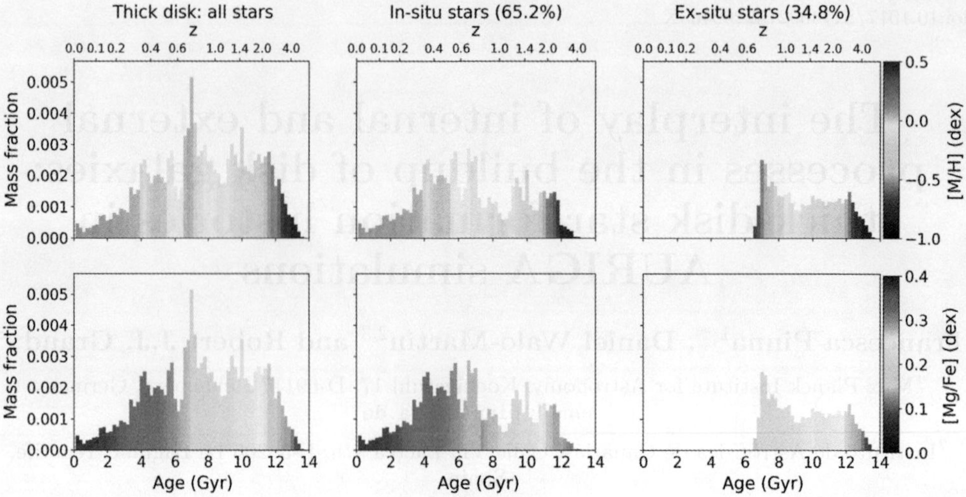

Figure 1. Star-formation history of the thick disk in AU5, color-coded by total metallicity (top panels) and [Mg/Fe] abundance (bottom panels). Left panels include all stars, while middle and right panels include respectively only stars formed in situ or ex situ, indicating their corresponding thick-disk mass fraction on top of each column.

from deep integral-field MUSE observations, have recently unveiled a combination of in-situ and ex-situ stellar populations, resulting from complex scenarios with different mechanisms at play (Pinna *et al.* 2019a, Pinna *et al.* 2019b, Martig *et al.* 2021).

Here we probe these complex scenarios with the use of numerical simulations.

2. Overview

Method. We used 24 Milky Way-mass galaxies from the AURIGA zoom-in cosmological simulation sample (Grand *et al.* 2017). We projected the galaxies in an edge-on view, to allow a similar analysis to what is usually done in integral-field spectroscopy observations. We performed a Voronoi binning to ensure a similar number of star particles in each spatial bin. We used for the analysis a region of radius the optical radius R_{opt} of the galaxy (Grand *et al.* 2017) and of height h_{scale}, the standard deviation of vertical positions of stars at R_{opt}. For each galaxy, we fitted vertical luminosity-density profiles, extracted in four radial bins, with two components associated with the thin and the thick disks, each one represented by a hyperbolic secant square function. The fits gave us the average height, for each galaxy, at which the thick disk starts to dominate.

Stellar populations. We mapped the stellar populations of the 24 galaxies and found a variety of ages, total metallicities ([M/H]) and [Mg/Fe] abundances in the thick disks, with average values respectively between 4 and 11 Gyr, -0.5 dex and solar values (0 dex), and between 0.15 and 0.23 dex (see Pinna et al., in prep.). Thick disks are in all cases older, more metal-poor and more [Mg/Fe] enhanced than thin disks. We extracted star-formation histories from the thin- and thick-disk-dominated regions, in terms of the mass fraction of stars of a specific age (following Pinna *et al.* 2019a, Pinna *et al.* 2019b and Martig *et al.* 2021). We color coded age bins according to their average metallicity and [Mg/Fe] abundance, to interpret these plots in terms of chemical evolution.

A representative galaxy: AU5. We show in Fig. 1 (left panels) the star-formation history of the thick disk in AU5, a barred spiral galaxy of stellar mass $\sim 7 \times 10^{10} M_{\odot}$ (see Pinna et al. in prep. for the full sample). At older ages, as expected from observations, we see a global increase of the star-formation rate (mass fraction in each age bin, from right to left), and a later decrease at young ages, with numerous star formation bursts across time. The colors display a global chemical enrichment, transitioning from metal-poor values and a strong [Mg/Fe] enhancement, to slightly subsolar metallicities and slightly supersolar [Mg/Fe] values. However, on top of the general trend, both [M/H] and [Mg/Fe] values oscillate at certain times across thick-disk evolution.

In-situ and ex-situ components. In order to understand the origin of these oscillations, which often correspond to star-formation bursts, we classified star particles as in situ or ex situ, by tracking them during the simulations. We mapped the mass fraction of accreted stars for the full galaxy sample, and found that they are mostly located in the thick-disk dominated regions (see maps in Pinna et al., in prep.). These regions host an accreted fraction between 7 and 61% of their mass, with a median fraction of $\sim$ 28%.

We decomposed the star-formation histories into in-situ and ex-situ components, and see that the oscillations in chemical abundances are driven by the combination of them. In AU5 (Fig. 1, right panels), a satellite contributed $\sim$ 35% of the thick-disk mass. This satellite had its own chemical evolution, and stars with high metallicity were formed during the star-formation burst at the final stage of the merger (at ages of about 7 Gyr, probably involving also some gas stripped from AU5; note that these higher metallicities are still much lower than values in the thin disk of AU5). The in-situ component (middle panels) is characterized by several star-formation bursts with lower metallicity values (and slightly higher [Mg/Fe]) than slightly younger and older ages. These drops in metallicity are driven by the accretion of gas from outside the galaxy.

Thick and thin disks in the full sample. Although each galaxy in the sample has peculiarities, the results derived from AU5 are representative of most thick disks. For thin disks, as expected from previous observations (e.g., Gallart *et al.* 2019 and Martig *et al.* 2021), we find a very low mass fraction of accreted stars, and the bulk of the mass was formed in recent times thanks to a large amount of accreted gas (for more details, see Pinna et al., in prep.).

3. Implications

Thick-disk formation. These results affirm that although thick disks are mainly formed in situ (but with a strong contribution of gas accreted at the time of frequent mergers), they host a significant contribution of accreted stars. This accreted fraction varies from 7 to 61%, on average 28% of the mass.

Galaxy evolution. Star-formation histories of thick and thin disks, thus of disk galaxies, result from the interplay of internal and external processes, in which the internal chemical enrichment is challenged by the accretion of more pristine gas and more metal-poor stars.

References

Abadi, M. G., Navarro, J. F., Steinmetz, M., & Eke, V. R., 2003, *ApJ*, 597, 21.

Brook, C. B., Kawata, D., Gibson, B. K., & Freeman, K. C., 2004, *ApJ*, 612, 894.

Di Matteo, P., Lehnert, M. D., Qu, Y., & van Driel, W., 2011, *A&A*, 525, L3.

Comerón, S., Salo, H., Knapen, J. H., 2018, *A&A*, 610, A5.

Comerón, S., Salo, H., Knapen, J. H., & Peletier, R. F., 2019, *A&A*, 623, A89.

Gallart, C., Bernard, E. J., Brook, C. B., Ruiz-Lara, T., Cassisi, S., Hill, V., & Monelli, M., 2019, *Nature Astronomy*, 3, 932.

Grand, R. J. J., *et al.*, 2017, *MNRAS*, 467, 179.

Kasparova, A. V., Katkov, I. Y., Chilingarian, I. V., Silchenko, O. K., Moiseev, A. V., & Borisov, S. B., 2016, *MNRAS*, 460, L89.

Martig, M., *et al.*, 2021, *MNRAS*, 508, 2458.

Pinna, F., *et al.*, 2019, *A&A*, 623, A19.

Pinna, F., *et al.*, 2019, *A&A*, 625, A95.

Yoachim, P., & Dalcanton, J. J., 2008, *ApJ*, 683, 707.

Resolving the Rise and Fall of Star Formation in Galaxies
Proceedings IAU Symposium No. 373, 2023
T. Wong & W.-T. Kim, eds.
doi:10.1017/S1743921322004616

Star formation feedback onto molecular clouds of KAGONMA sources using temperature distribution

N. Takeba[1], T. Handa[1], T. Murase[1], M. Kohno[2,3], T. Omodaka[1], M. Nakano[4], Y. Hirata[1], J. O. Chibueze[5,6] and R. A. Burns[7]

[1]Faculty of Science, Kagoshima University, 1-21-35 Korimoto, Kagoshima 890-0065, Japan.
email: n34.bamu.lss@gmail.com

[2]Astronomy Section, Nagoya City Science Museum, 2-17-1 Sakae, Naka-ku, Nagoya, Aichi 460-0008, Japan.

[3]Department of Physics, Graduate School of Science, Nagoya University, Furo-cho, Chikusa-ku, Nagoya, Aichi 464-8602, Japan.

[4]Faculty of Education, Oita University, 700 Dannoharu, Oita 870-1192, Japan.

[5]Centre for Space Research, Potchefstroom campus, North-West University, Potchefstroom 2531, South Africa

[6]Department of Phys. and Astron., Faculty of Physical Sci., Univ. of Nigeria, Carver Building, University road, Nsukka 410001, Nigeria

[7]National Astronomical Observatory of Japan (NAOJ), 2-21-1 Osawa, Mitaka, Tokyo 181-8588, Japan

Abstract. This paper reports on four of the sources observed in the KAGONMA (KAgoshima Galactic Object survey with the Nobeyama 45-m telescope by Mapping in Ammonia lines) project for which mapping observations have been completed (KAG35, KAG45, KAG64, and KAG71). In this study, we compiled the analysis results of four sources for which mapping observations were completed in the KAGONMA project and statistically investigated the range to which star formation activity affects the molecular gas. In order to investigate the affected range, we analyzed the heating range by focusing on the temperature distribution of the molecular cloud and found that it is within about 3 pc. This suggests that direct star formation feedback in molecular clouds is very spatially limited.

Keywords. ISM: clouds – HII regions

---

1. Introduction & Observations

The effects on the interstellar matter by star formation is critical to understand the next generation of star formation and galaxy evolution. This effect is expected by expansion of an HII region, bipolar outflow, intense ultraviolet radiation and shock wave, which produce shock waves and heat up the surrounding gas. Therefore, we have conducted a mapping observation project, named KAGONMA, to figure out the regions affected by star formation activities.

In the project we observed 72 molecular cloud cores with the Nobeyama 45-m radio telescope (NRO) since 2015 December. The NH_3 (J, K) = (1,1), (2,2), (3,3) and H_2O

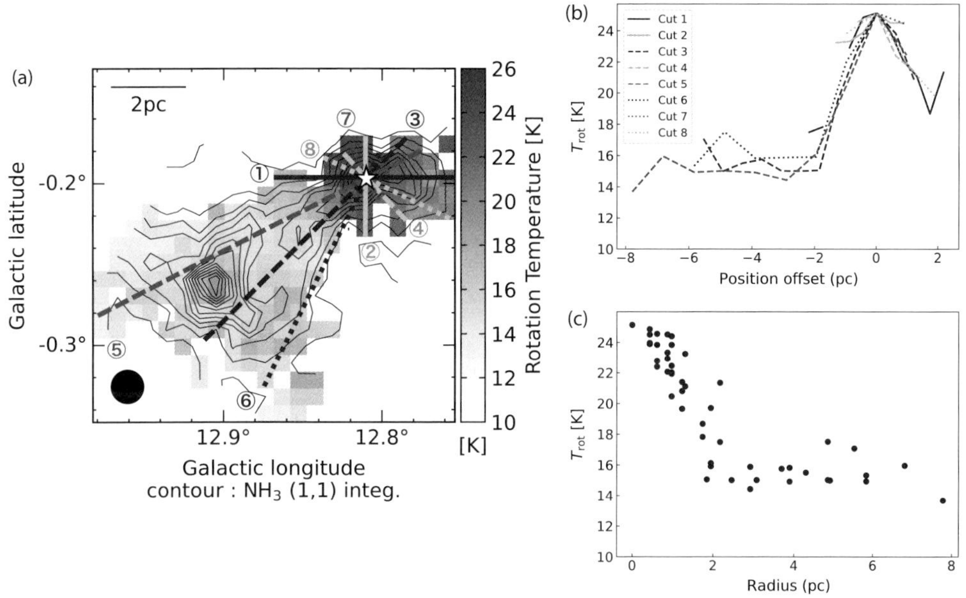

Figure 1. (a) The rotation temperature map of KAG64. The contours correspond to the $NH_3(1,1)$ integrated intensity map. The star symbol is reference point. The 8 lines are the cuts in 8 directions from the reference point. (b) $T_{\rm rot}$ along 8 directions from the reference point. Each color corresponds to the coded color of each cut in (a). (c) The radial distribution of estimated rotational temperature, which is turned around with respect to $x = 0$ at the reference point.

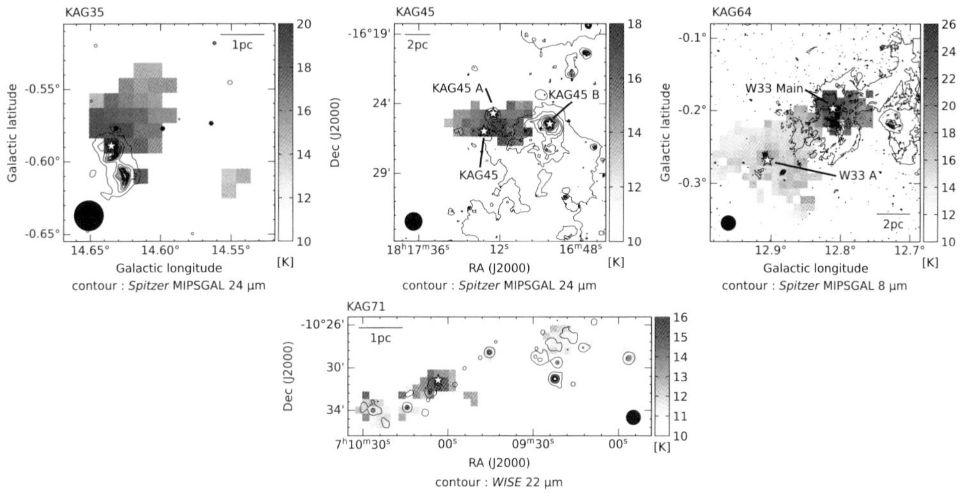

Figure 2. The map of rotation temperature (color coded cells) is overlaid on the infrared brightness distribution (contours). The infrared maps of KAG35 and KAG45 are used with *Spitzer* MIPSGAL 24μm. The maps of KAG64 and KAG71 are used with *Spitzer* GLIMPS 8μm and WISE 22μm, respectively.

maser lines were observed simulaneously. The KAGONMA project was performed over a wide area covering the entire molecular cloud. The telescope beam size was 75$''$ at 23 GHz. The pointing accuracy was checked using known H_2O maser sources and was better than 5$''$. The rms noise level was typically 0.04 K in T_a^* at each observed position.

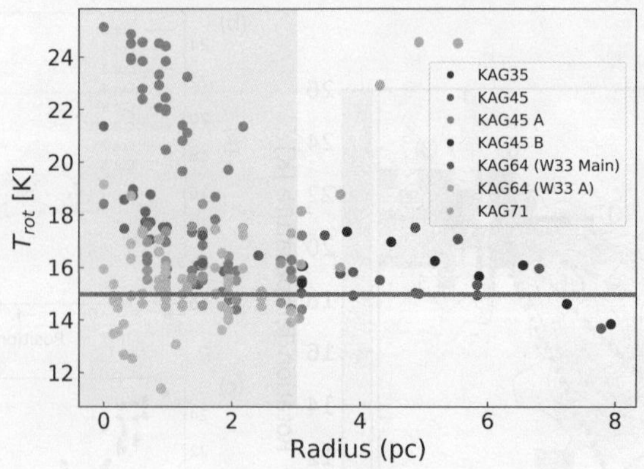

Figure 3. The radial distribution of the rotation temperature for all sources. The horizontal axis is the radial distance from each reference point. The colors of the plots indicate the source. The blue horizontal line is ~15 K, which is the typical temperature of interstellar molecular gas (Planck Collaboration et al. 2011), and the yellow shaded region is within 3 pc range from the reference point of each source.

2. Analysis Methods

To investigate the extent of the influence of star formation feedback on a molecular cloud, we focused on the temperature distribution of the molecular cloud and investigated the extent of the heated range. The star symbol in Fig. 1(a) indicates the reference point which is defined as the temperature peek position or identified excitation source, and cuts were made in eight directions from the reference point. The result is shown in Fig. 1(b). The radial distribution of estimated rotational temperature, which is turned around with respect to $x = 0$ at the reference point, is shown in Fig. 1(c).

3. Results & discussion

We show the preliminary results of four sources are shown in Fig. 3. For KAG35 (M17 southwest extension) we assigned a reference point at the peak in temperature, and the heated range is within 2 pc. For KAG45 we assigned three reference points at the peak in temperature or intense positions in 24μm, and the heated range is within 3 pc. Although KAG45 B shows a increase beyond 3 pc, it should be due to the influence of KAG45 A. For KAG64 (W33 star formation region) we assigned two reference points at the peak in temperature, and the heated range is within 2 pc. Although W33 A shows a increase beyond 3 pc, it should be due to the influence of W33 Main. The heated range of W33 Main is wider than that of W33 A. For KAG71 (CMa OB1 east filament) we assigned a reference point at the peak in temperature, and the heated range is within 1 pc.

From these results, we conclude that the heated range is within about 3 pc. This suggests that direct star formation feedback in molecular clouds is limited to a few pc.

Reference

Planck Collaboration 2011, *A&A*, 536, A25

Resolving the Rise and Fall of Star Formation in Galaxies
Proceedings IAU Symposium No. 373, 2023
T. Wong & W.-T. Kim, eds.
doi:10.1017/S1743921322004367

Investigating the Drivers of CO-to-H_2 Conversion Factor Variations in Nearby Galaxy Centers

Yu-Hsuan Teng and Karin Sandstrom

University of California San Diego,
9500 Gilman Dr, La Jolla, CA 92093, USA
email: yuteng@ucsd.edu

Abstract. The CO-to-H_2 conversion factor (α_{CO}) is crucial for accurate estimation of the amount and properties of molecular gas. However, α_{CO} is known to vary with environmental conditions, and previous kpc-scale studies have revealed lower α_{CO} in the centers of some barred galaxies, including NGC 3351, 3627, and 4321. We present ALMA Band 3, 6, and 7 observations toward the inner ~2 kpc of these galaxies tracing ^{12}CO, ^{13}CO, and C^{18}O lines at ~100 pc resolution. We show that dynamical effects resulting from turbulence/shear can lead to substantially lower α_{CO} in the bar-driven inflows of NGC 3351 due to lower optical depth. A clear, positive correlation between α_{CO} and ^{12}CO optical depth is seen in all three galaxy centers. We also find that the CO/^{13}CO(2–1) ratio mainly traces the ^{12}CO optical depth, and thus it may be a useful observable in predicting α_{CO} variation in galaxy centers.

Keywords. Barred spiral galaxies (136); CO line emission (262); Galaxy nuclei (609); Molecular gas (1073); Star forming regions (1565)

1. Introduction

Molecular gas plays an important role in star formation and galaxy evolution. To trace the cold molecular gas where stars are born, most studies use the low-J rotational lines of carbon monoxide (^{12}C^{16}O; hereafter CO). Because of this, the CO-to-H_2 conversion factor (α_{CO}) is the basis of measuring the amount and properties of molecular gas. α_{CO} is defined as the ratio of molecular gas mass to the CO 1–0 luminosity. While most studies assume a constant α_{CO} value similar to the Milky Way disk average, α_{CO} varies within and between galaxies, and it can vary by as much as orders of magnitude depending on environmental conditions such as metallicity, density, temperature, and opacity (Bolatto *et al.* (2013)). Thus, investigating the variation of α_{CO} and its relation to gas properties is critical to understanding molecular gas and star formation in galaxies.

Recent studies have shown that galaxy centers, especially those with spiral arms or stellar bars, tend to have lower α_{CO} than the typical Galactic value (Sandstrom *et al.* (2013); Israel (2020)). The specific environmental properties that drive such α_{CO} variations are not fully understood. With diverse gas conditions and/or altered gas dynamics, galaxy centers are ideal nearby laboratories for studying how α_{CO} depends on physical properties. In this work, we present a combination of results on three nearby galaxy centers: NGC 3351, 3627, and 4321. This will include our preliminary results on NGC 3627 and 4321, as well as the published results on NGC 3351 from Teng *et al.* (2022).

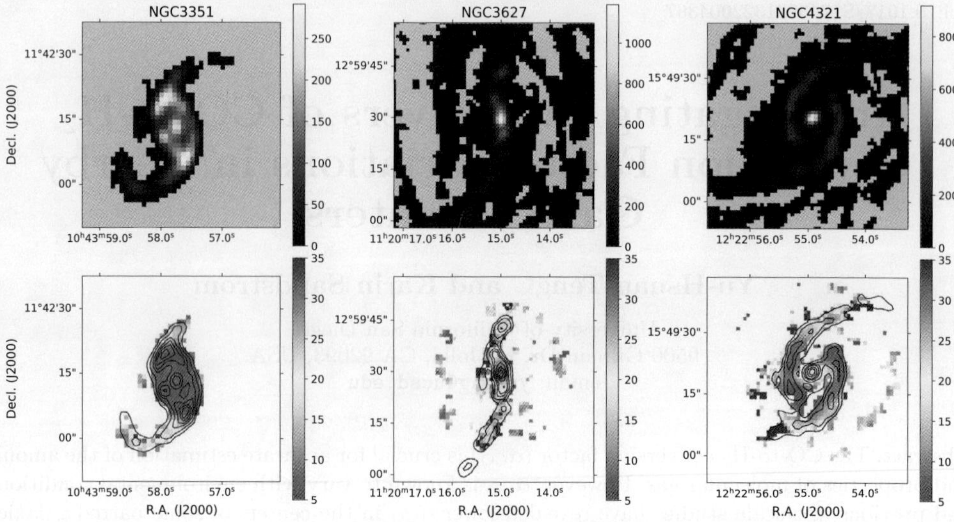

Figure 1. Maps of the CO 2–1 integrated intensity (top) and CO(2–1)/^{13}CO(2–1) line ratio (bottom). *Top:* The gray areas show regions with $< 3\sigma$ detection in CO 2–1. *Bottom:* The masking ensure pixels with S/N > 3 in both lines and 12-m/total flux recovered rate $> 70\%$. The overlaid contours represent the CO 2–1 integrated intensity shown in top panels.

2. ALMA Observations and Multi-line Modeling

To study what physical processes cause the α_{CO} variations in galaxy centers, we select three nearby barred galaxies that were found by previous kpc-scale observations to have lower-than-Galactic α_{CO} in their central few kpc. To accurately measure the gas properties and α_{CO}, we need optically-thin tracers like ^{13}CO and C^{18}O to estimate the optical depths in different lines. Therefore, we observe the inner $\sim$2 kpc regions of these galaxies with six low-J CO (1–0 and 2–1), ^{13}CO (2–1 and 3–2), and C^{18}O (2–1 and 3–2) lines using the Atacama Large Millimeter/submillimeter Array (ALMA).

Figure 1 shows the integrated intensity maps in CO 2–1 as well as the CO(2–1)/^{13}CO(2–1) line ratios with intensity units of K km s^{-1}. We match the beam sizes of all the lines to $\sim$100 pc, which corresponds to the scale of Giant Molecular Clouds. In the central $\sim$1 kpc of NGC 3351, we see a clear star-forming ring around the nucleus, and the ring is connected to two inflow arms driven by the stellar bar in the galaxy center. Interestingly, these bar-driven inflow arms also show a significantly higher CO/^{13}CO 2–1 line ratio, which could be due to variations in optical depths or CO isotopologue abundances. In NGC 3627 and 4321, the observations reveal a bright nucleus in the inner $\sim$300 pc of both galaxies, and those nuclei show distinct CO/^{13}CO line ratios from other regions.

To determine the molecular gas properties in different regions of these galaxies, we use the non-LTE radiative transfer code *RADEX* (van der Tak *et al.* (2007)) to jointly model the intensities of all six observed lines under various combination of CO column density per linewidth, kinetic temperature, H_2 volume density, CO/^{13}CO and ^{13}CO/C^{18}O abundances, and the beam-filling factor. Using a Bayesian likelihood analysis, we derive probability distributions for each parameter on a pixel-by-pixel basis (see Teng *et al.* (2022) for modeling details). We further derive α_{CO} distributions via the modeled parameters and CO 1–0 intensities, assuming a CO-to-H_2 abundance of 3×10^{-4}. We present our α_{CO} results and possible implications in the next section.

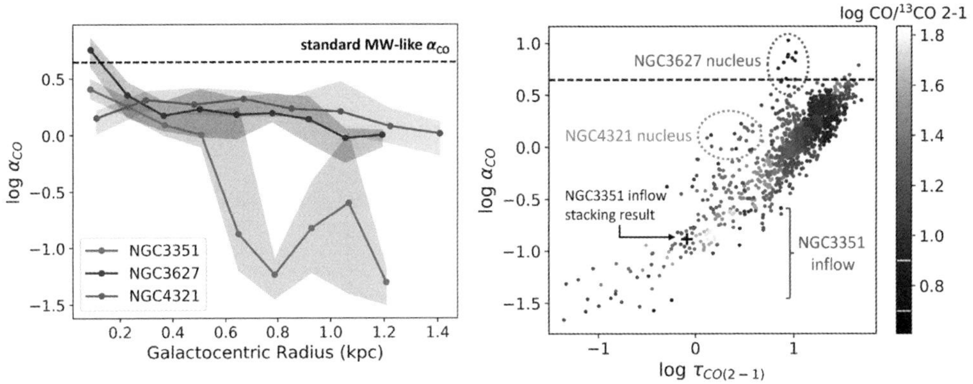

Figure 2. *Left*: Medians of the modeled α_{CO} within ~100 pc galactocentric radius bins in the central kpc of NGC 3351, 3627, and 4321. Shaded areas span the 25th and 75th percentile ranges. *Right*: Relation between the modeled α_{CO} and CO 2–1 optical depth from all three galaxies. Data points are color-coded by 2D-binned medians of the observed CO/^{13}CO 2–1 ratios. White lines on the color bar indicates the range of Galactic disk-like ratio.

3. Results and Discussion

Figure 2 shows how the derived α_{CO} varies with galactocentric radius and CO optical depths (τ_{CO}). Except for the NGC 3627 nucleus reaching a near-Galactic α_{CO}, all pixels have lower-than-Galactic α_{CO} values. The diverse α_{CO} trend in the nuclei is likely related to the denser and hotter gas condition as suggested by our modeling. In addition, substantially lower α_{CO} is found in the bar-driven inflows of NGC 3351, where τ_{CO} is low. We find a tight, positive correlation of α_{CO} with τ_{CO} in all three galaxy centers. Since τ_{CO} is proportional to CO column density per line width, this means that nuclear gas concentration and turbulence/shear effects play important roles in setting α_{CO} in these regions. Thus, the overall lower-than-Galactic α_{CO} may be explained by the higher velocity dispersion in barred galaxy centers which decreases the τ_{CO} (Sun *et al.* (2020)).

Furthermore, the right panel in Figure 2 shows that the CO/^{13}CO 2–1 ratio generally reflects τ_{CO} inversely. Using NGC 3351 as an example, the higher CO/^{13}CO 2–1 ratio we observe in the inflow arms is likely related to the significantly lower τ_{CO} and α_{CO}. Since the decreasing τ_{CO} can lead to more fraction of escaped CO emission, this would increase the CO/^{13}CO line ratio or the CO intensity per unit gas mass, and thus lower the α_{CO} by definition. With the strong dependence seen between α_{CO} and τ_{CO}, we conclude that the CO/^{13}CO ratio can be a useful observational tracer for α_{CO} variations, particularly in galaxy centers where optical depth effects are dominant.

Acknowledgements

This work would not have been possible without our collaborators: Jiayi Sun, Adam Leroy, Cliff Johnson, Alberto Bolatto, Diederik Kruijssen, Andreas Schruba, Antonio Usero, Guillermo Blanc, Brent Groves, Frank Israel, Daizhong Liu, Erik Rosolowsky, Eva Schinnerer, John-David Smith, Fabian Walter, and the PHANGS Team.

References

Bolatto, A. D., Wolfire, M., & Leroy, A. K. 2013, *ARAA*, 51, 207

Israel, F. P. 2020, *A&A*, 635, A131

Papadopoulos, P. P., van der Werf, P., Xilouris, E., Isaak, K. G., & Gao, Y. 2012, *ApJ*, 751, 10
Sandstrom, K. M., Leroy, A. K., Walter, F., *et al.* 2013, *ApJ*, 777, 5
Sun, J., Leroy, A. K., Schinnerer, E., *et al.* 2020, *ApJ* (Letters), 901, L8
Teng, Y.-H., Sandstrom, K. M., Sun, J., *et al.* 2022, *ApJ*, 925, 72
van der Tak, F. F. S., Black, J. H., Schöier, F. L., *et al.* 2007, *A&A*, 468, 627

Chapter 3:
The Decline of Star Formation: Feedback, Fuel shortage or Inefficiency?

Resolving the Rise and Fall of Star Formation in Galaxies
Proceedings IAU Symposium No. 373, 2023
T. Wong & W.-T. Kim, eds.
doi:10.1017/S1743921322004884

The spatially resolved view of star formation in galaxy clusters

Bianca M. Poggianti and the GASP team

INAF-Osservatorio Astronomico di Padova,
vicolo dell'Osservatorio 5, 35122 Padova, Italy
email: bianca.poggianti@inaf.it

Abstract. Integral field spectroscopic studies of galaxies in dense environments, such as clusters and groups of galaxies, have provided new insights for understanding how star formation proceeds, and quenches. I present the spatially resolved view of the star formation activity and its link with the multiphase gas in cluster galaxies based on MUSE and multi-wavelength data of the GASP survey. I discuss the link among the different scales (i.e. the link between the spatially resolved and the global star formation rate-stellar mass relation), the spatially resolved signatures and the quenching histories of jellyfish (progenitors) and post-starburst (descendants) galaxies in clusters. Finally, I discuss the multi-wavelength view of star-forming clumps both in galaxy disks and in the tails of stripped gas.

Keywords. Galaxy evolution; star formation; baryonic cycle; galaxy clusters

1. Introduction

When galaxies infall into clusters and move at high speed with respect to the hot intracluster medium (ICM), the ram pressure (RP) exerted by the ICM upon the galaxy interstellar medium (ISM) can remove the galaxy gas, wherever the RP exceeds the gravitation pull (Gunn & Gott 1972, Jaffe' et al. 2018). The hot halo gas is affected first, then the disk gas is removed starting from the outer disk regions and stripping works its way inward in the disk, until the galaxy remains eventually devoid of gas.

Neutral gas studies were the first direct evidence for gas stripping in clusters, revealing HI tails, truncated HI disks, HI disturbed morphologies and a progressively increasing HI deficiency towards the central regions of clusters (see Cortese et al. 2021 for a recent review). By now, observations of extraplanar tails of stripped material have been obtained with several different methods (HI line, integral-field spectroscopy, Hα imaging, X-ray emission, radio continuum, UV and optical imaging, see Boselli et al. 2022), showing tails of gas in different phases and even tails of stars formed in the stripped gas. It is important to realize that each of the methods for identifying galaxies affected by ram pressure provides only a partial view of this phenomenon.

The GASP (GAs Stripping Phenomena in galaxies) project investigates the physical mechanisms that remove gas from galaxies and their effects on the star formation activity and galaxy structure. This programme has been studying galaxies in low redshift clusters, groups, filaments and isolated galaxies, but in these proceedings I will focus on cluster galaxies only. Readers are refereed to Vulcani et al. (2021) for an overview of non-cluster galaxies.

GASP provides the only large sample of confirmed ram-pressure stripped galaxies with integral-field (IF) spectroscopy. It consists of 64 ram-pressure stripped galaxies with a wide range of stellar masses (10^9-$10^{11.5} M_{\odot}$), that were selected for showing unilateral

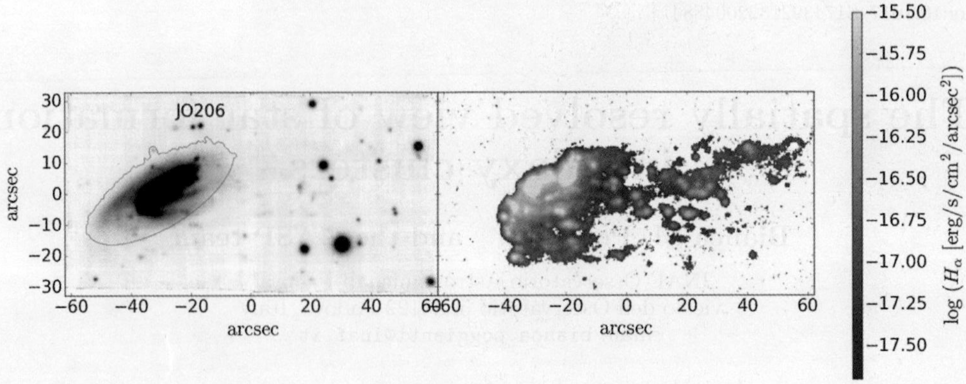

Figure 1. The jellyfish galaxy JO206, a massive ($9 \times 10^{10} M_\odot$) galaxy in a low mass cluster (velocity dispersion ~ 500km s^{-1}). Left. White-light MUSE image. Right. MUSE map of Hα emission. Note the tentacles of stripped ionized gas departing for ~ 80 kpc from the disk, to the west. Clearly visible in this tail are Hα-emitting, star-forming clumps. From Poggianti et al. (2017a).

debris in B-band images. They are hosted in 39 low redshift clusters (z = 0.04–0.07) whose velocity dispersions range from ~ 500 to ~ 1400km s^{-1}.† These galaxies have various stages and degrees of stripping, from weak or initial signs of stripping, to very strong stripping ("jellyfish galaxies", with long gas tails, Fig. 1), to the final stages of stripping (truncated disks with gas only left in the galaxy center). Their properties can be contrasted with the GASP control sample of 30 undisturbed galaxies. Recently, MUSE data of distant galaxy clusters have started to allow the first detailed studies of ram-pressure ionized gas tails also at higher redshifts (2 galaxies at z = 0.7, Boselli et al. 2019; 13 galaxies at z = 0.3–0.4, Moretti et al. 2022, Bellhouse et al. 2022).

The GASP survey is based on a MUSE ESO Large Program (spatial resolution $\sim$ 1kpc, galaxy coverage out to 7 effective radii on average), complemented by multi-wavelength follow-up programs with ALMA, APEX, JVLA, MeerKAT, ATCA, UVIT@ASTROSAT, LOFAR and Chandra. Thus, the GASP multi-wavelength coverage allows us to investigate all the main processes related to star formation, including gas in different phases, the stellar content and non-thermal processes.

2. Global and spatially resolved star formation rate-mass relation in the disks of galaxies

The existence of a relation between the integrated star formation rate (SFR) and the total stellar mass of galaxies (hereafter, "global" SFR-M relation) has been known for many years and has been observed from z = 0 to beyond z = 2. A correlation between the stellar mass surface density and the SFR surface density has been found in all integral-field spectroscopy surveys, even at higher redshifts, suggesting that the global relation originates from a perhaps more fundamental relation on small (1kpc) scales (Wuyts et al. 2013, Sanchez et al. 2013). The scatter in this relation is related to Hubble type (Gonzalez-Delgado et al. 2016) and variations in star formation efficiency (Ellison et al. 2020). Reaching larger galactocentric radii than any other large IF survey, GASP has shown that the resolved SFR-M correlation of undisturbed galaxies is broad, and that the scatter mainly arises from bright off-center star-forming knots. Moreover, each galaxy has a distinct resolved relation and the global relation appears to be driven by the existence of the size-mass relation (Vulcani et al. 2019).

† Cases of GASP ram-pressure stripped galaxies in groups and even in filaments have been presented in Vulcani et al. (2021) and references therein.

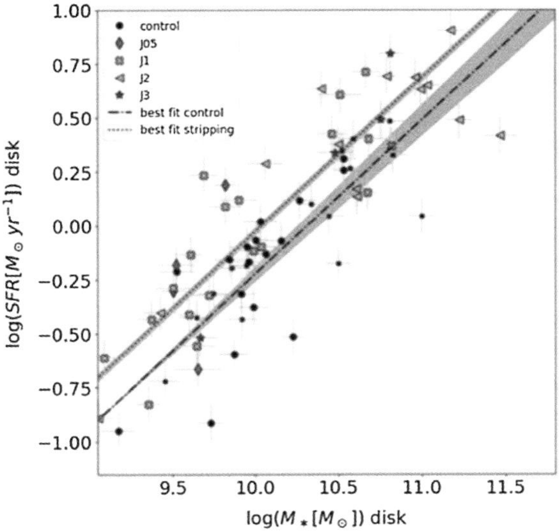

Figure 2. Global SFR-M relation for undisturbed galaxies (black points, grey line fit) versus RP stripped galaxies (colored points, light blue line fit). RP galaxies show a modest but significant enhancement in their SF activity. From Vulcani et al. (2018)

Ram-pressure stripped galaxies show a moderate enhancement of SFR in the disk for their stellar mass, lying on average above the global SFR-M relation of undisturbed galaxies by 0.2 dex (Fig. 2, Vulcani et al. 2018, Roberts & Parker 2020). Thus, during the initial and strongest stages of stripping, ram pressure enhances SF before leading to quenching. On spatially resolved scales, we find again an enhancement of SFR density at a given mass density at all galactocentric distances, consistent with being induced by RP compression waves (Vulcani et al. 2020b).

3. From stripping to quenching: post-starburst galaxies and AGN

RP stripping of the disk, and consequent quenching, proceeds mostly outside-in, or side-to-side if the galaxy is plunging edge-on into the ICM. It is therefore common to observe RP stripped galaxies still with ionized gas and intense star formation in their central regions as well as long tails of ionized gas, but having their outer disk regions already devoid of ionized gas and recently quenched (Gullieuszik et al. 2017, Poggianti et al. 2017a, 2019b, Bellhouse et al. 2017, 2019; Werle et al. 2022). These outer regions present typical post-starburst (PSB) spectra (Dressler & Gunn 1982, Poggianti et al. 1999), with no emission lines and strong Balmer lines in absorption, indicative of a local sudden truncation of the star formation activity at some point during the past ~ 0.5Gyr. It has been suggested that the strong decline of star formation is responsible for the global radio continuum excess of RP stripped galaxies compared to their ongoing SFR seen by LOFAR (Ignesti et al. 2022a,b), and that the high radio-to-Hα ratio in tails compared to disks is due to relativistic electrons advected from the disks.

If gas is totally stripped, the end product of this evolution is a PSB galaxy, with no ionized gas and no SF left anywhere in the disk. The GASP samples, both at low-z and in distant clusters, include by construction several PSB galaxies whose quenching history can be studied in details applying spectrophotometric codes (Fritz et al. 2017) to the MUSE spectroscopic datacube (Vulcani et al. 2020a, Werle et al. 2022). The stellar history maps confirm indeed mostly outside-in and side-to-side quenching, with characteristic signatures indicating that most cluster PSBs originate from RP stripping (Vulcani et al. 2020a, Werle et al. 2022). The MUSE spectra also reveal that generally,

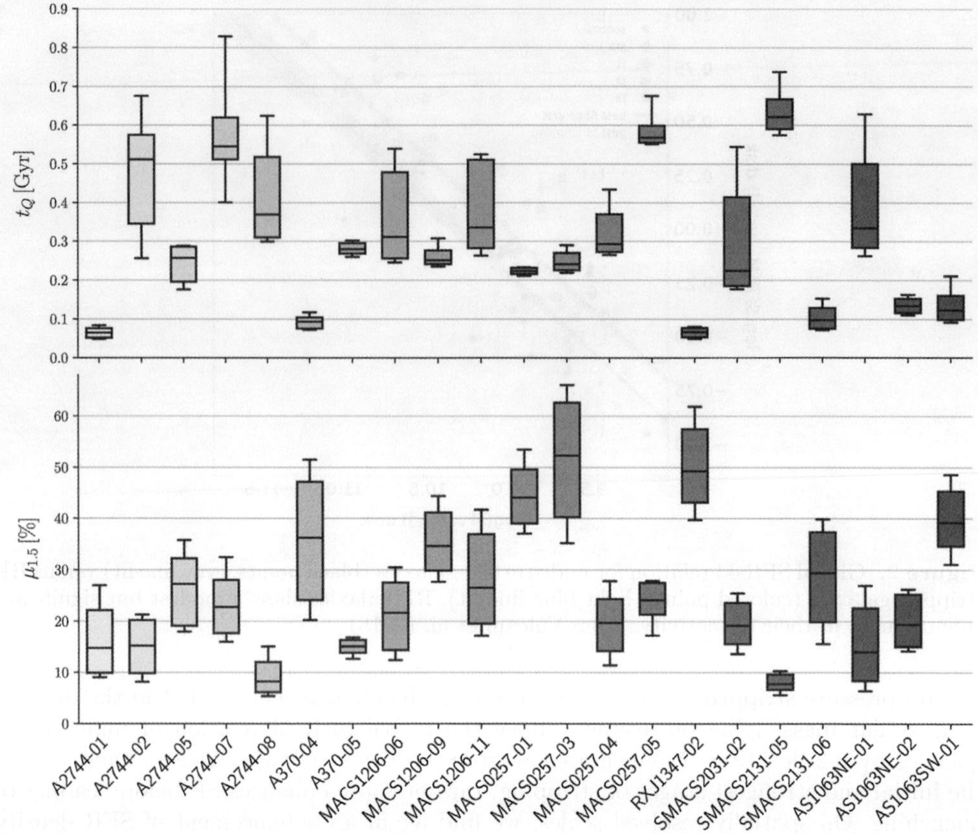

Figure 3. Characteristics of a sample of PSB galaxies in z = 0.3–0.4 clusters. Top. Time since quenching. Bottom. Fraction of mass assembled during the last 1.5 Gyr, used to roughly estimate the mass fraction involved in the burst prior to truncation. Boxes represent the 25th and 75th percentiles, the median is the horizontal line and the whiskers are the 15th and 85th percentiles. From Werle et al. (2022).

a strong local starburst takes place before the SF truncation, accounting for significant fractions of the mass formed (Fig. 3). Moreover, the quenching timescales and the total time it takes to quench all spaxels can be derived (Werle et al. 2022).

Interestingly, a small fraction of the PSB galaxies in distant clusters present an inside-out quenching, associated with AGN episodes which are still detectable from the MUSE emission lines in the centers (Werle et al. 2022). The quenching history in these cases appears to be dominated by AGN feedback, though this may be indirectly induced by RP. In fact, an usually high incidence of AGN has been found among strongly RP stripped galaxies (Poggianti et al. 2017), leading in some cases to AGN driven outflows of ionized gas (Radovich et al. 2019) and strong AGN feedback sweeping a large region around the galaxy center (George et al. 2019).

Recently, based on a combination of GASP galaxies and all available literature data for RP stripped galaxies, contrasted with a MANGA mass-matched undisturbed sample, Peluso et al. (2022) has confirmed that the incidence of AGN is higher in RP stripped galaxies. This phenomenon seems to occur preferably during the strongest phase of stripping, when the gas tails are longest. Hydrodynamical simulations reach similar conclusions (Ricarte et al. 2020; Farber et al. 2022). Simulations have previously shown that loss of angular momentum due to the interaction of the rotating ISM with the non-rotating ICM can potentially draw gas on lower orbits, and recent work has clarified the respective roles of mixing and torques from pressure gradients (Akerman et al. 2023 submitted).

4. Star formation in the tails of stripped gas and gas mixing with the intracluster medium

The fact that new stars can form in-situ in the stripped gas, outside of the galaxy disk and even far out in the stripped tails, has been shown by many studies using several different SF indicators (e.g. Yagi et al. 2007; Smith et al. 2010; Sun et al. 2010; Merluzzi et al. 2013; Fumagalli et al. 2014; Fossati et al. 2016; Consolandi et al. 2017; Boselli et al. 2018; Abramson et al. 2011; Kenney et al. 2014; George et al. 2018; Cramer et al. 2019).

In the extraplanar Hα emitting tails of GASP stripped galaxies the dominant ionization mechanism is indeed photoionization by young massive stars, as obtained by MUSE BPT diagrams (Poggianti et al. 2019a). This SF takes place in Hα-bright, dynamically cold star-forming clumps formed in-situ in the tails with luminosities similar to giant and super-giant HII regions (Fig. 1, Poggianti et al. 2019a).

The star formation occurring in the tail can produce from a negligible fraction to up to $\sim 20\%$ of the total SFR of the system (disk+tail) (Poggianti et al. 2019a), and the fraction of star formation in the tails roughly follows the fraction of gas that is stripped according to traditional analytical formulation (Gullieuszik et al. 2020): the SFR in the tail can thus be roughly predicted, in a statistical sense, knowing four main observable quantities: galaxy mass, cluster mass, galaxy line-of-sight velocity within the cluster and projected clustercentric distance.

With HST, using broad band filters spanning from UV to I-band and a narrow-band filter covering Hα, we can now study at higher spatial resolution (70pc) the star-forming clumps, in the disks, in the vicinity of the disk but "extraplanar", and in the tails. For a large sample of both Hα-selected (over 2400 clumps) and UV-selected (over 3700) clumps in GASP jellyfish galaxies we have been able to study sizes, luminosities, hierarchical structure as well as stellar masses, star formation histories and stellar ages (Giunchi et al. submitted, Werle et al. in prep.). Hα and UV luminosity and size distribution functions, together with luminosities-size relations, can be found in Giunchi et al. (2023, submitted).

The emerging picture is that basic star-forming clump properties such as luminosity and size do not depend strongly on whether the clumps are sitting in a galaxy disk or far out in the tail without an underlying disk. These characteristics of GASP clumps resemble more closely those in highly turbulent, Hα-bright galaxies (Fisher et al. 2017) than those in normal low-z spiral galaxies (see Giunchi et al. 2023 for details).

Interestingly, Hα and UV-selected clumps can be embedded in larger star-forming complexes emitting at optical wavelengths (V-band), whose total masses range typically from 10^4 to $10^7 M_\odot$ (Werle et al. in prep.). Thanks to these data, it is now possible to investigate the fate of these clumps, whether they are likely to remain intracluster isolated entities (Globular clusters? Ultracompact dwarf galaxies? Ultradiffuse galaxies?), or get dispersed and go to contribute to the general intracluster light.

Stripped material and intracluster material can exchange baryons in both directions. Evidence for mixing between the stripped interstellar medium (ISM) and the intracluster medium (ICM) has accumulated over the past few years, based e.g. on X-ray (Poggianti et al. 2019b, Campitiello et al. 2021, Sun et al. 2022; Bartolini et al. 2022) and on the analysis of the diffuse ionized gas (Tomicic et al. 2021a,b). The most direct evidence for ICM-ISM mixing has come from the gas metallicity in the stripped gas, which decreases along the tails, going away from the disk (Fig. 4, Franchetto et al. 2021, as predicted also in hydrodynamical simulations (Tonnesen & Bryan 2021).

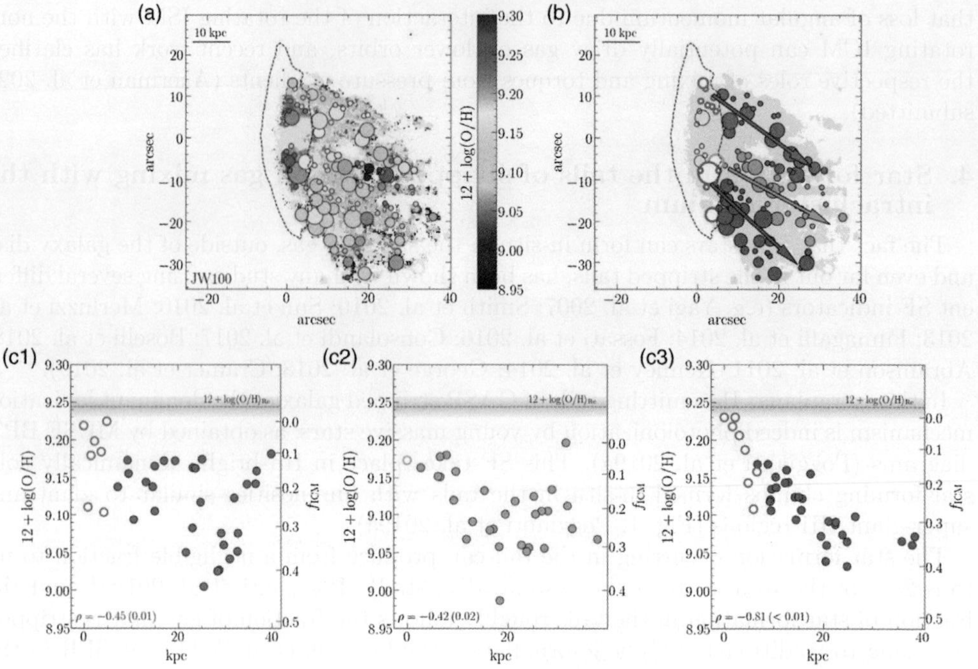

Figure 4. MUSE gas metallicity map of the galaxy JW100 (top left). Three main tails can be identified (top right). The metallicity strongly decreases along each tail (bottom panels). Assuming a typical ICM metallicity of 0.3 times solar, the fraction of gas provided by ICM mixing can be estimated (right Y-axes in bottom panels). From (Franchetto et al. 2021).

Overall, the data supports a scenario in which the stripped gas gets mixed with the ICM, accreting a significant fraction of mass, and still manages to cool and collapse to form new stars. Observations of the magnetic field aligned with the tail in the stripping direction in the JO206 GASP jellyfish galaxy favor a "magnetic draping" scenario (the galaxy moving within the magnetized ICM sweeping up the surrounding magnetic field, Dursi & Pfrommer 2008) and point out that the magnetic field may play a role favoring the SF in the tails by reducing the thermal conduction between ICM and ISM, hence the ISM evaporation (Mueller et al. 2021).

In order to complete the picture, there is another crucial stage in the star formation process: the molecular cloud formation from the neutral gas, as discussed below.

5. Neutral gas, molecular gas and star formation

As mentioned above, no selection method can provide the whole census of ram pressure stripped galaxies. It is thus of fundamental importance to obtain multi-wavelength information for samples selected in different ways. Only then we will be able to understand under what conditions multiphase tails are formed, and what is the evolutionary sequence among the observations of tails seen at different wavelengths. There are still relatively few RPS galaxies for which a detailed multi-lambda coverage is available (e.g. ESO137-001 and JW100).

For a subset of the GASP jellyfish galaxies, both HI and CO observations have been collected, in addition to MUSE. These GASP jellyfish galaxies are slightly HI deficient compared to similar non-stripped galaxies but have a significant excess of star formation for their HI content (Fig. 5, Ramatsoku et al. 2019, 2020; Deb et al. 2020, 2022, see also Luber et al. 2022 for an HI comparison of jellyfish and other cluster galaxies).

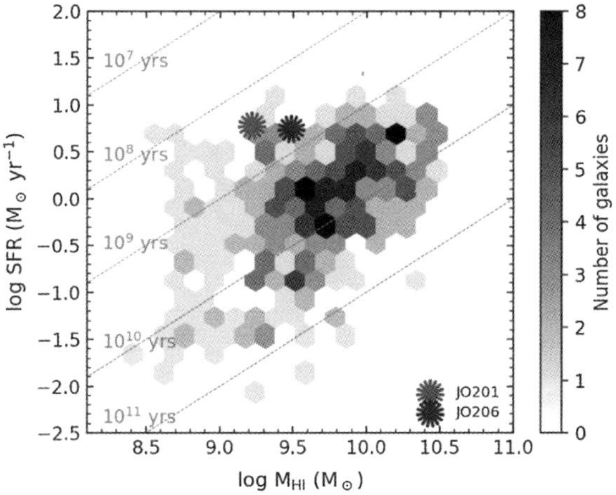

Figure 5. Global SFR - HI mass relation for two GASP jellyfish galaxies (JO201 and JO206), compared to a sample of normal spirals of similar masses (grey distribution). In jellyfish galaxies there is a clear excess of SF for their HI mass. From Ramatsoku et al. (2020), see also Deb et al. (2022).

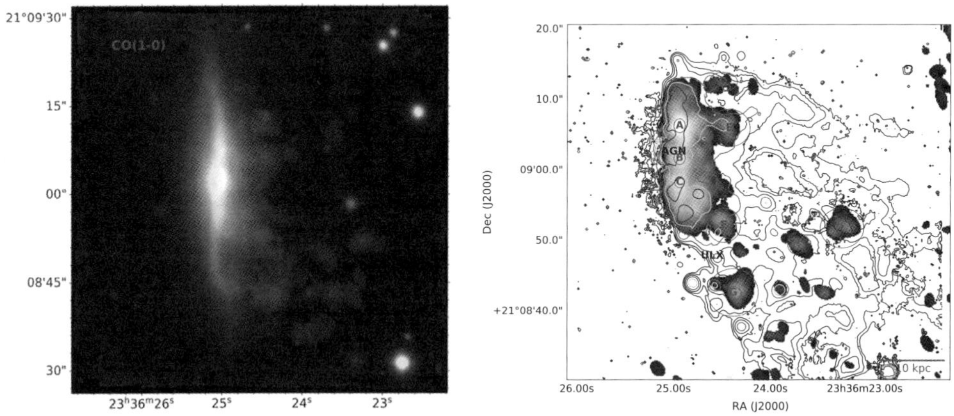

Figure 6. ALMA molecular gas (CO(1-0) emission in the jellyfish galaxy JW100 (red in the left panel, grey in the right panel). Left. There is a large amount of extraplanar CO, in clouds with molecular gas masses ranging from 10^6 to $10^9 M_\odot$. Right. Blue contours are the Hα emission. While molecular gas close to the disk may be partly stripped, the clouds far out in the tail must have formed in-situ. From Moretti et al. (2020b).

These same galaxies have very large amounts of molecular gas, as estimated from their CO(2-1) and CO(1-0) emission (Fig. 6, Jachym et al. 2017, 2019; Moretti et al. 2018b, 2020a,b). Their molecular gas mass at a given stellar mass is 4-5 times higher than in undisturbed galaxies (Fig. 7). The molecular to neutral gas ratio in their disks is between 4 and 100 times higher than normal. Surprisingly, overall, the total (molecular plus neutral) gas mass is similar to that in normal galaxies. These results, shown in Moretti et al. 2020b, are solid irrespective of the conversion factors used for CO to H_2 and they suggest a very efficient conversion of neutral gas into molecular gas in jellyfish galaxies.

As fas as the star formation efficiency is concerned (ratio between star formation rate and molecular gas mass) this appears to be significantly lower in the tails than in normal spirals, with long depletion times up to 10^{10} yr (Moretti et al. 2020b, and in prep.).

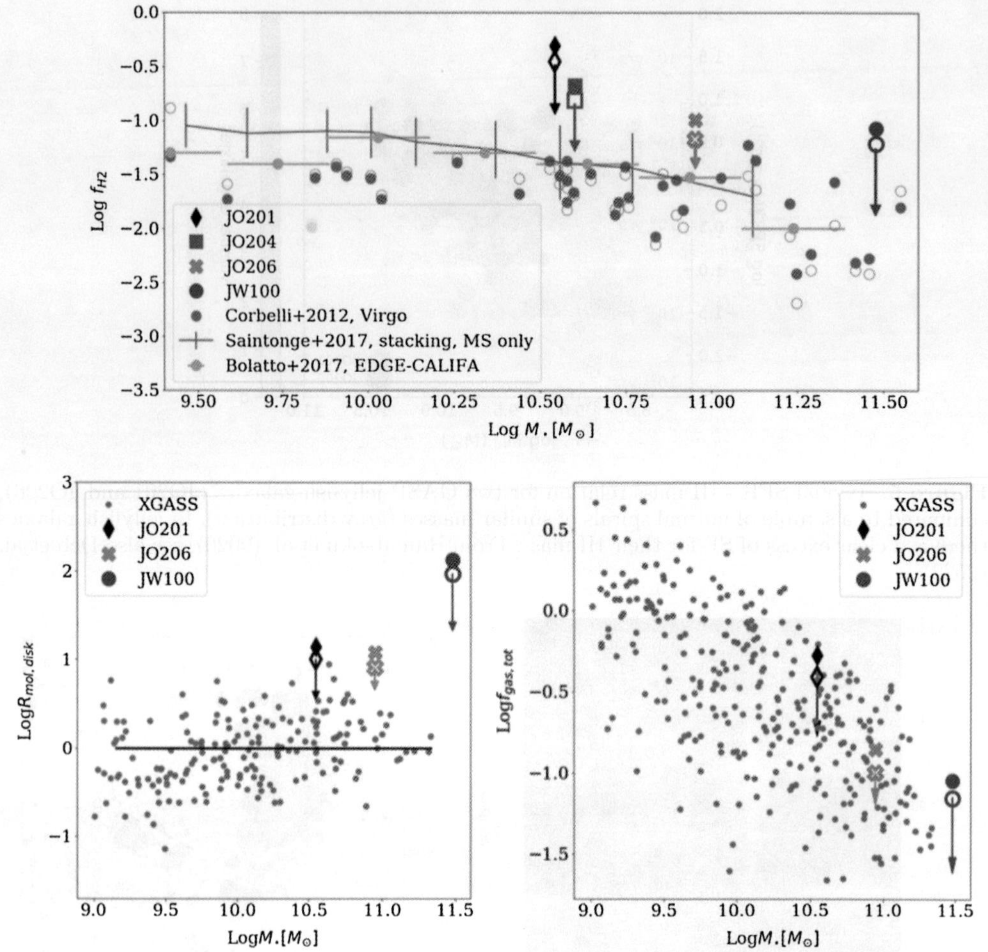

Figure 7. The molecular gas content of 4 GASP jellyfish galaxies obtained with ALMA (colored symbols) is compared with samples of normal galaxies (grey points and lines). Top. Ratio of molecular gas mass and galaxy stellar mass as a function of stellar mass. Bottom left. Molecular to neutral gas mass ratio versus stellar mass. Bottom right. Total (molecular+neutral) gas mass over stellar mass as a function of stellar mass. From Moretti et al. (2020b).

MeerKAT is soon to provide HI-selected samples for which additional multi-wavelength informations, including molecular gas estimates, have been secured.

6. Conclusions

Ram-pressure stripped galaxies are unique laboratories to study the star formation process and the baryonic cycle under unusual conditions. Gas clouds in stripped tails are embedded within the hot ICM and do not experience the influence of an underlying stellar disk, yet stars are commonly formed in stripped tails of multi-phase gas. Also in the galaxy disks, SF appears to be slightly enhanced globally, and strongly enhanced locally at any given time, where gas is still left. We witness an exceptionally efficient conversion of neutral to molecular gas in these galaxies, while the star formation efficiency ranges from usually normal in disks to much lower than normal in tails. PSB galaxies are the natural end-product of the stripping and subsequent quenching, and AGN activity possibly triggered by RP itself sometimes contributes in an inside-out manner to the quenching. I have presented only some of the highlights of the GASP project concerning

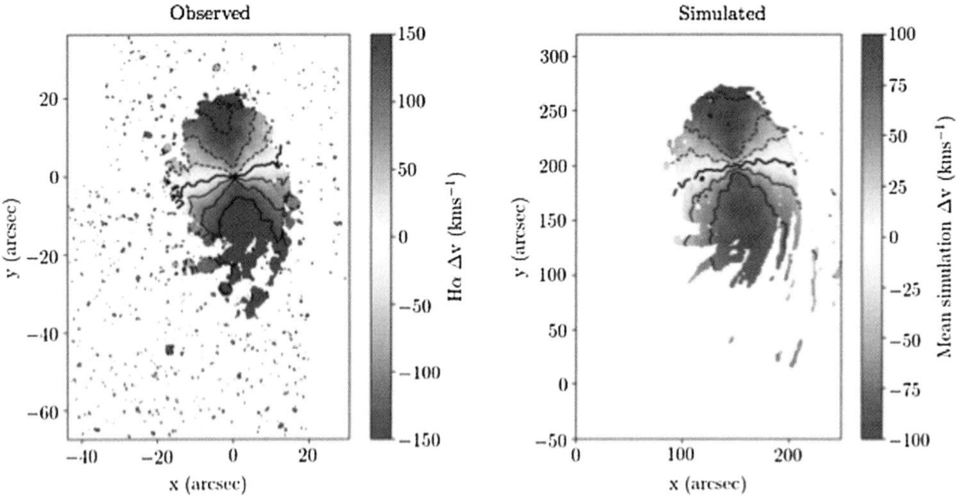

Figure 8. Example of unwinding arms in a RP stripped galaxy (JO200). The observed Hα velocity map (left) is compared with RP simulations (right). From Bellhouse et al. (2021).

the star formation activity and the baryonic cycle in cluster galaxies. I have not presented the consequences of the RP-induced SF on the galaxy structure: as an example, RP stripping can cause the unwinding of spiral arms without the contribution of tidal interactions (Fig. 8, Bellhouse et al. 2021). Interested readers can find the full list of GASP publications at `https://web.oapd.inaf.it/gasp/`.

Acknowledgements

Based on observations collected at the European Organization for Astronomical Research in the Southern Hemisphere under ESO programme 196.B-0578. This project has received funding from the European Research Council (ERC) under the European Union's Horizon 2020 research and innovation programme (grant agreement No. 833824).

References

Abramson, A., Kenney, J.D.P., Crowl, H.H., et al. 2011, *AJ*, 141, 164
Akerman et al. 2023, *ApJ* in press; arXiv 2301.09652
Bartolini, C., Ignesti, A., Gitty, M., et al. 2022, *ApJ*, 936, 74
Bellhouse, C., Jaffe', Y.L., Hau, G., McGee, S., Poggianti, B.M., et al. 2017, *ApJ*, 844, 49
Bellhouse, C., Jaffe', Y.L., McGee, S.L., et al. 2019, *MNRAS*, 485, 1157
Bellhouse, C., McGee, S., Smith, R., et al. 2021, *MNRAS*, 500, 1285
Bellhouse, C., Poggianti, B.M., Moretti, A., et al. 2022, *ApJ*, 937, 18
Boselli, A., et al. 2018, *A&A*, 614, A56
Boselli, A., Epinat, B., Contini, T., et al. 2019, *A&A*, 631, 114
Boselli, A., Fossati, M., Sun, M., et al. 2022, *A&ARv*, 30, 3
Campitiello, M.G., Ignesti, A., Gitti, M., et al. 2021, *ApJ*, 911, 144
Consolandi, G., Gavazzi, G., Fossati, M., et al. 2017, *A&A*, 606, A83
Cortese, L., Catinella, B., & Smith, R. 2021, *PASA*, 38, 35
Cramer, W.J., Kenney, J.D.P., Sun, M., et al. 2019, *ApJ*, 870, 63
Deb, T., Verheijen, M., Gullieuszik, M.,, et al. 2020, *MNRAS*, 494, 5029
Deb, T., Verheijen, M., Poggianti, B.M., et al. 2022, *MNRAS*, 516, 2683
Dressler, A., & Gunn, J.E. 1982, *ApJ*, 263, 533
Dursi, L.J., & Pfrommer, C. 2008, *ApJ*, 677, 993
Ellison, S., Thorp, M.D., Lin, L., et al. 2020, *MNRAS*, 493, 39
Farber, R.J., Ruszkowski, M., Tonnesen, S., & Holguin, F. 2022, *MNRAS*, 512, 5927

Fisher, D.B., Glazebrook, K., Damjanov, I., et al. 2017, *MNRAS*, 464, 491
Fossati, M., Fumagalli, M., Boselli, A., et al. 2016, *MNRAS*, 455, 2028
Franchetto, A., Tonnesen, S., Poggianti, B.M., et al. 2021, *ApJLetter*, 922, L6
Fumagalli, M., Fossati, M., Hau, G.K.T., et al. 2014, *MNRAS*, 445, 4335
Fritz, J., Moretti, A., Gullieuszik, M., et al. 2017, *ApJ*, 848, 132
George, K., Poggianti, B.M., Gullieuszik, M., et al. 2018, *MNRAS*, 479, 4126
George, K., Poggianti, B.M., Bellhouse, C., et al. 2019, *MNRAS*, 487, 3102
Giunchi, E., Gullieuszik, M., Poggianti, B.M., et al. 2023, *ApJ* in press arXiv2302.10615
Gonzalez-Delgado, R.M., Cid Fernandes, R., Perez, E., et al. 2016, *A&A*, 590, 44
Gullieuszik, M., Poggianti, B.M., Moretti, A., Fritz, J., Jaffe' Y.L., et al. 2017, *ApJ*, 846, 27
Gullieuszik, M., Poggianti, B.M., McGee, S., Moretti, A., Vulcani, B., et al. 2020, *ApJ*, 899, 13
Gunn, J.E., & Gott, J.R. 1972, *ApJ*, 176, 1
Ignesti, A., Vulcani, B., Poggianti, B.M., et al. 2022a, *ApJ*, 924, 64
Ignesti, A., Vulcani, B., Poggianti, B.M., et al. 2022b, *ApJ*, 937, 58
Jachym, P., Sun, M., Kenney, J.D.P., et al. 2017, *ApJ*, 839, 114
Jachym, P., Kenney, J.D.P., Sun, M., et al. 2019, *ApJ*, 883, 145
Jaffe', Y.L., et al. 2018, *MNRAS*, 482, 3454
Kenney, J.D.P., Geha, M., Jachym, P., et al. 2014, *ApJ*, 780, 119
Luber, N., Mueller, A., van Gorkom, J. et al. 2022, *ApJ*, 927, 39
Merluzzi, P., Busarello, G., Dopita, M.A., et al. 2013, *MNRAS*, 429, 1747
Moretti, A., Poggianti, B.M., Gullieuszik, M., et al. 2018a, *MNRAS*, 475, 4055
Moretti, A., Poggianti, B.M., Gullieuszik, M., et al. 2018b, *MNRAS*, 480, 2508
Moretti, A., Paladino, R., Poggianti, B.M., et al. 2020a, *ApJ*, 897, L30
Moretti, A., Paladino, R., Poggianti, B.M., Serra, P., Reodiger, E., et al. 2020b, *ApJ*, 889, 9
Moretti, A., Radovich, M., Poggianti, B.M., et al. 2022, *ApJ*, 925, 4
Mueller, A., Poggianti, B.M., Pfrommer, C., et al. 2021, *Nature Astrionomy*, 5, 159
Peluso, G., Vulcani, B., Poggianti, B.M., Moretti, A., Radovich, M., et al. 2022, *ApJ*, 927, 130
Poggianti, B.M., Smail, I., Dressler, A., et al. 1999, *ApJ*, 518, 576
Poggianti, B.M., Moretti, A., Gullieuszik, M.,, et al. 2017a, *ApJ*, 844, 48
Poggianti, B.M., Jaffe', Y.L., Moretti, A.,, et al. 2017b, *Nature*, 548, 304
Poggianti, B.M., Gullieuszik, M., Tonnesen, S., et al. 2019a, *MNRAS*, 482, 4466
Poggianti, B.M., Ignesti, A., Gitti, M., et al. 2019b, *ApJ*, 887, 155
Radovich, M., Poggianti, B.M., Jaffe, Y.L., et al. 2019, *MNRAS*, 486, 486
Ramatsoku, M., Serra, P., Poggianti, B.M., et al. 2019, *MNRAS*, 487, 4580
Ramatsoku, M., Serra, P., Poggianti, B.M., et al. 2020, *A&A*, 640, 22
Ricarte, A., Tremmel, M., Natarajan, P, & Quinn, T. 2020, *ApJL*, 895, L8
Roberts, I.D, & Parker, L.C. 2020, *MNRAS*, 495, 554
Sanchez, S., Rosales-Ortega, F.F.,Jungwiert, B., et al. 2013, *A&A*, 554, 58
Smith, R.J., Lucey, J.R., Hammer, D., et al. 2010, *MNRAS*, 408, 1417
Sun, M., Donahue, M., Roediger, E., et al. 2010, *ApJ*, 708, 946
Sun, M., Ge, C., Luo, R., et al. 2022, *Nature Astronomy*, 6, 270
Tomicic, N., Vulcani, B., Poggianti, B.M., et al. 2021b, *ApJ*, 922, 131
Tomicic, N., Vulcani, B., Poggianti, B.M., et al. 2021a, *ApJ*, 907, 22
Tonnesen, S., & Bryan, G.L. 2021, *ApJ*, 911, 68
Vulcani, B., Poggianti, B.M., Gullieuszik, M., et al. 2018, *ApJ*, 866, L25
Vulcani, B., Poggianti, B.M., Moretti, A., et al. 2019, *MNRAS*, 488, 1597
Vulcani, B., Fritz, J., Poggianti, B.M., Bettoni, D., Franchetto, A., et al. 2020a, *ApJ*, 892, 146
Vulcani, B., Poggianti, B.M., Tonnesen, S., McGee, S., Moretti, A., et al. 2020b, *ApJ*, 899, 98
Vulcani, B., Poggianti, B.M., Moretti, A., Franchetto, A., Bacchini, C., et al. 2021, *ApJ*, 914, 27
Vulcani, B., Poggianti, B.M., Smith, R., Moretti, A., Jaffe', Y.L., et al. 2022, *ApJ*, 927, 91
Yagi, M., Komiyama, Y., Yoshida, M., et al. 2007, *ApJ*, 660, 1209
Werle, A., Poggianti, B.M., Moretti, A., Bellhouse, C., Vulcani, B., et al. 2022, *ApJ*, 930, 43
Wuyts, S., Forster Schreiber, N.M., Nelson, E., et al. 2013, *ApJ*, 779, 135

Resolving the Rise and Fall of Star Formation in Galaxies
Proceedings IAU Symposium No. 373, 2023
T. Wong & W.-T. Kim, eds.
doi:10.1017/S1743921322004410

What Drives Galaxies from the Main Sequence to the Green Valley?

Lihwai Lin

Institute of Astronomy & Astrophysics, Academia Sinica, Taipei 10617, Taiwan

Abstract. Green valley galaxies (by selection) exhibit lower specific star formation rates and are thought to be in the transition from the active star-forming phase to the quiescent state. Physical mechanisms responsible for the depleted star formation in green valley galaxies, however, are still under debate. Using the ALMA-MaNGA Quenching and STar formation (ALMaQUEST) CO observations, we study the so-called 'resolved star formation scaling relations', which describe relationships among surface densities of star formation rate, stellar mass, and molecular gas mass. By comparing the kpc-scale scaling relations between the main sequence and green valley galaxies, we are able to quantify if the deficit of star formation in green valley galaxies is driven by depleted molecular gas or inefficient star formation. And finally, we present our recent ALMA dense gas (HCN and HCO^+) observations for a set of selected ALMaQUEST galaxies to discuss whether the green valley galaxies lack dense molecular gas or not.

Keywords. Galaxies; Molecular Gas; Galaxy Evolution

1. Introduction

The distributions of galaxies on the star formation rate (SFR) and the stellar mass (M_*) plane have been widely used to constrain the galaxy evolution models. In particular, the star-forming galaxies show a tight relation between the global SFR and M_*, referred to as the 'star-forming main sequence', across a wide range of redshifts (SFMS or MS; Noeske et al. 2007; Daddi et al. 2007; Lin et al. 2012; Whitaker et al. 2012; Speagle et al. 2014). Not only we want to know why there exists the star-forming main sequence, but also we want to understand under what kind of condition galaxies depart from the main sequence and become quiescent. To address theses questions, we focus on two issues, the molecular gas and the spatially resolved properties.

First, we need to bring the molecular gas into the whole picture because it is the direct fuel of star formation. Among the three variables, SFR, M_*, and molecular gas mass (M_{H_2}), we know that the relation between SFR and M_{H_2} has already been intensively studied, namely the Schmidt-Kennicutt Relation (SK relation; Schmidt 1959; Kennicutt 1998). This can be easily understood because the star formation is closely linked to how much gas is available. Compared to the SK relation, the origin of the relation between SFR and M_*, is less clear. In particular, M_* is simply the SFR integrated over time, and it is not physically intuitive why the current star formation has a strong dependence on its past history. On the other hand, the 3rd scaling relation between M_* and M_{H_2}, although not highlighted as intensively as the other two scaling relations, can be qualitatively understood as the consequence of the fact that both the gas and stellar components trace the underlying gravitational potential. Studying the gas content of galaxies is therefore essential if we want to gain a full picture about the link between star formation and M_*.

In addition to the gas, it has been recently acknowledged that the spatially-resolved observations offer a promising prospect to understand the quenching process. This is because different quenching mechanisms will lead to different patterns in the spatial distributions of gas and SF (e.g. Lin et al. 2017). For example, in the case of ram-pressure stripping, the quenching can be outside-in, and in the case of AGN, the quenching is likely to proceed from the inner part to the outer part of galaxies. Therefore, the combination of spatially-resolved observations in both gas and stars gives clues about the nature of the quenching mechanisms.

The ALMaQUEST survey (Lin et al. 2020), standing for the ALMA-MaNGA quenching and star formation, is designed to offer spatially resolved molecular gas observations that have matched spatial resolutions to the MaNGA galaxies. In total, we observe 46 MaNGA-selected galaxies in CO(1-0) at kpc scales. We target three classes of galaxies, from starburst, the main sequence, to the green valley galaxies, allowing us to probe the gas content across different populations. In addition to ALMaQUEST, there are other surveys that also offer statistical samples with spatially resolved observations, for example, the EDGE-CALIFA (Bolatto et al. 2017), COMING (Sorai et al. 2019), and PHANGS-ALMA (Leroy et al. 2021) surveys, etc., enabling detailed studies of the interplay between gas and stellar populations at kpc or subkpc scales. Compared to ALMaQUEST, the sample sizes of EDGE-CALIFA and COMING are two or even three times bigger, and the spatial resolution of PHANGS is almost 10 times better. However, the main strength of ALMaQUEST is that it samples a wider range in terms of sSFR of galaxies and galaxy properties. In addition to the main sample that covers from SB to GV mentioned here, we are now continuing to add other classes of samples like AGNs and mergers.

2. Star Formation Scaling Relation for Main Sequence Galaxies

Using the spatially resolved measurements, there have been many studies quantifying the scaling relations among the SFR, M_*, and M_{H_2} surface densities on local scales (e.g. Lin et al. 2019; Brownson et al. 2020; Morselli et al. 2020; Sánchez et al. 2021; Ellison et al. 2021; Pessa et al. 2021). For star-forming regions in MS galaxies, these three parameters are strongly correlated with each other in the 3-D space as shown in Fig. 1, and each pair forms a tight sequence, corresponding to the resolved star-forming main sequence (rSFMS), resolved Schmidt-Kennicutt (rSK) relation, and resolved molecular gas main sequence (rMGMS; Lin et al. 2019), respectively. In Lin et al. (2019) we have performed a variety of analyses, which demonstrate that the rSFMS is likely a consequence of the other two relations that are more physically motivated. For example, the correlation coefficient is weaker for rSFMS than for the other two relations. We have also performed the partial correlation, showing that the SFR is primarily driven by M_{H_2}, and once that is taken into account, there is no additional dependence on the stellar mass. Nevertheless, the interpretation of this is still under debate, and there are also other hypotheses proposed by other people Sánchez et al. (see 2021, for an alternative explanation).

3. Star Formation Scaling Relation for Green Valley Galaxies

The above relations are obtained based on well-selected star-forming spaxels of main-sequence galaxies. Then what about galaxies departing from the MS, for instance, the green valley (GV) galaxies? One thing worth paying attention to is that GVs consist of not only star-forming spaxels but also a significant fraction of non-star-forming spaxels, for example, the retired spaxels. Therefore, in addition to separating the MS and GV galaxies, in Lin et al. (2022) we also separate the star-forming and retired spaxels using their locations in the BPT diagram. The retired spaxels are defined as those with Hα equivalent width (EW) less than 3 angstroms. These are the regions that we believe are no longer with active star formation.

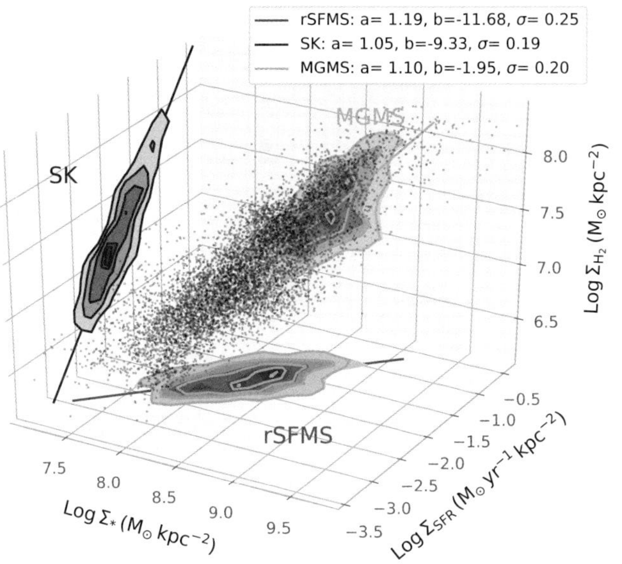

Figure 1. The 3D distribution between Σ_*, Σ_{H_2}, and Σ_{SFR}, computed for 5383 spaxels (black points) identified as star-forming regions taken from 14 ALMaQUEST (Lin et al. 2020) main sequence galaxies. The contours show the results projected on the 2D planes (red: rSFMS; blue: rSK; orange: rMGMS), with the contour levels corresponding to 20%, 40%, 60%, 80%, and 90% of the density peaks. The best-fit parameters and associated scatters (σ) based on ODR fitting are given in the legend. Figure taken from Lin et al. (2019).

Fig. 2 plot shows the resolved SFMS relation, left for the star-forming spaxels and right for the retired spaxels. The top panels are for the MS galaxies while the bottom panels are for the GV galaxies. The blue dotted line represents the original SFMS and is shown in every panel to guide our eyes. As expected (or by definition), the retired spaxels lie below the MS line. However, what is interesting here is that the relations between MS and GV are different even if we only focus on the star-forming spaxels of MS galaxies. Bearing in mind that the ratio of SFR to M_* is sSFR, this means that the sSFR of star-forming spaxels of GV galaxies is lower compared to that of MS galaxies Lin et al. (2022).

Next, we show the resolved SK relation in Fig. 3. As we reported in Lin et al. (2022), the retired spaxels lie well below the reference line, meaning that the star formation efficiency (SFE), defined as SFR divided by M_{H_2}, is pretty low in these retired regions as opposed to the star-forming spaxels. And how about the star-forming spaxels between MS and GV galaxies? If we compare the solid line, which is the best-fit for the GV galaxies, to the reference line from MS galaxies, you can see that the GV deviates from the MS galaxies toward a lower value of SFE.

Similarly, we can compare the resolved MGMS between MS and GV galaxies, as well as between the star-forming and retired spaxels as discussed in Lin et al. (2022). Fig. 4 shows that the retired spaxels also lie well below this reference line, in other words, a lower gas fraction (f_{H2}) compared to the star-forming spaxels. When comparing the star-forming spaxels between GV and MS galaxies, although the difference is small, the normalization of GV galaxies is still found to be 3-σ lower compared to that of the MS galaxies.

From the above analyses, we found that statistically GV galaxies show lower f_{H2} and lower SFE compared to the MS galaxies. On the other hand, we know that from the work by Ellison et al. (2021), each of the resolved scaling relations actually shows a lot of

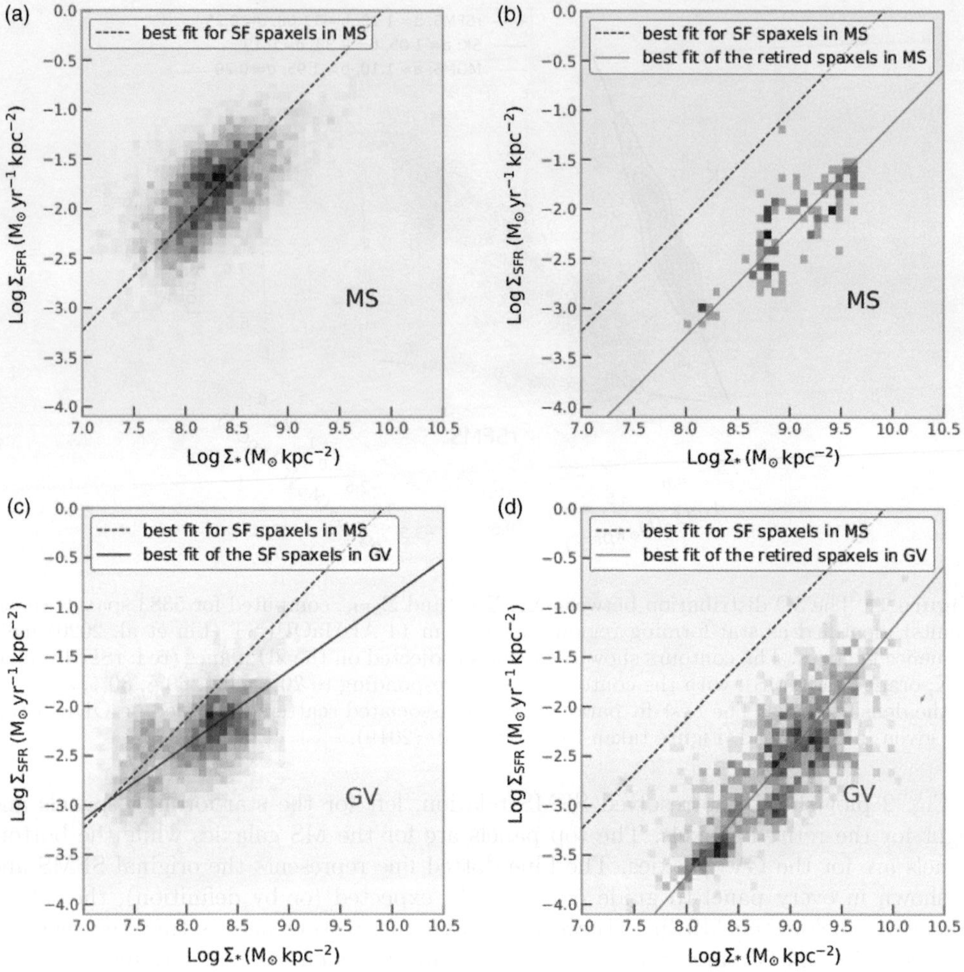

Figure 2. Number densities of star-forming (left panels; blue scales) and retired (right panels; red scales) spaxels on the rSFMS plane for MS (top panels) and GV (bottom panels) galaxies. The solid lines represent the best fit of the data points (blue for star-forming spaxels and red for retired spaxels) shown in each panel. The blue dashed lines are the same reference lines in each panel, corresponding to the best fit of the rSFMS derived for the star-froming spaxels in MS galaxies (i.e. a fit to the spaxels in the top left panel). Figure from (Lin et al. 2022).

galaxy-to-galaxy variation. So the next question we can ask is whether the deficit of gas and the inefficiency of star formation coexist in every single GV galaxy, or some galaxies show lower gas fractions whereas some others have lower SFE.

To investigate this issue, for each GV, we plot the spaxels with suppressed sSFR on the ΔSFE vs. Δf_{H2} plane (Fig. 6). An one-to-one line is plotted in each panel. If most spaxles lie above this line, it means that the low sSFR of this galaxy is under fuel-driven quenching. In contrast, if most spaxels lie below this line, it means that it is under efficiency-driven quenching. For the 15 GV galaxies that we are able to perform this classification, we found that roughly 40% are efficiency-driven, 33% are fuel-driven, and 27% have contributions from both factors. All these different situations are comparable, and there seems no dominant factor between fuel and efficiency driven that leads to the quenching in GV galaxies. Our results suggest that the mechanisms that cause the deficit of SFR in GV galaxies are likely case by case.

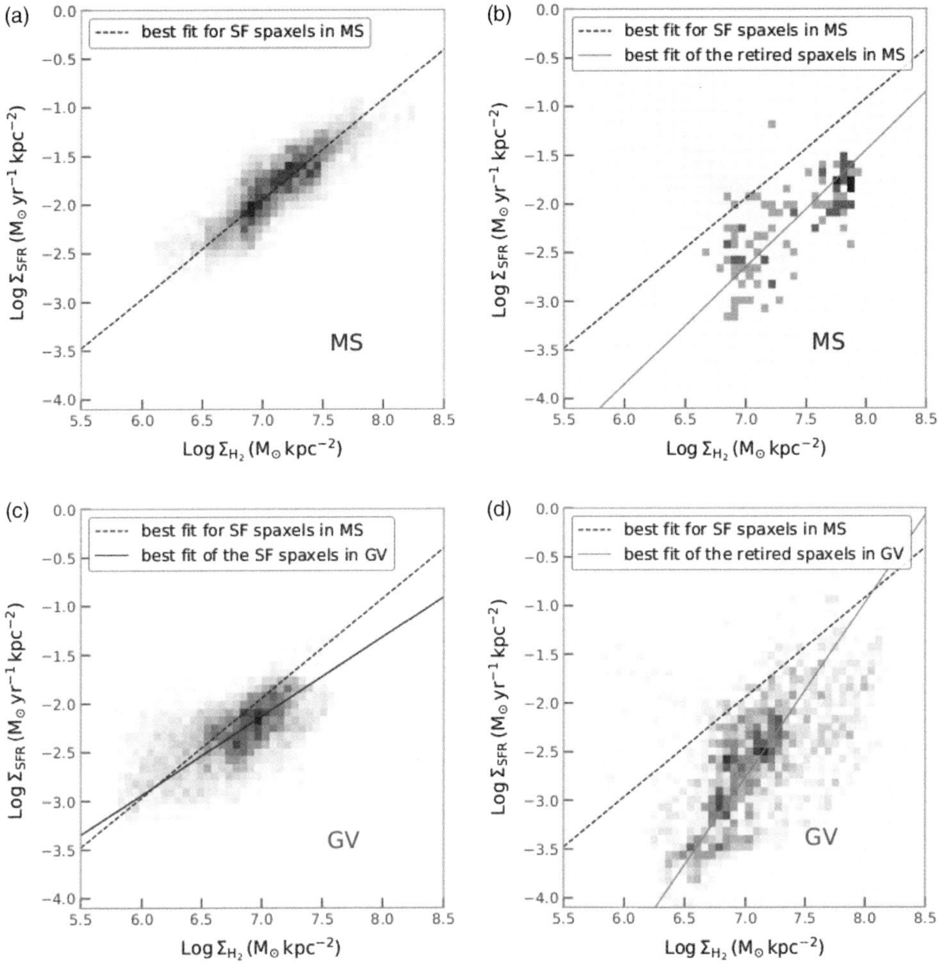

Figure 3. Similar to Fig. 2, but for the star formation rate surface density ($\Sigma_{\rm SFR}$) versus the molecular gas mass surface density ($\Sigma_{\rm H_2}$). Figure from Lin et al. (2022).

4. Dense gas

So far, our discussion was based on CO-based measurements. However, we know that it is the dense gas that is more directly relevant to star formation. The molecular gas-based SFE can be expressed as the dense gas SFE times the dense gas fraction (defined as the dense-to-molecular gas ratio), so the low CO-based SFE could be either due to the low dense gas star formation efficiency or the low dense gas fraction.

In order to distinguish these two effects, we follow up 5 ALMaQUEST galaxies with dense gas tracers, HCN and HCO^+ using ALMA. In Fig. 6, we show the CO-based SFE, versus the dense gas fraction. The blue dots are star-forming spaxels, and the red colors denote spaxels classified as retired regions. The grey symbols are unclassified spaxels. We can see that the CO-based star formation efficiency is considerably lower in retired and unclassified regions compared to that in star-forming spaxels as we discussed previously, but their dense fraction is actually comparable to those of star-forming regions. There is no obvious correlation between the CO-based SFE and the dense gas fraction. If this is true, then we expect that the dense gas-based SFE should be low for non-star forming regions.

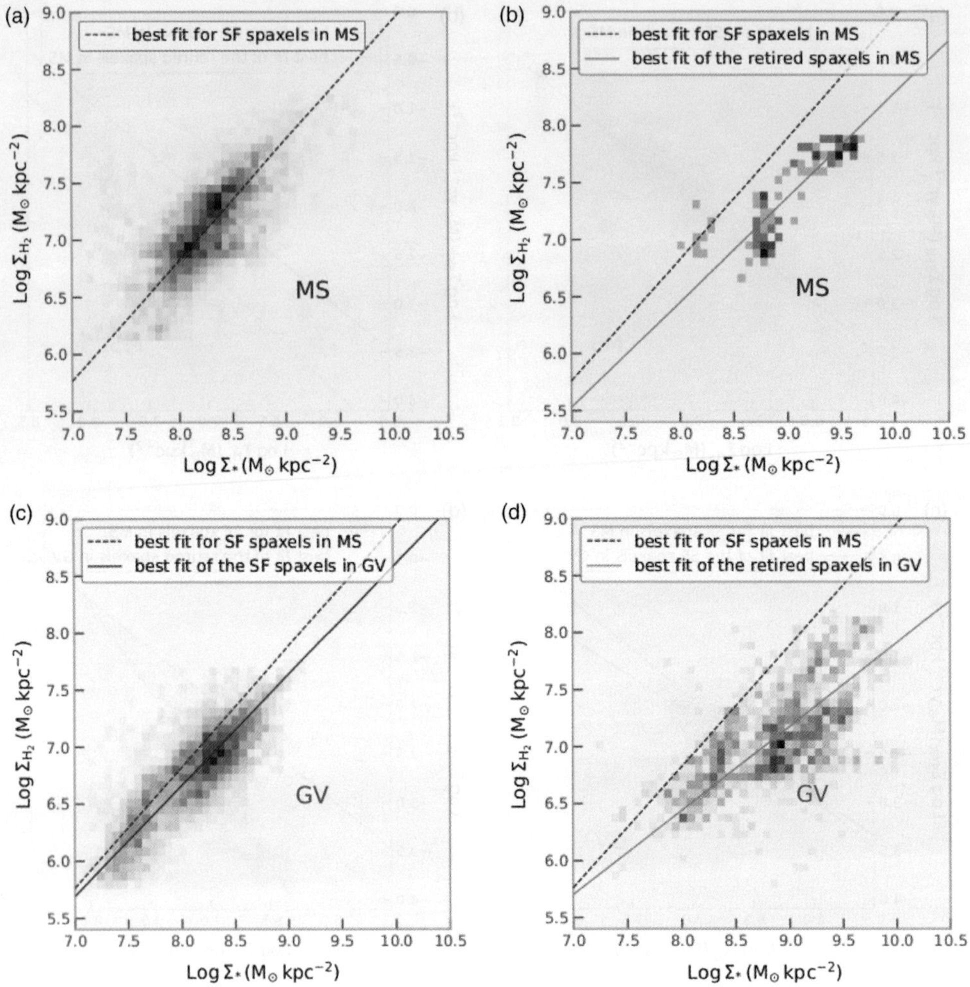

Figure 4. Similar to Fig. 2, but for molecular gas mass surface density (Σ_{H_2}) versus the stellar mass surface density (M_*) relation. Figure from Lin et al. (2022).

This is exactly what we see in the dense gas Schmidt-Kennicutt relation as shown in Fig. 7. We can see that while the star-forming spaxels form a clear correlation between the SFR and the dense gas surface density, the non-star-forming regions lie below the resolved SK relation formed by the star-forming spaxels and hence show lower dense gas SFE. In other words, there is some process that prevents even the dense gas from forming stars, leading to the overall low SFE (defined by the CO-based measurement), despite that they still have the dense gas.

5. Summary

The advent of spatially resolved stellar and gas observations has unveiled crucial links between the star formation rate, stellar mass, and molecular gas mass at (sub)kpc scales. These local relationships provide new insight to the understanding of global relationships. The major issues touched in this talk can be summarized as below:

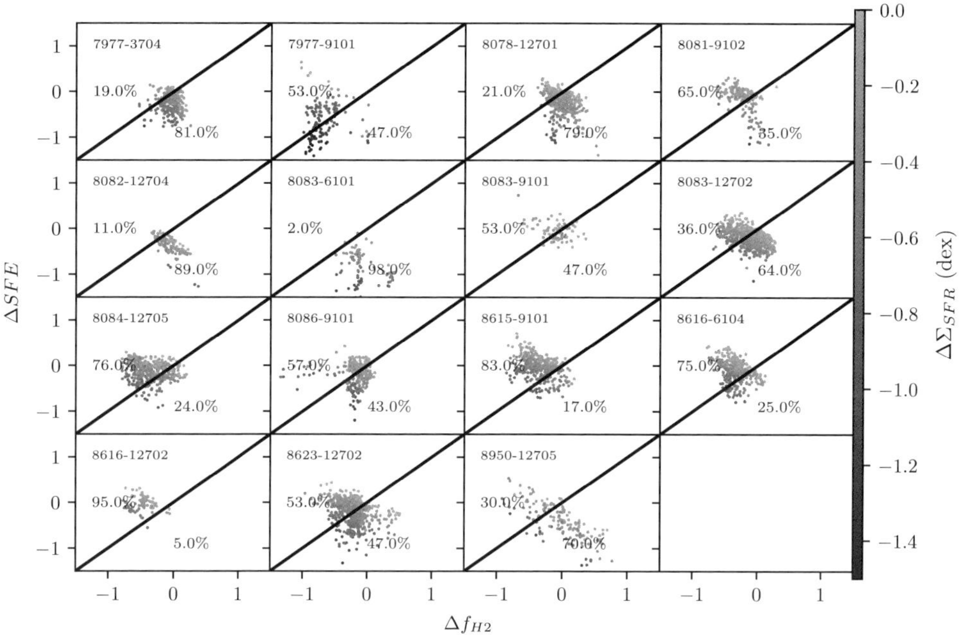

Figure 5. The spaxel-by-spaxel based ΔSFE versus Δf_{H2} in our GV sample, color coded by the ΔSFR. The fractions of spaxels locating on either side of the one-to-one solid line separating the SFE-driven and f_{H2} driven contributions are shown in each panel. Figure from Lin et al. (in prep.).

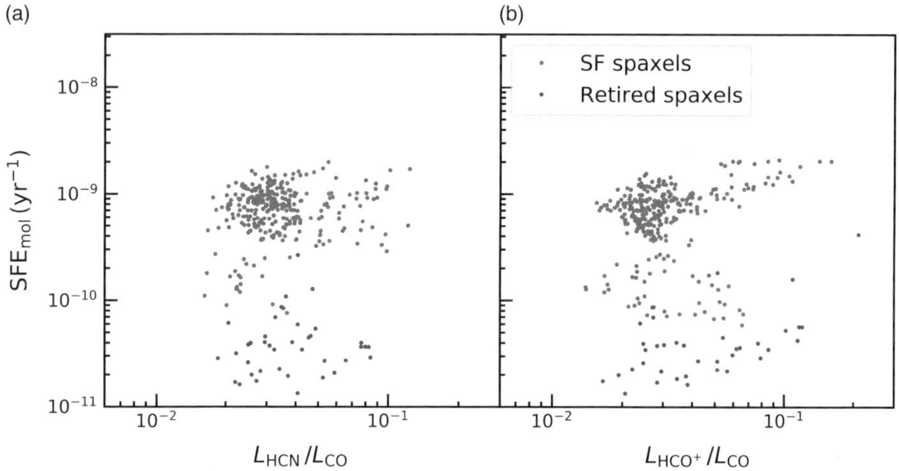

Figure 6. The CO-based star formation efficiency as a function of HCN-to-CO luminosity ratio (left panel) and HCO^+-to-CO luminosity ratio. The blue dots are star-forming spaxels, and the red colors denote spaxels classified as retired regions. The grey symbols are unclassified spaxels. Figure from Lin et al. (in prep.).

• Spatially resolved observations offer promising opportunities to probe the scaling relations of galaxies at kpc scales. Each pair of the 3 parameters, SFR, M_*, and M_{H_2} surface densities form a tight relation, referred to as rSFMS, rSK, and rMGMS, respectively. The rSFMS is suggested to be a by-product of the other two relations.

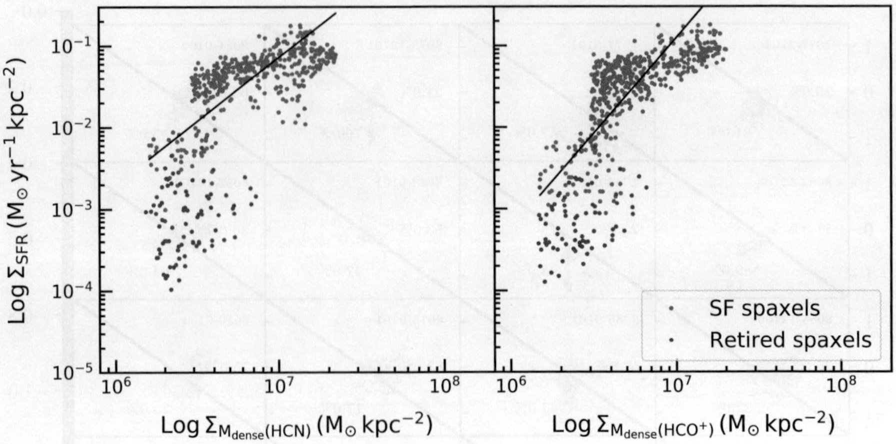

Figure 7. The dense gas version of the resolved Schmidt-Kennicutt relation in our sample. Left: HCN is used as the dense gas tracer. Right: HCO^+ is used as the dense gas tracer. The blue dots are star-forming spaxels, and the red colors denote spaxels classified as retired regions. The grey symbols are unclassified spaxels. Figure from Lin et al. (in prep.).

• The green valley galaxies show distinct scaling relations from the main sequence galaxies, having lower star formation efficiency and lower molecular gas fraction, even in the star-forming spaxels.

• The low CO-based star formation efficiency (SFE) of green valley galaxies is more associated with the low dense gas SFE, rather than the lower dense-to-molecular gas ratio.

Acknowledgements

LL thanks the conference organizers for their invitation. We also thank the ALMaQUEST members and other co-authors on the work described in this talk. A significant amount of the work presented here is based on the MaNGA survey and the precious ALMA observations.

References

Bolatto, A. D., Wong, T., Utomo, D., et al. 2017, ApJ, 846, 159
Brownson, S., Belfiore, F., Maiolino, R., et al. 2020, MNRAS, 498, L66
Daddi, E., Dickinson, M., Morrison, G., et al. 2007, ApJ, 670, 156
Ellison, S. L., Lin, L., Thorp, M. D., et al. 2021, MNRAS, 501, 4777
Kennicutt, R. C., Jr. 1998, ApJ, 498, 541
Leroy, A. K., Schinnerer, E., Hughes, A., et al. 2021, ApJS, 257, 43
Lin, L., Dickinson, M., Jian, H.-Y., et al. 2012, ApJ, 756, 71
Lin, L., Belfiore, F., Pan, H.-A., et al. 2017, ApJ, 851, 18
Lin, L., Pan, H.-A., Ellison, S. L., et al. 2019, ApJL, 884, L33
Lin, L., Ellison, S. L., Pan, H.-A., et al. 2020, ApJ, 903, 145
Lin, L., Ellison, S. L., Pan, H.-A., et al. 2022, ApJ, 926, 175
Morselli, L., Rodighiero, G., Enia, A., et al. 2020, MNRAS, 496, 4606
Noeske, K. G., Faber, S. M., Weiner, B. J., et al. 2007, ApJ, 660, L47
Pessa, I., Schinnerer, E., Belfiore, F., et al. 2021, A&A, 650, A134
Sánchez, S. F., Barrera-Ballesteros, J. K., Colombo, D., et al. 2021, MNRAS, 503, 1615
Schmidt, M. 1959, ApJ, 129, 243
Sorai, K., Kuno, N., Muraoka, K., et al. 2019, PASJ, 71, S14.
Speagle, J. S., Steinhardt, C. L., Capak, P. L., & Silverman, J. D. 2014, ApJS, 214, 15
Whitaker, K. E., van Dokkum, P. G., Brammer, G., et al. 2012, ApJ, 754, L29

Resolving the Rise and Fall of Star Formation in Galaxies
Proceedings IAU Symposium No. 373, 2023
T. Wong & W.-T. Kim, eds.
doi:10.1017/S1743921322004860

Osaka Feedback Model II: Modeling Supernova Feedback Based on High-Resolution Simulations

Yuri Oku[1], Kengo Tomida[2], Kentaro Nagamine[1,3,4], Ikkoh Shimizu[5] and Renyue Cen[1,6]

[1]Theoretical Astrophysics, Department of Earth & Space Science, Graduate School of Science, Osaka University, 1-1 Machikaneyama, Toyonaka, Osaka 560-0043, Japan
email: oku@astro-osaka.jp

[2]Astronomical Institute, Tohoku University, Aoba, Sendai, Miyagi 980-8578, Japan

[3]Kavli IPMU (WPI), The University of Tokyo, 5-1-5 Kashiwanoha, Kashiwa, Chiba 277-8583, Japan

[4]Department of Physics & Astronomy, University of Nevada, Las Vegas, 4505 S. Maryland Pkwy, Las Vegas, NV 89154-4002, USA

[5]Shikoku Gakuin University, 3-2-1 Bunkyocho, Zentsuji, Kagawa, 765-8505, Japan

[6]Department of Astrophysical Sciences, Princeton University, Peyton Hall, Princeton, NJ 08544, USA

Abstract. Feedback from supernovae (SNe) is an essential mechanism that self-regulates the growth of galaxies. We build an SN feedback model based on high-resolution simulations of superbubble and SN-driven outflows for the physical understanding of the galaxyCGM connection. Using an Eulerian hydrodynamic code `Athena++`, we find universal scaling relations for the time evolution of superbubble momentum, when the momentum and time are scaled by those at the shell-formation time. We then develop an SN feedback model utilizing Voronoi tessellation, and implement it into the `GADGET3-Osaka` smoothed particle hydrodynamic code. We show that our stochastic thermal feedback model produces galactic outflow that carries the metals high above the galactic plane but with weak suppression of star formation. Additional mechanical feedback further suppresses star formation. Therefore, we argue that both thermal and mechanical feedback is necessary for the SN feedback model of galaxy evolution when an individual SN bubble is unresolved.

Keywords. Galaxy formation, Numerical simulation, Stellar feedback, Supernovae, Galactic winds

1. Introduction

Supernova feedback (SN FB) is an essential mechanism that self-regulates the growth of galaxies, and a better understanding of SN FB is necessary to resolve the rise and fall of star formation in galaxies. Numerical simulations are a powerful tool to study the nonlinear evolution of galaxies on a cosmological timescale. However, it is challenging to resolve individual SN bubbles in galaxy simulations due to their low resolution of mass and space, which are not enough to resolve the cooling mass and length of the SN bubble. Therefore, it is necessary to model the effect of SN FB on sub-resolution scale, and many flavors of SN FB subgrid models have been developed (e.g., Stinson et al. 2006; Dalla Vecchia & Schaye 2012; Keller et al. 2014; Kimm & Cen 2014; Hopkins et al. 2018).

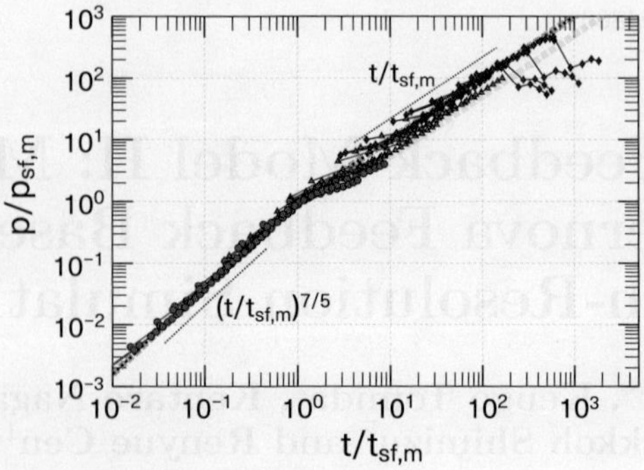

Figure 1. Evolution of momentum p normalized by $p_{\rm sf,m}$ vs. time t normalized by $t_{\rm sf,m}$. Red and blue points depict the runs with $\Delta t_{\rm SN} < 0.1 t_{\rm PDS}$ and $\Delta t_{\rm SN} > 0.1 t_{\rm PDS}$, respectively. Different symbols connected by lines correspond to runs at different SN intervals: $\Delta t_{\rm SN} = 0.01$ Myr (circle), $\Delta t_{\rm SN} = 0.1$ Myr (triangle), $\Delta t_{\rm SN} = 1$ Myr (diamond). The black dotted lines indicate the $(p/p_{\rm sf,m}) \propto (t/t_{\rm sf,m})^{7/5}$ and $(p/p_{\rm sf,m}) \propto (t/t_{\rm sf,m})$ power law, respectively.

In this work, we focus on two effects of SN FB: mechanical FB driving interstellar turbulence and thermal FB launching hot outflows. We show that models are necessary for both FB modes; mechanical FB model for spatially unresolved growth of superbubble, and stochastic thermal FB model for hot gas unresolved in mass. For the mechanical FB, we first investigate the momentum input by superbubble using an Eulerian hydrodynamic code `Athena++` and derive a formula to predict the input momentum for given density and metallicity of the interstellar medium (ISM).

This article is based on our published paper (Oku et al. 2022), and the interested readers are referred to the paper for further details.

2. `Athena++` simulation of superbubbles

We perform a suite of three-dimensional hydrodynamical simulations of a superbubble using `Athena++` (Stone et al. 2020). The SN energy of 10^{51} erg is injected ten times with a fixed time interval $\Delta t_{\rm SN}$ at the center of the simulation box. We investigate the dependence of the momentum of a superbubble on the density n and metallicity Z of the ISM and $\Delta t_{\rm SN}$. The parameter ranges investigated in this work are $n/{\rm cm}^{-3} \in \{0.1, 1, 10\}$, $Z/Z_\odot \in \{10^{-3}, 10^{-2}, 10^{-1}, 1\}$, and $\Delta t_{\rm SN}/{\rm Myr} \in \{0.01, 0.1, 1\}$, and the simulation suite consists of 36 simulations with different combinations of n, Z, and $\Delta t_{\rm SN}$. The size of the simulation box is $L/{\rm pc} = 135,\ 300,\ 900$ for cases of $n/{\rm cm}^{-3} = 0.1$, 1, 10, and the simulation box has periodic boundaries.

Figure 1 shows the time evolution of the momentum of the simulated superbubbles with x- and y-axis normalized by the time and momentum at the shell formation. We extend the analytic estimation of the shell formation time $t_{\rm sf,m}$ by Kim & Ostriker (2015) and Kim et al. (2017) by including the metallicity dependence:

$$t_{\rm sf,m} = 1.7 \times 10^4 \,{\rm yr}\, E_{51}^{0.28}\, \Delta t_{\rm SN,6}^{-0.28}\, n_0^{-0.71}\, \Lambda_{6,-22}^{-0.42}, \tag{2.1}$$

where $E_{51} = E_{\rm SN}/10^{51}$ erg, $\Delta t_{\rm SN,6} = \Delta t_{\rm SN}/1\,{\rm Myr}$, $n_0 = n/1\,{\rm cm}^{-3}$, and $\Lambda_{6,-22} = \max(1.9 - 0.85\, Z/Z_\odot, 1.05)\,(Z/Z_\odot) + 10^{-1.33}$. Here, $\Lambda_{6,-22}$ is the fitting function to the value at $T = 10^6$ K of the metal cooling function by Sutherland & Dopita (1993).

The normalized time evolution of the superbubble momentum show universality over a wide range of environments, indicating that the shell-formation time is the characteristic timescale of superbubble growth, which is consitent with the previous finding by Kim et al. (2017).

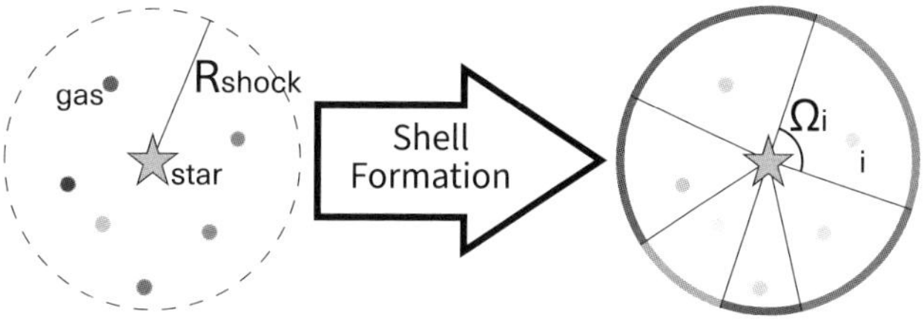

Figure 2. Schematic description of the spherical superbubble feedback model developed in this work. The left-hand side shows the particle distribution of gas and star particles when the supernova explosion takes place, whereas the right-hand side shows the superbubble shell formation, and how we split the shell using Voronoi tessellation based on the gas particle distribution inside the bubble.

We further derive the momentum of superbubble per SN assuming the initial star cluster mass function $dN/dM \propto M^{-2}$ (Krumholz et al. 2019):

$$\hat{p} = 1.75 \times 10^5\, M_\odot\, \mathrm{km\,s^{-1}} n_0^{-0.05}\, \Lambda_{6,-22}^{-0.17}. \tag{2.2}$$

The dependency on $\Lambda_{6,-22}$ indicates that the momentum input by superbubble is larger by a factor of three in a zero metallicity environment than in solar metallicity.

3. Construction of SN FB model

In this section, we construct an SN FB model for particle-based simulations of galaxy formation.

Mechanical FB Model. We model the mechanical effect of SN FB using Equation (2.2) to predict the amount of momentum input by SNe at the SN event site. We distribute the momentum input to surrounding gas particles using Voronoi tesselation as illustrated in Figure 2. The surrounding particles are projected to the surface of a sphere centered at the SN site, and the Voronoi tesselation is performed on the surface. We use the solid angle to the face on the Voronoi polyhedron Ω_i for weighting the SN input to the surrounding particles.

Stochastic Thermal FB Model. Superbubbles can grow to the size of galactic disk scale height, and then break out to produce energetic outflow (Fielding et al. 2018). We model this wind launching effect by SNe as stochastic thermal energy injection, similar to the model developed by Dalla Vecchia & Schaye (2012). We use the entropy $S = k_\mathrm{B}\, T\, n^{1-\gamma}$ as a controlling parameter. Hu (2019) showed that typical entropy of hot outflow to be $10^8 - 10^9\, k_\mathrm{B}\, \mathrm{K\,cm^2}$ using a high-resolution dwarf galaxy simulation, and we adopt $S_\mathrm{FB} = 10^8\, k_\mathrm{B}\, \mathrm{K\,cm^2}$ as the default value. We stochastically inject thermal energy ΔE_i to increase the entropy of a i-th particle to S_FB, and the probability of injecting thermal energy to i-th particle is $P_i = \Delta E_i/(\Omega_i E_\mathrm{FB,th})$, where $E_\mathrm{FB,th}$ denotes the total thermal energy input by the SN event, which is set to 70% of total energy input.

4. `GADGET3-Osaka` simulation of isolated galaxies

To explore the thermal and mechanical FB effects, we run five simulations with different FB settings summarized in Table 1. We implement the FB model described in the previous section to the smoothed particle hydrodynamics code `GADGET3-Osaka` (Shimizu et al. 2019) which includes radiative cooling, heating, and star formation models as described in Oku et al. (2022). The initial condition is an isolated Milky-way-mass galaxy, taken from the AGORA project (Kim et al. 2014), and we add gas halo following Shin et al. (2021). We evolve the system to 1 Gyr with subgrid physics turned on after evolving it adiabatically to 0.5 Gyr for relaxation. In the following, we show the time evolution of the simulated galaxies for 1 Gyr with subgrid physics.

Table 1. List of isolated galaxy simulations used in this work

Run Name	FB Type				Notes
	SNII-M[a]	SNII-T[b]	Stochastic[c]	SNIA & OB[d]	
Fiducial	✓	✓	✓	✓	Fiducial run with all feedback models
Non-stochastic	✓	✓		✓	Fiducial but thermal energy is distributed without stochastic treatment
SNII-kinetic	✓				Type II SN mechanical feedback only
SNII-thermal		✓	✓		Type II SN thermal feedback only
No-FB					No feedback

Notes. Model description of simulations used in this work. The leftmost column shows the name of run, the rightmost column shows notes, and the middle four columns show FB types activated in the simulation.
[a]Type-II SN mechanical FB
[b]Type-II SN thermal FB
[c]Stochastic thermal FB
[d]Type-Ia and OB star mechanical FB

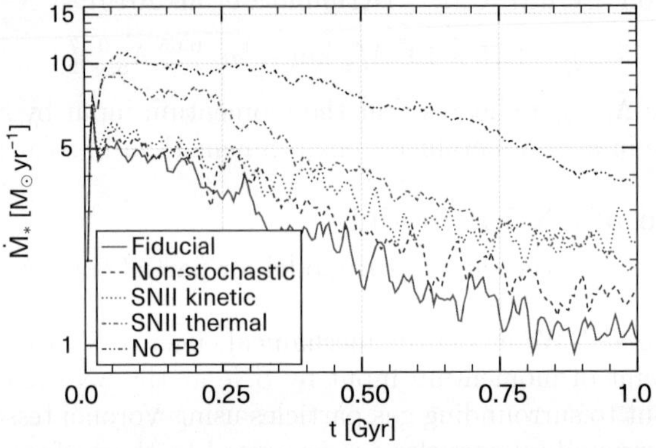

Figure 3. Star formation rate as a function of time for the isolated galaxy simulations given in Table 1. The No-FB run has the highest SFR, and the SNII-thermal run shows weak suppression of SFR. In other runs, SFR is suppressed by the mechanical SN FB efficiently.

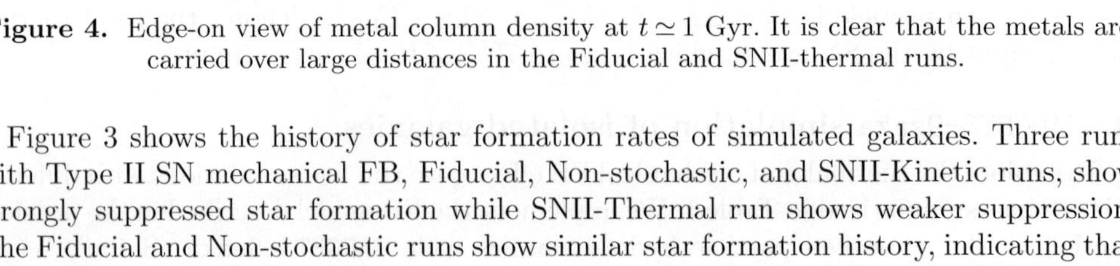

Figure 4. Edge-on view of metal column density at $t \simeq 1$ Gyr. It is clear that the metals are carried over large distances in the Fiducial and SNII-thermal runs.

Figure 3 shows the history of star formation rates of simulated galaxies. Three runs with Type II SN mechanical FB, Fiducial, Non-stochastic, and SNII-Kinetic runs, show strongly suppressed star formation while SNII-Thermal run shows weaker suppression. The Fiducial and Non-stochastic runs show similar star formation history, indicating that the stochastic thermal feedback has a minor effect on the suppression of star formation, while the Fiducial run shows a slightly lower star formation rate.

Figure 4 shows the edge-on view of metal column density of simulated galaxies at $t \simeq 1$ Gyr. One can clearly see that metals are carried over large distances in the Fiducial and SNII-Thermal runs, indicating strong outflow is produced by stochastic thermal FB as intended in our modeling.

5. Conclusion and Future Prospects

We have studied the environment dependence of the momentum input from superbubble by three-dimensional hydrodynamical simulation using `Athena++` code, and obtained Equation (2.2), which predicts the momentum input per SN for given density and metallicity of ISM. We then constructed the SN FB model for particle-based simulation using Voronoi tessellation as shown in Figure 2. We separately modeled the mechanical and thermal effect of SN FB based on results from high-resolution simulations. We have implemented the SN FB model into `GADGET3-Osaka` code, and performed isolated milky-way-mass galaxy simulations. We have demonstrated that the mechanical FB suppresses star formation while stochastic thermal FB drives strong metal outflows.

We have performed idealized isolated galaxy simulations in this work, and more realistic cosmological simulation is necessary to compare our result to observational data. We will investigate metal distribution in CGM and statistics of galaxies, e.g., galaxy stellar mass function and mass-metallicity relation, in cosmological simulations with our SN FB model developed in this work, and compare them with observations in our future work.

References

Dalla Vecchia, C. & Schaye, J. 2012, *MNRAS*, 426, 140
Fielding, D., Quataert, E., & Martizzi, D. 2018, *MNRAS*, 481, 3325
Hopkins, P. F., Wetzel, A., Kereš, D., *et al.* 2018, *MNRAS*, 477, 1578
Hu, C.-Y. 2019, *MNRAS*, 483, 3363
Keller, B. W., Wadsley, J., Benincasa, S. M., & Couchman, H. M. P. 2014, *MNRAS*, 442, 3013
Kim, J.-H., Abel, T., Agertz, O., *et al.* 2014, *ApJS*, 210, 14
Kim, C.-G. & Ostriker, E. C. 2015, *ApJ*, 802, 99
Kim, C.-G., Ostriker, E. C., & Raileanu, R. 2017, *ApJ*, 834, 25
Kimm, T. & Cen, R. 2014, *ApJ*, 788, 121
Krumholz, M. R., McKee, C. F., & Bland-Hawthorn, J. 2019, *ARAA*, 57, 227
Oku, Y., Tomida, K., Nagamine, K., Shimizu, I., & Cen, R. 2022, *ApJS*, 262, 9
Shimizu, I., Todoroki, K., Yajima, H., & Nagamine, K. 2019, *MNRAS*, 484, 2632
Shin, E.-J., Kim, J.-H., & Oh, B. K. 2021, *ApJ*, 917, 12
Stinson, G., Seth, A., Katz, N., Wadsley, J., Governato, F., & Quinn, T. 2006, *MNRAS*, 373, 1074
Stone, J. M., Tomida, K., White, C. J., & Felker, K. G. 2020, *ApJS*, 249, 4
Sutherland, R. S. & Dopita, M. A. 1993, *ApJS*, 88, 253

Resolving the Rise and Fall of Star Formation in Galaxies
Proceedings IAU Symposium No. 373, 2023
T. Wong & W.-T. Kim, eds.
doi:10.1017/S1743921322004240

On the H I Content of MaNGA Major Merger Pairs

Qingzheng Yu[1], Taotao Fang[1], Shuai Feng[2,3], Bo Zhang[4], C. Kevin Xu[4,5], Yunting Wang[1,6] and Lei Hao[7]

[1]Department of Astronomy, Xiamen University, Xiamen, Fujian 361005, China; email: fangt@xmu.edu.cn

[2]College of Physics, Hebei Normal University, 20 South Erhuan Road, Shijiazhuang, Hebei 050024, China

[3]Hebei Key Laboratory of Photophysics Research and Application, Shijiazhuang, Hebei 050024, China

[4]National Astronomical Observatories, Chinese Academy of Sciences (NAOC), Beijing 100101, China

[5]Chinese Academy of Sciences South America Center for Astronomy, National Astronomical Observatories, CAS, Beijing 100101, China

[6]Department of Physics and Astronomy, University of British Columbia, 6225 Agricultural Road, Vancouver, V6T 1Z1, Canada

[7]Shanghai Astronomical Observatory, Chinese Academy of Sciences, Shanghai 200030, China

Abstract. To study the role of H I content in galaxy interactions, we select galaxy pairs and control galaxies from the SDSS-IV MaNGA IFU survey, adopting kinematic asymmetry as a new effective indicator to describe the merger stage. With archival data from the HI-MaNGA survey and new observations from the Five-hundred-meter Aperture Spherical radio Telescope (FAST), we investigate the differences in H I gas fraction ($f_{\rm H\,I}$), star formation rate (SFR), and H I star formation efficiency ($\rm SFE_{H\,I}$) between pairs and controls. Our results suggest that on average the H I gas fraction of major-merger pairs is marginally decreased by $\sim 15\%$ relative to isolated galaxies, and paired galaxies during pericentric passage show weakly decreased $f_{\rm H\,I}$ (-0.10 ± 0.05 dex), significantly enhanced SFR (0.42 ± 0.11 dex), and $\rm SFE_{H\,I}$ (0.48 ± 0.12 dex). We propose the marginally detected H I depletion may originate from the gas consumption in fueling the enhanced H_2 reservoir of galaxy pairs.

Keywords. Galaxy interactions, Galaxy pairs, Galaxy mergers, Interstellar atomic gas, Star formation

1. Introduction

Interactions of galaxies provide an effective way to study the impact of H I gas during galaxy evolution, as recent observations and simulations have connected galaxy mergers with the enhancement of star formation, gas regulation, triggering of AGN, and starburst (Di Matteo et al. 2007; Cox et al. 2008; Ellison et al. 2008; Satyapal et al. 2014; Hani et al. 2018). Although many studies have focused on the gas conditions and star formation during galaxy interactions, the role of H I gas is still under debate. Using different samples of merging galaxies, some observations indicate that the H I gas fractions of galaxy mergers are enhanced relative to that of isolated galaxies (Casasola et al. 2004; Janowiecki et al. 2017; Ellison et al. 2018), while others find no significant difference in H I gas fractions

(Ellison et al. 2015; Zuo et al. 2018) or decreased H I content (Hibbard & van Gorkom 1996; Georgakakis et al. 2000). The confusing results may have been caused by several factors; for example, limited pair sample or lack of robust control sample (Moreno et al. 2019). However, the lack of rigorously defined merger stages is less explored in previous studies.

Considering the pair selection mainly relies on the projected separation (d_p) and the difference between line-of-sight velocities (Δv) (Ellison et al. 2008; Patton et al. 2013; Zuo et al. 2018), the current definition of the merger stage for galaxy pairs has several shortcomings. The projected separation (d_p) may not reveal the physical separation of member galaxies (Soares 2007) due to the projection effect. Furthermore, galaxy pairs with the same separation but at different merging stages could experience different interacting strengths (Torrey et al. 2012).

Recent integral field unit (IFU) surveys of nearby galaxies have revealed a possible connection between the asymmetry of gas kinematics with galaxy interactions (Barrera-Ballesteros et al. 2015; Bloom et al. 2018). Using the IFU data from Mapping Nearby Galaxies at APO (MaNGA) survey, Feng et al. (2020) investigated the kinematic asymmetry ($\overline{v}_\mathrm{asym}$) of the ionized gas in a large sample of paired galaxies. Derived from Hα velocity maps of galaxies (see details in Feng et al. 2020), the value of $\overline{v}_\mathrm{asym}$ describes the asymmetry of the velocity field contributed by the interaction-induced nonrotating motion. They find the star formation of paired galaxies with high kinematic asymmetries is significantly enhanced, while paired galaxies with low kinematic asymmetries show no significant enhancement of star formation rate (SFR) even at small projected separation. For pairs with high $\overline{v}_\mathrm{asym}$, the enhancement of SFR shows a tight anti-correlation with projected separation. These findings suggest that the kinematic asymmetry is an effective indicator of galaxy mergers.

The enhancement of star formation during galaxy-galaxy interactions requires sustaining gas supply, which can be further investigated by observations of cold gas (e.g., H I gas) in galaxy pairs. In this work, we compile a major-merger galaxy pairs sample selected from the MaNGA survey to study the H I content of merging galaxies, adopting the kinematic asymmetry and projected separation as indicators of the merging stage. We compare the H I gas properties and star formation of the galaxy pair sample with a robustly matched control sample. Our study can provide better constraints on the H I gas fraction in merging galaxies and study the HI gas depletion/replenishment in clearly defined interacting stages.

2. Samples

The parent sample (Feng et al. 2020) consists of 578 isolated galaxy pairs, which are selected following these criteria: (1) the projected separation for member galaxies: 5 h^{-1} kpc $\leq d_\mathrm{p} \leq$ 200 h^{-1} kpc, (2) the line-of-sight velocity difference: $|\Delta v| \leq$ 500 km s^{-1}, (3) each pair member only has one neighbor satisfying the above criteria, (4) at least one member galaxy of each pair has been observed in the MaNGA survey, and the member galaxy has more than 70% spaxels with Hα emission at S/N $>$ 5 within 1.5 effective radius (R_e), and (5) we only study star-forming galaxies ($\log(\mathrm{sSFR/yr}^{-1}) > -11$) in this work. To study major-merger pairs, we require the mass ratio $M_1/M_2 < 3$, where M_1 and M_2 represent the stellar masses of primary galaxies and companions, respectively. The H I data used in this work are either obtained by our PI programs with FAST or extracted from the HI-MaNGA survey (Stark et al. 2021). Constrained with the above requirements, the final pair sample consists of 66 galaxy pairs with H I detections at S/N $>$ 5.

Following Feng et al. (2020), we extract two subsamples from the pair sample based on their $\overline{v}_\mathrm{asym}$ values: low asymmetry ($0.007 < \overline{v}_\mathrm{asym} < 0.029$), and high asymmetry ($0.029 < \overline{v}_\mathrm{asym} < 0.316$). As revealed by recent simulations, paired galaxy after the

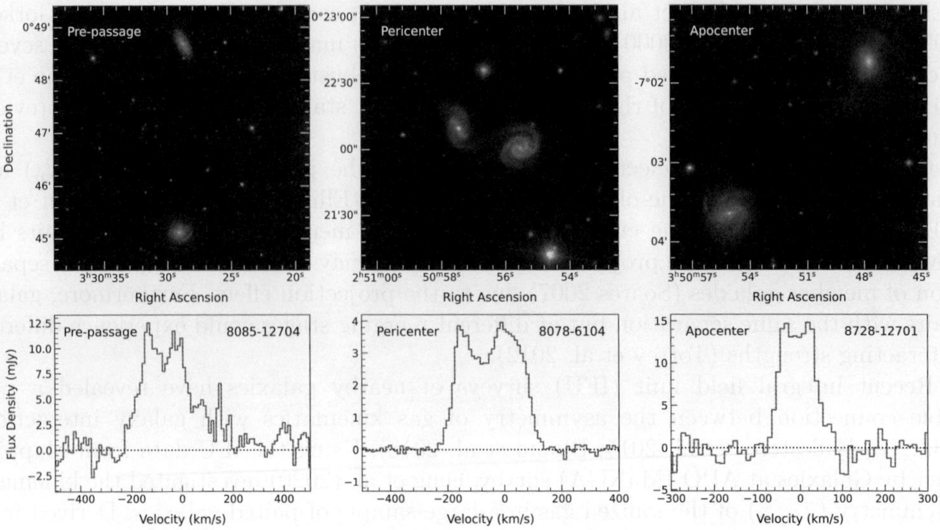

Figure 1. Examples of SDSS optical images and the corresponding H I line profiles of galaxy pairs at pre-passage, pericenter, and apocenter stages, respectively.

pericentric passage shows a more disturbed velocity field than that before the passage (Hung et al. 2016). Therefore, galaxy pairs with low $\overline{v}_{\rm asym}$ values are defined as pre-passage pairs, which indicates there are no significant interactions between member galaxies. Simulations suggest the physical separation constantly decreases before pericentric passage and increases when the paired galaxies are approaching apocenter (Torrey et al. 2012; Moreno et al. 2019). Hence, we use the projected separation ($d_{\rm p}$) as a reference to distinguish the pericenter and apocenter stages of merging galaxy pairs. We define the high $\overline{v}_{\rm asym}$-valued pairs with $d_{\rm p} < 50\ h^{-1}$ kpc to be at the stage of pericenter passage, and the high $\overline{v}_{\rm asym}$-valued pairs with $d_{\rm p} > 50\ h^{-1}$ kpc are approaching the apocenter passage. In Figure 1, we present examples of the SDSS optical images and corresponding H I line profiles of galaxy pairs at pre-passage, pericenter, and apocenter stages, respectively. We have compiled a control sample of isolated star-forming galaxies from MaNGA to compare the effect of galaxy interactions on star formation and H I gas properties.

3. Results

3.1. H I *Gas Properties*

We detect 6 H I emission lines and 1 absorption line in FAST observations of 8 galaxy pairs. For each detected source, we estimate the H I mass. For galaxy pairs in the HI-MaNGA survey, we adopt the H I mass from the HI-MaNGA DR2 catalog (Stark et al. 2021). Adopting the derived H I mass ($M_{\rm H\,I}$), we calculated the H I gas fraction ($f_{\rm H\,I}$) and the star formation efficiency of H I gas (SFE$_{\rm H\,I}$). The H I gas fraction $f_{\rm H\,I}$ is defined as $f_{\rm H\,I} = M_{\rm H\,I}/M_\star$. The mean H I gas fraction of the pair sample in logarithm is $\log f_{\rm H\,I} = -0.12 \pm 0.06$. In contrast, the mean H I gas fraction of the control sample ($\log f_{\rm H\,I} = -0.03 \pm 0.02$) is higher than that of the pair sample, which indicates the control galaxies are more gas-rich. However, the $f_{\rm H\,I}$ is tightly correlated to the global stellar mass ($M_\star$) for star-forming galaxies (Catinella et al. 2010), so we perform a galaxy-by-galaxy match between the pair sample and control sample in Section 3.2.

We use $\mathrm{SFE}_{\mathrm{H\,I}} = \mathrm{SFR}/\mathrm{M}_{\mathrm{H\,I}}$ to describe the star formation efficiency of H I gas for galaxies in pairs and controls. The mean log $(\mathrm{SFE}_{\mathrm{H\,I}}/\mathrm{yr}^{-1})$ values are -9.80 ± 0.06 and -9.96 ± 0.02 for galaxies in pairs and controls, respectively.

3.2. H I *Depletion and SFR Enhancement at Different Merger Stages*

To compare the H I gas fraction and other properties of galaxies in the pair sample and control sample along the merger sequence, we calculate the "offset" quantities following the method of Ellison et al. (2015, 2018). We match each paired galaxy in stellar mass and redshift with at least five isolated galaxies in the control pool. The matched galaxies are required to satisfy the tolerance of $|\Delta\log(M_\star/M_\odot)| < 0.2$ and $|\Delta z| < 0.01$ (Feng et al. 2020).

After matching the stellar mass and redshift, the offset of H I gas fraction, is calculated as

$$\Delta f_{\mathrm{H\,I}} = \log f_{\mathrm{H\,I,\,pair}} - \log \mathrm{median}(f_{\mathrm{H\,I,\,control}}), \qquad (1)$$

where the $\log f_{\mathrm{H\,I,pair}}$ is the H I gas fraction of paired galaxy, and the $\log \mathrm{median}(f_{\mathrm{H\,I,control}})$ represents the median H I gas fraction of its matched control galaxies in the logarithm scale. We apply the same method to compute the offsets of SFR (ΔSFR) and $\mathrm{SFE}_{\mathrm{H\,I}}$ ($\Delta\mathrm{SFE}_{\mathrm{H\,I}}$). The mean H I gas fraction offset of the galaxy pair sample is $\Delta f_{\mathrm{H\,I}} = -0.06 \pm 0.03$ dex, which indicates the average $f_{\mathrm{H\,I}}$ of paired galaxies is ~15% deficient compared to isolated galaxies. The SFR offset of the pair sample is $\Delta\mathrm{SFR} = 0.09 \pm 0.05$ dex, suggesting the SFR of paired galaxies and isolated galaxies on average have no significant difference. The enhanced SFR (~23%) is only marginally detected with large uncertainty. The $\mathrm{SFE}_{\mathrm{H\,I}}$ offset of the pair sample is 0.12 ± 0.06 dex, which indicates the average $\mathrm{SFE}_{\mathrm{H\,I}}$ in galaxy pairs is marginally enhanced by ~32% compared with control galaxies.

We present our main results in Figure 2. The histograms of the offsets of different galaxy properties are shown in top panels of Figure 2, and lower panels show the mean values of each distribution. Our results suggest mild H I depletion occurs during the merging process, especially when pairs are at the pericenter stage. In Figure 2(a) and Figure 2(d), we plot the distributions of H I gas fraction offsets for each subsample. Compared to the controls, the mean values of each distribution indicate that on average the H I gas fraction of pairs is marginally decreased at pericenter and apocenter stages for 0.10 ± 0.05 dex and 0.05 ± 0.04 dex (~26% and ~12%, respectively). At the pre-passage stage, the H I gas fraction offset of galaxy pairs is $\Delta f_{\mathrm{H\,I}} = -0.04 \pm 0.05$ dex, which is comparable to that of isolated galaxies. The cyan data points show the results of FIRE-2 simulation (Moreno et al. 2019), which predicts a ~ 4% enhancement of the cool gas mass on average during the galaxy-pair period. Comparing with their simulation results, our data indicate weak decreases of $f_{\mathrm{H\,I}}$ for pairs at pericenter and apocenter stages, respectively.

The offset of SFR reveals that the star formation is enhanced when galaxy pairs are encountering close interactions. In Figure 2(b), the ΔSFR in pairs during pericentric passage tends to distribute at positive values, which indicates enhanced SFR. As shown in Figure 2(e), on average the paired galaxies during pericenter passage have strong SFR enhancement $\Delta\mathrm{SFR} = 0.42 \pm 0.11$ dex, and the mean ΔSFR values of paired galaxies during pre-passage and apocenter passage are -0.01 ± 0.10 dex and 0.04 ± 0.06 dex, respectively.

Similar to SFR, our data suggest paired galaxies at the pericenter stage present significantly enhanced $\mathrm{SFE}_{\mathrm{H\,I}}$Ȧs shown in Figure 2(c), the distributions of $\Delta\mathrm{SFE}_{\mathrm{H\,I}}$ for each subsample are similar to those of ΔSFR. Our results in Figure 2(f) indicate that the average $\mathrm{SFE}_{\mathrm{H\,I}}$ of paired galaxies during pericenter passage is significantly enhanced by

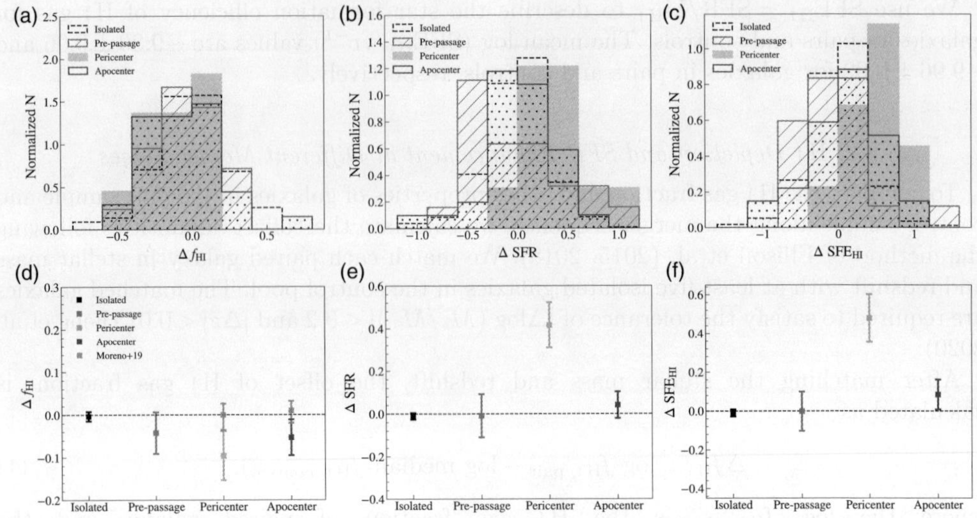

Figure 2. The top panels show histograms of different galaxy property offsets. Paired galaxies at different merger stages and the control sample are marked with blue, orange, red and black, respectively. (a) The distribution of $\Delta f_{\rm H\,I}$. (b) The distribution of ΔSFR. (c) The distribution of $\Delta \rm SFE_{H\,I}$. The middle panels show the mean value distributions for different galaxy property offsets. In each plot, the mean value is indicated by square point in the middle and the error of the mean as error bar for each distribution. (d) The mean value of $\Delta f_{\rm H\,I}$ of our results and the simulation results (cyan data points) from Moreno et al. (2019). (e) The mean value of ΔSFR. (f) The mean value of $\Delta \rm SFE_{H\,I}$.

0.48 ± 0.12 dex. In contrast, the paired galaxies at the pre-passage stage shows no difference of the $\rm SFE_{H\,I}$ compared with the controls ($\Delta \rm SFE_{H\,I} = 0.00 \pm 0.10$ dex). The mean $\rm SFE_{H\,I}$ offset of pairs at the apocenter stage is $\Delta \rm SFE_{H\,I} = 0.08 \pm 0.08$ dex.

4. Discussions and Summary

For galaxy pairs at the pre-passage stage, there are no significant interactions between member galaxies. Therefore, the H I gas fraction, SFR, and $\rm SFE_{H\,I}$ of pairs are similar to that of isolated galaxies, which has been presented by our results (Figure 2).

During the pericentric passage, our analysis suggest that the member galaxies show significantly enhanced SFR (a factor of ~ 2.6), which is consistent with previous studies of close galaxy pairs (Patton et al. 2013). As revealed by previous CO observations of close pairs, the molecular gas fraction is also significantly enhanced and correlated to the enhancement of SFR (Pan et al. 2018; Violino et al. 2018; Lisenfeld et al. 2019). Due to external pressure, the enhancement of molecular gas fraction may originate from an accelerated transition from atomic to molecular gas, which can arise in the early stage of the merger (Kaneko et al. 2017). In this scenario, the moderate decrease of $f_{\rm H\,I}$ revealed by our work can be explained as H I gas depletion in fueling the H_2 reservoir. Furthermore, the enhanced molecular-to-atomic gas mass ratio in close interacting pairs (Lisenfeld et al. 2019) may lead to the significantly enhanced $\rm SFE_{H\,I}$ indicated by our data, considering the $\rm SFE_{H\,I} = SFE \times (M_{H_2}/M_{H\,I})$, where the SFE is not enhanced compared with isolated galaxies (Casasola et al. 2004; Lisenfeld et al. 2019). Alternatively, some recent observations of close pairs found small (< a factor of 2) SFE enhancement (Pan et al. 2018; Violino et al. 2018), which may also be able to drive the enhanced $\rm SFE_{H\,I}$ revealed by our data.

At the apocenter stage, our data suggest the H I gas fraction of pairs remains suppressed compared to isolated galaxies. Compared to pairs during the pericentric passage, the H I gas reservoir seems to be mildly replenished after strong interaction. This can be explained by the cooling of hot/warm gas from CGM (Moster et al. 2011; Tonnesen & Cen 2012), considering the interaction-induced shocks at the pericenter stage can be gradually alleviated when pairs are approaching the apocenter. As for SFR, the star formation of pairs at the apocenter stage is decreased relative to the previously enhanced SFR during pericentric passage. The decrease of SFR enhancement is in agreement with the previous result that galaxy pairs with large projected separations present weak SFR enhancement (Scudder et al. 2012; Patton et al. 2013). Simulations also suggest the merger-induced SFR enhancement is gradually decreased when the paired galaxies are approaching the apocenter (Moreno et al. 2019). As discussed above, the change of $\mathrm{SFE}_{\mathrm{H\,I}}$ can be driven by SFE and $M_{\mathrm{H}_2}/M_{\mathrm{H\,I}}$. However, previous CO studies lack observations of pairs at the apocenter stage, which requires further CO observations of our pair sample to help determine the dominating factor for the observed suppression of $\mathrm{SFE}_{\mathrm{H\,I}}$ enhancement.

We conclude that our study reveals marginal detection of the H I gas depletion during the galaxy-pair period of the merger. The suppressed H I gas fraction may originate from the gas consumption in fueling the enhanced H_2 reservoir of galaxy pairs. As a new method to use kinematic asymmetry as the indicator of galaxy interaction, we can further continue this study as more and more IFU data become available. Since our results are based on the global properties of galaxies in pairs, spatially resolved H I observations will help us further explore the interplay between galaxy interactions and the H I gas. Combining observations of the molecular gas for our sample, we will be able to build a complete picture of the cold gas evolution during mergers in the local universe.

References

Barrera-Ballesteros J. K., et al., 2015, *A&A*, 582, A21

Bloom J. V., et al., 2018, *MNRAS*, 476, 2339

Casasola V., Bettoni D., Galletta G., 2004, *A&A*, 422, 941

Catinella B., et al., 2010, *MNRAS*, 403, 683

Cox T. J., Jonsson P., Somerville R. S., Primack J. R., Dekel A., 2008, *MNRAS*, 384, 386

Di Matteo P., Combes F., Melchior A. L., Semelin B., 2007, *A&A*, 468, 61

Ellison S. L., Patton D. R., Simard L., McConnachie A. W., 2008, *AJ*, 135, 1877

Ellison S. L., Fertig D., Rosenberg J. L., Nair P., Simard L., Torrey P., Patton D. R., 2015, *MNRAS*, 448, 221

Ellison S. L., Catinella B., Cortese L., 2018, *MNRAS*, 478, 3447

Feng S., Shen S.-Y., Yuan F.-T., Riffel R. A., Pan K., 2020, *ApJL*, 892, L20

Georgakakis A., Forbes D. A., Norris R. P., 2000, *MNRAS*, 318, 124

Hani M. H., Sparre M., Ellison S. L., Torrey P., Vogelsberger M., 2018, *MNRAS*, 475, 1160

Hibbard J. E., van Gorkom J. H., 1996, *AJ*, 111, 655

Hung C.-L., Hayward C. C., Smith H. A., Ashby M. L. N., Lanz L., Martínez-Galarza J. R., Sanders D. B., Zezas A., 2016, *ApJ*, 816, 99

Janowiecki S., Catinella B., Cortese L., Saintonge A., Brown T., Wang J., 2017, *MNRAS*, 466, 4795

Kaneko H., Kuno N., Iono D., Tamura Y., Tosaki T., Nakanishi K., Sawada T., 2017, *PASJ*, 69, 66

Lisenfeld U., Xu C. K., Gao Y., Domingue D. L., Cao C., Yun M. S., Zuo P., 2019, *A&A*, 627, A107

Moreno J., et al., 2019, *MNRAS*, 485, 1320

Moster B. P., Macciò A. V., Somerville R. S., Naab T., Cox T. J., 2011, *MNRAS*, 415, 3750

Pan H.-A., et al., 2018, *ApJ*, 868, 132

Pan H.-A., et al., 2019, *ApJ*, 881, 119

Patton D. R., Torrey P., Ellison S. L., Mendel J. T., Scudder J. M., 2013, *MNRAS*, 433, L59
Satyapal S., Ellison S. L., McAlpine W., Hickox R. C., Patton D. R., Mendel J. T., 2014, *MNRAS*, 441, 1297
Scudder J. M., Ellison S. L., Torrey P., Patton D. R., Mendel J. T., 2012, *MNRAS*, 426, 549
Smith B. J., Campbell K., Struck C., Soria R., Swartz D., Magno M., Dunn B., Giroux M. L., 2018, *AJ*, 155, 81
Soares D. S. L., 2007, *AJ*, 134, 71
Stark D. V., et al., 2021, *MNRAS*, 503, 1345
Tonnesen S., Cen R., 2012, *MNRAS*, 425, 2313
Torrey P., Cox T. J., Kewley L., Hernquist L., 2012, *ApJ*, 746, 108
Violino G., Ellison S. L., Sargent M., Coppin K. E. K., Scudder J. M., Mendel T. J., Saintonge A., 2018, *MNRAS*, 476, 2591
Zuo P., Xu C. K., Yun M. S., Lisenfeld U., Li D., Cao C., 2018, *ApJS*, 237, 2

Resolving the Rise and Fall of Star Formation in Galaxies
Proceedings IAU Symposium No. 373, 2023
T. Wong & W.-T. Kim, eds.
doi:10.1017/S1743921322004574

Evolution of Gas Flows over the Starburst to Post-Starburst to Quiescent Galaxy Sequence

Yang Sun[1], Gwang-Ho Lee[1,2], Ann I. Zabludoff[1], K. Decker French[3], Jakob M. Helton[1], Nicole A. Kerrison[1], Christy A. Tremonti[4] and Yujin Yang[2]

[1]Steward Observatory, University of Arizona, 933 North Cherry Avenue, Tucson, AZ 85721, USA

[2]Korea Astronomy and Space Science Institute, 776 Daedeokdae-to, Yuseong-gu, Daejeon, 305-348, Republic of Korea

[3]Department of Astronomy, University of Illinois, 1002 West Green Street, Urbana, IL 61801, USA

[4]Department of Astronomy, University of Wisconsin–Madison, 475 North Charter Street, Madison, WI 53703, USA

Abstract. Even though galactic winds are common in galaxies with starbursts or active galactic nuclei (AGN), the role of such gas flows in galaxy evolution remains uncertain. Here we examine how winds vary along a likely evolutionary sequence connecting starburst to post-starburst to quiescent galaxies. To detect the interstellar medium and measure its bulk flows, we examine the residual Na D absorption line doublet after the stellar contribution has been removed from each galaxy's spectrum. We discover that outflows diminish along this sequence, i.e., as star formation ends. We then focus on the wind behavior within the post-starburst sample, for which we have measured the time elapsed since the starburst ended (post-burst age) via detailed modeling of their star formation histories (French *et al.* 2018). Even within our post-starburst sample, the fraction of galaxies with significant winds and the average wind velocities decrease with post-burst age after controlling for stellar mass.

Keywords. ISM: jets and outflows – ISM: kinematics and dynamics – galaxies: evolution – galaxies: ISM

1. Introduction

In the local Universe, observational surveys suggest there are two main types of galaxies: star-forming and quiescent. The current understanding of this bimodality indicates that star-forming galaxies will eventually stop forming stars and evolve to the quiescent phase. Transitioning galaxies evolving between star-forming and quiescent are found in the "green valley". One rare class of transitioning galaxies is the post-starburst galaxy. This type of galaxy is believed to have experienced a burst of star formation that was halted in the past ~ 1 Gyr (Dressler & Gunn 1983; Couch & Sharples 1987). Their optical spectra show strong Balmer absorption lines suggesting a recent starburst has occurred but are without strong nebular emission lines indicating a low current star formation rate. Even though the star formation must decline along the evolutionary sequence from starburst to post-starburst to quiescent, the reasons why star formation has stopped are still unknown.

One likely mechanism for this quenching is feedback. For current cosmological simulations of galaxy formation and evolution, feedback is required to reproduce observed galaxy baryonic mass functions and star formation efficiency (Kereš *et al.* 2009; Hopkins *et al.* 2014; Somerville & Davé 2015; Naab & Ostriker 2017, and references therein). Massive stars, supernovae, and active galactic nuclei (AGN) may be sources of negative feedback through thermal or kinetic energy transferred to the interstellar medium (ISM) in galaxies. If negative feedback processes remove the cold gas in winds (see Veilleux *et al.* 2005; King & Pounds 2015; Veilleux *et al.* 2020), star formation could rapidly end due to the removal of the fuel for star formation.

The connections between galactic outflows and star formation have been well studied in the past decades. For instance, Chen *et al.* (2010) systematically analyzed the stacked Na D line of star-forming galaxies in the SDSS DR7 survey, finding strong ISM Na D absorption lines are only prevalent in disk galaxies with high star formation rates (SFR), high SFR per unit area (ΣSFR), or high stellar mass ($M_\star$). However, none have tested how winds evolve along the galaxy evolutionary sequence where star formation is declining. Based on previous findings that higher outflow detection rates and larger Na D EWs prefer to locate at higher SFR and $M_\star$ spaces, we expect that there should be decline in winds over this sequence. Therefore, this work aims to build a star formation declining sequence from starburst to post-starburst to quiescent galaxy in the local Universe ($0.010 < z < 0.325$) and tests for wind evolution along it. Moreover, French *et al.* (2018) (hereafter F18) modeled the star formation histories (SFHs) of their ~500 post-starburst galaxies with post-burst ages, burst mass fraction and burst durations. These ages allow us to reveal wind evolution with a more highly time-resolved view.

We use the spectroscopic sample of the SDSS DR12 survey (Alam *et al.* 2015) and analyze the Na I λλ5890, 5896 line to trace the bulk motion of the neutral gas of starburst, post-starburst and quiescent samples. In this text we report the current results on how wind properties change along the first sequence and with post-burst age.

2. Building the SFR-declining galaxy evolutionary sequence

Here we briefly introduce how we built the star formation declining galaxy evolutionary sequence from the SDSS galaxy.

F18 provided a sample of 522 post-starburst galaxies with $0.010 < z < 0.325$ that are drawn from the spectroscopic sample in the Sloan Digital Sky Survey Data Release 8 (DR8, Aihara *et al.* 2011). In the equivalent widths (EWs) of the Hα and the Lick Hδ absorption line index (H$\delta_{\rm A}$) diagram F18 selected the post-starburst sample following the selection criteria: H$\alpha < 3$ Å in the rest-frame, representing a lack of current star formation, and strong Balmer absorption line index $\mathrm{H}\delta_{\rm A} - \sigma(\mathrm{H}\delta_{\rm A}) > 4$ Å, where $\sigma(\mathrm{H}\delta_{\rm A})$ is the measurement error, representing recent starburst in the past $\sim 1 - 1.5$ Gyr.

Then we construct two different comparison samples of starburst galaxies and quiescent galaxies that represent the likely progenitors and descendants of post-starburst galaxies, respectively. We start with a parent sample of 593,807 galaxies drawn from the SDSS DR12 survey (Alam *et al.* 2015), which is at the same redshift range as post-starburst galaxies, has continua S/N≥ 10, and has galaxy properties and spectral line flux measurements from the MPA-JHU Catalogs (Kauffmann *et al.* 2003; Brinchmann *et al.* 2004; Tremonti *et al.* 2004). For the starburst galaxy sample, we require they are star-forming galaxies in the BPT diagram (Kewley *et al.* 2006), their SFRs are 5σ higher than the main-sequence and their stellar masses are within the proper range considering each post-starburst galaxy's current stellar mass and burst mass fraction. Quiescent galaxies are selected to have $\mathrm{H}\delta_{\rm A} + \delta(\mathrm{H}\delta_{\rm A}) < 1$ Å and Hα EW< 3 Å, with the intention of selecting galaxies with no current and recent star formation. We finalize the starburst sample with

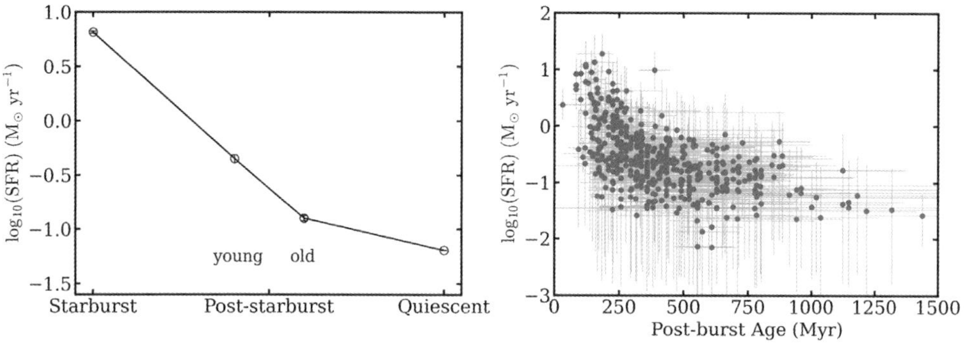

Figure 1. SFR evolves along the evolutionary sequence (left) and with post-burst age (right), which confirms that SFR declines along the evolutionary sequence we build. In the left panel, the post-starburst sample is divided into two subsamples: young (post-burst age < 400 Myr) and old (≥ 400 Myr). SFR measurements are obtained from MPA-JHU catalogs.

3306 galaxies and the quiescent sample with 72,251 galaxies. Figure 1 shows SFR of our samples declines along the whole sequence and with post-burst age.

3. Detecting neutral ISM bulk flows

In this section, we introduce how we detect neutral bulk flows and measure their properties in our three galaxy sample.

Firstly, the stellar continua are well fitted by the penalized pixel fitting (pPXF, Cappellari & Emsellem 2004; Cappellari 2017) Python code with MILES (Vazdekis *et al.* 2010) simple stellar population (SSP) models.

After obtaining the best-fit stellar continuum model, we measured the equivalent widths of both the total and the residual (after subtracting the best-fit stellar continuum model) of Na D absorption line ($EW_{NaD,tot}$ and $EW_{NaD,exc}$) using Monte Carlo simulation to derive the mean values and errors of $EW_{NaD,tot}$ and $EW_{NaD,exc}$.

The velocity of bulk flow motions (Δv) from the galaxy redshift is another important properties we measure. To do this, we apply a double Gaussian profile to fit the Na D residuals at the wavelength of 5850-5940 Å. we assume the widths of Na D are the same and the doublet flux ratio is in the range from 1 to 2, corresponding to the scenario of optically thin and thick, respectively. The shifted wavelength of the Na D excess is constrained as μ: $[-20, 20]$+5890 Å that corresponds to $\Delta v \sim \pm 1000$ km s^{-1}. Depending on the geometry, it is possible to see Na D emission in the case that absorbed photons are re-emitted along the line of sight. This scattering process can produce P-Cygni profiles of Na D. We do find a small fraction ($\sim 1\%$) of galaxies in our three samples showing a clear P-Cygni profile in their residual spectra. In this case, the overall profile should be fit by two double Gaussian components but Δv is measured from the blueshifted absorption component only. Then we use Monte Carlo simulation to obtain the Δv and its uncertainty, similar to what we do for EW measurement. A example of the Na D excess fitting for a post-starburst galaxy with the P-Cygni profile is shown in Figure 2.

Excluding samples without Na D ISM absorption or good velocity measurements, finally we obtain Δv measurements for 589 starburst, 202 post-starburst, and 9,711 quiescent galaxies. However, to avoid unreliable further analysis for very weak bulk flows, we further define flow-detected samples as samples with significant Na D excess ($EW_{NaD,exc} - 3\sigma(EW_{NaD,exc}) > 0$) or P-Cygni profile. The flow-detected sample includes 370 starburst galaxies, 164 post-starburst galaxies, and 4,598 quiescent galaxies. Hereafter we only focus on those samples with strong bulk flows.

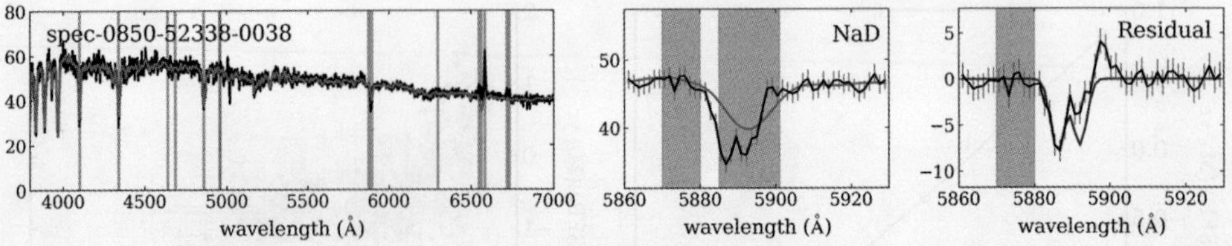

Figure 2. Example of galaxy spectral fitting. Left: the SDSS spectra of a post-starburst galaxy in the rest-frame (black) and their best-fit stellar continuum model obtained using pPXF (red). Middle: The zoom-in Na D spectra (black) with the best-fit stellar continuum models (red). Right: The Na D residual spectra (black), representing the ISM, and their best-fit double Gaussian profile (blue).The errorbars in three panels are the original flux uncertainties from the SDSS spectra. The masked regions are shown as gray areas. This post-starburst galaxy have the P-Cygni profile in the Na D residual which is composed of a blueshifted absorption component and a redshift emission component.

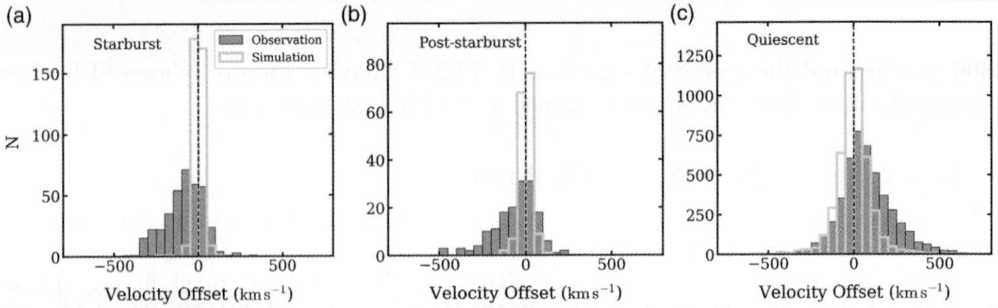

Figure 3. The histograms of Δv for the (a) starburst, (b) post-starburst and (c) quiescent sample. The observed wind distributions are displayed by gray filled histograms. The yellow open histograms represent simulated data expected from zero velocity wind and 1σ measurement error. The mean velocities and their errors of three galaxy samples from left to right are -84.6 ± 5.9 km s^{-1}, -71.8 ± 11.4 km s^{-1}, and 76.6 ± 2.3 km s^{-1}. For starburst and post-starburst samples, the distributions of observation shows significant negative velocity tails comparing with those of simulations. On the contrary, the quiescent sample has a significant positive tail.

4. Outflows in starburst and post-starburst galaxies

Assuming a possible evolutionary sequence from starburst to post-starburst to quiescent, Figure 3 shows the histograms of wind velocities for the three samples with this sequence (grey histograms). The histograms for the starburst and post-starburst samples are significantly skewed with a tail at high negative wind velocities. The mean values of wind velocities are -84.6 ± 5.9 km s^{-1} and -71.8 ± 11.4 km s^{-1} for the starburst and post-starburst samples, respectively. On the other hand, the quiescent galaxy's histogram is skewed to the positive velocities with the mean velocity of 76.6 ± 2.3 km s^{-1}. But the overall shape of the histogram is symmetric compared to the histograms for the two other samples.

To test whether these differences are statistically significant, we compare the distribution of observed wind velocities with that expected from measurement errors alone in each panel. We use the 1σ velocity measurements errors of samples and randomly made them half negative and half positive, then added to zero velocities, and compared the zero wind models (yellow histograms in Figure 3) with the observed wind distributions. For the starburst and post-starburst samples, the distributions of Δv show negative wind velocity tails exceeding the distributions expected from measurement errors alone. On the other hand, the histogram of the observed wind velocities for the quiescent galaxy sample

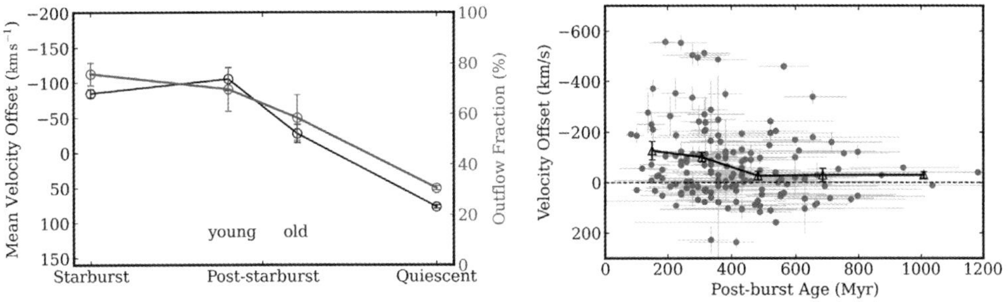

Figure 4. Left: The trends of outflow velocity (black) and outflow fraction (blue) with the evolutionary sequences, from starburst to post-starburst to quiescent galaxies. The post-starburst sample is divided into two subsamples: young (post-burst age < 400 Myr) and old (≥ 400 Myr). Outflow velocity and outflow fraction both decrease along the sequence. Right: The overall trend of wind velocity with post-burst age for the post-starburst sample without removing stellar mass dependency. The black triangles show the mean velocity in each post-burst age bin. Within the post-starburst phase, outflow velocities still decrease with post-burst age. More accurate partial-correlation analysis confirms the correlation between wind velocity and post-burst age controlling stellar mass is with 2.26σ significance. The figure is reproduced from Sun *et al.* (in prep).

shows a significant excess of positive wind velocities. Those differences are confirmed by Kolmogorov-Smirnov test (fails a KS test).

Assuming Na D excess represents the average properties from the foreground ISM in the light of sight, negative and positive wind velocities represent bulk flows of the ISM in opposite direction: negative velocities indicate outflows, while positive velocities indicate inflows. Therefore, the exceeding negative velocity tails of the starburst and post-starburst samples show that these two groups have significant strong outflows, while the positive tails of the quiescent sample suggests the presence of strong inflows.

5. Trends along the evolutionary sequence

In this section, we introduce how wind properties we derived (i.e., outflow fraction and velocities) evolve along the assumed evolutionary sequence (starburst - post-starburst - quiescent), which is the main result of Sun *et al.* (in prep). For the post-starburst sample, we further divided it into two groups: the young post-starburst group (post-burst age < 400 Myr) and the old post-starburst group (post-burst age ≥ 400 Myr).

The outflow fraction is defined as the number of galaxies with outflows as a percentage of the number of galaxies with flow. The outflow fractions for the starburst, post-starburst and quiescent sample are $76 \pm 5\%$, $65 \pm 6\%$, and $31 \pm 1\%$, respectively. The left panel of Figure 4 shows the overall trends of outflow fractions with the sequence. Dividing the post-starburst into two age groups, the outflow fraction of the young group is $70 \pm 9\%$ and that of the old one is $58 \pm 9\%$. As we mentioned in Section 4, The wind velocity decreases from the starburst sample (-84.6 ± 5.9 km s^{-1}) to the post-starburst sample (-71.8 ± 11.4 km s^{-1}), and then changes to a positive value in the quiescent stage (76.6 ± 2.3 km s^{-1}). The left panel of Figure 4 shows the velocity trend with the sequence, but here the post-starburst sample are divided into the young and old group as well. The mean wind velocity of the young post-starburst galaxy sample (-105.1 ± 16.6 km s^{-1}) is slightly higher than the value of the starburst galaxy sample, but are consistent within 1σ error. The value of the old post-starburst galaxy sample (-27.6 ± 13.0 km s^{-1}) is lower than that of the young group. Even though we do not only focus on outflows, since overall wind velocities change from more negative values to less negative values to positive values, we represent this trend as: outflows diminish along the evolutionary sequence.

We also test the outflow velocity evolution at the post-starburst stage. The right panel of Figure 4 visualizes the average trend of wind velocity and post-burst age. After controlling the stellar mass effect, the final Spearman partial-correlation coefficient for the trend is 0.177, with 2.26σ significance. The correlation implies that, outflows diminish with the evolution of post-starburst phase as well.

6. Conclusions

In this work, we use the Na D absorption line to analyze bulk gas motions in the ISM of ∼80,000 galaxies from the SDSS DR12 survey who follow a SFR declining galaxy evolutionary sequence, from starburst to post-starburst, to quiescent galaxies. The bulk flow are detected in all three galaxy samples, but are mainly located at high mass end ($\log(M_*/M_\odot) \geq 10$). We find a small fraction ($\sim 1\%$) of starburst, post-starburst, and quiescent with bulk flows show the P-Cygni profile. The starburst and post-starburst sample show statistically significant negative velocity tails, which means many of them have outflows. Even though the quiescent sample has positive velocity tail which represents significant inflows, inflows are weak due to the small EW of Na D excess. Outflow fractions and outflow velocities similarly decrease from starburst ($76 \pm 5\%$, $-84.6 \pm 5.9\,\mathrm{km\,s^{-1}}$) to young ($70 \pm 9\%$, $-105.1 \pm 16.6\,\mathrm{km\,s^{-1}}$) and old post-starburst ($58 \pm 9\%$, $-27.6 \pm 13.0\,\mathrm{km\,s^{-1}}$) to quiescent ($31 \pm 1\%$, $76.6 \pm 2.3\,\mathrm{km\,s^{-1}}$). Even within the post-starburst phase, outflows diminish with post-burst age. It indicates that neutral gas outflows we detected diminish with star formation over time.

References

Aihara, H., Allende Prieto, C., An, D., *et al.* 2011, *ApJS*, 193, 29
Alam, S., Albareti, F. D., Allende Prieto, C., *et al.* 2015, *ApJS*, 219, 12
Brinchmann, J., Charlot, S., White, S. D. M., *et al.* 2004, *MNRAS*, 351, 1151
Cappellari, M. 2017, *MNRAS*, 466, 798
Cappellari, M. & Emsellem, E. 2004, *PSAP*, 116, 138
Chen, Y.-M., Tremonti, C. A., Heckman, T. M., *et al.* 2010, *AJ*, 140, 445
Couch, W. J. & Sharples, R. M. 1987, *MNRAS*, 229, 423
Dressler, A. & Gunn, J. E. 1983, *ApJ*, 270, 7
French, K. D., Yang, Y., Zabludoff, A. I., *et al.* 2018, *ApJ*, 862, 2
Hopkins, P. F., Kereš, D., Oñorbe, J., *et al.* 2014, *MNRAS*, 445, 581
Kauffmann, G., Heckman, T. M., White, S. D. M., *et al.* 2003, *MNRAS*, 341, 33
Kereš, D., Katz, N., Davé, R., *et al.* 2009, *MNRAS*, 396, 2332
Kewley, L. J., Groves, B., Kauffmann, G., *et al.* 2006, *MNRAS*, 372, 961
King, A. & Pounds, K. 2015, *ARA&A*, 53, 115
Naab, T. & Ostriker, J. P. 2017, *ARA&A*, 55, 59
Somerville, R. S. & Davé, R. 2015, *ARA&A*, 53, 51
Sun, Y., Lee, G., Zabludoff, A. I., *et al.* in prep
Tremonti, C. A., Heckman, T. M., Kauffmann, G., *et al.* 2004, *ApJ*, 613, 898
Vazdekis, A., Sánchez-Blázquez, P., Falcón-Barroso, J., *et al.* 2010, *MNRAS*, 404, 1639
Veilleux, S., Cecil, G., & Bland-Hawthorn, J. 2005, *ARA&A*, 43, 769
Veilleux, S., Maiolino, R., Bolatto, A. D., *et al.* 2020, *A&AR*, 28, 2

Resolving the Rise and Fall of Star Formation in Galaxies
Proceedings IAU Symposium No. 373, 2023
T. Wong & W.-T. Kim, eds.
doi:10.1017/S1743921322003945

Evidence for supernova feedback sustaining gas turbulence in nearby star-forming galaxies

C. Bacchini[1,2,3], F. Fraternali[2], G. Pezzulli[2], G. Iorio[4], A. Marasco[1] and V. Nipoti[3]

[1]INAF - Astronomical Observatory of Padova, Vicolo dell'Osservatorio 5, IT-35122, Padova, Italy
email: cecilia.bacchini@inaf.it

[2]Kapteyn Astronomical Institute, University of Groningen, Landleven 12, 9747 AD Groningen, The Netherlands

[3]Department of Physics and Astronomy, University of Bologna, via Gobetti 93/2, IT-40129, Bologna, Italy

[4]Department of Physics and Astronomy, University of Padova, Vicolo dell'Osservatorio 3, IT-35122, Padova, Italy

Abstract. HI and CO observations indicate that the cold gas in galaxies is very turbulent. However, the turbulent energy is expected to be quickly dissipated, implying that some energy source is needed to explain the observations. The nature of such turbulence was long unclear, as even the main candidate, supernova (SN) feedback, seemed insufficient. Other mechanisms have been proposed, but without reaching a general consensus. The key novelty of our work is considering that the gas disc thickness and flaring increase the dissipation timescale of turbulence, thus reducing the energy injection rate required to sustain it. In excellent agreement with the theoretical expectations, we found that the fraction of the SN energy (a.k.a. SN coupling efficiency) needed to maintain the cold gas turbulence is $\sim 1\%$, solving a long-standing conundrum.

Keywords. Turbulence, stellar feedback, supernovae, ISM structure, star-forming galaxies.

1. What drives the turbulence in the cold interstellar medium?

In star-forming galaxies, the cold gas velocity dispersion derived from the broadening of emission lines exceeds the values expected from the gas thermal motion. This additional broadening is typically ascribed to turbulence. Turbulence in the interstellar medium (ISM) is well described by the Kolmogorov's theory (Kolmogorov 1941). In this framework, the turbulent energy is injected into the ISM at a certain driving scale and then is transferred to the smaller scales through the so-called Kolmogorov's cascade. In practice, the turbulent eddies break down into smaller eddies, transferring to the kinetic energy to smaller scales. This energy is conserved until the dissipation scale is reached, then the turbulent energy is dissipated by the gas viscosity. Since we observe turbulence, this process must be in a stationary state and turbulence must be maintained by some source of new energy. This fresh turbulent energy must be provided with a certain injection rate depending on the dissipation timescale of turbulence. This latter is $\tau_{\rm diss} \equiv L/v_{\rm turb}$, with L and $v_{\rm turb}$ being the physical scale of the system and the turbulent velocity, respectively. In the case of galaxies, L is typically assumed proportional to the gas disc scale height h.

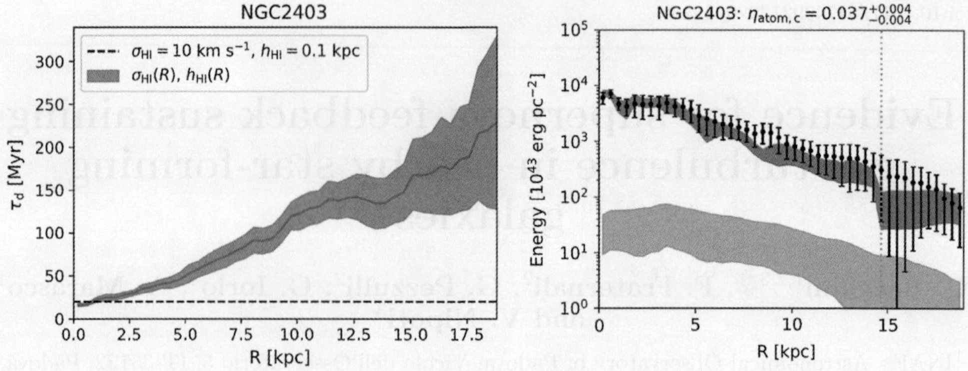

Figure 1. *Left:* Dissipation timescale of the atomic gas turbulence in the spiral galaxy NGC 2403. The dashed black line is for a thin disc with no flaring. The red curve is for a flaring disc in VHE (the red area shows the uncertainty). *Right:* Observed kinetic energy density per unit area of the atomic gas in NGC 2403 (black points). Blue and red areas are respectively the gas thermal energy and the turbulent energy injected by SNe with the efficiency $\eta_{\rm atom,c}$ reported above. The green area is the total model energy. The grey dotted vertical line indicates the outermost radius with measured $\Sigma_{\rm SFR}$; upper limits are used for larger radii.

Supernova (SN) explosions are the best candidates for being the engine of turbulence in galaxy discs, as they inject a huge amount of energy into the ISM. The energy flux injected by SNe can be estimated from the observed SFR ($\Sigma_{\rm SFR}$) using $\dot{E}_{\rm SNe} = \eta \Sigma_{\rm SFR} f_{\rm cc} E_{\rm SN}$, where η is SN efficiency (or coupling efficiency), $E_{\rm SN} = 10^{51}$ erg is the energy of a single SN, and $f_{\rm cc}$ is the fraction of stars exploding as SN core-collapse given an initial mass function. The key parameter is η, which is the fraction of the total SN energy that goes into feeding turbulence. Analytical and numerical models of SN remnant evolution predict $\eta \lesssim 0.1$, as most of the SN energy is radiated away (Chevalier 1974; Martizzi et al. 2015). To understand whether SNe can sustain the cold gas turbulence in galaxies, the expected SN energy must be compared to the observed kinetic energy density of the gas $E_{\rm obs} = \frac{3}{2} \Sigma_{\rm obs} \sigma^2_{\rm obs}$, with $\Sigma_{\rm obs}$ and $\sigma_{\rm obs}$ being the gas surface density and velocity dispersion. Tamburro et al. (2009) used this approach to investigate the origin of the atomic gas turbulence in nearby galaxies, assuming $L = 100$ pc and $v_{\rm turb} = 10$ km s^{-1}(i.e. $\tau_{\rm diss} = 10$ Myr) across the disc. They found that the SN energy can explain $E_{\rm obs}$ in the innermost regions of their galaxies, while very high ($\eta \gg 0.1$) or even non-physical ($\eta > 1$) efficiencies were needed in the outer parts where $\Sigma_{\rm SFR}$ is low (see also Utomo et al. 2019). Other mechanisms have been proposed, but no consensus has been reached about the main engine of turbulence (Mac Low and Klessen 2004).

2. Slow dissipation in flaring discs

The key improvement in our work (Bacchini et al. 2020a) consists in taking into account that the dissipation of turbulence is slower in thick and flaring discs than in thin discs without the flaring. Gas discs in galaxies are indeed expected to be in vertical hydrostatic equilibrium (hereafter VHE), meaning that the gas pressure, which tends to inflate the disc, balances the gravitation force towards the midplane, which tends to flatten the disc. Since gravity weakens with increasing distance from the galaxy centre, the gas layer thickens at large galactocentric radii (flaring). The natural driving scale of turbulence in galaxy discs is $L = 2h$, with h being the gas disc scale height. As shown in the left panel of Fig. 1, the turbulence dissipation timescale in the outer regions of a flaring gas disc is one order of magnitude longer than in the case of a thin disc without the flaring, which is the case assumed by Tamburro et al. (2009).

We collected a sample of 10 nearby galaxies made of 2 dwarfs and 8 spirals. For each galaxy, we measured the distribution and kinematics of the atomic and the molecular gas from 21-cm and CO(2–1) emission line datacubes using the software ^{3D}BAROLO (Di Teodoro and Fraternali 2015), which performs a tilted-ring model fitting on the observations. Then, we calculated the radial profile of $E_{\rm obs}$ for both the gas phases. We derived $\tau_{\rm diss}$ using the gas scale heights based on the VHE. To this task, we used the Python tool `Galpynamics` (Iorio 2018). We adopted a simple model for the kinetic energy density per unit area ($E_{\rm mod}$) that includes the gas thermal energy and the turbulent energy expected from SNe. This latter depends on $\tau_{\rm diss}$ and the free parameter η, which we assumed to be constant across the disc. For the atomic gas, we explored two single-phase cases (assuming either cold or warm neutral medium), and the more realistic scenario of a two-phase gas. We fitted $E_{\rm obs}$ with $E_{\rm mod}$ to find η. The right panel in Fig. 1 shows the example of the atomic gas in NGC 2403 assuming a single-phase, cold neutral gas. This case requires the maximum SN energy, as the gas thermal energy is essentially negligible. Nonetheless, $E_{\rm obs}$ is very well reproduced by our model, notably with very low efficiency. We find the same results for all the galaxies and for both the atomic gas and the molecular gas, which is remarkable considering that we assumed η to be independent from the galactocentric radius. Considering the galaxy sample as a whole and the systematic uncertainties (e.g. on gas temperature and $\Sigma_{\rm SFR}$), we obtained $\eta = 0.02^{+0.02}_{-0.01}$ for the atomic gas and $\eta = 0.003^{+0.007}_{-0.002}$ for the molecular gas.

3. Concluding remarks

The take-home message of this proceeding is that SNe can sustain the cold gas turbulence in nearby galaxies with only a few percent ($\lesssim 2\%$) of their energy. Hence, no additional source of energy is compulsorily required. The η values obtained in this work can be useful to implement sub-grid physics in some numerical simulations of galaxy discs with limited resolution.

In a broader context, our results can be interpreted as empirical evidence for the self-regulating cycle of star formation (Ostriker et al. 2010). Bacchini et al. (2019a,b, 2020) found a tight correlation, called volumetric star formation (VSF) law, between the volume densities of the cold gas and the SFR of nearby disc galaxies. These volume densities were obtained from the observed surface densities using the gas scale height derived in the same way as here. The VHE implies that the gas scale height depends on its velocity dispersion, which includes the contributions of thermal motions and turbulence. The results presented here show that turbulence depends in turn on the SFR. Hence, we can envision the following self-regulating cycle. In a quasi-steady state, higher SFR means more turbulence (thicker disc) and lower gas volume density. According to the VSF law, the SFR declines and turbulence follows, causing the gas volume density to grow again and bringing to a new phase of high SFR.

References

Bacchini, C., Fraternali, F., Iorio, G., & Pezzulli, G. 2019a, *A&A*, 622a, A64.

Bacchini, C., Fraternali, F., Iorio, G., Pezzulli, G., Marasco, A., & Nipoti, C. 2020a, *A&A*, 641a, A70.

Bacchini, C., Fraternali, F., Pezzulli, G., & Marasco, A. 2020b, *A&A*, 644b, A125.

Bacchini, C., Fraternali, F., Pezzulli, G., Marasco, A., Iorio, G., & Nipoti, C. 2019b, *A&A*, 632b, A127.

Chevalier, R. A. 1974, *ApJ*, 188, 501–516.

Di Teodoro, E. M. & Fraternali, F. 2015, *MNRAS*, 451, 3021–3033.

Iorio, G. 2018,. PhD Thesis, University of Bologna, (2018).

Kolmogorov, A. 1941, *Akademiia Nauk SSSR Doklady*, 30, 301–305.

Mac Low, M.-M. & Klessen, R. S. 2004, *Reviews of Modern Physics*, 76(1), 125–194.
Martizzi, D., Faucher-Giguère, C.-A., & Quataert, E. 2015, *MNRAS*, 450(1), 504–522.
Ostriker, E. C., McKee, C. F., & Leroy, A. K. 2010, *ApJ*, 721, 975–994.
Tamburro, D., Rix, H.-W., Leroy, A. K., Mac Low, M.-M., Walter, F., Kennicutt, R. C., Brinks, E., & de Blok, W. J. G. 2009, *AJ*, 137, 4424–4435.
Utomo, D., Blitz, L., & Falgarone, E. 2019, *ApJ*, 871(1), 17.

Resolving the Rise and Fall of Star Formation in Galaxies
Proceedings IAU Symposium No. 373, 2023
T. Wong & W.-T. Kim, eds.
doi:10.1017/S1743921322003921

Testing the quenching subgrid physics in Green Valley galaxies

Ignacio Ferreras[1,2], James Angthopo[3,4] and Andrea Negri[1]

[1]Instituto de Astrofísica de Canarias, C/ Vía Láctea s/n, E38205, La Laguna, Tenerife, Spain

[2]Dept. of Physics and Astronomy, University College London, London WC1E 6BT, UK

[3]MSSL, University College London, Holmbury St Mary, Surrey, RH5 6NT, UK

[4]INAF, Osservatorio Astronomico di Brera, Via Brera 28, 20121, Milano, Italy

Abstract. In the title of this Symposium: "The rise and fall of star formation in galaxies", the "falling" stage is mostly represented by so-called Green Valley galaxies. In this phase, quenching mechanisms operate, concerning the evolution from star formation towards quiescence. Therefore, GV galaxies are ideal laboratories to test cosmological simulations. This contribution focuses on the application of a novel, dust-independent, definition of the GV, to two of the most recent simulations: EAGLE and Illustris-TNG. We present some of the results, concerning the excess fraction of quenched galaxies in simulations, with respect to observational data from SDSS. We suggest possible causes for the mismatch.

Keywords. galaxies: evolution – galaxies: stellar content – methods: numerical

1. Introduction

A standard statistical methodology to understand the formation and evolution of galaxies involves the use of bivariate plots where the distribution of various observational measurements are compared. One of the most important of these diagrams is arguably the relation between the mass of a galaxy (in its various guises) and a parameter describing the stellar content: either colour, a more targeted spectral indicator or star formation rate. A bimodality is evident in the distribution, with two major components: the blue cloud (BC) made up of star-forming galaxies, and the red sequence (RS) dominated by passively evolving galaxies (see, e.g., Strateva et al. 2001; Baldry et al. 2004). In between these two lie the so-called green valley (GV) populated by galaxies that evolve from one group to the other (e.g., Salim 2014): most often from BC to RS (quenching) but with a second channel from RS to BC (rejuvenation). The GV therefore holds the key to understanding the fundamental processes operating during the evolution of galaxies. Cosmological simulations of galaxy formation are typically calibrated to the scaling relations of general populations of galaxies (see, e.g., Crain et al. 2015), such as the stellar mass function, the Tully-Fisher relation, or the mass-metallicity relation. While those are the lowest order fundamental constraints, it is important to assess whether this calibration manages to explain more focused phases of evolution. The GV, as a transitioning region where a substantial fraction of the quenching population is found, represents an ideal laboratory to test the details incorporated into the models beyond its resolution ("sub-grid physics").

Angthopo et al. (2019), as part of the doctoral thesis of the first author, proposed a novel definition of the GV based on the 4000Å break, a spectral indicator that is less affected by systematics concerning dust. Details of this definition on a sample of galaxy spectra from SDSS are presented in Angthopo, Ferreras, & Silk (2020). This contribution focuses on how GV galaxies can be used to test cosmological models of galaxy formation (based on Angthopo et al. 2021). We refer interested readers to that paper for details on the material presented here.

2. GV galaxies and the quenching of star formation

We make use of the latest galaxy formation models from the EAGLE (Schaye et al. 2015) and Illustris-TNG (Nelson et al. 2019) collaborations. We adopt the runs defined inside L=100 Mpc boxes (RefL0100N1504 for EAGLE and TNG100 for Illustris-TNG). We extract the formation histories given by the time evolution of the stellar particles, and combine them with population synthesis models (Vazdekis et al. 2015) to produce synthetic spectra at the same resolution as SDSS (see Negri et al. 2022, for details). We also homogeneise the samples, so that these distributions are equivalent in stellar mass – we match the samples differently in EAGLE and TNG100 as they have different mass distributions. In Angthopo et al. (2021) we show that the $D_n(4000)$ vs stellar mass diagram is comparable between SDSS and simulations, although the latter produce stronger values of the 4000Å break and noticeably flatter trends with stellar mass. The fraction of star-forming, AGN and quiescent galaxies in the GV overall matches the observational constraints. While there is an acceptable level of agreement, we find a discrepancy in the GV concerning the fraction of quiescent galaxies at the massive end. This is the focus of this contribution.

Fig. 1 (left/top) shows the fraction of quenched galaxies in the GV as a function of stellar mass for EAGLE (red, left) and TNG100 (blue, right), as dashed lines. This sample corresponds to the mid-GV (mGV), defined as the mid tercile of the distribution of the 4000Å break in GV galaxies, selected in bins of stellar mass. The mGV therefore gives a more focused representation of transitioning galaxies, as it targets the trough of the bimodality between BC and RS. The observational constraint from SDSS is shown as shaded areas. Note the sizeable excess of quenched galaxies at $\log M_s/M_\odot \gtrsim 10.5$, especially in TNG100. Such an excess must be related to the way the simulations implement the evolution from star formation to quiescence. In order to look for differences between these two models, Fig. 1 (left/bottom) shows the cumulative stellar mass growth of the galaxies in the two most massive bins, where the discrepancy is largest. Note the difference between both simulations, especially in the most massive bin: while EAGLE assembles the stellar mass from early times in a relatively fast way, followed by rapid quenching at around $\sim$5 Gyr, TNG100 builds up the stellar content more gradually, extending to later times, producing the quenching at later cosmological times. Moreover, in this model, once the galaxy is quenched, the system no longer experiences subsequent episodes of star formation – unlike EAGLE, that allows for later phases of star formation within the last $\sim$1 Gyr. The TNG100 data could be explained if the quenching subgrid prescriptions need to be strong enough to produce the z=0 results that match the observations. This is clearly a signature of the imposed AGN feedback, given that the biggest discrepancy occurs in the most massive galaxies.

Fig. 1 (right) compares the central black hole mass versus the stellar mass of the galaxy in both simulations, at zero redshift, as labelled. At fixed galaxy mass, TNG100 features systematically higher BH masses. Since the AGN feedback depends on M_{BH}, it is expected that TNG100 will inject more energy into the ISM during the AGN phase. Moreover, the prescription for AGN feedback in TNG100 includes a kinetic term that applies to the massive end ($\gtrsim 10^{10.5} M_\odot$, Weinberger et al. 2017). Therefore, TNG100 galaxies in the GV quench at later times and more efficiently. In contrast, a larger fraction of GV galaxies in EAGLE show later episodes of star formation, producing lower quenched fractions that are more consistent with the observational constraints. Independendent studies

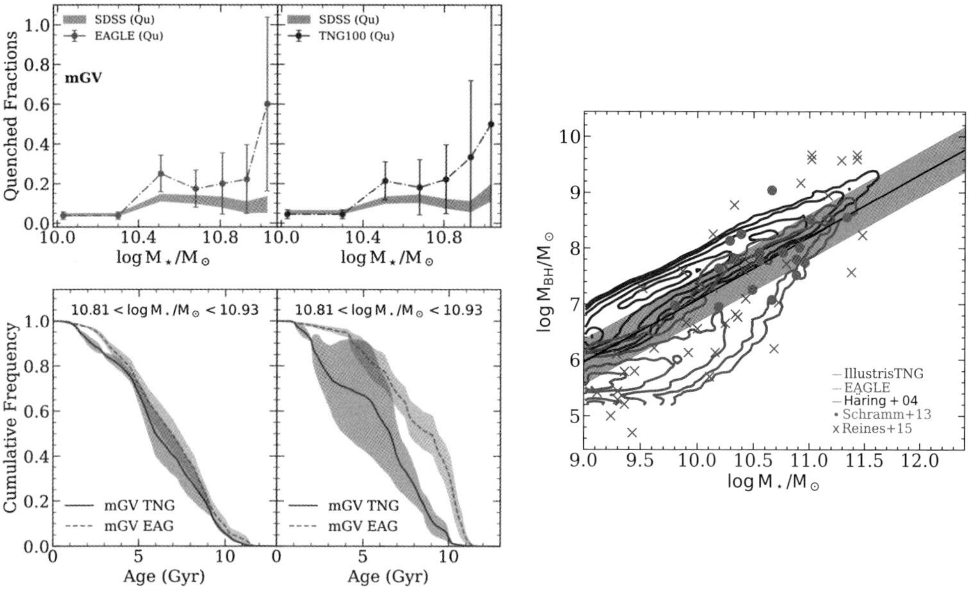

Figure 1. *Left:* The top panels show the fraction of quenched GV galaxies as a function of stellar mass for EAGLE (red, left) and TNG100 (blue, right), shown as dashed lines with error bar accounting for the 1σ uncertainties. The equivalent observed fraction in SDSS is shown as a shaded region. The sample is defined in the mid tercile of the GV selection. The bottom panels show the cumulative stellar mass growth in the two most massive bins of the sample, for EAGLE (red) and TNG100 (blue). *Right:* The central black hole mass vs stellar mass is shown for EAGLE (red) and TNG100 (blue), with data points from observational constraints. Note the higher black hole mass at fixed stellar mass in the TNG100 simulation. Adapted from Angthopo et al. (2021).

have noted the strong quenching nature of kinetic feedback (see, e.g., Terrazas et al. 2020; Davies et al. 2020), along the lines of our comparisons, but the definition of the sample is important, as other approaches may produce more consistent results with the observations (Donnari et al. 2021). We emphasize that our $D_n(4000)$-based definition of the GV provides a robust, dust-independent way of selecting transitioning galaxies. The difference shown in Fig. 1 between the black hole masses of EAGLE and TNG100 galaxies could be down to the systematic difference in seeding the central black holes that eventually act as AGN: Both simulations inject a "seed" BH at the centre, once the dark matter halo reaches a predetermined threshold mass. TNG100 impose a higher BH seed mass, and a higher halo mass threshold (a factor $\sim 8\times$ higher than EAGLE in both). Along with the kinetic feedback prescription, this choice would statistically induce later AGN-related quenching, which might explain our comparisons with the observational data.

References

Angthopo J., Ferreras I., Silk J., 2019, *MNRAS Lett.*, 488, L99

Angthopo J., Ferreras I., Silk J., 2020, *MNRAS*, 495, 2720

Angthopo J., Negri A., Ferreras I., de la Rosa I. G., Dalla Vecchia C., Pillepich A., 2021, *MNRAS*, 502, 3685

Baldry I. K., Glazebrook K., Brinkmann J., Ivezić Ž., Lupton R. H., Nichol R. C., Szalay A. S., 2004, *ApJ*, 600, 681

Crain R. A. et al., 2015, *MNRAS*, 450, 1937

Davies J. J., Crain R. A., Oppenheimer B. D., Schaye J., 2020, *MNRAS*, 491, 4462

Donnari M., Pillepich A., Nelson D., Marinacci F., Vogelsberger M., Hernquist L., 2021, *MNRAS*, 506, 4760

Negri A., Dalla Vecchia C., Aguerri J. A. L., Bahé Y., 2022, *MNRAS*, 515, 2121

Nelson D., Springel V., Pillepich A., Rodriguez-Gomez V., Torrey P., Genel S., Vogelsberger M., et al., 2019, *Comp. Astronomy & Cosmology*, 6, 2

Salim S., 2014, *Serb. Astron. J.*, 189, 1

Schaye J., Crain R. A., Bower R. G., Furlong M., Schaller M., Theuns T., Dalla Vecchia C., et al., 2015, *MNRAS*, 446, 521

Strateva I., Ivezić Ž., Knapp G. R., Narayanan V. K., Strauss M. A., Gunn J. E., Lupton R. H., et al., 2001, *AJ*, 122, 1861

Terrazas B. A. et al., 2020, *MNRAS*, 493, 1888

Vazdekis A., Koleva M., Ricciardelli E., Rčk B., Falcón-Barroso J., 2016, *MNRAS*, 463, 3409

Weinberger R., Springel V., Hernquist L., Pillepich A., Marinacci F., Pakmor R., Nelson D., et al., 2017, *MNRAS*, 465, 3291

Resolving the Rise and Fall of Star Formation in Galaxies
Proceedings IAU Symposium No. 373, 2023
T. Wong & W.-T. Kim, eds.
doi:10.1017/S1743921322003775

Is star formation in gas-rich bars suppressed?

Fumiya Maeda

Institute of Astronomy, Graduate School of Science, The University of Tokyo, 2-21-1 Osawa, Mitaka, Tokyo 181-0015, Japan

Abstract. Whether the star formation efficiency (SFE) in the bar region is lower than those in the other regions in a barred galaxy has recently been debated. We statistically investigate the SFEs along the bars in nearby gas-rich massive star-forming barred galaxies by distinguishing the center, bar-end, and bar regions for the first time. The molecular gas surface density is derived from archival CO(1–0) and/or CO(2–1) data and the star formation rate surface density is derived from a linear combination of far-ultraviolet and mid-infrared intensities. To distinguish the three regions, we targeted 18 galaxies with a large apparent bar length ($\geq 75''$). The resulting SFE in the bars is about $0.6 - 0.8$ times lower than that in the disks, which suggests the star formation in the bars tends to be systematically suppressed.

Keywords. galaxies: ISM – galaxies: structure

1. Introduction

It has been commonly known that massive star formation in bars of barred galaxies is suppressed. A low star formation efficiency (SFE $= \Sigma_{\rm SFR}/\Sigma_{\rm mol}$) in the bar region has been reported for many barred galaxies such as NGC 1300 (Maeda et al. 2020), NGC 4303 (Yajima et al. 2019), and NGC 5236 (Hirota et al. 2014). However, in contrast to the studies that observed the individual barred galaxies, recent statistical studies have suggested that the SFE in the bars is not systematically lower than those in other regions and is, in fact, environmentally independent (Muraoka et al. 2019; Díaz-García et al. 2021; Querejeta et al. 2021).

What makes this inconsistency? Two methodological differences exist. First difference is the definition of the bar. While the galactic center and bar-end are distinguished as other environments from the bar in most of the previous studies focused on individual galaxies, the definition of the bar in recent statistical studies includes (part of) center and bar-end, which may make the difference of SFEs between bars and disks small because of the high SFE in the center and bar-end. Second difference is the spatial resolution. The recent statistical studies included galaxies in the sample whose apparent bar major axis is as small as several times the angular resolution of the images they used. In this case, the center, bar, and bar-end cannot be distinguished, which may also smooth the differences in the SFE between the environments.

We aim to statistically determine whether the SFE in the bar region is lower than those in other regions by distinguishing the galactic center, bar-end, and bar, similar to previous studies that focused on individual barred galaxies. To distinguish between these environments and avoid the smoothing the differences in the SFE, we focus on 18 gas-rich galaxies whose bar major axis is at least five times larger than the angular resolution of the images we used ($15''$).

2. Sample Selection and Data Analysis

We first select 35 nearby barred galaxies with a large apparent bar major axis from the HyperLeda database according to the following criteria: (1) The morphological type is SAB or SB and the Hubble T stage is in $T = 0 - 7$. (2) The recessional velocity is < 2000 km s^{-1} ($\sim$ 27 Mpc). (3) The inclination is $< 70°$. (4) The Galactic latitude (b) is $|b| > 10°$. (5) GALEX FUV and WISE 22 μm data available. The $\Sigma_{\rm SFR}$ is derived from a linear combination of them. (6) CO(1–0) or/and CO(2–1) data cubes available. Here, we refer to the catalogs of previous CO survey projects (NRO CO atlas, COMING, HERACLES, and PHANGS-ALMA) and search the ALMA archival data sets. (7) Bar semi-major axis ($R_{\rm bar}$) is $\geq 37.''5$. This selection is based on the requirement that the major axis of the bar ($2R_{\rm bar}$) must be at least five times larger than the angular resolution of the WISE 22 μm ($15''$) to distinguish between the center, bar, and bar-end.

Then, we select 18 galaxies with gas-rich bar from the 35 galaxies. We define a gas-rich bar as a bar region where more than 40% area is $\Sigma_{\rm mol} \geqq 5\ M_\odot$ pc^{-2}. Here, we calculate the $\Sigma_{\rm mol}$ by assuming the Galactic CO-to-H_2 conversion factor and a CO(2–1)/CO(1–0) ratio of 0.65. All CO data cubes are used after convolving them to a beam size of $15''$. We define the center, bar, and bar-end regions of the galaxy based on the stellar bar structure defined by the cataloged ellipse. We define a rectangle with a width of $2 \times 1.25 \times R_{\rm bar}$, height of $2R_{\rm bar}(1 - \epsilon_{\rm bar})$, and position angle of the bar (Figure 1(a)). Here, the $\epsilon_{\rm bar}$ is ellipticity. In this rectangle, we define the center, bar, and bar-end as the regions where the distance to the minor axis of the ellipse (R) is $0.0 - 0.25R_{\rm bar}$, $0.25R_{\rm bar} - 0.75R_{\rm bar}$, and $0.75R_{\rm bar} - 1.25R_{\rm bar}$, respectively. The region outside of this rectangle and inside the FoV of the CO data cube is defined as a disk. Most of the 18 galaxies are massive ($M_{\rm star} \geqq 10^{10}\ M_\odot$) and located in the upper side of the main sequence. The angular resolution is $15''$, which corresponds to $0.3 - 1.8$ kpc.

3. Results

Figure 1(b) and (c) display the SFE profiles from $R/R_{\rm bar} = 0.0$ to 1.25 that are normalized by the SFE in the disk of the sample galaxies. We find the SFE in the bar to be systematically lower than that in the disk regardless of whether $\Sigma_{\rm mol}$ is measured using CO(1–0) or CO(2–1). The median normalized SFE (SFE/SFE$_{\rm disk}$) is $0.6 - 0.8$ in $R/R_{\rm bar} = 0.25 - 0.75$. The SFE/SFE$_{\rm disk}$ is at a minimum at around $R/R_{\rm bar} = 0.5$. These results suggest that massive star formation in the bar region is systematically suppressed in comparison with the disk region in massive ($M_{\rm star} \geqq 10^{10}\ M_\odot$) high sSFR (i.e., upper side of the main sequence) galaxies with gas-rich bar and disk. Although the SFE in the bar region is systematically suppressed, its scatter is approximately 0.5 dex; some areas possess SFEs comparable to those in the disk, whereas others possess significantly lower SFEs than those in the disk. Additionally, the degree of suppression of star formation appears to vary among galaxies and within a galaxy.

Our results are inconsistent with the results of nonenvironmental dependence on the SFE by similar recent statistical studies (Muraoka et al. 2019; Díaz-García et al. 2021; Querejeta et al. 2021). The possible causes of this inconsistency are the differences in the definition of the bar region, spatial resolution, the CO-to-H_2 conversion factor, and sample galaxies.

4. Discussion: Relationship between the velocity width and SFE

The degree of the star formation suppression seems to vary among galaxies and within a galaxy. What determines the degree of the suppression? One promising parameter is the strength of noncircular motion in the bar region. Star formation has been arguably suppressed by some dynamical effects, such as strong shock, large shear, and fast cloud-cloud collisions, which are caused by the noncircular motion. Here, we investigate the

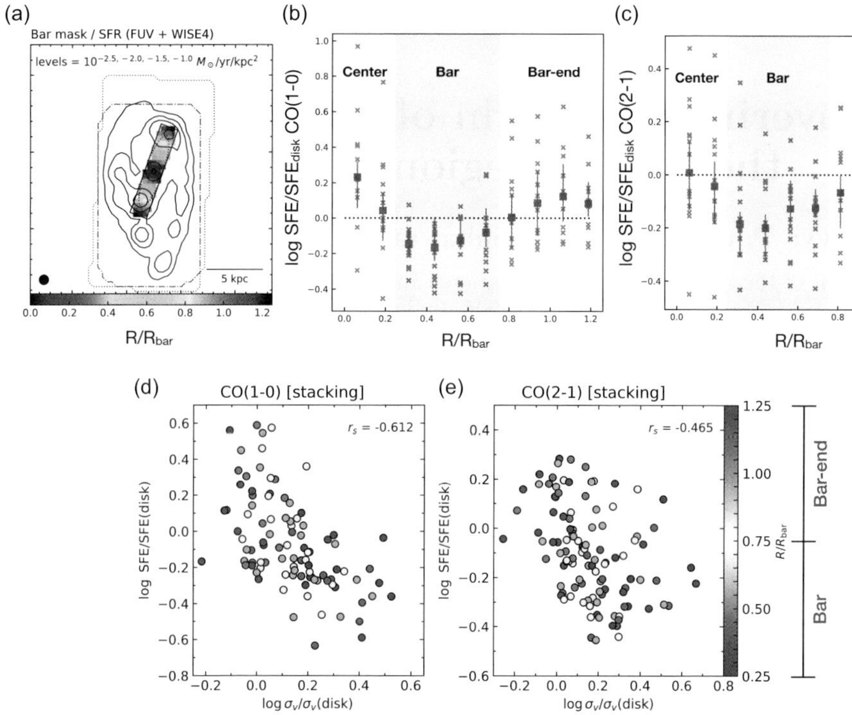

Figure 1. (a) Environmental mask (NGC 3627). The contours show the $\Sigma_{\rm SFR}$. The FoVs of the CO(1–0) and CO(2–1) observations are represented as a red dotted and blue dash-dotted lines, respectively. The black ellipse is the cataloged bar structure. Color map shows the normalized distance to the minor axis of the ellipse. (b-c) SFE profiles as a function of the $R/R_{\rm bar}$ derived from CO(1–0) and CO(2–1). The data points of each galaxy is shown by gray crosses and the median value and IQR of all sample galaxies in each bin are shown as a red square and bar, respectively. For each galaxy, we show the median SFE in each bin normalized by the median SFE in the disk. (e-f) Relationship between normalized velocity width of the CO spectrum and normalized SFE in the bar and bar-end regions. The velocity width, which is derived from the stacking profile, is normalized by that in the disk region. The Spearman's correlation rank (r_s) is given in the top right corner.

relationship between the velocity width of the CO spectrum and SFE in the bar, bar-end, and disk regions. Figure 1(d-e) show the relationship between the normalized velocity width of the CO spectrum and normalized SFE in the bar and bar-end regions. The velocity width is normalized by that in the disk region of the galaxy. We find negative correlations between the normalized velocity width and SFE. The σ_v in the bar-end is $\sim 0.8 - 1.2$ times larger than that in the disk, whereas that in the bar is is $\sim 1.2 - 4.0$ times larger than that in the disk. This negative correlation would support the idea that the larger the noncircular motion, the lower the SFE.

References

Díaz-García S. et al. 2021, *A&A*, 654, A135
Hirota A. et al. 2014, *PASJ*, 66, 46
Maeda F. et al. 2020, *MNRAS*, 495, 3840
Muraoka K. et al. 2019, *PASJ*, 71, S15
Querejeta M. et al. 2021, *A&A*, 656
Yajima Y. et al. 2019, *PASJ*, 71, S13

Resolving the Rise and Fall of Star Formation in Galaxies
Proceedings IAU Symposium No. 373, 2023
T. Wong & W.-T. Kim, eds.
doi:10.1017/S1743921322004161

Recovering the origin of star formation in the central region of I Zw 81

Divya Pandey[1], Kanak Saha[2] and Ananta C. Pradhan[1]

[1]Department of Physics and Astronomy, National Institute of Technology, Rourkela, Odisha 769008, India

[2]Inter-University Centre for Astronomy & Astrophysics, Postbag 4, Ganeshkhind, Pune 411007, India

Abstract. We have studied the star formation properties of a massive void galaxy - I Zw 81. We performed 2D structural decomposition on *Canada France Hawaii Telescope (CFHT)* g- and r-band observation of I Zw 81 using GALFIT. The galaxy consists of an unresolved small bulge, a bar, an inner ring, and a truncated disk. We have used far-ultraviolet (FUV) and near-UV (NUV) observation of Ultraviolet Imaging Telescope (UVIT) onboard *AstroSat* for our analysis. The NUV−r color map of the lenticular galaxy illustrates a shallow positive color gradient in the profile, implying that the bar and inner ring are more star-forming than the outer disk. The FUV emission is mainly concentrated in the central region of the galaxy. A tidal tail-like feature is detected in the CFHT observations. We infer that bar and minor mergers-like interactions enhance the gas inflow and drive star formation in the center of I Zw 81.

Keywords. galaxies: elliptical and lenticular, ultraviolet: galaxies, galaxies: interactions

1. Introduction

Local galaxies disparate themselves into distinct classes, i.e., star-forming and quiescent (Kauffmann et al. 2003; Baldry et al. 2004). This separation is an outcome of various physical transformation processes that act on galaxies over their cosmic lifetime. It is well established that the environment and stellar mass play crucial role in determining the properties of galaxies (Peng et al. 2010). The low-mass galaxies are affected mainly by their surrounding environment, while at the high stellar mass range, several mass quenching mechanisms get activated that cause star formation cessation in galaxies (Alpaslan et al. 2015; Peng et al. 2010).

We are interested in studying the properties of a massive ($M_\star \approx 10^{10.9}$ $M_\odot$) early-type galaxy, I Zw 81 present in the Bootes void (Kirshner et al. 1987). Our previous analysis using Ultraviolet Imaging Telescope (UVIT) data reveals that the I Zw 81 is star-forming, shows blue color and strong ultraviolet (UV) emission (Pandey et al. 2021). The relatively rare class of massive blue early-type galaxies similar to I Zw 81 are thought to be an outcome of major merger events (Kannappan et al. 2009). However, such interactions are improbable in voids (Penny et al. 2015). In this work, we have attempted to understand the physical mechanisms responsible for the present state of massive void galaxy, I Zw 81.

2. Central star formation and FUV emission in bar

We have primarily used *Canada France Huwaii Telescope (CFHT)* g- and r-band and UVIT near-UV (NUV) and far-UV (FUV) imaging data to study the star formation and perform the structural analysis of I Zw 81. We performed the structural decomposition

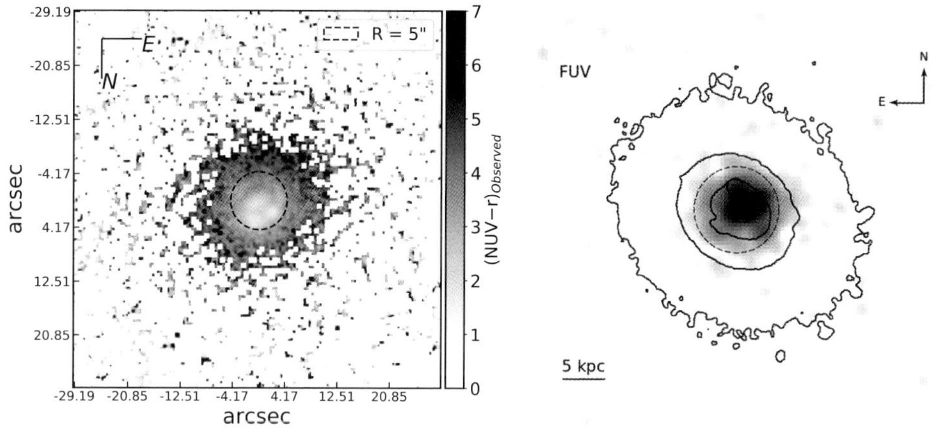

Figure 1. Left panel: NUV−r color map of I Zw 81; Right panel: FUV observation of I Zw 81. The *CFHT*-g band contours shown by black solid lines are overlayed. The contour comprises of 5 surface brightness levels = [17.37, 19.24, 21.11, 22.94, 24.58] mag/arcsec^2. The star-forming clumps are shown in black circles on the second contour. The black dashed circles (r = 5″) shown in both panels enclose the bar and ring.

of I Zw 81 using GALFIT (Peng et al. 2002) on *CFHT* observation, and found that the galaxy consists of an unresolved small bulge, a bar, an inner ring, and a truncated disk.

We studied the mass assembly of the galaxy using the NUV−r color map shown in the left panel of Figure 1. The central region comprising the bar and inner ring is represented by a dashed black circle of radius = 5″. Note that the point source function (PSF) matching on *CFHT* r-band data was performed prior to the color map construction. The color map shows that the central region of the galaxy is bluer than its outskirts. The shallow positive gradient displayed by I Zw 81 is generally observed in low-mass galaxies and is thought to be the product of environmental disturbances (Pan et al. 2015). The right panel of Figure 1 shows FUV observation of the galaxy with *CFHT* g-band contours overlayed on the image. The FUV emission traces recent star formation in a galaxy averaged to a timescale of 100 Myr (Kennicutt et al. 1998). We found that the FUV emission is localized in the central region of the galaxy enclosing the bar. The FUV emission in the bar region of I Zw 81 is similar to that reported by Saha et al. 2021 in the Malin 1 galaxy whose central region resembles lenticular morphology.

The bar generally transfers the gas from galactic outskirts to the inner region. As a result, shock waves and gravitation torque rise inside the bar which lead to decrease the ongoing star formation activity (Khoperskov et al. 2018). Massive lenticular galaxies similar to I Zw 81 show FUV emission only in the nuclear region of bars (Diaz-Garcia et al. 2020). On the other hand, the FUV emission in I Zw 81 is spread across the entire bar and the inner ring implying that galaxy is sufficiently gas-rich. We investigate the source of the gas supply and observed star formation in the next section.

3. Role of mergers and interactions

Mergers and interactions between galaxies strongly influence their star formation properties (Alonso et al. 2012; Kaviraj 2014). Since the galaxies present in voids are unlikely to go through major merger interaction in their lifetime (Penny et al. 2015), we only discuss the effect of minor mergers and tidal interactions on I Zw 81. We used deep *CFHT* g-band observation to look for any signatures of interactions. We found a tidal tail-like feature in the galactic outskirt shown in Figure 2. We detect a companion galaxy WISEA

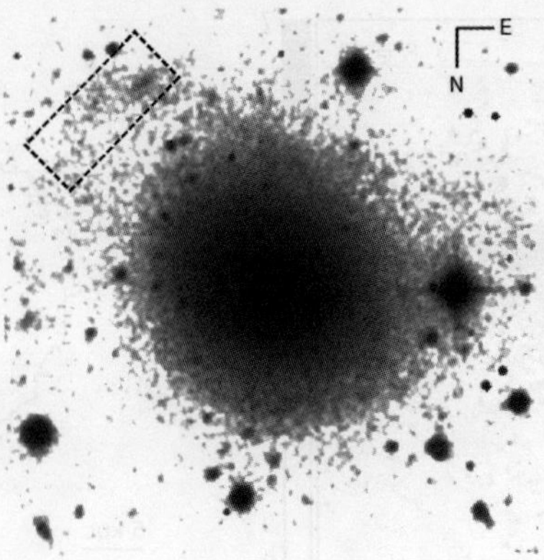

Figure 2. The black dashed rectangular box highlights the tidal tail-like feature observed in *CFHT*-g band image of the galaxy.

J140811.38+485344.2 at a distance of 120 kpc from I Zw 81. The galaxy is not massive enough to enhance the ongoing star formation in I Zw 81 via tidal interaction or flyby.

Minor merger interactions in the past seem to be the most favorable scenario to explain the rapid star formation. Such interactions can ignite star formation activity in the galaxies by supplying fresh gas to the systems (Bournaud et al. 2007). Galaxies in voids can sustain their gas supplies for longer time due to secular evolution. Hence, the low-density environment of I Zw 81 may play a part in enhancing the star formation induced by the interactions. The bar transfers the available gas to the galactic center which lead to the observed star formation. The results are explained in detail in (Pandey et al. 2022).

References

Alonso, S., Mesa, V., Padilla, N., et al. 2012, *AAP*, 539, A46
Alpaslan, M., Driver, S., Robotham, A. S. G., et al. 2015, *MNRAS*, 451, 3249
Baldry, I. K., Glazebrook, K., Brinkmann, J., et al. 2004, *ApJ*, 600, 681
Bournaud, F., Jog, C. J., & Combes, F. 2007, *AAP*, 476, 1179
Diaz-Garcia, S., Moyano, F. D., Comerón, S., et al. 2020, 896, *A&A*, 644, A38
Kannappan, S. J., Guie, J. M., & Baker, A. J. 2009, *AJ*, 138, 579
Kauffmann, G., Heckman, T. M., White, S. D. M., et al. 2003, *MNRAS*, 341, 33
Kaviraj, S. 2014, *MNRAS*, 440, 2944
Kennicutt, R. C. 1998, *AARA*, 36, 189
Khoperskov, S., Haywood, M., Di Matteo, P., Lehnert,932 M. D., & Combes, F. 2018, *A&A*, 609, A60
Kirshner, R. P., Oemler, A., Schechter, P. L., et al. 1987, *ApJ*, 314, 493
Pan, Z., Li, J., Lin, W., et al. 2015, *ApJL*, 804, L42
Pandey, D., Saha, K., & Pradhan, A. C. 2021, *APJ*, 919, 101
Pandey, D., Saha, K., Pradhan, A. C., & Kaviraj, S. 2022, *APJ*, 941, 128
Peimbert, M. & Torres-Peimbert, S. 1992, *AAP*, 253, 349
Peng, C. Y., Ho, L. C., Impey, C. D., & Rix, H.W. 2002, 973, *AJ*, 124, 266
Peng, Y.-. jie., Lilly, S. J., Kovač, K., et al. 2010, *ApJ*, 721, 193
Penny, S. J., Brown, M. J. I., Pimbblet, K. A., et al. 2015, *MNRAS*, 453, 3519
Saha, K., Dhiwar, S., Barway, S., et al. 2021, *JApA*, 42, 59

Chapter 4:
The Rise and Fall of Star Formation Across Cosmic Time

Resolving the Rise and Fall of Star Formation in Galaxies
Proceedings IAU Symposium No. 373, 2023
T. Wong & W.-T. Kim, eds.
doi:10.1017/S1743921322003866

The dust properties of star-forming galaxies in the first billion years

Elisabete da Cunha

International Centre for Radio Astronomy Research, University of Western Australia, 35 Stirling Hwy, Crawley, WA 6009, Australia
ARC Centre of Excellence for All Sky Astrophysics in 3 Dimensions (ASTRO 3D)

Abstract. The Atacama Large Millimetre/Sub-millimetre Array (ALMA) is obtaining the deepest observations of early galaxies ever achieved at (sub-)millimetre wavelengths, and detecting the dust emission of young galaxies in the first billion years of cosmic history, well in the epoch of reionization. Here I review some of the latest results from these observations, with special focus on the REBELS large programme, which targets a sample of 40 star-forming galaxies at $z \simeq 7$. ALMA detects significant amounts of dust in very young galaxies, and this dust might have different properties to dust in lower-redshift galaxies. I describe the evidence for this, and discuss theoretical/modelling efforts to explain the dust properties of these young galaxies. Finally, I describe two additional surprising results to come out of the REBELS survey: (i) a new population of completely dust-obscured galaxies at $z \simeq 7$, and (ii) the prevalence of spatial offsets between the ultraviolet and infrared emission of UV-bright, high-redshift star-forming galaxies.

Keywords. galaxies: evolution – galaxies: high-redshift – galaxies: ISM – dust, extinction

1. Introduction

Interstellar dust plays a critical role in galaxy assembly. Dust grains are a site of efficient molecule formation, specifically H_2, the fuel for new star formation (e.g., Hirashita & Ferrara 2002). Dust grains also affect how we observe galaxy assembly: the cosmic infrared background is equivalent in power to the cosmic optical background (e.g., Dole et al. 2006), implying that about half of all the UV/optical light produced by stars in the Universe is reprocessed by dust. In fact, we know that dust-obscured star formation contributes significantly to the total star formation rate (SFR) density of the Universe (e.g., Madau & Dickinson 2014). Recent studies show that dust-obscured star formation dominates the cosmic SFR density out to at least $z \simeq 3$ (e.g., Zavala et al. 2021), and it still provides an important contribution at higher redshifts, although the exact amount of dust-obscured star formation at the earliest redshifts ($z \gtrsim 6$, i.e., in the epoch of reionization, or 'cosmic dawn') is still uncertain.

The main limitation in measuring obscured star formation at the highest redshifts has been the difficulty to detect and measure the dust emission of individual high-redshift galaxies using far-infrared/sub-millimetre facilities such as the *Herschel Space Observatory* and single-dish sub-mm observatories. Due to their poor resolution and sensitivity (see, e.g., figure 10 of Casey et al. 2014), observations with these facilities suffer from important confusion noise, which limited the study of high-redshift sources to only the brightest objects, such as quasars (e.g., Bertoldi et al. 2003) and sub-millimetre galaxies (SMGs; e.g., Walter et al. 2012; Riechers et al. 2013). Alternatively, several studies attempted to overcome the confusion limits and detect fainter galaxies using stacking (e.g., Schreiber 2015) and de-blending techniques (e.g., Liu et al. 2018; Jin et al. 2018), but these approaches have limitations. With its unprecedented sensitivity and spatial resolution at (sub-)millimetre wavelengths, the Atacama Large Millimetre/Submillimetre Array (ALMA), has revolutionised this field, allowing us to directly detect, for the first time, the dust (and gas) emission from individual, normal star-forming galaxies at $z > 6$ (see Hodge & da Cunha 2020 for a recent review). In this paper, I briefly review what recent ALMA observations are revealing about the interstellar dust properties of star-forming galaxies in the first billion years.

2. ALMA observations of Lyman-break galaxies

2.1. *Different dust properties at high-redshift?*

Before direct dust detections with ALMA, high-redshift studies routinely corrected the observed UV emission of distant galaxies for dust attenuation using the so-called 'IRX-β' relation between the ultraviolet spectral slope (β) and the infrared excess (IRX, defined as the ratio of infrared to ultraviolet luminosity). Local starburst galaxies follow a tight relation in the IRX vs β diagram (Meurer et al. 1999), which can be explained if stars are behind a screen of dust with properties similar to those of dust in the Milky Way (Calzetti et al. 1994). Before the advent of ALMA, high-redshift galaxies were assumed to follow a similar relation. However, early ALMA follow-up of known Lyman-break galaxies (LBGs) at $z > 4$ in Band 6 ($\simeq 1$ mm) found infrared luminosities that were lower than those predicted by the local starburst IRX-β relation. This was found in a variety of targeted observations and stacks of $z > 4$ LBGs (e.g., Capak et al. 2015; Bouwens et al. 2016; Fudamoto et al. 2017; Barisic et al. 2017; Bowler et al. 2018; Schouws et al. 2022): the lower IR excesses were more consistent with a steeper dust extinction curve similar, to the one observed in the Small Magellanic Cloud. Taken at face value, this result would imply that interstellar dust in EoR galaxies has fundamentally different properties than dust in low-redshift galaxies. This led to speculation that perhaps these galaxies have different dust properties due to their lower metallicities. This also implied that dust properties evolve rapidly in the early universe, since studies at $z < 4$ seemed to show properties already consistent to Milky Way dust (e.g., Fudamoto et al. 2020). However, these results are plagued by the large uncertainty in measuring the total infrared luminosity of a galaxy from a single 1-mm ALMA measurement without knowing the dust temperature (see Fig. 1 and discussion in section 3.3 of Hodge & da Cunha 2020).

2.2. *The need for robust dust temperatures*

In all the aforementioned studies, a dust temperature had to be assumed to extrapolate the single-band ALMA measurement to a total IR luminosity, and most studies assumed temperatures around 40 K, similar to local and intermediate-redshift star-forming galaxies. Faisst et al. (2017) argued that local analogues to high-redshift LBGs have higher dust temperatures ($T_{\rm dust} \sim 60 - 90$ K), and that adopting such temperatures would bring the ALMA detections (and upper limits) more in agreement with Milky-Way dust in the IRX-β relation. Theoretical work also seems to support hotter dust at high redshifts (e.g., Behrens et al. 2018; Narayanan et al. 2018; Ma et al. 2019; Liang et al. 2019; McAlpine et al. 2019; Sommovigo et al. 2020).

These higher dust temperatures need to be confirmed through multi-frequency observations with ALMA, preferably sampling close to the peak of the dust emission, to measure $T_{\rm dust}$. For galaxies at $z \gtrsim 6$, such observations are time consuming even with ALMA, but they have been achieved for a handful of galaxies so far. Faisst et al. (2020) followed-up four $z \sim 5.5$ galaxies in Band 8 (sampling rest-frame $\sim 100\mu$m, close to the peak of dust emission), which, combined with existing Bands 6 and 7 observations, allowed them to

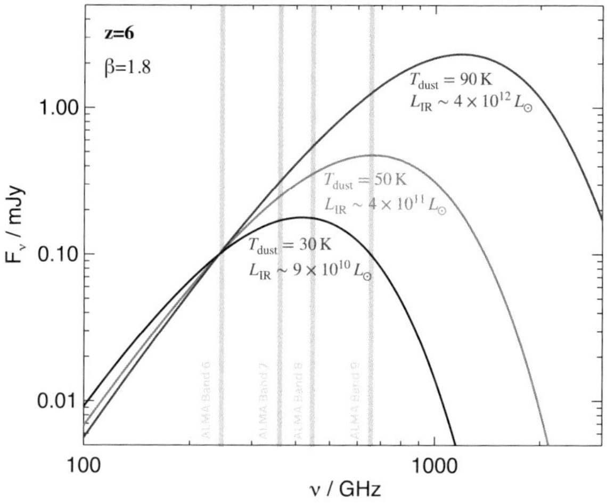

Figure 1. Three model dust emission spectral energy distributions (SEDs) at $z = 6$ normalized at the same ALMA Band 6 (230 GHz) flux of 0.1 mJy. This illustrates how sources with the same Band 6 flux can have inferred infrared luminosities, $L_{\rm IR}$, that vary by almost two orders of magnitude depending on the adopted dust temperature, $T_{\rm dust}$. To constrain the temperatures and luminosities, we need multi-band observations at higher frequencies (e.g., in the ALMA bands shown in grey) sampling the dust SED closer to its peak.

measure dust temperatures $T_{\rm dust} \sim 30 - 40$ K. In another recent study, Bakx et al. (2021) combined Band 9 observations with observations in three other bands (Bands 6, 7, and 8), to constrain a dust temperature of $T_{\rm dust} = 42$ K for a galaxy at $z = 7.13$.

It is worth noting that having ALMA high-frequency observations that sample the peak of the dust emission (or even beyond the peak) is critical for robust constraints on the dust temperature (and hence, luminosity). But even with multi-band observations, potential degeneracies with the dust emissivity index and dust opacity assumptions need to be taken into account (e.g., da Cunha et al. 2021; Jin et al. 2022; Witstok et al. 2022). For example, in a detailed study of the dust emission of SMGs, da Cunha et al. (2021) show that with only three ALMA bands sampling the dust emission SED, it is very difficult to distinguish between optically thin dust emission and dust that it optically thick at far-infrared wavelengths – but important systematics can be introduced in the derived dust temperatures and masses (Fig. 2).

2.3. *Is dust in galaxies really getting hotter with redshift?*

As mentioned before, theoretical simulations predict an increase of the dust temperature with redshift, which is generally explained by several factors such as stronger radiation fields, lower dust masses, and more compact dust-emitting regions at high redshifts. Empirically, an increase of dust temperature with redshift has also been observed, e.g., by Schreiber et al. (2018), though we note that this is still somewhat controversial as it could be a result of selection effects: since there is a positive correlation between dust temperature and dust luminosity, and flux-limited samples detecting higher luminosity sources at higher redshifts (e.g., Dudzevičiūtė et al. 2020). In fact, the direct temperature measurements of sources at $z \gtrsim 6$ with multi-band ALMA observations described above (Faisst et al. 2020; Bakx et al. 2021) do not follow the extrapolation of the temperature-redshift relation of Schreiber et al. (2018): their temperatures are about 10 to 20 K lower than predicted by that relation. If these temperatures are typical for high-redshift galaxies, then the dust temperatures might not be high enough to reconcile the IR excesses

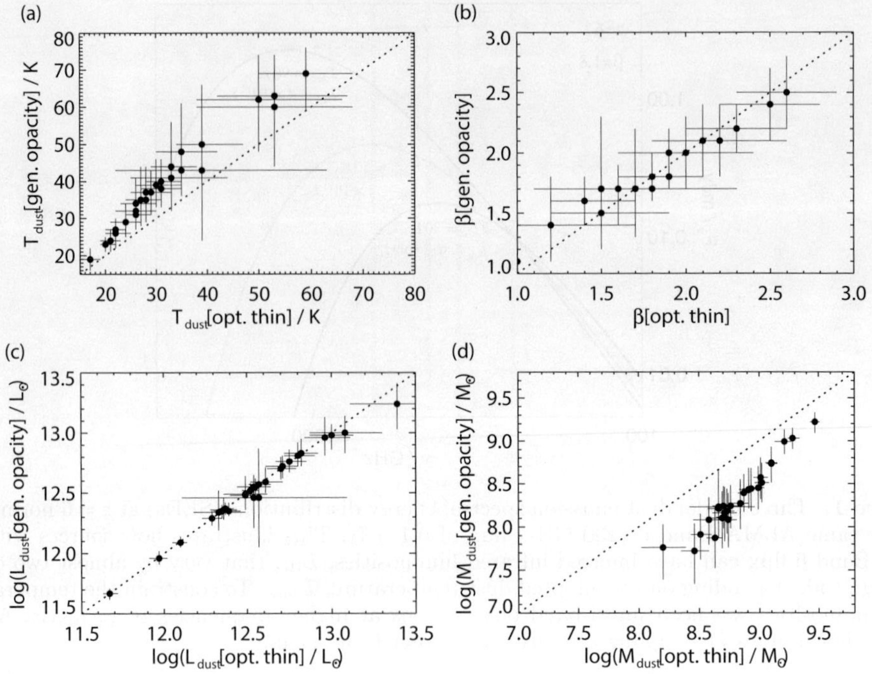

Figure 2. Comparison between dust physical properties derived from fitting the dust emission with simple isothermal models when using an optically-thin assumption (x-axes) vs a general opacity model (y-axes): (a) dust temperature; (b) dust emissivity index; (c) dust luminosity; (d) dust mass. The points shown are 27 SMGs with well-sampled dust SEDs studied in da Cunha et al. (2021). While the dust luminosity and emissivity index are quite robust to optical depth assumptions, assuming the wrong optical depth can introduce biases in the derived dust temperatures and dust masses (see more details in da Cunha et al. 2021). Figure first published in da Cunha et al. (2021).

with Milky Way-like dust attenuation in the IRX-β relation, which would indicate that indeed galaxies at the EoR have different dust properties. However, it is to be expected that there is a spread in dust temperatures at any given redshift (depending on different galaxy properties), so larger samples of galaxies t $z \gtrsim 6$ with ALMA continuum observations are needed. One such samples comes from the REBELS survey described below.

3. Results from the REBELS survey

3.1. *Continuum detections in REBELS*

The Reionization Epoch Bright Emission Line Survey (REBELS) is an ALMA Large Programme targeting [OIII] and [CII] lines and their underlying dust continuum (at ~ 90 and 160μm) in 40 UV-bright LBGs at $z > 6.5$ (Bouwens et al. 2022). From this sample, 18 sources have been detected in the continuum so far (Inami et al. 2022), which more than tripled the previous number of dust continuum detections at $z > 6.5$. The inferred infrared luminosities of these continuum-detected sources imply that about half of their star formation is dust-obscured. One of the detections, REBELS-25, is noteworthy as it is an ultra-luminous infrared galaxy (ULIRG, with $L_{\rm IR} > 10^{12}\,L_{\odot}$) at $z = 7.4$ (Hygate et al., submitted).

So far the REBELS sources are only continuum-detected in one ALMA band, therefore we cannot directly measure the dust temperature from fitting the dust SEDs.

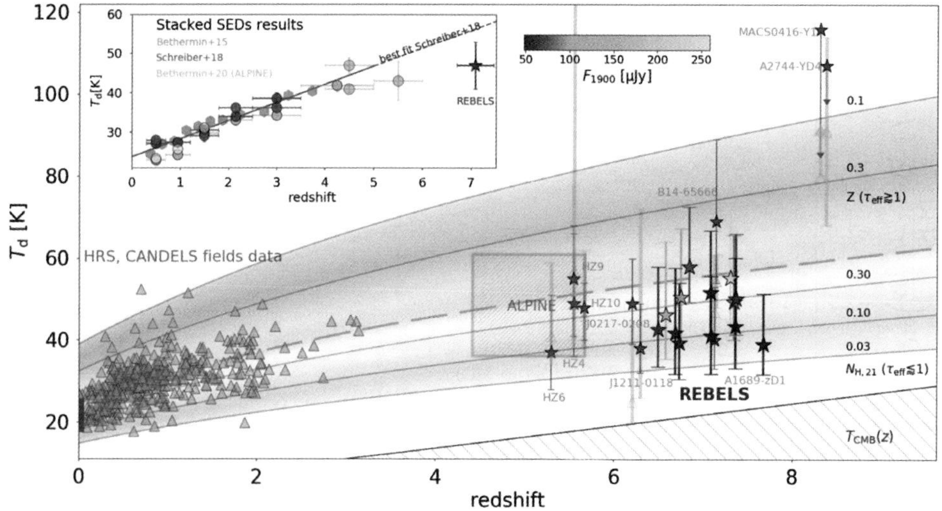

Figure 3. Empirical relation between the redshift and the dust temperature of galaxies. The REBELS sources are plotted as colourful stars. The inset shows that the average dust temperature of REBELS sources fall short of the extrapolation of the $T_{\rm dust}$-redshift relation of Schreiber et al. (2018). Figure first published in Sommovigo et al. (2022).

But Sommovigo et al. (2022) estimate the temperatures of REBELS sources using a novel analytical method that uses only the [CII] line and its underlying dust continuum (Sommovigo et al. 2021). They find an average $T_{\rm dust} \simeq 46\,{\rm K}$ for the REBELS sources (Fig. 3) – broadly consistent with the more direct multi-band measurements of Faisst et al. (2020); Bakx et al. (2021) for galaxies at similar redshifts.

3.2. *Dust-obscured star formation and its contribution to the cosmic SFR density at $z \simeq 7$*

The dust temperature estimates of REBELS sources by Sommovigo et al. (2022) are used to extrapolate from the single continuum measurements to obtain their total infrared luminosities, and infer their dust-obscured star formation rates. Algera et al. (2022) find that between 30% and 60% of the star formation rate in these galaxies is obscured. They find that this fraction depends on the stellar mass of the sources, similar to what is found at lower redshifts (Whitaker et al. 2017; the fraction of dust-obscured SFR increases with stellar mass). They also use these results to quantify the contribution of dust-obscured star formation to the cosmic star formation rate density at $z \sim 7$. Since REBELS is not a blind, volume-limited survey, but a targeted survey, infrared luminosity functions cannot be directly directly derived. Their method involves first finding a relation between the dust luminosity and stellar mass, and then integrating across the (known) stellar mass function at $z = 7$ (Algera et al. 2022; see also, Barrufet et al., submitted). Algera et al. (2022) find that dust-obscured star formation at high redshifts is more important than previously assumed: it can contribute up to a third of the total cosmic SFR density at $z \simeq 7$ (Fig. 4). This implies also that the cosmic SFR density does not decrease as fast as previously predicted at $z > 3$. While there are caveats attached to this result (stellar mass and dust luminosity uncertainties, sample selection effects, etc.), it could have a profound meaning: that galaxies in the epoch of reionization already have significant amounts of dust, and indeed perhaps more dust than we were expecting to observe in such young galaxies. This in turn might challenge our understanding of how metals and dust evolve in the interstellar medium of early galaxies, as I discuss in the next section.

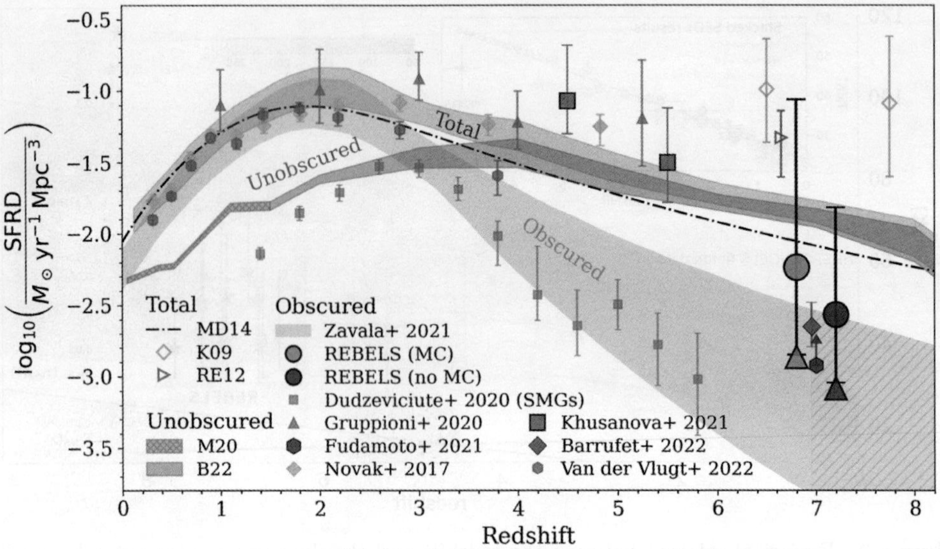

Figure 4. Redshift evolution of the cosmic star formation rate density. The purple and orange filled circles/triangles show the estimated contribution by the REBELS sources at $z \simeq 7$. Figure first published in Algera et al. (2022).

4. Can we explain so much dust at early times?

Having large amounts of dust in early galaxies, when the Universe was less than a billion years old, could be a challenge for our current understanding of how dust forms and evolves. The lifecycle of dust in intimately related with the lifecycle of stars (Fig. 5). In the local Universe, one of the main environments for stardust formation is the cool envelope of AGB stars, however, the timescales for these stars to evolve are of the order of 1 billion years or more (i.e., longer than the Hubble time at $z > 6$). Based on stellar evolution timescales, at early ages, the main channel for stardust production has to be supernova dust, however the effective dust yields in supernova are still uncertain even in the local Universe. After being formed in stellar envelopes/supernova remnants, dust grows through coagulation and mantle accretion in dense molecular clouds, and although the timescale and efficiency for this process is still debated, models predict that this ISM growth is only efficient once the gas reaches a 'critical metallicity' ($Z_{\rm crit} \sim 0.05 - 0.3\,Z_\odot$; e.g., Asano et al. 2013a, see also Galliano et al. 2018 for a review). For all these reasons, in galaxies at the epoch of reionization (which presumably have young ages and low metallicities), the main source of dust should be supernova stardust (e.g., Gall et al. 2011).

The different channels of dust production and growth also affect the grain size distributions, which in turn affect the dust extinction curves in galaxies (e.g., Asano et al. 2013b; Aoyama et al. 2020). This could explain the location of high-redshift galaxies in the IRX-β diagram. Mancini et al. (2016) post-processed a hydrodynamical galaxy simulation with a detailed chemical and dust evolution model and find that lower stellar mass galaxies at high redshift, which are less chemically evolved, follow a steep SMC-like attenuation curve because their dust is mostly contributed by stellar sources, with less efficient ISM growth than in more metal-rich massive galaxies (see also, Graziani et al. 2020).

Regarding the production of the large dust masses found at high-redshift (REBELS and other ALMA observations find dust masses $M_{\rm dust} \gtrsim 10^7\,M_\odot$ at $z \simeq 7$), models have not reached a consensus yet. Some models find that the dust masses can be explained

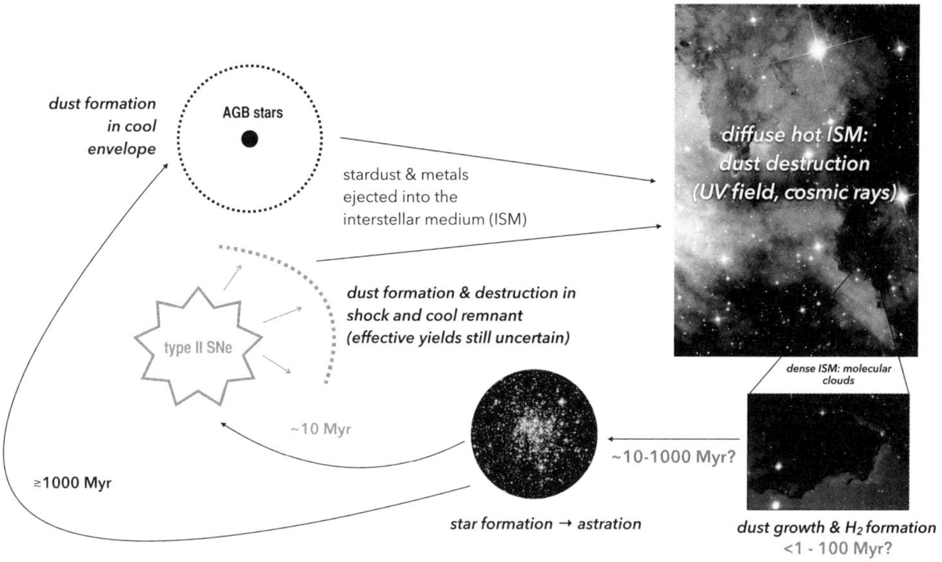

Figure 5. A schematic representation of the lifecycle of dust in galaxies (after, e.g., Dwek 1998; Zhukovska et al. 2008; Dwek & Cherchneff 2011; Asano et al. 2013a; Zhukovska 2014).

by high supernova rates and/or yields, or efficient ISM dust growth (e.g., Popping et al. 2017), while others find that efficient ISM growth is needed (e.g., Mancini et al. 2015). The need for efficient ISM dust growth in early galaxies may be problematic: Ferrara et al. (2016) point out that at high-redshifts, icy dust grain mantles formed in the cold ISM do not survive in the hot diffuse ISM. Recently, Dayal et al. (2022) showed that the DELPHI semi-analytic model can reproduce the dust masses of REBELS sources with minimal ISM dust growth, coupling metal and dust enrichment, and using new stellar yields from Kobayashi et al. (2020).

This is not a resolved issue. There are still large uncertainties in modelling ISM dust growth and stellar yields. On the observational side, the dust mass measurements of most sources are still uncertain as they rely on limited ALMA continuum measurements. The stellar masses, gas-phase metallicities, and star formation histories of these sources are also critical ingredients in the dust formation and evolution modelling, but so far these measurements are either non-existent (in the case of metallicities) or very uncertain (in the case of the stellar masses and star formation histories; Topping et al. 2022). Future observations of the rest-frame optical emission of EoR galaxies with the JWST will be critical to constrain these properties (e.g., Schaerer et al. 2022; Katz et al. 2022).

5. A few surprises/challenges from recent ALMA observations

5.1. *A hidden population of dusty galaxies in the EoR?*

A surprising result to come out of the REBELS survey was the serendipitous detection of two dust-obscured galaxies at $z \simeq 7$ that have no optical counterparts in deep HST imaging (Fudamoto et al. 2021) – this implies SEDs and dust attenuations similar to those of the dustiest known galaxies, sub-millimetre galaxies, but in galaxies at much higher redshifts and lower stellar masses than typical SMGs. ALMA observations are not only revealing that there is a variety of dust obscuration at high-redshifts, but that there might be a hidden population of dusty galaxies at the EoR that we did not know about before (similar to optically-dark galaxies seen at lower redshifts; e.g., Wang et al. 2019; Smail et al. 2021). Fudamoto et al. (2021) estimate that these dusty galaxies could contribute 10 to 25% of the total cosmic SFR density at $z > 6$. This intriguing population needs further study, and these sources are prime targets for JWST follow-up.

5.2. *Spatial offsets between UV/optical and infrared emission*

Some of the REBELS sources show large spatial offsets between the observed ultraviolet and infrared emission – up to 1.5 arcsec (or 7.6 kpc; Inami et al. 2022). Similar offsets have also been observed in high-resolution observations of star-forming galaxies at lower redshifts (e.g., Hodge et al. 2019; Cochrane et al. 2021). These offsets could be a result of the multi-phase ISM – dust and stars are not homogeneously distributed, with opaque star-forming clumps and a more transparent diffuse component. This could be a challenge for SED fitting methods that assume energy balance (e.g., da Cunha et al. 2008), as well as for modelling the IRX-β relation (Ferrara et al. 2022). More high-resolution follow-up of the dust (with ALMA) and stellar (with JWST) emissions in these sources will be key to better understand these offsets.

6. Summary & Open Questions

Thanks to ALMA we can now detect the dust emission of galaxies all the way into the epoch of reionization. Dust at those early times seems to have different properties than dust in star-forming galaxies at lower redshifts, with some indication of steeper attenuation curves, resembling dust in the SMC rather than in the Milky Way. However, this result depends on assumptions on the dust temperature (used to extrapolate single-band continuum measurements to obtain the total infrared luminosity). Dust in high-redshift galaxies seems to be warmer than dust in lower-redshift galaxies, but not enough to reconcile their infrared excesses with a Milky Way-like attenuation curve, and not as hot as predicted by theoretical models or by extrapolating the temperature-redshift relation calibrated at lower redshift. More direct measurements of the dust emission SEDs in multiple bands for larger samples of galaxies are needed to confirm this result.

The contribution of dust-obscured star formation rate to the cosmic SFR density at $z \gtrsim 6$ is non-negligible, and it could be as high as 30% based on recent REBELS results. This however needs to be confirmed with larger, preferably, blind surveys of dust emission at those redshifts.

ALMA observations show that many $z > 6$ UV-selected galaxies have copious amounts of dust. The production of high dust masses in the early Universe still pose a challenge to models. For now there seems to be a consensus in the literature that high supernova yields and/or significant ISM dust growth are needed, but the exact contribution of each process is still debated. Going forward, we need to improve the measurements of the main stellar and dust parameters that need to be compared with models, which at the moment have large uncertainties. JWST will yield more robust stellar masses, star formation histories, and metallicities; ALMA multi-band observations will constrain the dust masses more accurately.

The REBELS survey also finds a new population of dust-obscured galaxies in the epoch of reionization with no optical counterparts – how significant is this population and what are the properties of these galaxies? Searches for larger samples of these sources with ALMA, and JWST follow-up to observe their stellar emission are needed to address these questions.

Finally, REBELS also provides the first opportunity to compare the UV and IR emission in early galaxies in a systematic way. This shows that spatial offsets between these two emissions are common, which begs the question: how should we deal with decoupled stellar and dust emissions? This may involve revising how we model the SEDs of galaxies.

High-resolution ALMA and JWST follow-up of these sources will be crucial to further characterise their ISM and stellar morphologies and study the impact of these offsets.

Acknowledgements. I would like to thank the organizers of this symposium for the invitation to present this talk and my colleagues from the REBELS team for their excellent work in producing the results I reviewed. I also gratefully acknowledge the Australian Research Council as the recipient of a Future Fellowship (project FT150100079) and the ARC Centre of Excellence for All Sky Astrophysics in 3 Dimensions (ASTRO 3D; project CE170100013). I acknowledge that this paper was written on Noongar land, and pay my respects to its traditional custodians and elders, past and present.

References

Algera, H., Inami, H., Oesch, P., et al. 2022, arXiv e-prints, arXiv:2208.08243
Aoyama, S., Hirashita, H., & Nagamine, K. 2020, MNRAS, 491, 3844
Asano, R. S., Takeuchi, T. T., Hirashita, H., & Inoue, A. K. 2013a, Earth, Planets, and Space, 65, 213
Asano, R. S., Takeuchi, T. T., Hirashita, H., & Nozawa, T. 2013b, MNRAS, 432, 637
Bakx, T. J. L. C., Sommovigo, L., Carniani, S., et al. 2021, MNRAS, 508, L58
Barisic, I., Faisst, A. L., Capak, P. L., et al. 2017, ApJ, 845, 41
Behrens, C., Pallottini, A., Ferrara, A., Gallerani, S., & Vallini, L. 2018, MNRAS, 477, 552
Bertoldi, F., Carilli, C. L., Cox, P., et al. 2003, A&A, 406, L55
Bouwens, R. J., Aravena, M., Decarli, R., et al. 2016, ApJ, 833, 72
Bouwens, R. J., Smit, R., Schouws, S., et al. 2022, ApJ, 931, 160
Bowler, R. A. A., Bourne, N., Dunlop, J. S., McLure, R. J., & McLeod, D. J. 2018, MNRAS, 481, 1631
Calzetti, D., Kinney, A. L., & Storchi-Bergmann, T. 1994, ApJ, 429, 582
Capak, P. L., Carilli, C., Jones, G., et al. 2015, Nature, 522, 455
Casey, C. M., Narayanan, D., & Cooray, A. 2014, Physics Reports, 541, 45
Cochrane, R. K., Best, P. N., Smail, I., et al. 2021, MNRAS, 503, 2622
da Cunha, E., Charlot, S., & Elbaz, D. 2008, MNRAS, 388, 1595
da Cunha, E., Hodge, J. A., Casey, C. M., et al. 2021, ApJ, 919, 30
Dayal, P., Ferrara, A., Sommovigo, L., et al. 2022, MNRAS, 512, 989
Dole, H., Lagache, G., Puget, J. L., et al. 2006, A&A, 451, 417
Dudzevičiūtė, U., Smail, I., Swinbank, A. M., et al. 2020, MNRAS, 494, 3828
Dwek, E. 1998, ApJ, 501, 643
Dwek, E., & Cherchneff, I. 2011, ApJ, 727, 63
Faisst, A. L., Fudamoto, Y., Oesch, P. A., et al. 2020, MNRAS, 498, 4192
Faisst, A. L., Capak, P. L., Yan, L., et al. 2017, ApJ, 847, 21
Ferrara, A., Viti, S., & Ceccarelli, C. 2016, MNRAS, 463, L112
Ferrara, A., Sommovigo, L., Dayal, P., et al. 2022, MNRAS, 512, 58
Fudamoto, Y., Oesch, P. A., Schinnerer, E., et al. 2017, MNRAS, 472, 483
Fudamoto, Y., Oesch, P. A., Magnelli, B., et al. 2020, MNRAS, 491, 4724
Fudamoto, Y., Oesch, P. A., Schouws, S., et al. 2021, Nature, 597, 489
Gall, C., Andersen, A. C., & Hjorth, J. 2011, A&A, 528, A13
Galliano, F., Galametz, M., & Jones, A. P. 2018, ARA&A, 56, 673
Graziani, L., Schneider, R., Ginolfi, M., et al. 2020, MNRAS, 494, 1071
Hirashita, H., & Ferrara, A. 2002, MNRAS, 337, 921
Hodge, J. A., & da Cunha, E. 2020, Royal Society Open Science, 7, 200556
Hodge, J. A., Smail, I., Walter, F., et al. 2019, ApJ, 876, 130
Inami, H., Algera, H. S. B., Schouws, S., et al. 2022, MNRAS, 515, 3126
Jin, S., Daddi, E., Liu, D., et al. 2018, ApJ, 864, 56
Jin, S., Daddi, E., Magdis, G. E., et al. 2022, A&A, 665, A3
Katz, H., Saxena, A., Cameron, A. J., et al. 2022, MNRAS
Kobayashi, C., Karakas, A. I., & Lugaro, M. 2020, ApJ, 900, 179
Liang, L., Feldmann, R., Kereš, D., et al. 2019, MNRAS, 489, 1397
Liu, D., Daddi, E., Dickinson, M., et al. 2018, ApJ, 853, 172

Ma, X., Hayward, C. C., Casey, C. M., et al. 2019, MNRAS, 487, 1844
Madau, P., & Dickinson, M. 2014, ARA&A, 52, 415
Mancini, M., Schneider, R., Graziani, L., et al. 2016, MNRAS, 462, 3130
—. 2015, MNRAS, 451, L70
McAlpine, S., Smail, I., Bower, R. G., et al. 2019, MNRAS, 488, 2440
Meurer, G. R., Heckman, T. M., & Calzetti, D. 1999, ApJ, 521, 64
Narayanan, D., Davé, R., Johnson, B. D., et al. 2018, MNRAS, 474, 1718
Popping, G., Somerville, R. S., & Galametz, M. 2017, MNRAS, 471, 3152
Riechers, D. A., Bradford, C. M., Clements, D. L., et al. 2013, Nature, 496, 329
Schaerer, D., Marques-Chaves, R., Barrufet, L., et al. 2022, A&A, 665, L4
Schouws, S., Stefanon, M., Bouwens, R., et al. 2022, ApJ, 928, 31
Schreiber, C., Elbaz, D., Pannella, M., et al. 2018, A&A, 609, A30
Schreiber, C., Pannella, M., Elbaz, D., et al. 2015, A&A, 575, A74
Smail, I., Dudzevičiūtė, U., Stach, S. M., et al. 2021, MNRAS, 502, 3426
Sommovigo, L., Ferrara, A., Carniani, S., et al. 2021, MNRAS, 503, 4878
Sommovigo, L., Ferrara, A., Pallottini, A., et al. 2020, arXiv e-prints, arXiv:2004.09528
—. 2022, MNRAS, 513, 3122
Topping, M. W., Stark, D. P., Endsley, R., et al. 2022, MNRAS, 516, 975
Walter, F., Decarli, R., Carilli, C., et al. 2012, Nature, 486, 233
Wang, T., Schreiber, C., Elbaz, D., et al. 2019, Nature, 572, 211
Whitaker, K. E., Pope, A., Cybulski, R., et al. 2017, ApJ, 850, 208
Witstok, J., Smit, R., Maiolino, R., et al. 2022, MNRAS, 515, 1751
Zavala, J. A., Casey, C. M., Manning, S. M., et al. 2021, ApJ, 909, 165
Zhukovska, S. 2014, A&A, 562, A76
Zhukovska, S., Gail, H. P., & Trieloff, M. 2008, A&A, 479, 453

Resolving the Rise and Fall of Star Formation in Galaxies
Proceedings IAU Symposium No. 373, 2023
T. Wong & W.-T. Kim, eds.
doi:10.1017/S1743921322003854

The cold gas supply through cosmic time: insights on the galaxy assembly at early epochs

Manuel Aravena

Núcleo de Astronomía, Facultad de Ingeniería y Ciencias,
Universidad Diego Portales, 8370191 Santiago, Chile
email: manuel.aravenaa@mail.udp.cl

Abstract.
Remarkable progress has been made in the last few years in understanding the global properties of galaxies and how they evolve through cosmic time. Major focus has been given to studies of how the availability of molecular gas regulates star-forming activity and galaxy growth, the eventual quenching of star formation, and how these mechanisms evolve through cosmic time. Most of these advances have been made thanks to ALMA and the upgraded capabilities of NOEMA. In this contribution, I briey review the latest constraints on the molecular gas content based on dierent tracers of the interstellar medium (ISM; dust continuum and CO, [CI] and [CII] line emission), including recent determinations of the molecular gas fraction, gas depletion timescales, and molecular gas cosmic density provided by the recent ALMA programs out to $z \sim 7$. Finally, I concentrate on recent and ongoing studies aiming to spatially and kinematically resolve the cold ISM and star formation activity down to kpc scales in galaxies out to $z \sim 6 - 7$, which represent an unprecedented view of the galaxy assembly and feedback processes in the early universe.

Keywords. galaxies: high-redshift, galaxies: evolution, galaxies: ISM

1. Rise and fall of star formation in galaxies

One of the main results of galaxy evolution in the past decades has been the determination of the cosmic star formation rate (SFR) density of the universe out to redshifts of 10 (Madau & Dickinson 2014). Three main epochs are defined, including an epoch where the first galaxies appeared and the star formation steadily increased ($z > 6$), an epoch of galaxy assembly where most of the stars in the universe were formed($1 < z < 4$), and the epoch of galaxy quenching ($z < 1$).

A key result has been the determination that most star-forming galaxies appear to form a tight relation between the SFRs and their stellar masses at all redshifts. The galaxies in this so-called "main sequence" appear to have the predominant contribution to the cosmic SFR density (e.g. Elbaz et al. 2007). Understanding what drives this evolution and how galaxies build up their stellar mass (as a function of their location in the SFR vs. stellar mass plane) is among the most fundamental question in galaxy evolution studies. To describe the physical processes involved in the galaxy's growth, it is critical to measure their cold interstellar medium (ISM) molecular gas content, as it represents the fuel through which stars form.

2. Measuring the molecular ISM in distant galaxies

Several methods have been used to measure the molecular gas mass in high-redshift galaxies. These include measurements of the continuum and line emission from the rest-frame far-infrared and submillimeter regimes.

2.1. *Dust continuum emission*

This is the most commonly used method to measure the molecular gas mass regarding the number of sources targeted. Two approaches have been used.

The first one requires the acquisition of photometry of the target galaxy through the full infrared regime, with at least three data points, enabling to conduct spectral energy distribution (SED) fitting with standard routines, or simply approximating the dust SED as a multi-component gray body spectrum. Various physical parameters can be obtained, including the dust temperature, mass, emissivity index, and size. Using the dust mass and assuming a dust-to-gas mass ratio (typically $\sim 20 - 100$ for galaxies with close to solar metallicities), an estimate of the molecular gas mass can be obtained. The main caveat is that it can be particularly observationally expensive to target statistical samples of sources as this method needs observations in several far-infrared and submillimeter bands. Furthermore, obtaining photometry in the far-infrared regime, at wavelengths shorter than 350μm is almost prohibited from the ground, requiring space observatories, which are currently lacking after decommissioning *Spitzer* and *Herschel*.

The second approach corresponds to a simplified version, in which an estimate of the dust mass can be obtained from a single far-infrared data point in the Rayleigh-Jeans (RJ) limit. This method has received particular interest since it is easier to detect observationally and benefits from a negative K-correction out to high redshifts (e.g. Scoville et al. 2014). This method essentially uses the fact that the dust emission in the RJ regime ($> 200\mu$m) is most sensitive to the dust mass. This method has been calibrated for massive star-forming galaxies and has been widely used to measure dust and molecular gas masses in large samples of galaxies as it requires a single photometric data point at relatively long wavelengths (typically 850μm or 1.2mm). The main disadvantage of this method is that it is not clear that it would apply to galaxies with conditions like low stellar masses or metallicities and further calibrations are needed.

2.2. *CO, [CI] and [CI] line emission*

The method to measure the molecular gas mass through cold molecular gas tracers like the CO and [CI] lines is among the most traditional methods. This approach assumes that these tracers are directly linked to the bulk of molecular gas, H_2, and relies on the assumption of a line luminosity to gas mass conversion factor α. The latter has been observed to vary through different environments and metallicity (Bolatto et al. 2013). For starburst and "normal" galaxies typically values of ~ 1 and 4.6 $M_\odot$ [K km s^{-1} pc^2]$^{-1}$ are assumed.

The main advantage of this method is that line emission can provide significantly more information on dynamics, particularly estimates of the dynamical mass of the system. Similarly to the full dust SED method, it is more observationally expensive than simply observing a single continuum data point. For CO line emission, the observation of higher rotational transitions is less sensitive to the total molecular gas mass. It thus has the additional requirement of assuming line ratios to convert the emission into the lower J CO transitions.

The [CII] has recently been proposed as an estimator of the molecular gas mass. The [CII] line emission arises from various environments (neutral, ionized, photodissociation regions in molecular clouds). Thus its emission has been calibrated to measure the amount of neutral (HI) and molecular gas (H_2) mass (Heintz et al. 2021; Zanella et al. 2018). As with the CO and [CI] line emission, the [CII] molecular gas mass estimates require the assumption of a (calibrated) conversion factor between the [CII] line luminosity. Based on the Zanella et al. results, $\alpha_{[CII]} \sim 30\ M_\odot\ L^{-1}$. While it is under debate whether the [CII] method is applicable in different contexts, the main advantage is to enable access to a higher redshift regime not accessible to CO and/or dust emission due to their general faintness, and large cosmological distances, as [CII] is the among the brightest far-infrared emission lines in galaxies.

3. ISM surveys of star-forming galaxies at high redshift

3.1. *Targeted surveys of the ISM at high redshift*

Submillimeter galaxies. Significant efforts have been devoted in the last 20+ years to conducting blank-field sub-millimeter surveys over large sky areas with bolometer cameras in single-dish telescopes (e.g. Smail et al. 1997; Hughes et al. 1998). These surveys discovered an important population of dust-rich starburst galaxies at high redshift that contributed a significant fraction of the cosmic SFR density (e.g. Casey et al. 2014).

Given their selection wavelength, by construction, every (sub)millimeter continuum survey traces the cold dust content out to $z < 3$, and similarly for surveys at 1.2mm out to $z \sim 6$. However, detailed studies of this population have only been enabled through the identification of multiwavelength counterparts through interferometric (sub)millimeter follow-up with facilities like the ALMA, NOEMA, and the SMA. Despite the significant improvement in our understanding of the physical properties of SMGs in the last decade, only a few dedicated surveys have been performed to measure the molecular gas content in CO line emission for both unlensed (Bothwell et al. 2013; Sharon et al. 2016; Wardlow et al. 2018; Birkin et al. 2021) and lensed samples (Aravena et al. 2016; Harrington et al. 2018; Yang et al. 2017; Cañ ameras et al. 2018).

Main sequence galaxies and starbursts. Since these (sub)millimeter surveys are only able to grasp the brightest dusty star-forming galaxies in the sky, pinpointing the typical population of galaxies at $z \sim 1-3$, with SFRs $\sim 50-200\ M_\odot\ \mathrm{yr}^{-1}$, has required targeted observations of dust continuum and molecular gas (CO and [CI]) emission in galaxies that have been pre-selected through their optical colors and/or faint IR emission detected in *Herschel* IR surveys (e.g. Tacconi et al. 2020; Valentino et al. 2018).

Quiescent galaxies. The advent of sensitive ALMA capabilities has enabled access to the molecular gas emission from quiescent galaxies through observations of dust continuum and CO line emission at $z < 3$ (e.g. Gobat et al. 2018; Whitaker et al. 2021; Williams et al. 2021). Such observations have allowed studying the evolution of these populations in the context of more active star-forming main-sequence galaxies and put into question the validity of the scaling relations for these galaxies.

[CII] emission at $z > 4$. The [CII] line has been used extensively to target the ISM in massive star-forming galaxies at high redshift. When redshifted, this line falls into the accessible atmospheric windows from the ground, making it an ideal tracer of the ISM conditions and dynamics at $z > 3$. This has been exploited by various studies (e.g., Capak et al. 2015; Smit et al. 2018; Bakx et al. 2020), and most recently by two ALMA large programs, ALPINE (Le Fèvre et al. 2020) and REBELS (Bouwens et al. 2022), targeting galaxies at $z = 4-6$ and $6.5-9$, respectively.

Archival surveys. A different approach has been to use the ALMA archival data to detect all the dusty and line-emitting sources in the extragalactic pointing observations (Liu et al. 2019). This extremely successful approach significantly increased the number of detected objects at high redshift.

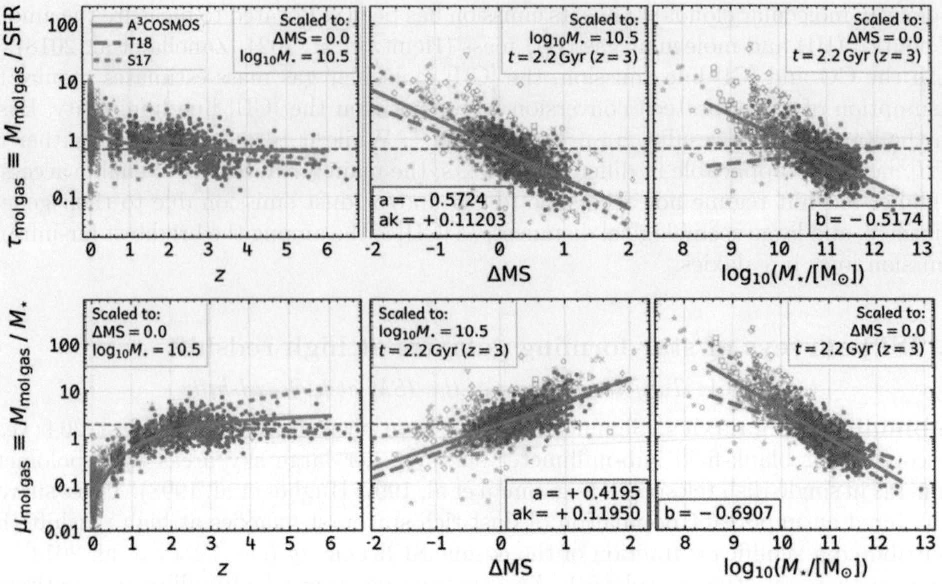

Figure 1. Scaling relations of the molecular gas to stellar mass ratio and molecular gas depletion timescale as a function of redshift, distance to the main sequence, and stellar mass (adapted from Liu et al. 2019). The molecular gas to stellar mass ratio changes by $\sim 5\times$ in the redshift range $0-2$, while the gas depletion timescale evolves only mildly. Galaxies below/above the main sequence have lower/higher gas fractions and longer depletion timescales.

3.2. *Interferometric cosmological deep fields*

Recent deep millimeter continuum surveys over increasingly large contiguous areas with ALMA toward cosmological deep fields area revolutionizing galaxy evolution studies by accessing the dust emission from the faint star-forming galaxy population.

These surveys have followed a "wedding cake" approach, covering large areas at shallower depths and narrow areas deeper and concentrating in the 1-mm band as it is more efficient in reaching good depth and area coverage (Franco et al. 2018; Hatsukade et al. 2018; Umehata et al. 2018; Dunlop et al. 2017; Hatsukade et al. 2016); and most recently two ALMA large programs, The ALMA Spectroscopic Survey in the HUDF (ASPECS; Walter et al. 2016; Gonzalez-Lopez et al. 2020) and the ALMA Lensing Cluster Survey (ALCS; Fujimoto et al. 2021). In the particular case of ASPECS, it provided a complementary approach to pure continuum millimeter-deep field observations, simultaneously observing CO/[CI] and [CII] line emission at all redshifts, along with continuum emission at 3mm and 1mm over the full Hubble Ultra Deep Field (HUDF). The availability of ancillary data has been key to providing a detailed characterization of the identified galaxies.

4. ISM evolution at high redshift

The surveys described above enabled the determination of key scaling relations between molecular gas fraction and depletion timescales, with redshift, stellar mass, and distance to the main sequence, which describes how galaxies regulate the stellar mass growth (Fig. 1; e.g. Tacconi et al. 2020).

Surveys of quiescent galaxies revealed stringent limits on gas fraction, with gas fractions $<6\%$ for galaxies $>3x$ below the main sequence, being still unclear the actual mechanisms for star formation quenching, whether through truncation of gas accretion

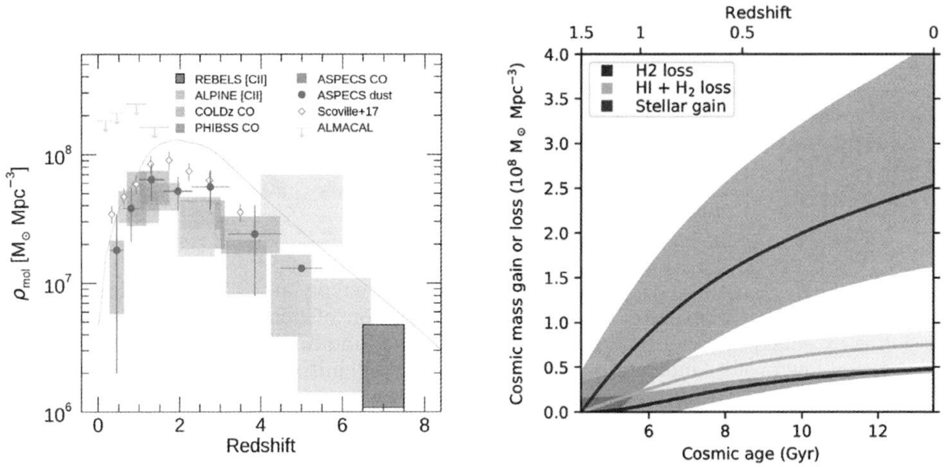

Figure 2. *(Left:)* Cosmic density of molecular gas as seen by various surveys for CO, dust, and [CII] line emission (from Aravena et al. in preparation; see also Decarli et al. 2020). The evolution of the cosmic molecular gas density appears to mimic the cosmic SFR density out to $z \sim 7$. *(Right:)* Cumulative gain of stellar mass since $z = 1.5$ to $z = 0$, compared to the loss of ISM gas, including H_2 and HI. This comparison suggests that gas accretion is needed to explain the difference between the amount of mass gained in stars and loss in gas.

truncation (gas depletion), or associated to low mass halos with strong feedback, or efficient gas consumption (Williams et al. 2021; Whitaker et al. 2021). An interesting individual example is Spilker et al. (2022), where resolved CO observations in a post-starburst galaxy at $z = 0.7$ showed the possibility of gas removal from the galaxy center through a recent merger.

Similarly, these surveys enabled the measurement of the cosmic molecular gas density (see Fig. 2). This seems to mimic the evolution of the cosmic SFR density, indicating a close link between both quantities and relatively constant efficiencies out to $z \sim 5 - 6$ (Decarli et al. 2020; Aravena et al. in preparation). Indeed by accounting for the contribution of the stellar, HI, and H2 cosmic mass densities, Walter et al. (2020) were able to argue that the significant gap between the number of stars created in the universe and the available gas (HI+H2) that can be used for star formation could only be explained by gas accretion from the galaxies circumgalactic medium (Fig. 2).

5. The next step? kpc-scale tomography of galaxies at high redshift

Understanding the physical processes of star formation within individual galaxies requires higher angular resolution. So far, the focus has been on the optical regime, centered on Hα and [OIII]5007 imaging at $z \sim 2$ using AO IFU and rest-UV using HST. The picture suggests that star formation occurs in massive clumps within turbulent disks (e.g. Genzel et al. 2008; Soto et al. 2017). However, such studies might be biased due to global obscuration across the sources, particularly in massive galaxies (Simpson et al. 2017; Hodge et al. 2019), based on the offsets seen between rest-UV and FIR components and the different extent and structure at the different wavelengths.

Resolved ISM observations exist only in a handful of objects. They have been mostly focused on size measurements from unresolved (Puglisi et al. 2019) and resolved continuum FIR observations (e.g. Barro et al. 2016; Tadaki et al. 2020; Chen et al. 2020; Cochrane et al. 2021). Current results indicate that the FIR emission is almost always more compact than optical and stellar mass sizes and that the central regions in galaxies can be starbursting, while outer disks are in the MS. So far, only a handful of sources

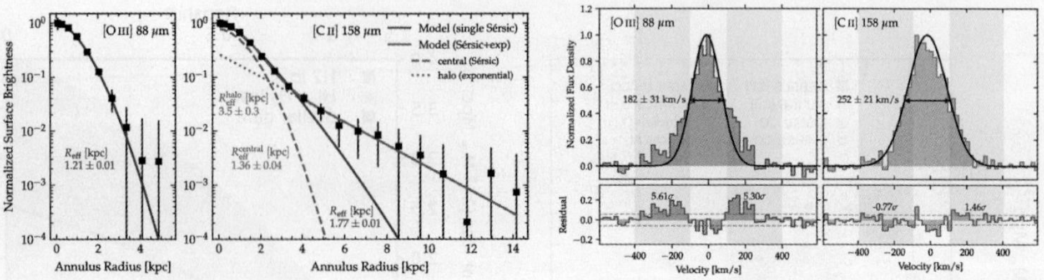

Figure 3. [OIII] 88 μ and [CII] 158μm line emission in a galaxy at $z = 7.1$ (adapted from Akins et al. 2022). *(Left:)* Comparison of the radial distribution of both lines shows that the [CII] emission is extended and can be described by a compact and an extended component. *(Right:)* The emission line profiles indicate a wider [CII] line profile, but with a double component [OIII] profile, which can be attributed to an outflow of ionized gas. Outflows have become a possibility to explain the extended [CII] structures seen recently in distant galaxies.

have resolved CO line emission observations both in unlensed (e.g. Spilker et al. 2019) and lensed objects (Dessauges-Zavadsky et al. 2019).

6. Feedback in action at high redshift?

The recent ALPINE and REBELS [CII] surveys of galaxies at $z > 4$ have found interesting results. Intriguingly, ALPINE finds evidence of extended [CII] emission, or "[CII]" halos around $z \sim 4-6$ galaxies. In particular, they find a 30-kpc scale halo in a multiple merger system at $z \sim 4$ (Ginolfi et al. 2020), while clear evidence that the [CII] radial distribution extends beyond rest-UV in massive galaxies in this redshift range both in individual cases and using stacking (Fujimoto et al. 2020). It is still unclear what the origin of the "halos" might be, with current possibilities being outflows, inflows, tidal tails, or mergers.

These surveys found key supporting evidence for the existence of rotating disks already at $z \sim 7$ (Smit et al. 2018). Similarly, ALPINE observations find that $\sim 40\%$ of galaxies at $z = 4-6$ show signs of rotating disks (Jones et al. 2020). [CII] size measurements from the REBELS and ALPINE surveys reveal little size evolution from $z \sim 4$ to 7, despite the decreasing UV-sizes at higher redshifts (Fudamoto et al. 2022a)

Resolved [CII] studies with ALMA toward a few galaxies at $z \sim 5-7$ have found contrasting results. Herrera-Camus et al. (2022) find evidence for a rotating disk and an outflow in a massive star-forming galaxy at $z = 5.5$. However, only mild evidence is found for extended [CII] emission, similar to the system studied by Posses et al. (2022) at $z = 6.8$. Akins et al. (2022) find an extended [CII] "halo" toward a galaxy at $z = 7.1$ when compared to the dust and rest-frame UV distributions and clear signatures for outflowing gas, indicating that outflows can be associated with the [CII] "halo' (Fig. 3). Finally, Lambert et al. (2022) find an extended [CII] emission compared to the rest-frame UV. However, the kinematics rather suggest a merger as the origin for the extended "halo".

Clearly, high-resolution multi-wavelength observations are needed to unveil the nature of distant galaxies and the associated build-up and feedback processes.

6.1. *The ALMA CRISTAL large program survey*

CRISTAL is a recently approved Cycle 8 ALMA Large Program (139 hr) that will spatially resolve the [CII] 158μm line and dust continuum emission in 19 mass-selected star-forming galaxies at $z \sim 4-5$ (Fig. 4). All CRISTAL galaxies have HST rest-frame UV data, and more than half of them will have Cycle 1 JWST rest-frame optical observations. With the combination of ALMA, HST, and JWST, CRISTAL will produce the

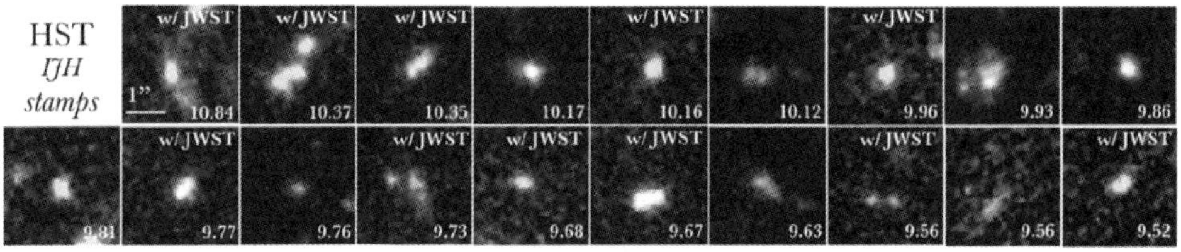

Figure 4. HST Postage stamps in the optical/NIR toward the mass-selected sample of 19 CRISTAL galaxies. Approximately half of the sample is expected to ave JWST imaging during its first 2 years of observations (Credit: R. Herrera-Camus, N. Förster-Schreiber).

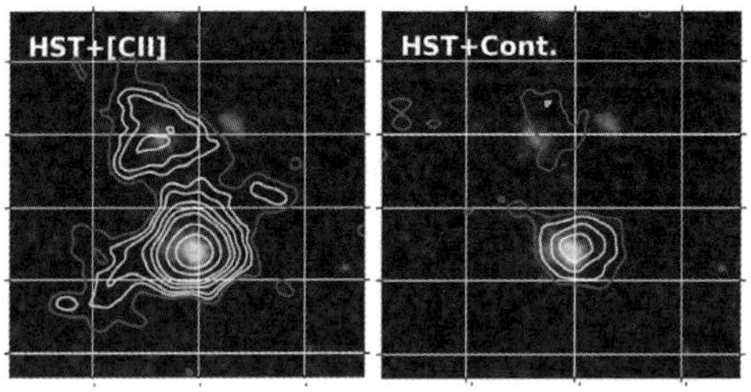

Figure 5. Example of the [CII] line and FIR continuum emission in one of the CRISTAL galaxies. The background image was created from the HST imaging, and the contours represent the [CII] line and FIR continuum emission, respectively. It is evident the difference in structure and extent seen in the ALMA and HST images.

first systematic census of gas, dust, and stars on $\sim 1-2$ kpc scales in typical star-forming galaxies when the Universe was only 1 Gyr old. This will allow us to connect the properties of the ISM, circumgalactic medium, star formation, and feedback in the form of outflows (Fig. 5).

7. Concluding remarks

• Observations of the ISM of galaxies at high redshift has provided a foundational framework for galaxy evolution and the transformation of gas to stars in galaxies. Recent studies of quiescent galaxies indicate that current scaling relations might not be valid for quenched objects.

• Measurements of the cosmic density of molecular gas indicate a simultaneous and similar evolution to that of the cosmic density of SFR. The current observations imply the need for cosmic gas accretion to match the number of stars created in the last 1.5 Gyr of cosmic time.

• Resolved observations of the ISM in high redshift galaxies are still scarce and focused mostly on the dust continuum emission. There is still debate about the mechanisms for quenching at $z < 2$.

• Recent ALMA large programs and follow-up higher resolution observations have suggested the existence of rotating disks and [CII] halos in galaxies at $z = 4-7$. The ALMA CRISTAL large program will provide insights on the origin and reality of these findings.

Acknowledgements

MA acknowledges support from FONDECYT grant 1211951, ANID+PCI+INSTITUTO MAX PLANCK DE ASTRONOMIA MPG 190030, ANID+PCI+REDES 190194 and ANID BASAL project FB210003.

References

Aravena, M., Spilker, J. S., Bethermin, M., et al. 2016, *MNRAS*, 457, 4406
Birkin, J. E., Weiss, A., Wardlow, J. L., et al. 2021, *MNRAS*, 501, 3926
Bakx, T. J. L. C., Tamura, Y., Hashimoto, T, et al. 2020, *MNRAS*, 493, 4294
Barro, G., Kriek, M., Pérez-González, P. G., et al. 2016, *ApJ*, 827, 32
Bolatto, A., Wolfire, M., Leroy, A. et al. 2013, *ARA&A*, 51, 207
Bothwell, M. S., Smail, I., Chapman, S. C., et al. 2013, *MNRAS*, 429, 3047
Bouwens, R. J., Smit, R., Schouws, S., et al. 2022, *ApJ*, 931, 160
Capak, P. L., Carilli, C. L., Jones, G., et al. 2015, *Nature*, 522, 455
Cañameras, R., Yang, C., Nesvadba, N. P. H., et al. 2018, *A&A*, 620, 61
Casey, C. M., Narayanan, D., Cooray, A., et al. 2014, *PhR*, 541, 45
Chen, C. C., Harrison, C. M., Smail, I. et al. 2020, *A&A*, 635, 119
Cochrane, R. K., Best, P. N., Smail, I. et al. 2021, *MNRAS*, 503, 2622
Decarli, R., Aravena, M., Boogaard, L. et al. 2020, *ApJ*, 902, 110
Dessauges-Zavadsky, M., Richard, J., Combes, F. et al. 2019, *NatAs*, 3, 1115
Dunlop, J. S., McLure, R. J., Biggs, A. D. et al. 2017, *MNRAS*, 466, 861
Elbaz, D., Daddi, E., Le Borgne, D., et al. 2007, *A&A*, 468, 33
Franco, M., Elbaz, D., Béthermin, M. et al. 2018, *A&A*, 620, 152
Fudamoto, Y., Smit, R., Bowler, R. A. A. et al. 2022, *ApJ*, 934,144
Fujimoto, S., Silverman, J. D., Béthermin, M. et al. 2020, *ApJ*, 900, 1
Fujimoto, S., Oguri, M., Brammer, G. et al. 2021, *ApJ*, 911, 99
Genzel, R., Burkert, A., Bouché, N. et al. 2008, *ApJ*, 687, 59
Ginolfi, M., Jones, G. C., Bethermin, M. et al. 2020, *A&A*, 643, 7
Gobat, R., Daddi, E., Magdis, G., et al. 2018, *NatAs*, 2, 239
González-López, J., Decarli, R., Pavesi, R. et al. 2019, *ApJ* accepted, arxiv:1903.09161
Harrington, K. C., Yun, M. S., Magnelli, B. et al. 2018, *MNRAS*, 474, 3866
Heintz, K. E., Watson, D., Oesch, P. et al. 2021, *ApJ*, 922, 147
Hatsukade, B., Kohno, K., Umehata, H. et al. 2016, *PASJ*, 68, 36
Hatsukade, B., Kohno, K., Yamaguchi, Y. et al. 2018, *PASJ*, 70, 105
Herrera-Camus, R., Forster-Schreiber, N., Price, S. H. et al. 2022, *A&A*, 665, 8
Hodge, J. A., Smail, I., Walter, F., et al. 2019, *ApJ*, 876, 130
Hughes, D. H., Serjeant, S., Dunlop, J., et al. 1998, *Nature*, 394, 241
Jones, G. C., Bethermin, M., Fudamoto, Y. 2019, *MNRAS*, 491, 18
Lambert, T. S., Posses, A., Aravena, M. et al. 2022, *ArXiv:2210.10023*
Le Fèvre, O., Béthermin, M., Faisst, A., et al. 2020, *A&A*, 643, A1
Liu, D., Lang, P., Magnelli, B. et al. 2019, *ApJS*, 244, 40
Madau, P. & Dickinson, M. 2004, *ARA&A*, 52, 415
Posses, A. C., Aravena, M., González-López, J. et al. 2022, *ArXiv:2206.13770*
Sharon, C. E., Riechers, D. A., Hodge J., et al. 2016, *ApJ*, 827, 18
Scoville, N., Aussel, H., Sheth, K. et al. 2014, *ApJ*, 783, 84
Scoville, N., Lee, N., Vanden Bout, P., et al. 2015, *ApJ*, 837, 150
Simpson, J. M., Smail, I., Swinbank, A. M. et al. 2017, *ApJ*, 839, 58
Smail, I., Ivison, R. J., & Blain, A. W. 1997, *ApJL*, 490, 5
Smit, R., Bouwens, R. J., Carniani, S. et al. 2018, *Nature*, 553, 178
Soto, E., de Mello, D. F., Rafelski, M. et al. 2017, *ApJ*, 837, 6
Spilker, J. S., Bezanson, R., Weiner, B. J. et al. 2019, *ApJ*, 883, 81
Spilker, J. S., Suess, K. A., Setton, D. et al. 2022, *ApJ*, 936, 11
Tadaki, K., Belli, S., Burkert, A., et al. 2020, *ApJ*, 901, 74
Tacconi, L. J., Genzel, R., Saintonge, A., et al. 2018, *ApJ*, 853, 179
Tacconi, L. J., Genzel, R., Sternberg, A. 2020, *ARA&A*, 58, 157
Umehata, H. Hatsukade, B., Smail, I. et al. 2018, *PASJ*, 70, 65
Valentino, F., Magdis, G. E., Daddi, E. et al. 2018, *ApJ*, 869, 27
Walter, F., Decarli, R., Aravena, M. et al. 2016, *ApJ*, 833, 67
Walter, F., Carilli, C. L., Neeleman, M. et al. 2020, *ApJ*, 902, 111

Wardlow, J. L., Simpson, J. M., Smail, I. et al. 2018, *MNRAS*, 479, 3879
Whitaker, K. E.., Williams, C. C., Mowla, L. et al. 2021, *Nature*, 579, 485
Williams, C. C., Spilker, J., Whitaker, K. E. et al. 2021, *ApJ*, 908, 54
Yang, C., Omont, A., Beelen, A. et al. 2017, *A&A*, 608, 144
Zanella, A., Daddi, E., Magdis, G. et al. 2018, *MNRAS*, 481, 2

Resolving the Rise and Fall of Star Formation in Galaxies
Proceedings IAU Symposium No. 373, 2023
T. Wong & W.-T. Kim, eds.
doi:10.1017/S174392132200429X

The Progressive Integral Step Method (PrISM) for Wide Field 3D Spectral Imaging of Nearby Galaxies: an Overview of the TYPHOON Survey

Kathryn Grasha

Research School of Astronomy and Astrophysics, Australian National University, Canberra, ACT 2611, Australia; email: kathryn.grasha@anu.edu.au
ARC Centre of Excellence for All Sky Astrophysics in 3 Dimensions (ASTRO 3D), Australia

Abstract. The TYPHOON program is producing an atlas of spectroscopic data cubes of 44 large-angular-sized galaxies with complete spatial coverage from 3650–9000 Å. This survey provides an unparalleled opportunity to study variations in the interstellar medium (ISM) properties within individual H II regions across the entire star-forming disks of nearby galaxies. This can provide key insights into the spatial distribution and *resolved* properties of the ISM to understand how efficiently metals are mixed and redistributed across spirals and dwarf galaxies. In this Proceeding, we present early science results from six nearby spiral galaxies as part of the TYPHOON program from Grasha et al. (2022). We use `HIIPhot` to identify the H II regions within the galaxy based on the surface brightness of the Hα emisison line and measure variations of the H II region oxygen abundance. In this initial work, we find that while the spiral pattern plays a role in organizing the ISM, it alone does not establish the relatively uniform azimuthal variations we observe across all the galaxies. Differences in the metal abundances are more likely driven by the strong correlations with the local physical conditions. We find a strong and positive correlation between the ionization parameter and the local abundances as measured by the relative metallicity offset Δ(O/H), indicating a tight relationship between local physical conditions and their localized enrichment of the ISM. These variations can be explained by a combination of localized, star formation-driven self-enrichment and large-scale mixing-driven dilution due to the passing of spiral density waves.

Keywords. galaxies: abundances — galaxies: ISM — galaxies: spiral — H II regions — ISM: abundances — ISM: evolution

1. Introduction

The 'baryon cycle' — the cycle of gas condensation, the resulting star formation and stellar feedback, and the re-injection of matter, energy, and momentum into the interstellar medium (ISM) — is the engine for (secular) galaxy evolution and metal enrichment. This cycle of the star formation process is an inherently multi-phase and multi-scale process, dictating the assembly history and properties of galaxies by controling the flows of gas in and out of galaxies. Equally important is information on the internal gas dynamics and connecting the small-scale cycle of baryons with the larger galactic and circum-galactic scales. These processes drive the evolution of galaxies in the large-scale cosmological context, fueling the diversity of galaxies we see in the Universe today. At the core of these processes is the small-scale physics of star formation, which is intrinsically linked to and controlled by the feedback from massive stars (Hopkins et al.

2013; Krumholz et al. 2019). A key unsolved challenge and area of modern astrophysics research, for both observations and theoretical models, is linking the observed global galaxy properties over cosmic time with the sub-parsec-scale physics of the star formation process.

Progress on the topic of galaxy evolution has rapidly moved into the domain of threedimensional (3D) spectroscopy. Integral field spectroscopy (IFS) enables the spatially-resolved measurement of stellar populations, ionized gas, on-going star formation, star formation histories, kinematics, dust distributions, and chemical abundance maps. Numerous 3D Integral Field Units (IFUs) are now in operation and several more are in development for the next generation of telescopes. Moreover, a number of important IFU based galaxy surveys exist, such as SAURON (de Zeeuw et al. 2002), PINGS (Rosales-Ortega et al. 2010), CALIFA (Sánchez et al. 2012), SAMI (Croom et al. 2012), MaNGA (Bundy et al. 2015), SIGNALS (Rousseau-Nepton et al. 2019), and PHANGS (Emsellem et al. 2022). Optical spectroscopy is a powerful tool to place the process of star formation in the context of its galactic host because it probes the ionized ISM that is intimately connected with young, massive stars. These surveys are making enormous progress in shedding light on the physical processes associated with star formation and variation with local conditions within local $z \sim 0$ galaxies at the scales of individual star-forming regions (e.g., Kawamura et al. 2009; Sun et al. 2020). The ideal spatial scale to study is the size of individual star-forming products in galaxies (individual gas clouds, associations and clusters of young stellar objects, and discrete star-forming photoionized regions), which are typically a few tens of parsec in size (e.g., Grasha et al. 2017, 2018; Ryon et al. 2017; Sun et al. 2022).

The production and build-up of heavy elements from stellar nucleosynthesis over cosmic time is one of the key physical properties for understanding galaxy evolution. The ISM oxygen abundance (i.e., the gas-phase metallicity) and distribution are built up by various physical processes throughout a galaxy's evolutionary history. The abundance of heavy elements is usually constrained by measuring the current gas phase oxygen abundance (metallicity) within H II regions (Kennicutt & Garnett 1996). The gas-phase metallicity sets the balance between processes that enrich gas, such as star formation, and processes that dilute or remove metals, such as inflows of pristine gas from the intergalactic medium, galactic winds, and outflows and drive changes in fundamental local ISM properties (e.g., gas-to-dust ratio, CO-to-H2 conversion factor). Although these physical processes are often coupled in complicated ways, chemical evolution models have been able to place quantitative constraints on the key processes driving galaxy evolution (e.g., Kobayashi et al. 2007; Torrey et al. 2019). Thus H II regions are strictly connected with the star formation process as they are powered by high mass ($>8\ M_\odot$) stars and are fundamental in driving the evolution of galaxies. H II regions are an important tool to investigate the star formation process and the properties of the ionized gas and the current chemical abundances of the gas from which stars are currently being produced.

With the complete spatial coverage delivered by IFS, recent studies have ushered in a new era of characterizing oxygen abundance variations, where predominantly negative radial trends in metallicity, well established across large samples of galaxies, traces the inside-out growth of galaxy disks (Boissier & Prantzos 1999). However, moving past simple 1D radial distributions and understanding the broader degree of chemical inhomogeneity in star-forming galaxies is key to making use of stellar metallicities within the field of Galactic archaeology and chemical tagging studies to assess accretion events and the formation history of galaxies (Buder et al. 2022).

The large FOV coverage of the TYPHOON data cubes (18' on one side, corresponding to deprojected galactocentric coverage of up to 35 kpc of the star-forming disk at the distances of the galaxies) provides us with an unparalleled sample to study variations in

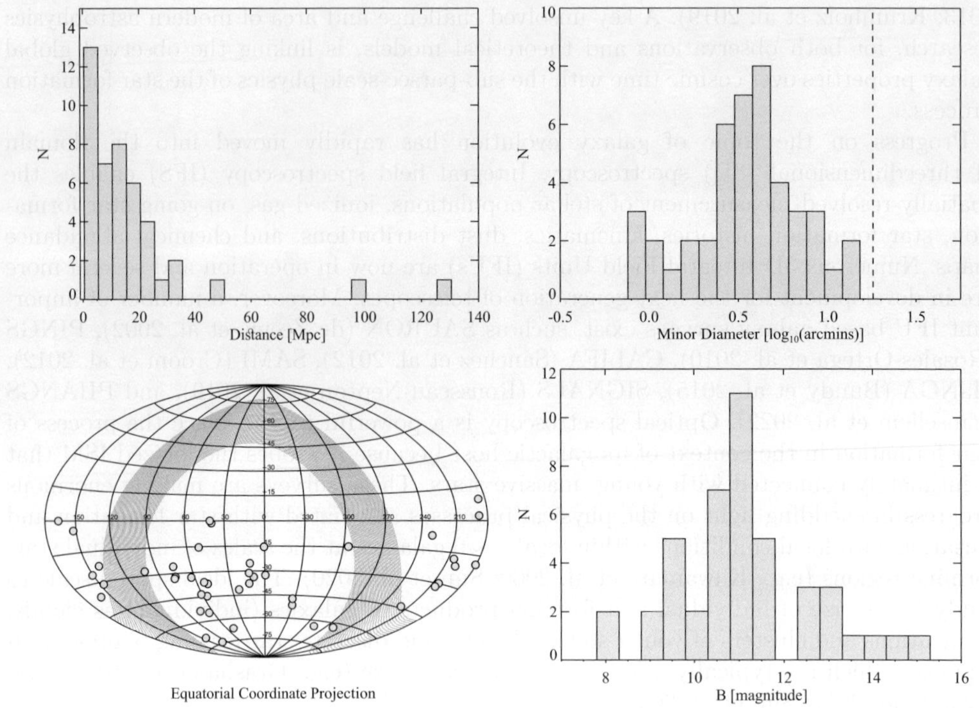

Figure 1. Description of the PrISM/TYPHOON survey sample. The full sample is listed in Table 1.

oxygen abundance in individual H II regions across entire star-forming disks. The spatial coverage of 3650–9000 Å enables constraints on the ISM pressure, density, and ionization parameter to provide an additional measure of the ISM conditions in these galaxies. Multiple diagnostics of the ISM conditions allows for the creation of simple chemical evolution model to understand the underlying physical processes governing observed variations in the chemical in-homogeneity of the ISM.

2. Introduction to the TYPHOON Survey

We are producing an atlas of spectroscopic data cubes of 44 large-angular-sized galaxies (Figure 1 and Table 1). In the world of IFU applications, this is a largely overlooked (but still critically important) sample; to date they have been difficult to observe with classical IFUs as current IFU technology typically has small fields of view. The TYPHOON sample fills a niche in the 3D spectroscopy domain: current IFU surveys are optimized to observe galaxies at redshifts greater than $z \sim 0.01$, and have spaxel sizes that resolve target galaxies at a physical scale of around 1 kpc. The simple method of PrISM is optimized (and only truly efficient) for bright large angular size galaxies at low redshift. This dataset of 44 local group galaxies within ∼20 Mpc and observable from the Southern Hemisphere will allow us to resolve at a median physical scale of 50 pc — the physical scales of H II regions (Azimlu et al. 2011). Early science results using TYPHOON galaxies have investigated the radial and azimuthal metallicity variations in select galaxies (Ho et al. 2017, 2018), the contribution of diffuse ionized gas to radial metallicity measurements (Poetrodjojo et al. 2019), and the starburst-active galactic nuclei mixing sequence (D'Agostino et al. 2018). A survey description paper will be forthcoming (Seibert et al.).

Table 1. The TYPHOON Sample.

ID	RA	Dec	Morph.	D(Mpc)	ID	RA	Dec	Morph.	D(Mpc)
WLM	00:01:58.16	-15:27:39.3	IB(s)m	1.04	NGC 2442	07:36:23.84	-69:31:51.0	SAB(s)bc	21.01
ESO 350–G 040	00:37:41.11	-33:42:58.8	S pec (Ring)	130	NGC 2835	09:17:52.91	-22:21:16.8	SAB(rs)c	10.36
NGC 0253	00:47:33.12	-25:17:17.6	SAB(s)c	3.21	NGC 2997	09:45:38.70	-31:11:27.9	SA(s)c	9
NGC 0289	00:52:42.36	-31:12:21.0	SAB(rs)bc	19.16	NGC 3109	10:03:06.88	-26:09:34.5	SB(s)m	1.33
NGC 0300	00:54:53.48	-37:41:03.8	SA(s)d	1.94	Sextans A	10:11:00.80	-04:41:34.0	IBm	1.43
NGC 0625	01:35:04.63	-41:26:10.3	SB(s)	3.78	NGC 3354	10:43:02.90	-36:21:44.6	S: pec	49
NGC 1015	02:38:11.56	-01:19:07.3	SB(r)a	35.1	NGC 3521	11:05:48.58	-00:02:09.1	SAB(rs)bc	11.39
NGC 1068	02:42:40.71	-00:00:47.8	(R)SA(rs)b	10.58	ARP 244	12:01:53.30	-18:52:37.0	HII	20.94
NGC 1313	03:18:16.05	-66:29:53.7	SB(s)d	4.13	NGC 4424	12:27:11.61	+09:25:14.4	SB(s)a	14.11
NGC 1309	03:22:06.56	-15:24:00.2	SA(s)bc	28.13	NGC 4536	12:34:27.05	+02:11:17.3	SAB(rs)bc	15.04
NGC 1316	03:2?:41.72	-37:12:29.6	SAB(s)	19.18	NGC 5068	13:18:54.81	-21:02:20.8	SB(s)d	5.88
NGC 1365	03:33:36.37	-36:08:25.4	SBb(s)b	16.56	NGC 5236	13:37:00.95	-29:51:55.5	SAB(s)c	6.51
NGC 1448	03:44:31.92	-44:38:41.4	SAcd	16.43	NGC 5247	13:38:03.04	-17:53:02.5	SA(s)bc	9.35
NGC 1512	04:03:54.28	-43:20:55.9	SB(r)ab	12.49	NGC 5253	13:39:55.96	-31:38:24.4	Im pec	3.65
NGC 1532	04:12:04.33	-32:52:27.2	SB(s)b	17.89	NGC 5643	14:32:40.74	-44:10:27.9	SAB(rs)c	11.44
NGC 1566	04:20:00.42	-54:56:16.1	SAB(rs)bc	9.95	NGC 5917	15:21:32.57	-07:22:37.8	Sb	31
NGC 1705	04:54:13.50	-53:21:39.8	SA0- pec	5.32	NGC 6300	17:16:59.47	-62:49:14.0	SB(rs)b	12.26
NGC 1744	04:59:57.80	-26:01:20.0	SB(s)d	9.99	NGC 6744	19:09:46.10	-63:51:27.1	SAB(r)bc	7.24
NGC 1800	05:06:25.72	-31:57:15.2	IB(s)m	7.71	NGC 6822	19:44:57.74	-14:48:12.4	IB(s)m	0.52
UGCA 106	05:11:59.32	-32:58:21.4	SAB(s)m	10.2	IC 5152	22:02:41.51	-51:17:47.2	IA(s)m	2.16
ESO 034–IG 011	06:43:06.10	-74:13:35.0	(S0-) + Ring	99	NGC 7213	22:09:16.31	-47:09:59.8	SA(s)	22
NGC 2280	06:44:49.11	-27:38:19.0	SA(s)cd	22.9	NGC 7793	23:57:49.83	-32:35:27.7	SA(s)d	3.83

Coordinates (RA/Dec), morphological type, and distances (Mpc) are taken from the NASA/IPAC Extragalactic Database.

2.1. *Observational Techniques*

The TYPHOON Program† uses the Progressive Integral Step Method also know as "step-and-stare" or "stepped-slit" technique to construct 3D data of nearby galaxies. The PrISM data are the result of an observing methodology rather than a new instrument. We progressively step the slit of a conventional long-slit spectrograph across an object, building up a full data cube via a "step and stare" technique. We use PrISM with the Wide Field reimaging CCD (WFCCD) on the 2.5m du Pont telescope at Las Campanas Observatory. The WFCCD has a field of view of 25' with the 3D data cubes constructed using a custom long-slit (18' × 1.65"; 0.5 arcminute^2) which is typically placed along the northsouth direction and progressively scanned across each individual galaxy through stepping and staring. Each pointing position is integrated for 600 s before the slit was moved by one slit width in the eastwest direction for the next integration. After each exposure, the slit is precisely stepped 1.65" perpendicular to the spatial direction until the majority of the optical disk of each galaxy is covered.

Images are taken at the beginning and end of each sequence of spectra, allowing us to align our observations from night to night. To minimize slit loss, observations are performed only when the seeing is less than the slit width of 1.65". Spectrophotometric flux standards are observed each night.

2.2. *Data Reduction and Cube Construction*

Each individual frame is reduced using our custom-built pipeline. The spectra are bias-subtracted and flat-fielded using calibration frames taken each night. Exposures of a HeNeAr arc lamp are used to calculate a 2d wavelength solution for each spectrum, and each frame is rectified using a spatial solution generated with a Hartmann mask. Each frame thus results in a single 1.65" × 18' "slice" of the full datacube. We bin each slice along the spatial direction of the slit from the native 0.484" pixel size to 1.65". The wavelength calibration has a typical root-mean square value 0.05 Å for the entire data set. Flux calibration is accurate to 2% at the spaxel scale over the range of 45007500 Å. WCS solutions are individually obtained for single-night sections of the datacube using the astrometry.net software based on sources selected from GAIA DR2 or DR3. The reduced long-slit 2D spectra are tiled together to form the 3D data cube and we achieve astrometric solutions that are accurate to within 1 spaxel (offsets<1.65"). The final reduced data cube of each galaxy covers a wavelength range of 3650–9000 Å at a spectral resolution of R∼800, with spectral and spatial samplings of 1.5 Å and 1.65", respectively (Figure 2).

2.3. *Datacube Processing – LZIFU and HIIphot*

The emission-line fluxes are measured using LZIFU (Ho et al. 2016). The emission lines are fitted as a single Gaussian component using the LevenbergMarquardt least-squares method and the stellar continuum is modeled and subtracted for each spaxel. Further details on the LZIFU fitting procedure can be found in Ho et al. (2017).

We use continuum subtracted Hα emission line maps to identify the HII regions using HIIphot (Thilker et al. 2000). HIIphot morphologically isolates HII regions from background diffuse ionized gas (DIG), and provides a 2D mask that identifies separate regions within the map. Figure 3 displays the H II region catalog for NGC 5068 and further details can be found in Grasha et al. (2022). The H II region catalog paper will be forthcoming (Grasha et al.).

† `https://typhoon.datacentral.org.au`

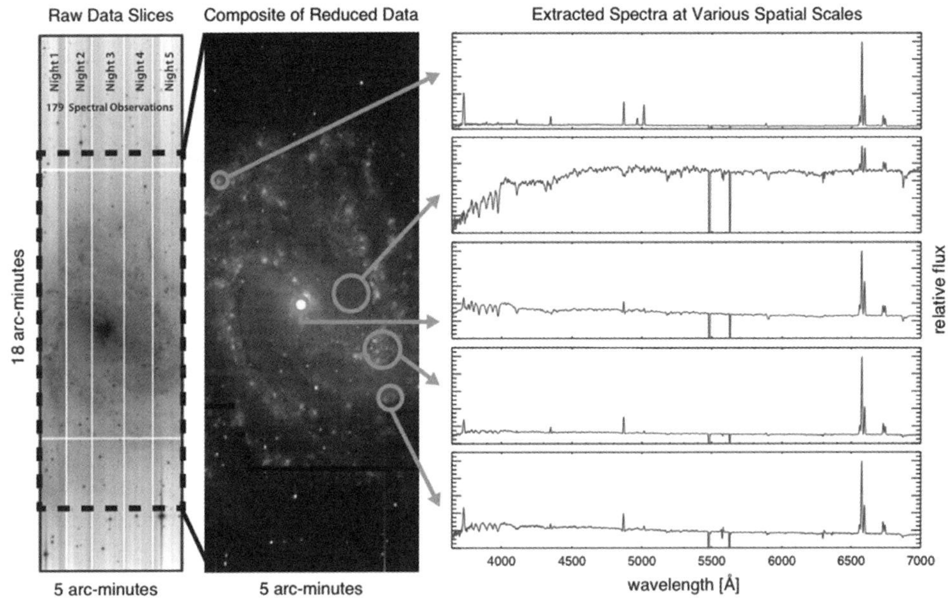

Figure 2. Schematic visualization of NGC 5236 and extracted spectra at various spatial scales and locations.

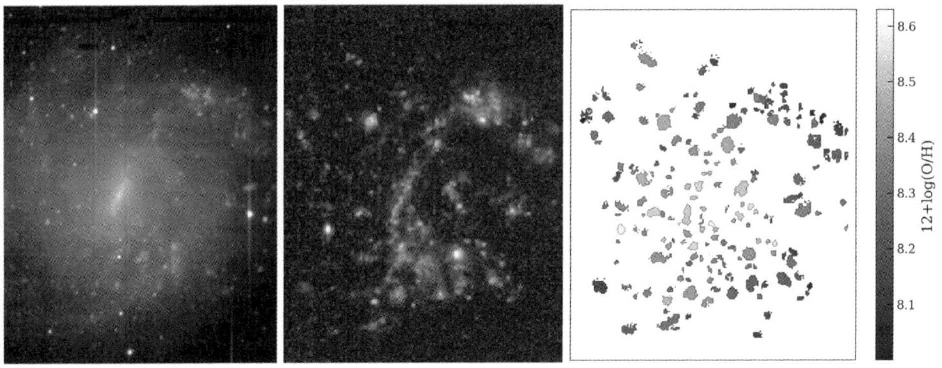

Figure 3. A composite image shows the stellar continuua of NGC 5068 (left), the Hβ, [OIII]λ5007, and Hα composite emission line image (middle), and the extracted HII regions colored by their gas-phase metallicity 12+log(O/H) (right, Grasha et al. 2022).

3. Early Science Demonstration of Grasha et al. (2022)

In a study of the first six spiral galaxies of TYPHOON, Grasha et al. (2022) found 'typical' negative radial metallicity gradients but also a flattening in the gradient in several of the galaxies at large galactocentric radius, suggesting that we have sufficient resolution and sensitivity to detect the metallicity plateau, set by direct accretion of gas from the circumgalactic medium. The measured metallicity floor informs on the chemical enrichment of the galaxy and the outflow of metals to the outer regions of the disk from the enriched inner region of the galaxy disk through radial flows.

We additionally find a lack of azimuthal variations in the oxygen abundance in the spirals in the survey. While the spiral pattern plays a role in organizing the ISM, it alone does not establish the relatively uniform azimuthal variations we observe. Differences in the

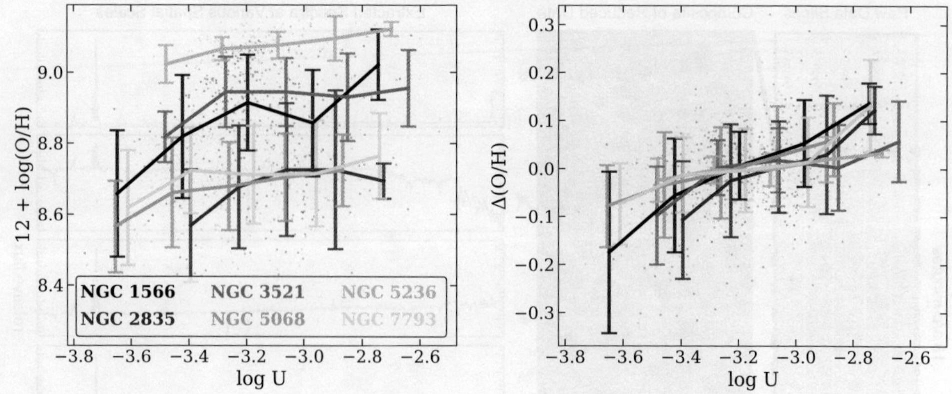

Figure 4. Left: 12 + log(O/H) metallicity of each H II region vs. the ionization parameter U colored by each individual galaxy. Right: the metallicity offset from the linear radial gradient $\Delta(O/H)$ vs. the ionization parameter U. There is a significant positive relationship between the residual metallicity $\Delta(O/H)$ and the ionization parameter U, where H II regions with higher metallicity offsets are association with higher ionization parameters.

metal abundances are more likely driven by the strong correlations with the local physical conditions. We find a strong and positive correlation between the ionization parameter U and the local abundances as measured by the relative metallicity offset $\Delta(O/H)$, indicating a tight relationship between local physical conditions and their localized enrichment of the ISM (Figure 4). This is in agreement with expectations from a combination of localized, star formation-driven self-enrichment and large-scale mixing-driven dilution due to the passing of spiral density waves (Ho et al. 2017).

References

Azimlu M., et al. 2011, AJ, 142, 139
Boissier S., et al. 1999, MNRAS, 307, 857
Buder S., et al. 2022, MNRAS, 510, 2407
Bundy K., et al. 2015, ApJ, 798, 7
Croom S. M., et al. 2012, MNRAS, no
D'Agostino J. J., et al. 2018, MNRAS, 479, 4907
Emsellem E., et al. 2022, A&A, 659, A191
Grasha K., et al. 2017, ApJ, 840, 113
——. 2018, MNRAS, 481, 1016
——. 2022, ApJ, 929, 118
Ho I.-T., et al. 2016, Astrophys Space Sci, 361, 280
——. 2017, ApJ, 846, 39
——. 2018, A&A, 618, A64
Hopkins P. F., et al. 2013, MNRAS, 432, 2647
Kawamura A., et al. 2009, ApJS, 184, 1
Kennicutt R. C. J., et al. 1996, ApJ, 456, 504
Kobayashi C., et al. 2007, MNRAS, 376, 1465
Krumholz M. R., et al. 2019, ARAA, 57, 227
Poetrodjojo H., et al. 2019, MNRAS, 487, 79
Rosales-Ortega F. F., et al. 2010, MNRAS
Rousseau-Nepton L., et al. 2019, MNRAS, 489, 5530
Ryon J. E., et al. 2017, ApJ, 841, 92
Sánchez S. F., et al. 2012, A&A, 538, A8

Sun J., et al. 2020, ApJ, 892, 148
——. 2022, AJ, 164, 43
Thilker D. A., et al. 2000, AJ, 120, 3070
Torrey P., et al. 2019, MNRAS, 484, 5587
de Zeeuw T. P., et al. 2002, MNRAS, 329, 513

Resolving the Rise and Fall of Star Formation in Galaxies
Proceedings IAU Symposium No. 373, 2023
T. Wong & W.-T. Kim, eds.
doi:10.1017/S1743921323000339

The Star Formation History and Dust Production in Andromeda IX

Hedieh Abdollahi[1], Atefeh Javadi[2], Mohammad Taghi Mirtorabi[1], Elham Saremi[2,3,4], Habib Khosroshahi[2], Jacco Th. van Loon[5], Iain McDonald[6,7], Elahe Khalouei[2], Sima T. Aghdam[2] and Maryam Saberi[8]

[1]Physics Department, Faculty of Physics and Chemistry, Alzahra University, Vanak, 1993891176, Tehran, Iran
email: Abdollahi.hedieh2013@gmail.com

[2]School of Astronomy, Institute for Research in Fundamental Sciences (IPM), Tehran, 19568-36613, Iran

[3]Instituto de Astrofisica de Canarias, C/ Vìa Làctea s/n, 38205 La Laguna, Tenerife, Spain

[4]Departamento de Astrofisica, Universidad de La Laguna, 38205 La Laguna, Tenerife, Spain

[5]Lennard-Jones Laboratories, Keele University, ST5 5BG, UK

[6]Jodrell Bank Centre for Astrophysics, Alan Turing Building, University of Manchester, M13 9PL, UK

[7]Department of Physical Sciences, The Open University, Walton Hall, Milton Keynes, MK7 6AA, UK

[8]Rosseland Centre for Solar Physics, University of Oslo, P.O. Box 1029, Blindern, NO-0315, Oslo, Norway

Abstract. Local Group (LG), the nearest and most complete galactic environment, provides valuable information on the formation and evolution of the Universe. Studying galaxies of different sizes, morphologies, and ages can provide this information. For this purpose, we chose the And IX dSph galaxy, which is one of the observational targets of the Isaac Newton Telescope (INT) survey. A total of 50 long-period variables (LPVs) were found in And IX in two filters, Sloan i' and Harris V at a half-light radius of 2.5 arcmin. The And IX's star formation history (SFH) was constructed with a maximum star formation rate (SFR) of about 0.00082 ± 0.00031 $M_\odot$ yr^{-1}, using LPVs as a tracer. The total mass return rate of LPVs was calculated based on the spectral energy distribution (SED) of about 2.4×10^{-4} $M_\odot$ yr^{-1}. The distance modulus of $24.56^{+0.05}_{-0.15}$ mag was estimated based on the tip of the red giant branch (TRGB).

Keywords. stars: AGB and LPV – stars: formation – stars: mass-loss– galaxies: evolution – galaxies: star formation – galaxies: individual: And IX

1. Introduction

Answering questions about the formation and evolution of the universe requires studying dwarf galaxies. While dwarf galaxies are the simplest system known, they are the most diverse in terms of star formation and chemical enrichment (Mateo 1998).

A comprehensive survey (up to now) in the optical bands has been launched to explore the evolution of dwarf galaxies using the Isaac Newton Telescope (INT) (Saremi et al.

2017, 2020). Probing long-period variable (LPV) stars, deriving the star formation history (SFH), and estimating the mass-loss of stars in galaxies are among the reasons for initiating the INT project. As a result of this survey, we can compare the SFHs of different galaxy types and study the evolution and quenching of dwarf galaxies (Hamedani Golshan et al. 2017; Saremi et al. 2019; Navabi et al. 2021). The method to reconstruct SFH was first applied by Javadi et al. (2011b) to rebuild the SFH of M 33.

The asymptotic giant branch (AGB) stars were selected because they are at their most luminous stage ($\sim 10^4$ $L_\odot$ (van Loon et al. 2005; Yuan et al. 2018)) and are easier to detect. AGBs are evolved stars and their luminosity is related to their birth mass (Rezaei et al. 2014; Hashemi et al. 2019). Moreover, their high mass-loss ($10^{-7} < \dot{M} \leq 10^{-3}$ $M_\odot$ yr^{-1} (van Loon et al. 1999; Javadi et al. 2013, 2019; Boyer et al. 2017)) contributes to enhancing the ISM.

The spheroidal dwarf galaxy And IX, located $\sim 39^{+5}_{-2}$ kpc (Weisz et al. 2019) along the major axis of M 31 galaxy, was selected as our candidate for study.

2. Photometry and detection of LPVs

2.1. *Observations and photometry*

Nine observations were made from 2015 – 2017 using the 2.5-m wide field camera (WFC) at INT in the Sloan i', Harris V, and RGO I filters. Data reduction is done using THELI (Transforming Heavenly Light into Image), an image processing pipeline adapted to multi-chip cameras. The DAOPHOT package performs the photometry procedures, and ADDSTAR task can evaluate the completeness of the photometry (Stetson 1987). Fully described details of the photometric procedure can be found in Saremi et al. (2020).

2.2. *Detection of LPVs as SFH tracers*

We detected LPVs using the method introduced by Stetson (1996). In this method, each star is assigned a variability index, which is a function of its magnitude and magnitude error. In each magnitude interval, a threshold for variability index is determined to separate variable stars from non-variables. More than 90% of the stellar population in this magnitude range should have variability indexes higher than the threshold.

To construct the SFH of And IX based on the LPVs, Milky Way foreground contamination should be excluded from candidates with an acceptable threshold. Omitting contamination was done by cross-correlating the INT catalog with the *Gaia* DR3 (Gaia Collaboration et al. 2021) and TRILEGAL simulation of Milky Way populations (Girardi et al. 2005).

We estimated the AGBs' birth mass through the birth mass-luminosity relation using PADOVA evolutionary tracks (Marigo et al. 2017). AGBs are at the end of their evolutionary path, and their luminosity is a function of their birth mass (Javadi et al. 2017). Age and pulsation duration are also calculated by mass-age and mass-pulsation relations (Javadi et al. 2011b; Saremi et al. 2021). The star formation rate (SFR) based on the mass, age, and pulsation duration over time represents the SFH. Fig. 1 presents SFH in two half-light radii (5 arcmin) of And IX in three metallicities. The highest peak of SFR reached at a level of 0.00082 ± 0.00031 $M_\odot$ yr^{-1} in 6 Gyr ago at $Z = 0.0001$. The metallicity of galaxies varies during their evolution, so in this study we consider metallicity $Z = 0.0002$ as well as a more metal-rich estimate $Z = 0.0003$ in addition to the main metallicity $Z = 0.0001$ (Collins et al. 2010; McConnachie et al. 2012; Kirby et al. 2013; Wojno et al. 2020).

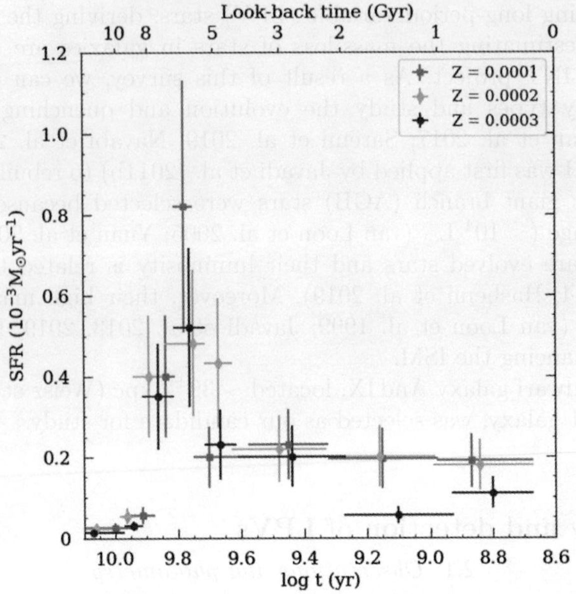

Figure 1. SFHs of And IX for metallicities of $Z = 0.0001$ (pink), $Z = 0.0002$ (orange), and $Z = 0.0003$ (blue) within two half-light radii (~ 0.022 deg^2).

3. Probing of dust in And IX

We modeled the spectral energy distribution (SED) by DUSTY code to estimate the mass-loss ratio of LPVs (Ivezi & Elitzur 1997). DUSTY assumes radiatively driven wind to solve hydrodynamic equations for AGB stars. Output parameters should be scaled for And IX as DUSTY solved equation by default parameters (gas-to-dust mass ratio $\psi_\odot = 200$, L $= 10^4$ L$_\odot$, and $\rho_{dust} = 3$ g cm^{-3}).

4. Results and conclusions

According to the different age gradients of the population in the inner and outer parts of the galaxy, the outside-in star formation scenario could be a galaxy evolution scenario.

The total mass return rate of our detected LPVs at Z = 0.0003, is estimated to be $\sim 10^{-4}$ M$_\odot$ yr^{-1}. By decreasing metallicity to Z = 0.0002, the total mass return increased by 50%. The total mass return rate is about 2.4×10^{-4} M$_\odot$ yr^{-1} in And IX at Z = 0.0001. The average mass-loss rate by C-rich stars is more than 80% of the total mass return rate in three metallicities.

Following are papers that attempt to estimate the total mass return rate of the variables and the SFH observed by the INT survey (Saremi et al. 2017, 2020) for other dwarf galaxies.

References

Boyer et al. 2017, ApJ, 851, 152
Collins M. L. M., et al. 2010, MNRAS, 407, 2411
Gaia Collaboration et al. 2021, A&A...649A...1G
Girardi, L., Groenewegen, M. A. T., Hatziminaoglou, E., da Costa, L., 2005, A&A, 436, 895G
Hamedani Golshan R., Javadi A., van Loon J. Th., et al. 2017, MNRAS, 466, 1764
Hashemi S. A., Javadi A., van Loon J. Th., 2019, MNRAS, 483, 4751
Ivezi e., Elitzur M., 1997, Monthly Notices of the Royal Astronomical Society, 287, 799–811
Javadi A., van Loon J. Th., Mirtorabi M. T., 2011b, MNRAS, 414, 3394

Javadi A., van Loon J. Th., Khosroshahi H., Mirtorabi M. T., 2013, MNRAS, 432, 2824
Javadi A., van Loon J. Th., Khosroshahi H., et al. 2017, MNRAS, 464, 2103
Javadi A., van Loon J. Th., 2019, IAUS, 343, 283
Kirby, E. N., Cohen, J. G., Guhathakurta, P., Cheng L., Bullock, J. S., Gallazzi, A., 2013, ApJ, 779, 102
McConnachie, A. W., 2012, AJ, 144, 4M
Marigo P., Girardi L., Bressan A., et al. 2017, ApJ, 835, 77
Mateo, Mario L. 1998, ARA&A..36..435M
Navabi M., Saremi E., Javadi A., et al. 2021, ApJ, 910, 127
Rezaei Kh S., Javadi A., Khosroshahi H., van Loon J. Th., 2014, MNRAS, 445, 2214
Saremi E., et al. 2017, Journal of Physics: Conference Series, 869, article id. 012068
Saremi E., Javadi A., van Loon J. Th., Khosroshahi H. G., Rezaeikh S., Hamedani Golshan R., Hashemi S. A., 2019, Proceedings of IAU Symposium, 344, 125
Saremi E., et al. 2020, ApJ, 894, 135S
Saremi E., Javadi A., Navabi M., et al. 2021, ApJ, 923, 164
Stetson P. B., 1987, PASP, 99, 191
Stetson P. B., 1996, PASP, 108, 851
van Loon J. Th., Groenewegen M. A. T., de Koter A., 1999, A&A, 351, 559
van Loon J. Th., Cioni M.-R. L., Zijlstra A. A., Loup C., 2005, A&A, 438, 273
Weisz D. R., et al. 2019, MNRAS, 489, 763W
Wojno J., Gilbert K. M., Kirby E. N., Escala I., Beaton R. L., Tollerud E. J., Majewski S. R., Guhathakurta P., 2020, ApJ, 895, 78W
Yuan W., et al. 2018, AJ, 156, 11

Resolving the Rise and Fall of Star Formation in Galaxies
Proceedings IAU Symposium No. 373, 2023
T. Wong & W.-T. Kim, eds.
doi:10.1017/S1743921322003982

Variable metallicity yields as tracers of inflows

Artemi Camps-Fariña[1], Patricia Sánchez-Blázquez[1], Santi Roca-Fàbrega[1] and Sebastián F. Sánchez[2]

[1]Departamento de Física de la Tierra y Astrofísica, Universidad Complutense de Madrid, Pl. Ciencias, 1, Madrid, 28015, Spain; email: arcamps@ucm.es

[2]Instituto de Astronomía, Universidad Nacional Autónoma de México, Circuito Exterior, Ciudad de México, 04510, México

Abstract. Pristine gas accretion is expected to be the main driver of sustained star formation in galaxies. We measure the required amount of accreted gas at each moment over a galaxy's history to produce the observed metallicity at that time given its star-forming history. More massive galaxies tend to have higher accretion rates and a larger drop of the accretion rate towards the present time. Within the same mass bin galaxies that are currently star-forming or in the Green Valley have similar, sustained, accretion histories while retired galaxies had a steep decline in the past. Plotting the T80 of the individual accretion histories, a measure of how sustained they are, versus the stellar mass and current sSFR we see a distribution such that currently star-forming galaxies have sustained or recent accretion and retired galaxies have declined accretion histories.

Keywords. galaxies: abundances, galaxies: evolution, techniques: spectroscopic

1. Introduction

The chemical enrichment history (ChEH) of a galaxy is the result of the net effect of the competing processes of enrichment and dilution that act on the interstellar medium over its lifetime. The enrichment is driven by the stellar population as they produce the elements and disperse them as they evolve and die. This input is balanced by the ISM interacting with its surroundings via accretion of pristine gas from the surroundings and outflows driven by stellar winds and AGN (Lilly et al. 2013).

As a simplified model, we can therefore consider the ChEH to be the result of the star-formation history (SFH) and the dilution history. We have measured the ChEH and SFH of MaNGA galaxies in a previous work (Camps-Fariña et al. 2022) which allows us to measure the dilution history by comparing the input of metals from the SFH to the measured ChEH. Pristine gas accretion is considered to be the key mechanism sustaining star-formation in galaxies (Sánchez Almeida et al. 2014).

2. Analysis

The ChEHs and SFHs were obtained by fitting the stellar spectra of galaxies in the MaNGA sample with a stellar population synthesis code, PyPipe3D (Lacerda et al. 2022). The MaNGA survey (Bundy et al. 2015) consists of IFU observations of ∼10,000 galaxies in the local universe providing spatially resolved spectroscopic information with a typical spectral resolution of ∼2000 over a wavelength range 3600-10300Å and an effective spatial resolution of ∼2.5". The fitting code uses templates of spectra of populations of stars of

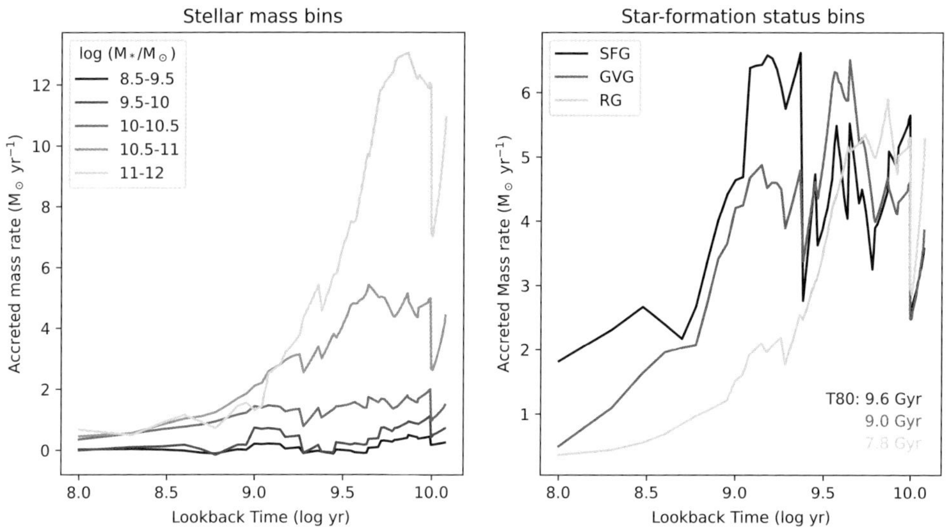

Figure 1. Pristine gas accretion histories for the galaxies in our sample. In the left panel the galaxies are divided into stellar mass bins. In the right panel a single stellar mass bin is considered: $10^{10.5-11}$ $M_{\odot}$ and the galaxies are further subdivided into current star-formation status mass bins using $EW_{H\alpha}$: Star-forming (SFG), Green Valley (GVG) and Retired (RG) galaxies. In the bottom right of the right panel we show the T80 of the accretion history defined as the time it takes to accrete 80% of the total accreted gas mass.

known age and metallicity values to recover the weights of each population allowing for the determination of the ChEH and SFH (Camps-Fariña et al. 2022).

The conversion from SFH to metal injection over time is done using a modified version of the galaxy chemical evolution code from Roca-Fàbrega et al. (2021) which was used to explain the distribution of Li abundance in the Milky Way as a result of a particular SFH and accretion history. In this work we treat the ChEH as an input rather than an output.

3. Results

In Figure 1 we show the resulting pristine gas accretion histories for the galaxies in the MaNGA sample divided into mass and star-formation status bins. The methodology to obtain these histories is to take the averaged ChEH and SFH of each stellar mass and star-forming status bins and calculate the necessary accretion history using the chemical evolution code. The histories correspond to those of Figures 3 and 5 from Camps-Fariña et al. (2022). The stellar mass bins are selected from current values of the mass and the current star-formation status bins are defined from the $EW_{H\alpha}$ with 3Å and 10Å as the boundary values.

The accretion histories show how more massive galaxies tend to accrete more gas over their lifetime compared to less massive ones. The former also show a more pronounced peak in the accretion rate such that they have suffered a significant decline in their accretion rate compared to early epochs.

If we control for stellar mass, however, we also see a difference in the accretion histories of galaxies depending on their current state of star-formation. In the right panel we show the histories within a single mass bin, $10^{10.5-11}$ $M_{\odot}$ but further divided into bins defined by their current star formation. Retired galaxies show a declined accretion history which is different from star-forming and Green Valley (GV) ones. The latter two have very similar histories until about 500 Myr ago (log LBT $\sim$8.7) when Green Valley galaxies

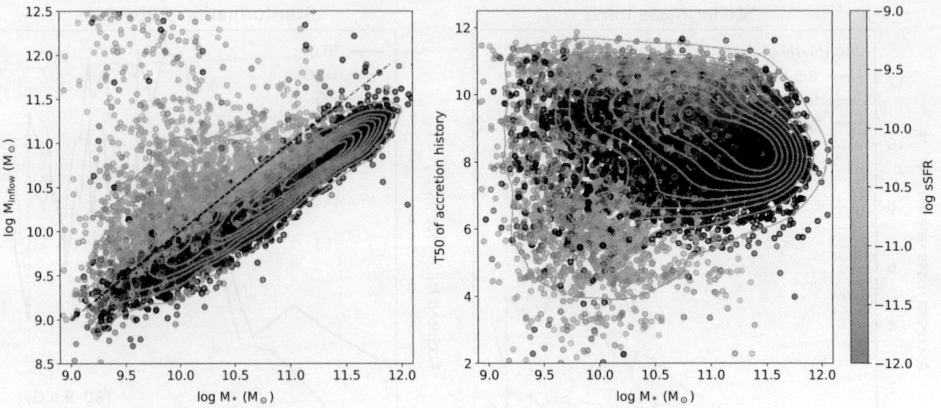

Figure 2. In the left panel, the correlation between the stellar mass and the total accreted gas mass. In the right panel, the correlation between the stellar mass and the T80 of the accretion history, defined as the time it takes to accrete 80% of the total accreted gas mass. In both panels the color of the points shows the sSFR and the orange contours the density of points.

have a drop in accretion rate. The similarity of star-forming and GV galaxies fits very well with our expectations as the latter are only now stopping their star-formation so they should have a similar history until recently.

3.1. *Individual histories*

The previous results were obtained using averaged ChEHs and SFHs but we can also calculate the gas accretion histories of each galaxy in the sample and study what correlations arise from their defining parameters.

In Figure 2 we show the correlation between stellar mass, sSFR, and either the total accreted gas mass or the T80 of the accretion history. The latter corresponds to the time it takes for 80% of the total mass to be accreted and is an indicator of whether the accretion is sustained or recent (high T80) or whether it has declined (low T80).

The left panel shows a tight correlation between the stellar mass and the accreted gas mass with currently star-forming galaxies located in an upper envelope of the accreted gas mass such that currently star-forming galaxieas have accreted more pristine gas.

In the right panel the main region shows a correlation such that for each stellar mass galaxies which are currently star-forming have more sustained or recent accretion. This figure is similar to the SFR-M_* plane, with a distribution of high-mass, retired galaxies and a tighter sequence of star-forming galaxies. This allows us to predict whether a galaxy is currently star-forming based on the shape of its chemical balance.

Given the relative simplicity of the approach this method could become a valuable tool to evaluate and validate galaxy evolution models and simulations, especially if the chemical evolution code is improved to include more complex dilution mechanisms and the effect of outflows.

References

Bundy, K., Bershady, M. A., Law, D. R., et al. 2015, ApJ, 798, 7.
Camps-Fariña, A., Sánchez, S. F., Mejía-Narváez, A., et al. 2022, ApJ, 933, 44.
Lacerda, E. A. D., Sánchez, S. F., Mejía-Narváez, A., et al. 2022, NewA, 97, 101895.
Lilly, S. J., Carollo, C. M., Pipino, A., et al. 2013, ApJ, 772, 119.
Roca-Fàbrega, S., Llorente de Andrés, F., Chavero, C., et al. 2021, A&A, 656, A64.
Sánchez Almeida, J., Elmegreen, B. G., Muñoz-Tuñón, C., et al. 2014, A&ARv, 22, 71.

Resolving the Rise and Fall of Star Formation in Galaxies
Proceedings IAU Symposium No. 373, 2023
T. Wong & W.-T. Kim, eds.
doi:10.1017/S1743921322003878

GALEX UV Catalog of Low-redshift Galaxies for Estimating Transient Rates

Jeeun Hwang, Myungshin Im, Hyeonho Choi, Gregory S.H. Paek and IMSNG team

SNU Astronomy Research Center, Astronomy Program, Dept. of Physics & Astronomy, Seoul National University, Seoul, Korea
email: jehwang@snu.ac.kr

Abstract. Detection of transients such as supernovae (SNe) and kilonovae (KNe) in early phase has recently become important for understanding the progenitor properties and multi-messenger astronomy. Predicting which galaxy has the higher probability of hosting the transient events would help detect the early phase of the events and get information on their progenitors. The SN and KN rates are known to be a function of star formation rate (SFR) and stellar mass of the host galaxy. The SFR of a galaxy can be estimated from ultraviolet (UV) luminosity. However, the UV magnitudes have been derived carefully only for a limited number of nearby galaxies. Here, we introduce GALEX galaxy catalog of all-sky UV brightness of low redshift galaxies. To do so, we derive the UV photometry of galaxies in the GLADE catalog using the GALEX AIS images, supplemented by GALEX NGS and MIS data. From the near-UV (NUV) and far-UV (FUV) magnitudes, we calculate the SFRs of the galaxies, which will further be useful for estimating the SN and KN rate. The results are compared with previous GALEX UV catalog of galaxies. There will be an updated catalog based on this catalog for calculating KN rate of the galaxies in the future work.

Keywords. GALEX, UV, transient, catalog, local universe, star formation rate, galaxy

1. Introduction

Transients are astronomical phenomena that last for a limited amount of time. Their timescales are short, from miliseconds to days, weeks, and several years. The example of transients are supernovae, kilonovae, fast radio bursts, gamma ray bursts, and tidal disruption events. Since they unexpectedly pop up and rapidly gets faint, immediate follow-up observation is necessary to study the nature of transients. For instance, IMSNG (Intensive Monitoring Survey of Nearby Galaxies); Im et al. (2019) performs a high cadence monitoring survey to catch early light curves of supernovae.

There are several researches that investigated the relation between the transient and its host galaxy properties. Smith et al. (2012) suggested the parameterization of supernova type Ia rate on host galaxy properties, namely, stellar mass and SFR. Artale et al. (2020a) investigated the properties of the galaxies hosting the compact binary merger events with simulations and showed that the binary merger rate per galaxy is strongly correlated with the stellar mass of the host galaxy. In both papers, it is clear that the host galaxy properties are closely related to the occurrence of transient events. Based on these results, we try to predict which galaxy has a higher probability of hosting transients. To achieve the goal of detecting the early phase of transients, therefore, estimating the stellar mass and SFR of the galaxy is important.

Figure 1. Galaxy mask of NGC 6962.

SFR of a galaxy can be estimated with ultraviolet (UV) continuum luminosity. The photometric catalog of galaxies in UV band is useful for estimating the SFR. Gil de Paz et al. (2007) presented the GALEX photometry results of 1034 nearby galaxies. Bai et al. (2015) offered the extended GALEX photometry results of 4038 nearby galaxies. Bianchi et al. (2017) provided more extended GALEX photometry including 15659 results for galaxies. However, galaxies are often split into several sources in the all sky point-source catalog of Bianchi et al. (2017), and some of the point-source photometry are inadequate for extended sources like galaxies. Therefore, we are performing careful galaxy UV photometry using GALEX data. Here, we present results from our initial attempt of constructing the UV catalog for galaxies at < 50 Mpc.

2. Photometry

Sample selection. To obtain the all-sky SFR map of galaxies, the sample galaxies are selected from Dálya et al. (2022). The 31457 galaxies within 50 Mpc were selected.

Galaxy mask. Since star-forming regions tend to be patchy, galaxies detected in UV are often split into multiple sources. Making a galaxy mask will help capture the whole extent of the galaxy, minimizing the loss of UV photons during photometry. The mask is made from images taken in optical bands (Figure 1), when the extents of galaxies in UV are too patchy or too small in comparison to optical images.

Tool. Galaxy masks were built by performing the Source Extractor dual mode, using optical images as detection images and GALEX images as photometry images. The Source Extractor provides a segmentation image and all the sources except for a target galaxy are masked to solely include the pixels of the galaxy.

Galactic Extinction correction. The effect of Milky Way extinction was corrected using the Galactic reddening map of Schlegel et al. (1998). The reddening law of Cardelli et al. (1989) is used to convert E(B - V) to the UV extinction.

3. Results

GALEX galaxy UV catalog. The GALEX galaxy UV catalog is composed of galaxy name, right ascension, declination, distance, magnitude in FUV band, magnitude in NUV band, and star formation rate.

Star formation rate. The SFR of the galaxy is calculated from the UV flux. The following conversion equation is the calibration of Madau et al. (1998) to Salpeter (1955)'s initial mass function with mass limits 0.1 and 100 $M_{\odot}$. It is valid for the wavelength interval of 1500-2800 Å.

$$SFR(UV) = 1.4 \times 10^{-28} L_{\nu} [erg\, s^{-1} Hz^{-1}]$$

Sample SFR map of the sky. The sample of 340 galaxies located at right ascension $0° - 180°$ has been plotted over the Mollweide projection of the sky in Figure 2. The color bar indicates the SFR in $M_{\odot}$ of each galaxies marked as different colors of points.

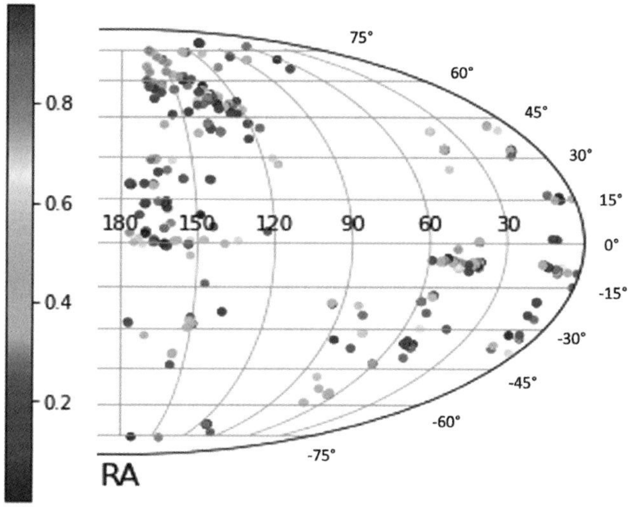

Figure 2. Sample SFR map of the sky.

4. Discussion

AGN contamination. The observed UV luminosity of a galaxy can be originated not only from stars, but also from Active Galactic Nucleus (AGN). Hence, the star formation rate calculated using UV luminosity can be contaminated by these types of AGNs. Therefore, we flag AGNs in this work.

UV as a SFR tracer in transient science. Star formation in a galaxy can be traced by various indicators, such as UV luminosity, recombination lines, forbidden lines, and far-infrared luminosity Kennicutt et al. (1998). Among them, UV luminosity is selected as a SFR tracer in this work. However, since the UV light can be extincted by dust, we plan to make corrections for the internal extinction, and use IR data to more accurately measure SFRs.

Stellar mass and more sample. Stellar mass is another key parameter for estimating SN and KN rates. We plan to derive stellar mass for galaxies and extend our analysis to galaxies at > 50 Mpc.

Acknowledgments

This work was supported by the NRF grants, No. 2020R1A2C3011091 and No. 2021M3F7A1084525.

References

Artale M. C., Marpelli, M., Bouffanais, Y., et al. 2020, *MNRAS*, 491, 3419
Bai, Y., Zou, H., Liu, J., & Wang, S. 2015, *ApJS*, 220, 6
Bianchi, L., Shiao, B., & Thilker, D. 2017, *ApJS*, 230, 24
Cardelli J. A., Clayton G. C., Mathis J. S. 1989, *ApJ*, 345, 245
Dálya et al. 2022, *MNRAS*, 514, 1403
Gil de Paz A. et al. 2007, *ApJS*, 173, 185
Im M. et al. 2019, *J. Korean Astron. Soc.*, 52, 11
Kennicutt, R. C., Jr. 1998, *ARAA*, 36, 189
Madau, P., Pozzetti, L., & Dickinson, M. 1998, *ApJ*, 498, 106
Salpeter E. E. 1955, *ApJ*, 121, 161
Schlegel D. J., Finkbeiner D. P., Davis M. 1998, *ApJ*, 500, 525
Smith M. et al. 2012, *ApJ*, 755, 61

Resolving the Rise and Fall of Star Formation in Galaxies
Proceedings IAU Symposium No. 373, 2023
T. Wong & W.-T. Kim, eds.
doi:10.1017/S1743921322003842

Tug of War, or Cohabitation? : Star Formation and AGN Activities within Type 1 AGN Host Galaxies

Ji Hoon Kim[1], Myungshin Im[1,2], Hyunsung D. Jun[1] and Dohyeong Kim[3]

[1]SNU Astronomy Research Center, Seoul National University, Republic of Korea

[2]Department of Physics and Astronomy, Seoul National University, Republic of Korea

[3]Department of Earth Science Education, Pusan National University, Republic of Korea

Abstract. The interplay between star formation (SF) activity and active galactic nuclei (AGN) governs the co-evolution of supermassive black holes (SMBHs) and their host galaxies. AGN feedback has been hailed as the de facto process to suppress, or even shut down SF within the framework of hierarchical galaxy merger based on the current ΛCDM paradigm. However, it is unclear what physical processes regulate the growth of SMBHs and how SMBHs and their evolution are interconnected with their host galaxies when SMBHs and host galaxies are of hugely different physical scales. In fact, there has been no observational evidence to show that AGN feedback works, but rather some evidence to speculate that the more powerful AGNs reside in the more actively star-forming host galaxies. While it is difficult to measure the amount of SF from AGN host galaxies, polycyclic aromatic hydrocarbon (PAH) emission features emerged as good proxies for this purpose. Although having several caveats as SFR indicators, such as metallicity dependency, and non-SF contribution from evolved stellar populations, or AGNs, PAH emissions have been utilized to investigate SF activity of AGN host galaxies with varying results. Utilizing the slitless spectroscopic apability of the AKARI Infrared Camera, we obtained the spectra in the wavelength range of 2∼5 μm from extended regions of 79 type 1 AGN host galaxies to detect and measure the 3.3 μm (PAH) emission feature as star formation rate proxy. Based on 18 sample galaxies, we found that the luminosity of the 3.3 μm PAH emission feature is strongly correlated with AGN luminosity, except for ultra-luminous infrared galaxies (ULIRGs). Therefore, we suggest that host galaxies with stronger AGN activities have stronger star formation activities. However, it is still unclear why ULIRGs deviate from the correlation, not to mention why the detection rate of the 3.3 μm emission feature is so low. High spatial resolution imaging not only for the circumnuclear region of AGN host galaxies, but also for entire galaxies should help the cause. We present the prospective studies to diagnose SF regulation for AGN host galaxies with various space telescope facilities, such as JWST, and SPHEREx.

Keywords. galaxies: star formation, galaxies: ISM, galaxies: Seyfert, infrared: galaxies

---

1. Introduction

The formation and evolution of supermassive black holes (SMBHs) and their host galaxies are closely linked. While SMBHs reside at the centers of most bulge-dominated galaxies, their masses have tight correlations with the various properties of their host galaxies. While there have been enormous efforts to understand the connection between SMBHs and their host galaxies, it is still uncertain what physical processes govern the growth of SMBHs and their host galaxies, and how their evolutionary phases are interconnected. Numerous theoretical models have been devised to explain the co-evolution of SMBHs and their host galaxies albeit a huge gap in physical scales. Notably, active galactic nucleus (AGN) feedback is a common feature of these models. AGN feedback was originally devised to reconcile the "overcooling" problem of massive halos within the frame work of hierarchical galaxy merger based on the ΛCDM paradigm. While the primary function of AGN feedback is believed to suppress, or to slow down star formation (SF) activity of host galaxies by removing cold gas within them, some recent models suggest that AGN feedback can enhance SF activity by plowing gas into denser clumps. Then, more sophisticated models implement scenarios with a time offset between SF and AGN activities within the same event, or regulated growths of BHs and host galaxies through multiple events of interplays between them.

Therefore, measuring SF and AGN activities to probe the connection between them provides a crucial information on how AGN feedback, either "positive", or "negative", works on the growth of host galaxies. However, it is not trivial to measure SF activity in AGN host galaxies. In general, the radiation from AGN overpowers SF activity of host galaxies. Especially within the ultraviolet and visible wavelength regimes, AGN powers SF rate (SFR) proxies, such as far-ultraviolet emission, and hydrogen recombination lines, more than SF activity. Hence, there have been numerous efforts to measure SF activity using the longer wavelength regimes, such as infrared (IR), and sub-mm wavelength regimes.

Within these longer wavelength regimes, polycyclic aromatic hydrocarbon (PAH) emission features have emerged as good candidates to provide SFR for AGN host galaxies thanks to various space missions, such as *Spitzer* and *AKARI* space telescopes; a handful of studies showed strong correlations between SF and AGN activities by measuring SF activity with various PAH emission features. However, the 3.3 μm PAH emission feature which is the only PAH emission feature accessible for high-z (z > 4) galaxies with *JWST* space-telescope has not been studied due to its weak nature.

2. Data

This study present the results based on the Quasar Spectroscopic Observation with NIR Grism (QSONG). The sample of QSONG consists of two large samples of AGNs; a low redshift (z < 0.5) sample called low redshift QSONG (LQSONG) and high redshift (z > 3.0) sample (HQSONG). Chosen to achievee high S/N with IRC onboard *AKARI* space-telescope, the LQSONG are composed mainly from two differenct sources; 35 bright type 1 AGNS of which BH masses are measured by reverberation mapping method, and 69 PG QSOs. Between them,we observed 31 out of 35 and 48 out of 69 from the reverberation sample and the PG QSO sample, respectively and used for the analysis.

3. Detection of the 3.3 μm PAH Emission

We detected the 3.3 μm PAH emission from 18 out of 79 sample galaxies (34 % detection rate). The detection rates for the subsamples are 42% and 10% for the reverberation sample and the PG QSO sample, respectively. The differences of the detection rates are not be attributed to physical origins, since the only difference is the measurement methods of their SMBH masses.

On the other hand, the detection rates of the 3.3 μm PAH emission feature depend on the strength of AGN activity; the stronger AGN activity means the lower detection rate. However, stronger AGNs are at higher redshifts in general. Therefore, we may not rule out that the decreasing detection rate for stonger AGNs are simply due to the sample selection and/or the detection limit.

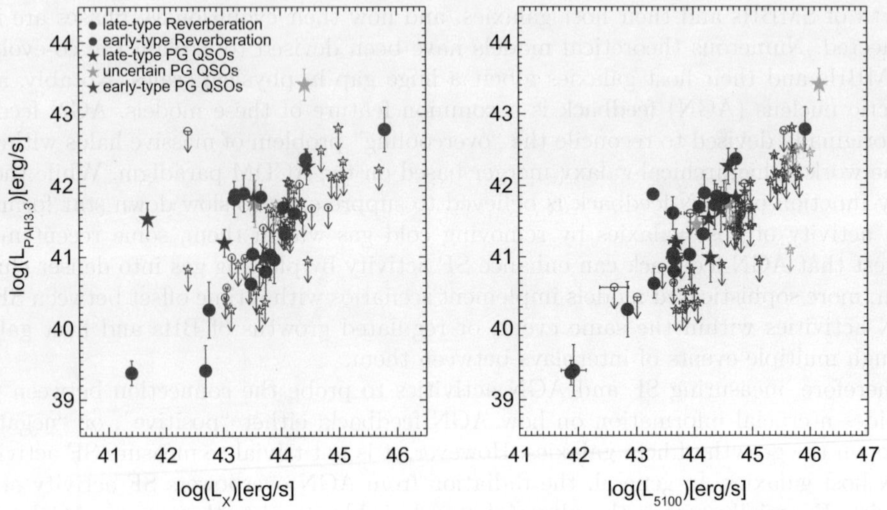

Figure 1. Top two panels of Figure 3 from Kim *et al.* (2019) $L_{\mathrm{PAH3.3}}$ against nuclear activity probes. Symbols represent the subsamples; circles are of the reverberation mapping sample, while stars of the PG QSO sample. Colors represent the host galaxy morphology; red symbols are of early-type, blue symbols are of late-type, and green symbols are of the unclear. All the empty symbols represent upper limits.

4. SF versus AGN

The strength of AGN activity can be probed by various properties; $\lambda L_{5100,\mathrm{AGN}}$ and L_{X} are among the most used ones. Figure 1 shows the correlations of $L_{\mathrm{PAH3.3}}$ with $\lambda L_{5100,\mathrm{AGN}}$ and L_{X}. $L_{\mathrm{PAH3.3}}$ has strong correlations with both L_{X} and $L_{5100,\mathrm{AGN}}$, although there are several outliers against L_{X} more than against $L_{5100,\mathrm{AGN}}$.

Moreover, there is no systematic dependence on the morphology of host galaxies. Regardless of their host galaxy morphology, the correlations between $L_{\mathrm{PAH3.3}}$ and L_{X}, or $L_{5100,\mathrm{AGN}}$ persist for the sample galaxies. Therefore, if $L_{\mathrm{PAH3.3}}$ represents SFR of galaxies, there is a strong positive correlation between star formation activity of AGN host galaxies also correlates with AGN activity. Still, it is hard to rule out that the correlations of $L_{\mathrm{PAH3.3}}$ with L_{X} and $L_{5100,\mathrm{AGN}}$ result from massive objects being luminous.

Several studies suggested that the 3.3 μm PAH emission correlates with various properties of AGN host galaxies. However, the correlations found by Kim *et al.* (2019) are the closet to the unity.

The linear fit between $L_{\mathrm{PAH3.3}}$ and $L_{5100,\mathrm{AGN}}$ follows as;

$$log L_{\mathrm{PAH3.3}} = 0.88 \times log L_{5100,\mathrm{AGN}} + 5.81. \tag{1}$$

The linear fit between $L_{\mathrm{PAH3.3}}$ and L_{X} follow as;

$$log L_{\mathrm{PAH3.3}} = 0.90 \times log L_{\mathrm{X}} + 5.20. \tag{2}$$

5. Discussion

It is clear that there are strong correlations between $L_{\mathrm{PAH3.3}}$ and nuclear activities. There are several caveats, though. First, the PAH emission features are absent for ultraluminous infrared galaxies (ULIRGs; galaxies with $\mathrm{L_{IR}} > 10^{12}\ \mathrm{L_{\odot}}$) (Kim *et al.* 2012). Although it is well known that not only the 3.3μm PAH emission feature, but also all the longer wavelength PAH emission features are absent for ULIRGs, it is unclear why these PAH emission features absent only for ULIRGs regardless of the presence of AGNs. Secondly, the abundance of PAH molecules are believed to depend on metallicity, which may impact the overall calibration of luminosity of PAH emission features as SFR proxy. Now that JWST started its operation, there will be more opportunities to measure $L_{PAH3.3}$ from SF regions of AGN host galaxies at various redshift regimes.

References

Kim, J. H., Im, M., Lee, H. M., et al. 2012, ApJ, 760, 120

Kim J. H., Im M., Kim D., Woo J.-H., Karouzos M., Lee H. M., Lee M. G., et al., 2019, PASJ, 71, 25

Resolving the Rise and Fall of Star Formation in Galaxies
Proceedings IAU Symposium No. 373, 2023
T. Wong & W.-T. Kim, eds.
doi:10.1017/S1743921322003787

High [OIII] luminosities from star formation and shocks in z∼6 quasars

Bomee Lee[1] and **Ranga-Ram Chary**[2]

[1]Korea Astronomy and Space Science Institute, Daedeokdae-ro 776, Yuseong-gu, Daejeon 34055, Republic of Korea; email: bomee@kasi.re.kr

[2]MS314-6, Infrared Processing and Analysis Center, California Institute of Technology, Pasadena, CA 91125, USA

Abstract. We use archival WISE and Spitzer photometry to derive optical emission line fluxes for a sample of distant quasars at z∼6. We find evidence for exceptionally high equivalent width [OIII] emission (rest-frame EW∼400Å) similar to that inferred for star-forming galaxies at similar redshifts. The median Hα and Hβ equivalent widths are derived to be ∼400Å and ∼100Å respectively, and are consistent with values seen among quasars in the local Universe, and at $z \sim 2$. After accounting for the contribution of photoionization in the broad line regions of quasars, we suggest that the narrow [OIII] emission likely arises from feedback due to massive star-formation in the quasar host. Forthcoming mid-infrared spectroscopy with the James Webb Space Telescope will help constrain the physical conditions in quasar hosts further.

Keywords. high redshifts, quasars, emission lines, galaxy formation/evolution

1. Introduction

Emission lines from ionized gas, particularly Hα, Hβ and [OIII] are a tracer of ISM conditions, particularly the magnetic field strength, the shock velocity, gas density and metallicity (Allen et al. 2008). At high redshifts, these diagnostic lines are redshifted to mid-infrared wavelengths. Although this will be remedied imminently through spectroscopic observations with the *James Webb Space Telescope*, it has not been possible thus far, to place constraints on the line strengths. In the case of star-forming, emission-line galaxies at high redshift, a pioneering approach which utilized the excess emission in broad, mid-infrared bandpasses yielded strong evidence for high equivalent width (EW) nebular emission but these studies were limited by signal to noise ratio (SNR; e.g. Chary et al. 2005; Faisst et al. 2016).

In comparison, quasars, by being the most luminous sources in the distant Universe, have multi-wavelength photometry that has much higher SNR than for most star-forming galaxies. In this study, we leverage this high quality photometry to derive rest-frame optical emission-line properties of a sample of $z \sim 6$ quasars.

2. Estimates of the rest-frame optical emission line strengths

Our initial sample consists of a list of 520 quasars at $z > 5$ that have been spectroscopically confirmed (Ross & Cross 2020). We collect multi-wavelength photometric data between the optical $y-$band ($\sim 0.9\,\mu$m) and 8 μm from archival imaging data on those $z > 5$ quasars including UKIRT/WFCAM and VISTA/VIRCAM, Pan-STARRS1 DR2, *Spitzer*/IRAC, and WISE. We select 53 quasars at $5.03 < z < 6.3$ having SNR> 5 in both IRAC 1 (3.6 μm) and 2 (4.5 μm) channels (see Lee & Chary (2022) for more details).

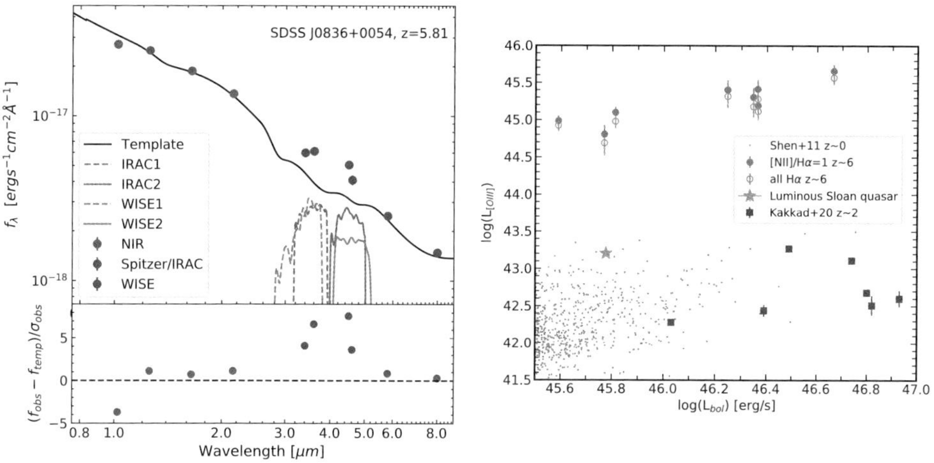

Figure 1. Left: An example fit of SDSS J0836+0054, a quasar at z = 5.81 with its observed photometry. Black line represents the continuum template best-fitted to near-IR bands and IRAC 5.8 and 8μm bands. The transmission curves of IRAC 3.6 and 4.5μm (orange) and WISE 3.3 and 4.6μm bands (light blue) are overplotted. Bottom panel: The difference between observed flux (f_{obs}) and continuum flux (f_{temp}) as a function of wavelength. One can clearly see the bumps between 3 and 5 μm due to [OIII] and Hβ in the $\sim 3.5\mu$m bandpasses and Hα+[NII] in the $\sim$4.5μm bandpasses. **Right**: Bolometric Luminosity vs our derived [OIII] luminosity of 8 quasars at z$\sim$6 (blue and orange), AGNs from SDSS DR7 (grey), and z$\sim$2 quasars from VLT/SINFONI (red). Empty orange and filled blue circles represent the range in [OIII] luminosities for the two broad-line region scenarios discussed in the text. A local AGN having similar [OIII]/Hβ ratio as $z \sim 6$ quasars is the green star.

The entrance of redshifted emission lines, such as Hα, [NII], Hβ, and [OIII], in $z > 5$ galaxies, can introduce a photometric excess in certain bands. These lines introduce a significant difference between the observed photometry and the underlying continuum flux density, if the EW of the lines is high. Thus, if we can fit for the underlying continuum flux density using templates, we can obtain an estimate of the emission line fluxes using the excess emission in the different bands. As an example, for the z = 5.81 quasar shown in the left of Figure 1, the Hα line flux can be estimated from the photometry in the IRAC 2 and WISE 2 bands, while Hβ+[OIII] can be estimated with IRAC 1 and WISE 1 bands. For typical quasars, the Hydrogen Balmer lines arise from photoionization in the broad line region (BLR). As a result it is difficult to separate the [NII] and [OIII] narrow-line emission from the broad Hydrogen lines. In order to assess if there is a component of narrow-line emission in the photometry, we adopt two scenarios: 1) where the BLR Hα emission accounts for half the IRAC2 and WISE2 photometric excess (with the NLR accounting for the rest), and 2) where the BLR accounts for all the excess. Since we cannot fit spectra to measure line widths, we fix FWHM = 3400km/s for all lines. FWHM = 3400km/s is chosen as a median FWHM of the MgII line in 82 quasars at $4.5 < z < 6.4$ (see Lee & Chary (2022) for references) and assumes a correlation of FWHM between $MgII$ and Hα lines (Jun et al. 2015).

With these assumptions and a Gaussian profile for the line, we derive the $H\alpha$ line fluxes using the excess emission at IRAC2 and/or WISE 2. Hβ is derived by assuming an Hα/Hβ= 2.86 (i.e. no dust). We derive [OIII] after subtracting the Hβ contribution from the excess emission in IRAC 1 and/or WISE 1. Depending on the redshift, the final line flux is the SNR-weighted line flux from both the IRAC 1 (IRAC2) and WISE 1 (WISE2) excess, when available. We use a Monte Carlo simulation to produce robust uncertainty estimates on the fitted emission lines. As a result, we are able to constrain Hβ and [OIII] line fluxes for only a subset of 33 quasars at $5.03 < z < 6.3$ due to large photometric uncertainties in most of cases, and inadequacy of the templates in a small subset of cases.

3. Optical emission line constraints at z ∼ 6

To investigate the physical characteristics of z > 5 quasars, we further restrict our sample to those which have $SNR > 3$ in $H\alpha$ and [OIII]. We have a final sample of 8 quasars at $5.03 < z < 6.23$ with a median redshift of $z = 5.94$ and compute the [OIII] luminosity of them (Figure 1 right panel). The median bolometric luminosity of this sample of quasars is 2×10^{46} erg/s with a range of $3.9 \times 10^{45} < L_{AGN}\ [erg/s] < 4.6 \times 10^{46}$. The derived median rest-frame EWs of Hα, Hβ and [OIII] are 400Å, 100Å and 440Å respectively. The median derived [OIII] luminosity is 2×10^{45} erg/s. [OIII] luminosities of $z \sim 6$ quasars are about two orders of magnitude higher than luminous local quasars (Shen et al. 2011) or z∼ 2 quasars studied with VLT/SINFONI (Vietri et al. 2020) (Figure 1). Moreover, we find rest-frame Hα EWs of ∼300-800Å at luminosities of $\sim 10^{46}$ erg s^{-1}. This is in good agreement with the values derived for $z > 5$ quasars, using a similar methodology by Leipski et al. (2014), and similar to the Hα EWs of Sloan quasars (Shen et al. 2011). The median Hβ rest-frame EW of the local SDSS quasars is 100Å (Shen et al. 2011), similar to our sample, implying that much of the excess in the bluer bandpass is indeed because of the [OIII] doublet.

4. Physical Implication for $z \sim 6$ Quasars

The photometric excess that we are measuring is the sum of emission from the narrow-line region (NLR), the broad-line region (BLR) and from stellar/ISM emission. Given the luminosity of the sample, the BLR is likely dominating the Hydrogen Balmer emission-line luminosities. For instance, in Shen et al. (2011), the local quasars show broad-line luminosities that are ∼20 times the narrow-line luminosities for Hα and Hβ. Much of the broad line emission is powered by photoionization by the central source. However, local quasar hosts are relatively quiescent with low-levels of ambient star-formation compared to our high-z galaxy sample (Zakamska et al. 2016). Our measurement of the enhanced [OIII] emission is surprising, implying a significant component arising from the NLR and from star-formation (Vietri et al. 2020). In $z \sim 2$ quasars of comparable bolometric luminosities as the sources studied here, the NLR [OIII] line luminosities are $\sim 10^{43}$ erg s^{-1} (Kakkad et al. 2020). The narrow component of [OIII] in those systems account for about 35% of the total [OIII] emission. In contrast, the median [OIII] luminosity of our sample is $10^{45.3}$ erg s^{-1}. Since the EWs of the Balmer lines are consistent between low- and high-z quasars, the implication is that the high [OIII] luminosity of our sample arises from an additional component, likely star-formation, rather that the quasar-powered ionized outflows noted by Kakkad et al. (2020). Spatially resolved measurements of the kinematics of [OIII] emission are required to confirm this explanation although the compactness of the stellar population at high-z (<1 kpc) will make it challenging to resolve stellar emission against the brightness of the quasar.

The large [OIII] EW and luminosity of $z \sim 6$ quasar hosts compared to $z \sim 2$, is similar to what has been inferred for star-forming galaxies (Endsley et al. 2021) who present evidence that at redshifts ~ 7, 50% of star-forming galaxies have rest-frame [OIII] EWs greater than 500Å. Since it is challenging to obtain such high [OIII]/Hβ ratios purely from photoionization models (Baskin & Laor 2005), strong star-formation-driven outflows powering radiative shocks are required to explain the emission. Future, mid-infrared spectroscopy with JWST, will be crucial to measure the kinematics of these lines to distinguish the contribution of the BLR, and NLR and to accurately constrain the metallicity, and magnetic field strength.

References

Allen, M. G., Groves, B. A., Dopita, M. A., Sutherland, R. S., Kewley, L. J. 2008, ApJs, 178, 20

Baskin, A. and Laor, A. 2005, MNRAS, 358, 1043

Chary, R.-R., Stern, D., & Eisenhardt, P. 2005, ApJL, 635, L5. doi:10.1086/499205

Endsley, R., Stark, D. P., Chevallard, J., et al. 2021, MNRAS, 500, 5229. doi:10.1093/mnras/staa3370

Faisst, A. L., Capak, P., Hsieh, B. C., et al. 2016, ApJ, 821, 122. doi:10.3847/0004-637X/821/2/122

Jun, H. D., Im, M., Lee, H. M., Ohyama, Y. et al, 2015, ApJ, 806, 109

Kakkad, D., Mainieri, V., Vietri, G., et al. 2020, *A&A*, 642, A147. doi:10.1051/0004-6361/202038551

Lee, B., & Chary, R., 2022, arXiv:2207.07290

Leipski, C., Meisenheimer, K., Walter, F., et al. 2014, ApJ, 785, 154. doi:10.1088/0004-637X/785/2/154

Ross, N. P., Cross, N. J. G. 2020, MNRAS, 494, 789

Shen, Y., Richards, G. T., Strauss, M. A., et al. 2011, ApJs 194, 45. doi:10.1088/0067-0049/194/2/45

Vietri, G., Mainieri, V., Kakkad, D., et al. 2020, *A&A*, 644, A175. doi:10.1051/0004-6361/202039136

Zakamska, N. L., Lampayan, K., Petric, A., et al. 2016, MNRAS, 455, 4191. doi:10.1093/mnras/stv2571

Resolving the Rise and Fall of Star Formation in Galaxies
Proceedings IAU Symposium No. 373, 2023
T. Wong & W.-T. Kim, eds.
doi:10.1017/S1743921322004409

Star-formation Property of High Redshift Galaxies in Clusters: Perceptive View from Observation and Simulation

Seong-Kook Lee[1,2], Myungshin Im[1,2], Eunhee Ko[1,2], Changbom Park[3], Juhan Kim[3], Jaehyun Lee[3] and Minhee Hyun[4]

[1]SNU Astronomy Research Center, Seoul National University, 1 Gwanak-ro, Gwanak-gu, Seoul 08826, Republic of Korea

[2]Astronomy Program, Department of Physics & Astronomy, Seoul National University, 1 Gwanak-ro, Gwanak-gu, Seoul 08826, Republic of Korea

[3]Korea Institute for Advanced Study, 85 Hoegi-ro, Dongdaemun-gu, Seoul 02455, Republic of Korea

[4]Korea Astronomy and Space Science Institute, Daejeon 34055, Republic of Korea

Abstract. The evolution of star formation properties of galaxies depends on the environment where galaxies reside, and generally star formation of galaxies in dense environment decreases more quickly. Interestingly, the star formation property of high-redshift galaxies clusters vary largely even though they are at similar redshift. We have found that the large-scale environment surrounding each galaxy cluster can contribute to make this cluster-by-cluster variation. This correlation is found in the results from observational data as well as in the simulations of galaxy formation. We suggest the 'Web-feeding model' to explain this trend. Star-forming galaxies falling into the galaxy cluster from surrounding large-scale structure make the quiescent galaxy fraction of the cluster lower than relatively isolated clusters.

Keywords. star formation, galaxy cluster, high redshift, large scale structure

1. Introduction

Galaxies are building blocks of the Universe, and understanding the evolution of galaxies is an interesting and important subject in modern astronomy. Among various properties of galaxies, the evolution of star formation activity of galaxies is one of the matters of substance, because it is the main channel of galaxy growth (modulo merging).

For this reason, there have been many observational efforts to understand the star formation properties of galaxies. While there are many issues which are still in controversy, one major uncontroversial finding is that the Universe experienced a decrease of an order of magnitude in the star formation rate density within it from $z \sim 2$ to now (e.g., Madau & Dickinson 2014). Also, many observational results have shown that average star formation rate (SFR) of $z \sim$ 1-2 galaxies is much higher than local.

Here, we briefly summarize our current understanding about the evolution of star formation activity of galaxies and present the results from our recent study on the environmental dependence of star formation properties of galaxies.

2. Star formation activity of high redshift galaxies

2.1. *Evolution and quenching of Star formation*

As stated above, the average SFR has been changed throughout the history of the Universe. Especially, it has decreased significant during the last 10 billion year from $z \sim 2$ to $z \sim 0$. There have been many mechanism or causes which can cut down or stop the star formation activity of galaxies. These can be categorized into two: (1) internal or nature or mass-quenching and (2) external or nurture or environmental quenching.

The suggested candidates for internal quenching includes shock heating of gas, stellar and AGN feedback and morphological transformation (such as bulge formation). These effects show somehow the dependence on stellar mass of galaxies, which is the reason this is called as 'mass-quenching'.

Environmental quenching is associated with processes occur mainly in dense environment such as galaxy clusters. These include ram pressure stripping (Gunn & Gott 1972) and strangulation (or starvation; Larson *et al.* 1980).

2.2. *Environmental Effects: Evolution of Environmental Dependence*

While average SFRs galaxies have been known to decrease since $z \sim 2$, this decrease is more significant in dense environment (e.g., Scoville *et al.* 2013). The evolution of relative importance of 'mass-quenching' and 'environmental quenching' has been studied by many authors (e.g., Peng *et al.* 2010; Lee *et al.* 2015).

This is the subject for which many researches are still being done, but most results from these studies suggest the relative importance of environmental quenching increases with decreasing redshift with crossing redshift occurs at $z \sim 1.4$-1.5, and environmental effects are manifested more significantly for low mass galaxies (e.g., Lee *et al.* 2015).

3. Galaxy Clusters in Different Large-scale Enivronment

One of interesting findings in Lee *et al.* (2015) was that the star formation property (measured by the quiescent galaxy fraction) shows a large variation among clusters residing at similar redshfit. The reason for this cluster-by-cluster variation among high redshift galaxy clusters was investigated in our subsequent works as summarized below.

3.1. *QF-Environment correlation in UDS and COSMOS*

We studied properties of galaxy clusters at $z \sim 0.65$-0.8 in the UDS, using the data from the multi-object spectroscopy (MOS) obtained by the Inamori-Magellan Areal Camera Spectrograph (IMACS) on the Magellan telescope (Lee *et al.* 2019). In this study, we found that the quiescent galaxy fraction of galaxy clusters shows a dependence on the large-scale environment where each galaxy cluster resides, in a sense that galaxy clusters located in more dense large-scale environment have on average lower quiescent galaxy fraction.

We parametrized the large-scale environment of galaxy clusters using 'FoF fraction', which is defined as the ratio of dense surface area within the 10 Mpc radius vicinity of each galaxy cluster. The galaxy clusters in the UDS show a anti-correlation between the quiescent galaxy fraction and the FoF fraction: galaxy clusters with large FoF fraction have low quiescent galaxy fraction. We have also found a similar trend between the quiescent galaxy fraction and cluster's large scale environment in the study done in the COSMOS field (Ko, E. et al. in preparation).

3.2. *Web Feeding Model*

To explain this correlation, we suggested the 'Web-feeding model' (Lee *et al.* 2019). The assumption of the web-feeding model is that new material, such as galaxies and gas, keeps infalling into the central galaxy clusters from surrounding overdense structures. This infalling material makes the quiescent galaxy fractions to be maintained low, comapred to relatively isolated clusters.

These aussmptions have been checked using Illustris TNG300 simulation and found to be true: (1) if there are filamentary structures connected to the galaxy clusters, galaxies are falling into clusters from the connected filametary structure, and (2) these infalling galaxies have lower level of quiescent galaxy fraction than cluster galaxies (Ko, E. et al. in preparation).

3.3. *Web Feeding and QF-FoF correlation in Galaxy Formation Models*

The correlation between the quiescent galaxy fraction of galaxies and the large-scale environment (e.g. FoF fraction) is also found in the galaxy formation models. For this, we analysed about 300 groups and clusters at $z \sim 0.9$ from the Horizon Run 5 (HR5) galaxy formation simulation (Lee *et al.* 2021).

These clusters from the HR5 simulation also shows a large range of quiescent galaxy fraction, similarly with the galaxy clusters found in the UDS and COSMOS. From the analysis of these mock groups and clusters, we have found that there are several factors which can affect the quiescent galaxy fraction of these groups and clusters. Among these factors, the large-scale environment of clusters, parametrized as FoF fraction, is found to affect and has correlation with the quiescent galaxy fraction of galaxy clusters, assuring our observational finding (Lee, S.-K. et al. in preparation).

4. Conclusion

In this proceeding, we present a brief summary of our interesting findings about the star formation properties of high redshift galaxy clusters. These findings are done utilizing observational survey data, follow-up spectroscopic data, as well as the up-to-date simulations of galaxy formation.

Our findings can be summarized as follows.

(1) The quiescent galaxy fraction of high redshift galaxy clusters increases with decreasing redshift.
(2) There is a cluster-by-cluster in the quiescent galaxy fraction among galaxy clusters even though they reside at similar redshift.
(3) The quiescent galaxy fraction shows a dependence on the large-scale environment surrounding each galaxy cluster.
(4) To explain this correlation, we suggest the 'Web-feeding model', in which infalling fresh star-forming galaxies from surrounding environment can keep the quiescent galaxy fraction of galaxy cluster to be lower compare to isolated clusters.

Acknowledgement

S.L. acknowledges support from a National Research Foundation of Korea (NRF) grant (2020R1I1A1A01060310) funded by the Korean government (MIST).

References

Gunn, J. E. & Gott, J. R. 1972, *ApJ*, 176, 1
Larson, R. B., Tinsley, B. M. & Caldwell, C. N. 1980, *ApJ*, 237, 692
Lee, J., Shin, J., Snaith, O. N., *et al.* 2021, *ApJ*, 908, 11

Lee, S.-K., Im, M., Kim, J.-W., *et al.* 2015, *ApJ*, 810, 90
Lee, S.-K., Im, M., Hyun, M., *et al.* 2019, *MNRAS*, 490, 135
Madau, P. & Dickinson, M. 2014, *ARA&A*, 52, 415
Peng, Y.-J., Lilly, S. J., Kovac, K., *et al.* 2010, *ApJ*, 721, 193
Scoville, N., Arnouts, S., Aussel, H., *et al.* 2013, *ApJS*, 206, 3

Resolving the Rise and Fall of Star Formation in Galaxies
Proceedings IAU Symposium No. 373, 2023
T. Wong & W.-T. Kim, eds.
doi:10.1017/S1743921322004951

From evolved stars to the formation and evolution of galaxies

Hamidreza Mahani[1], Atefeh Javadi[1], Jacco Th. van Loon[2], Habib Khosroshahi[1], Elham Saremi[3,4], Roya Hamedani Golshan[5], Mahdieh Navabi[1], Seyed Azim Hashemi[6], Mahtab Gholami[1] and Sima Taefi Aghdam[1]

[1]School of Astronomy, Institute for Research in Fundamental Sciences (IPM), Tehran, 19568-36613, Iran
email: mahani@ipm.ir

[2]Lennard-Jones Laboratories, Keele University, ST5 5BG, UK

[3]Instituto de Astrof'ısica de Canarias, C/ V'ıa L'actea s/n, 38205 La Laguna, Tenerife, Spain

[4]Departamento de Astrof'?sica, Universidad de La Laguna, 38205 La Laguna, Tenerife, Spain

[5]Physikalisches Institut der Universität zu Köln, Zülpicher Str. 77, D-50937 Köln, Germany

[6]Department of Physics and Astronomy, University of California Riverside, CA 92521, The USA

Abstract. Due to observational constraints, our detailed knowledge of stellar populations, formation, and evolution of galaxies is limited to a few dozen galaxies located in the Local Group. The Local Group of galaxies offers a unique opportunity to construct the formation histories and probe the structure and dynamics of many dwarf galaxies surrounding the Milky Way and Andromeda and of isolated dwarf galaxies. In this regard, we monitored the majority of galaxies in the Local Group, including the M33 galaxy and satellites galaxies surrounding the Milky Way and Andromeda galaxy, as well as isolated dwarf galaxies. We identified stellar populations and based on light curve analysis, the cool evolved stars pulsating in the fundamental mode were identified. In this paper, first, we will present the results we obtained for SFH and dust production rate in individual galaxies separately to answer how different types of galaxies have been formed and evolved over cosmic time. Then, we will discuss whether the mass return from dusty evolved stars can provide enough gas reservoirs to sustain the star formation or even rejuvenate the dwarf galaxy, as some seem to harbor relatively young stars.

Keywords. stars: AGB and post-AGB – stars: carbon – stars: mass-loss – supergiants – galaxies: evolution

1. Introduction

The Local Group members provide us with a unique opportunity to study and learn more about the structure and history of galaxies. One method to calculate the star formation rate (SFR) and star formation history (SFH) of these composite stellar populations is to use stars that have left the main sequence (Javadi et al. 2011). Highly evolved stars with low to intermediate initial masses that have moved through the helium-core burning phase are known as asymptotic giant branch (AGB) stars.

Finding variable AGB stars in the last stages of their evolution is an effective mechanism for retracing the history of star formation in a galaxy. Since they have undergone more evolution than less evolved AGB stars, their luminosity is more closely tied to their birth mass. The term long-period variables (LPVs) refers to variable stars that typically exhibit periodic variations in brightness with periods of a month to a few thousand days and amplitudes ranging from a few tenths to around ten magnitudes in the visual bands (Whitelock, Feast, & Catchpole 1991; Wood 2000). Another application of variable stars is studying the dust they spew into the interstellar medium (ISM). When investigating the structure and evolution of galaxies and the entire universe, ISM and star formation are inextricably linked and follow one another (van Loon, Marshall, & Zijlstra 2005). The ISM is further energized and accelerated by the stellar winds of evolved stars and the violent deaths of the most massive stars, which cause turbulence and galactic fountains as superbubbles explode as they approach the 'surface' of the galactic disc.

We studied a variety of galaxies, including dwarf spheroidals and irregulars, dwarf transition galaxies, and globular clusters, using the UK Infrared and Isaac Newton Telescope monitoring survey of Local Group members (Javadi et al. 2011; Parto et al. 2021). The primary goals of this study are as follows: locating every LPV in the LG dwarf galaxies that can be seen in the northern hemisphere; producing the SFHs from the present-day luminosity function of the LPVs; Modeling their SEDs and examining the relationship between stellar characteristics including mass, luminosity, metallicity, and pulsation amplitude and mass-loss (Saremi et al. 2020).

2. Methodology

In order to calculate the SFH by employing LPVs, we take use of the fact that LPVs are at the very last phases of evolution and that their brightness can be converted into their birth mass by using theoretical evolutionary tracks or by isochrones (Marigo et al. 2008; Javadi et al. 2011). The SFR behaves as a function of elapsed time to describe the SFH. The amount of gas transformed into stars per year is measured by this function. Javadi et al. 2017 showed that SFR, $\xi(t)$, could be derived from the following equation.

$$\xi(t) = \frac{dn'(t)}{\delta t} \frac{\int_{m_{min}}^{m_{max}} f_{imf}(m) m dm}{\int_{m(t)}^{m(t+dt)} f_{imf}(m) dm} \tag{2.1}$$

In equation 2.1, f_{imf} represents an initial mass function (e.g., canonical IMF proposed by Kroupa 2001), δt is the pulsation duration, and $dn'(t)$ is the number of LPVs observed in the age interval.

Investigating the mass-loss, we used `DUSTY`, the publicly available code (Ivezic & Elitzur 1997), to model the SEDs of all variable stars having observations in at least two near-IR and two mid-IR (Spitzer Space Telescope) bands. Simulations demonstrated that the mass-loss of the stars could be determined using the existing correlations between the color of the stars and their optical depth (Javadi et al. 2013).

3. Conclusion

We have utilized the novel method of Javadi et al. 2011 to determine the SFH and dust production in nearby galaxies and local group objects using long-period variable AGB stars. Following, we refer to the data acquired for a few of these galaxies.

• **M33:** Javadi et al. 2011 discovered that the disc of M33 was formed around 6 Gyr ago when more than 80% of this galaxy's previous star formation occurred. Since then, shorter epochs of star formation may have taken place; the most recent significant instance of such an occurrence was 250 Myr ago and contributed less than 6% to the star formation of the galaxy. They developed a two-dimensional map of the mass-return rate, which reveals a radial decrease in the rate and local increases caused by clusters of massive stars. They predict a total mass-loss rate of $0.004 - 0.005\ \mathrm{M}_{\odot} yr^{-1} kpc^{-2}$, which rises

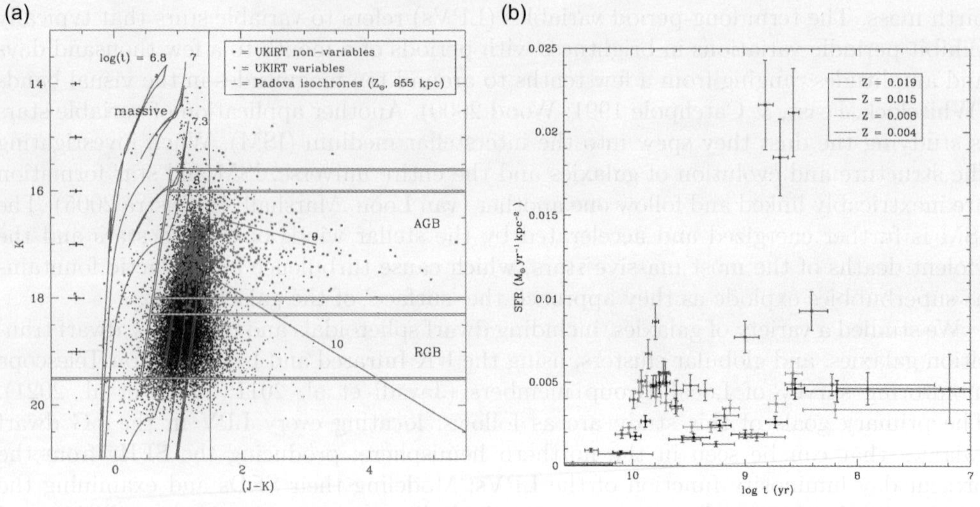

Colour–magnitude diagram of UKIRT catalogue

SFH compared for four choices of metallicity

Figure 1. As seen in subplot (a), the isochrones designated by their logarithmic ages, appear as significant trends towards red colors; carbonaceous dust has a distinct reddening slope at log t = 9 than oxygeneous dust does at log t = 10. Subplot (b) shows the SFH in M33's center that was generated by a near-IR AGB star and red-supergiant variable. Plots are adopted from Javadi et al. (2011).

to 0.006 $M_\odot yr^{-1} kpc^{-2}$ when eruptive mass-loss (such as supernovae) is taken into consideration. Comparing this to the present rate of star formation, they conclude that the center area of M33 can only support star formation if gas is accreted from circumgalactic regions or further out in the disc.

• **NGC 147 and NGC 185:** Regarding two of the most massive satellites of the Andromeda galaxy, Hamedani Golshan et al. 2017 showed that star formation began sooner in NGC 185 than in NGC 147, reaching its peak in NGC 185, 8.3 Gyr, and in NGC 147, 7 Gyr ago. In NGC 185, star formation persisted for the previous 6 Gyr, albeit at a much slower rate, in the center until 200 Myr ago, whereas it stopped much earlier in the edge. However, While star formation was prominent in NGC 147 between 3-6 Gyr ago, no star formation has been observed for the last 300 Myr.

• **And I:** Saremi et al. 2021 indicated that, in And I, a significant phase of star formation peaked around 6.6 Gyr ago, reached about 0.0035 $M_\odot yr^{-1}$, and only gradually declined until 1-2 Gyr ago. This galaxy's presence of a few dusty LPVs is consistent with a slight rise in recent star formation that peaked around 800 Myr ago. They estimated that And I is a late-quenching dSph and that the quenching period was around 4 Gyr ago.

• **And VII:** Through multi-epoch imaging with the Isaac Newton Telescope in the i and V bands, Navabi et al. 2021 discovered 55 LPV candidates within two half-light radii. The star formation rate of And VII was calculated as a function of cosmic time based on their birth mass function. They showed that star formation began in its primary phase 6.2 Gyr ago, with an SFR of $0.006 \pm 0.002\, M_\odot yr^{-1}$. They discovered slow star formation throughout the previous 6 Gyr, $0.0005 \pm 0.0002\, M_\odot yr^{-1}$, which persisted until 500 Myr ago.

• **IC 1613:** In IC 1613, Hashemi, Javadi, & van Loon 2018 noticed 53 highly evolved stars, eight of which are supergiants. Using stellar evolution models, they determined these stars' age and birth mass to recreate the SFH and reported that during the previous

Gyr, the average star formation rate was $\sim 0.0003\,\mathrm{M}_{\odot} yr^{-1} kpc^{-2}$. They demonstrated that the absence of a dominant era of star formation over the past 5 Gyr shows that IC 1613 has evolved in isolation for that prolonged, uninfluenced by other galaxies in the Local Group.

• **IC 10:** In order to calculate the luminosity function and rebuild the SFH for IC 10, Gholami et al. 2022 first discovered 430 LPVs within two half-light radii of the galaxy's center. After correcting for star extinction, they calculated the galaxy's current star formation rate, $\xi(0) = 0.32\,\mathrm{M}_{\odot} yr^{-1}$, for a constant metallicity Z = 0.0008, indicating that the galaxy is actively undergoing star formation.

• **NGC 5128:** According to Taefi Aghdam et al. 2022, NGC 5128 has three significant star formation episodes that occurred at 800 Myr, 3.2 Gyr, and 10 Gyr in the past. The star formation rate at about 800 Myr ago is consistent with earlier research suggesting the galaxy underwent a merger at that time. In addition, the galaxy has recently had a lower star formation rate, which may have been caused by jet-induction star formation, multiple AGN outbursts, a minor merger 400 Myr ago, and multiple AGN outbursts in this galaxy.

References

Gholami M., Javadi A., van Loon J. T., Khosroshahi H., Saremi E., 2022, Manuscript in preparation.

Hamedani Golshan R., Javadi A., van Loon J. T., Khosroshahi H., Saremi E., 2017, MNRAS, 466, 1764.

Hashemi S. A., Javadi A., van Loon J. T., 2019, MNRAS, 483, 4751.

Ivezic Z., Elitzur M., 1997, MNRAS, 287, 799.

Javadi A., van Loon J. T., Mirtorabi M. T., 2011, MNRAS, 411, 263.

Javadi A., van Loon J. T., Khosroshahi H., Mirtorabi M. T., 2013, MNRAS, 432, 2824.

Javadi A., van Loon J. T., Khosroshahi H. G., Tabatabaei F., Hamedani Golshan R., Rashidi M., 2017, MNRAS, 464, 2103.

Kroupa P., 2001, MNRAS, 322, 231.

Marigo P., Girardi L., Bressan A., Groenewegen M. A. T., Silva L., Granato G. L., 2008, A&A, 482, 883.

Navabi M., Saremi E., Javadi A., Noori M., van Loon J. T., Khosroshahi H. G., McDonald I., et al., 2021, ApJ, 910, 127.

Parto T., Dehghani S., Javadi A., Saremi E., van Loon J. T., Khosroshahi H., Taghi Mirtorabi M., et al., 2021, ApJ submitted.

Saremi E., Javadi A., Th. van Loon J., Khosroshahi H., Molaeinezhad A., McDonald I., Raouf M., et al., 2020, ApJ, 894, 135.

Saremi E., Javadi A., Navabi M., van Loon J. T., Khosroshahi H. G., Bojnordi Arbab B., McDonald I., 2021, ApJ, 923, 164.

Taefi Aghdam S., Javadi A., Hashemi S., van Loon J. T., Khosroshahi H., Hamedani Golshan R., Saremi E., Saberi M., 2022, Manuscript in preparation.

van Loon J. T., Marshall J. R., Zijlstra A. A., 2005, A&A, 442, 597.

Whitelock P., Feast M., Catchpole R., 1991, MNRAS, 248, 276.

Wilson C. D., Scoville N., Rice W., 1991, AJ, 101, 1293.

Wood P. R., 2000, PASA, 17, 18.

Resolving the Rise and Fall of Star Formation in Galaxies
Proceedings IAU Symposium No. 373, 2023
T. Wong & W.-T. Kim, eds.
doi:10.1017/S1743921323000200

The JWST/NIRSpec GTO programme "The Physics of Galaxy Assembly: IFS observations of high-z galaxies"

Michele Perna, on behalf of the NIRSpec Galaxy Assemby IFS GTO team

Centro de Astrobiología, (CAB, CSIC–INTA), Departamento de Astrofísica, Cra. de Ajalvir Km. 4, 28850 – Torrejón de Ardoz, Madrid, Spain
email: mperna@cab.inta-csic.es

Abstract. We present an overview of the project "The Physics of Galaxy Assembly: IFS observations of high-z galaxies", a Guaranteed Time Observations (GTO) programme of the James Webb Space Telescope (JWST). It an ambitious project aimed at investigating the internal structure of distant galaxies with the NIRSpec integral field spectrograph (IFS), having allocated 273 hours of JWST prime time. The NIRSpec capability will provide us with spatially resolved spectroscopy in the 1-5 μm range of a sample of over forty galaxies and Active Galactic Nuclei in the redshift range $3 < z < 9$. IFS observations of individual galaxies will enable us to investigate in detail the most important physical processes driving galaxy evolution across the cosmic epoch. More in detail, the main specific objectives are: to trace the distribution of star formation, to map the resolved properties of the stellar populations, to trace the gas kinematics (i.e. velocity fields, velocity dispersion) and, hence, determine dynamical masses and also identify non-virial motions (outflow and inflows), and to map metallicity gradients and dust attenuation.

Keywords. James Webb Space Telescope – galaxies: formation – galaxies: evolution – galaxies: high-redshift – galaxies: star formation

---

1. Introduction

The James Webb Space Telescope (JWST) promises to reveal a new view of galaxy formation in the early Universe. Thanks to its unprecedented sensitivity and spectroscopic capability in the near/mid-infrared (NIR/MIR) wavelengths, for the very first time, the rest-frame optical nebular emission lines (e.g., Hβ, [O III]$\lambda\lambda$4959,5007, Hα and [N II]$\lambda\lambda$6549,6585) of star-forming galaxies (SFGs) and active Galactic Nuclei (AGN) can be directly detected and resolved across the cosmic epoch, from the cosmic noon ($z \sim 2-3$; e.g. Förster Schreiber & Wuyts 2020) to the Epoch of Reionization (EoR, $z \lesssim 15$; e.g. Robertson 2022). Early Release Observations (EROs; Pontoppidan et al. 2022) results have clearly demonstrated the power of JWST's spectroscopic observations (e.g. Brinchmann 2022; Curti et al. 2022; Tacchella et al. 2022), promising many exciting discoveries to be made over the coming years.

The focus of this paper is to present an overview of the project "The Physics of Galaxy Assembly: IFS observations of high-z galaxies" (Leads: Santiago Arribas and Roberto Maiolino), a Guaranteed Time Observations (GTO) programme for the JWST NIRSpec instrument (Jakobsen et al. 2022; Ferruit et al. 2022). We will describe the sample, the observational design, and the general goals of the project.

2. Programme overview

The project "The Physics of Galaxy Assembly: IFS observations of high-z galaxies" (GA-IFS hereinafter) is based on the use of the NIRSpec's Integral Field Spectroscopy (IFS) mode (Böker et al. 2022). It allows to cover the wavelength range $0.6 - 5.3\mu$m at spectral resolutions of R$\approx$ 100, 1000, and 2700; more in detail, low resolution (R100) observations cover the full wavelength range, while for R1000 and R2700 multiple exposures with different bands are required to cover the entire range (see Table 3 in Böker et al. 2022). The field of view (FOV) of the IFS aperture is a contiguous $3.1'' \times 3.2''$ sky area, with a sampling of $0.1''$/spaxel and a spatial resolution even lower than $0.1''$ at shortest wavelengths (Böker et al. 2022).

The GA-IFS sample consists of 42 individual systems to be observed in Cycle 1 (230 hours of JWST time), plus a number of additional targets to be selected for Cycles 2 and 3 (43 hours). The sample contains a good representation of the most massive and extended SFGs in the redshift range $3 < z < 9$, selected to increase the leverage of IFS capabilities. For instance, at $z \sim 6$, the NIRSpec IFS FOV corresponds to a physical area of $\sim 20 \times 20$ kpc, hence large enough to cover the entire galaxy emission with a sub-kpc sampling. The GA-IFS sample also comprises AGN and quasars (QSOs) in the redshift range $3 < z < 7.5$, to test the IFS capabilities in detecting extended emission associated to their host galaxy component, and map virial and non-virial motions (e.g. kpc-scale outflows and inflows). A plethora of selection criteria have been used to identify these systems: for instance, most SFGs and AGN have been selected from the COSMOS and GOODS-S fields, on the basis of their available multi-wavelength information; the sample also contains Ly-α and [O III]88μm emitters, submillimeter-bright galaxies, and the brightest and most distant QSOs (up to z$\sim$ 7.5).

For the SFGs sub-samples, we request observations at low (R100) and high (R2700) spectral resolution, to better study the continuum and line emission, respectively. For AGN and QSOs, only R2700 observations are requested, as we are interested in studying only their emission lines properties (but see Sect. 3). The specific band for the high-resolution observations mainly depends on the redshift of the individual galaxies, and it is aimed at including in a single band the most important rest-frame emission lines (e.g. covering the Hβ– Hα range, redshifted in the NIR). The exposure time for each target is settled to map all prominent rest-frame optical lines, according to flux predictions obtained from different diagnostics (taking advantage of available multi-wavelength information). No extra background exposures are requested: for R100 observations, we expect to have a relatively large number of spaxels free from galaxy emission, useful to estimate the background level. High resolution observations are instead dedicated to the study of emission line features, less affected by background emission.

In Table 1 we summarise the general properties of the selected targets, as well as the observational design for each class of sources in our sample. For more details, we refer to the APT of each program of the GTO: PID 1215, 1216, 1217, 1218, 1219, 1220, 1222, 1262, 1263, 1264. The GTO project has allocated 273 hours of JWST prime time; most of the program will be executed during the first three JWST cycles (2022-2024).

3. Addressing the study of galaxy formation and assembly with NIRSpec IFS

GA-IFS high resolution (R2700) observations, with a spectral range encompassing the main rest-frame optical emission lines, allows a plethora of diagnostic tools to study the structural, chemical, and dynamical properties of the ionised gas with a sub-kpc sampling. More in detail, we will:

- trace the gas kinematics (i.e. velocity fields, velocity dispersion) and, hence, determine dynamical masses and also identify non-virial motions (outflow and inflows);

Table 1. Observations and targets.

#	Class and type of selection criteria	redshift range	Average exp. time and configuration(s) per target
14	SFGs: mainly optical and UV-selected; a few objects are IR-selected	$3 < z < 6.3$	$\sim 4-5$ hr (R2700) ~ 1 hr (R100)
9	SFGs at the EoR: Ly-α, [O III]88μm emitters	$6.5 < z < 9.1$	$\sim 3-5$ hr (R2700) ~ 1 hr (R100)
13	AGN: X-ray, optical and/or MIR selected	$3 < z < 5$	~ 1 hr (R2700)
6	type 1 AGN/QSOs	$6.3 < z < 7.5$	~ 3 hr (R2700, all targets) ~ 1 hr (R100, two targets*)

Notes: Column (1): Number of targets; (2): Class of targets and selection criteria; (3): redshift range; (4): average exposure time and spectral configuration per individual target.
*dedicated PSF star observations will be acquired, and used to model and subtract the QSO component and reveal the stellar emission.

- map metallicity gradients (e.g. using strong-line diagnostics) and dust attenuation (using the Balmer decrement);
- map the distribution of star formation traced by the Balmer emission.

Moreover, the R100 continuum emission (complemented with available R2700 emission line fluxes) will allow us a full characterisation of the SFGs in our sample, for instance tracing the distribution of dust, and mapping the resolved properties of the stellar populations (e.g. star formation history, stellar mass, age, and metallicity). R100 observations will be also obtained for two QSOs at $z > 6$: dedicated PSF star observations will be used to model and subtract the nuclear component and reveal the stellar emission.

To summarise, the GA-IFS sample has been chosen to maximize the novelty and the impact of the very first NIRSpec IFS outcomes; our project will also pave the way for more efficient observational strategies in future NIRSpec IFS programs.

References

Brinchmann, J. 2022, *MNRAS*, arXiv:2208.07467
Böker, T., Arribas, S., Lützgendorf, N., & Zeidler P. 2022, *A&A*, 661, 82
Curti, M., D'Eugenio, F., Carniani, S., & Wallace, I., E., B. 2022, *MNRAS*, arXiv:2207.12375
Ferruit, P., Jakobsen, P., Giardino, G., & Zeidler, P. 2022, *A&A*, 661, 81
Förster Schreiber, N. M. & Wuyts, Stijn, 2020 *ARA&A*, 58, 661
Jakobsen, P., Ferruit, P., Alves de Oliveira, C., & Zincke, C. 2022, *A&A*, 661, 80
Pontoppidan, K. M., Barrientes, J., Blome, C., & Nota, A. 2022, *ApJ*, 936, 14
Robertson, Brant E. 2022, *ARA&A*, 20, 121
Tacchella, S., Johnson, B. D., Robertson, B. E., & Witstok, J. 2022, *MNRAS*, arXiv:2208.03281

Resolving the Rise and Fall of Star Formation in Galaxies
Proceedings IAU Symposium No. 373, 2023
T. Wong & W.-T. Kim, eds.
doi:10.1017/S1743921322004793

New era of LSST data: Estimating the physical properties of main-sequence galaxies

G. Riccio

National Centre for Nuclear Research, Astrophysics division
ul. Pasteura 7, 02-093 Warszawa, Poland
email: gabriele.riccio@ncbj.gov.pl

Abstract. The main goal of the Vera C. Rubin observatory is to perform the 10 year *Legacy Survey of Space and Time* (LSST). This future state-of-art observatory will open the new window to study billions of galaxies from Local Universe as well as the high redshift objects. In this work we employ simulated LSST observations and uncertainties, based on the 50 385 real galaxies within the redshift range $0 < z < 2.5$ from the ELAIS-N1 and COSMOS fields of the *Herschel Extragalactic Legacy Project* (HELP) survey, to constrain the physical properties of normal star-forming galaxies, such as their star formation rate (SFR), stellar mass (Mstar), and dust luminosity (Ldust). We fit their spectral energy distributions (SEDs) using the Code Investigating GALaxy Emission (CIGALE). The stellar masses estimated based on the LSST measurements agree with the full UV to far-IR SED, while we obtain a clear overestimate of the dust-related properties (SFR, Ldust) estimated with LSST. We investigate the cause of this result and find that it is necessary to employ auxiliary rest-frame mid-IR observations, simulated UV observations, or the far-UV attenuation (AFUV)-Mstar relation to correct for the overestimate.

Keywords. SED, LSST, Extragalactic Astrophysics, SFR

1. Introduction

The total spectral energy distribution (SED) of galaxies is the result of a complex interplay of several components, such as old and young stars, stellar remnants, the interstellar medium, dust, and super- massive black holes (Walcher 2011, Conroy 2013). For this reason, the study of the multi-wavelength emission of galaxies from X-rays to radio was found to be necessary to properly analyse the physical properties of galaxies, as only the panchromatic view can give the full information about their physical properties. For example, in the literature (i.e. Kennicutt 1998, Schreiber 2015) it has been shown that the ultraviolet (UV) to infrared (IR) SED contains important information about the star formation activity of galaxies. In fact, young stars emission is dominant in the UV band, while the dust, that surrounds the star forming regions, thermally absorb this light and re-emits it in the IR band. However, it is rare to have available the full SED and considering only restricted wavelength ranges can cause problem in the estimation of the physical parameters, due to degeneration issues.

The upcoming *Legacy Survey of Space and Time* (LSST, Ivezic 2019) from the Vera C. Rubin Observatory, will provide high-quality optical to near-IR images in the *ugrizy* bands of about 20 billion expected galaxies over ten years of observations. This large dataset will raise multi-fold questions, such as how we can use only LSST observations to obtain estimates of the main physical properties of galaxies, and how realistic and reliable they would be. We investigate these topics by performing a simulation of LSST

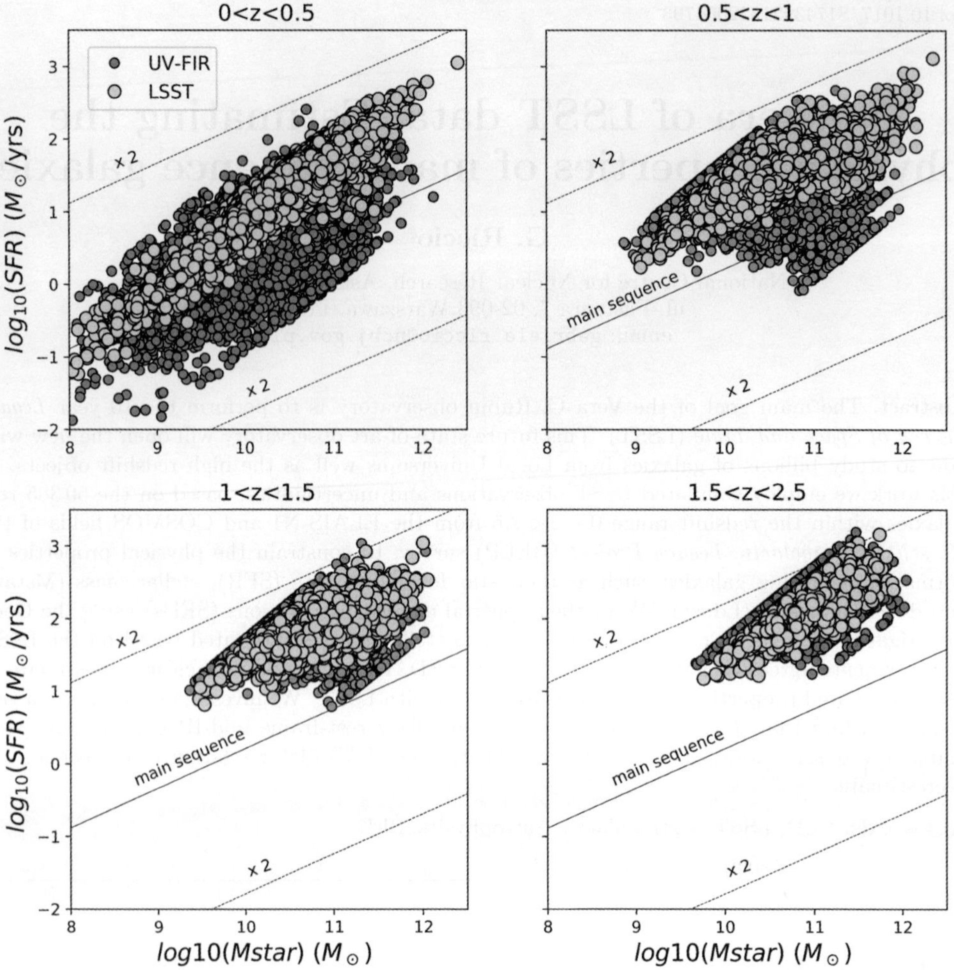

Figure 1. Main-sequence (SFR vs M_{star}) relation for the ELAIS-N1 and COSMOS fields in four redshift bins. In magenta we show the LSST-like sample, and in grey the real sample. The solid black line represents the MS by Speagle (2014), and the dashed lines mark the loci two times above and below the MS.

observations of main-sequence (hereafter MS) galaxies to estimate the main physical properties using an SED fitting method.

2. Data and methodology

The data adopted in this work comes from the *Herschel Extragalactic Legacy survey* (HELP, Shirley 2021), that provides extremely valuable multi-wavelength data over the HerMES (Oliver 2012) and the H-ATLAS survey fields (Eales 2010) and other relevant *Herschel* fields. We use two of the HELP fields: the European Large Area ISO Survey North 1, hereafter ELAIS-N1 (Oliver 2000), and the COSMOS field (Laigle 2016). In order to select MS galaxies, we remove all possible starburst and passive galaxies from the catalogue. We then simulate the LSST observations according to the technical description of the future survey (Ivezic 2019) and perform the SED fitting. The detailed description of the methodology, used parameters, and test can be found in Riccio (2021).

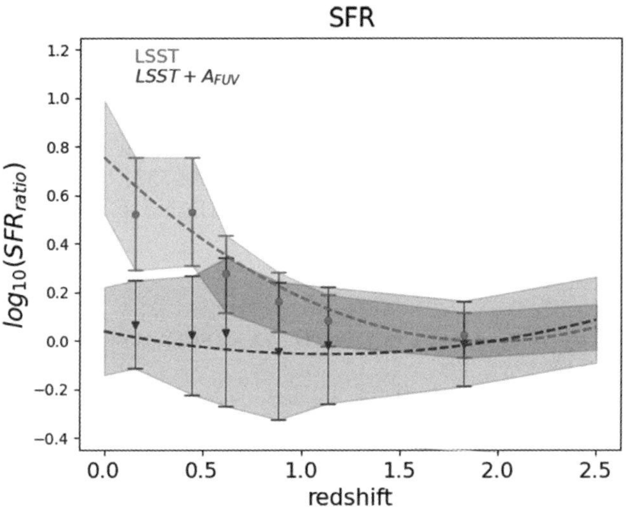

Figure 2. SFR ratio as a function of redshift for the ELAIS-N1 and COSMOS fields. Blue curve represents the result adding the A_{FUV} prior. The ratio is defined as $SFR_{ratio} = SFR_{LSST}/SFR_{UV-FIR}$, The points are the median values in each redshift bin, with the median absolute deviation as errors.

3. Results

We estimate the main galaxy physical properties (SFR, M_{star}, L_{dust}, M_{dust}) for the LSST sample and we compare the results with the full UV-FIR estimates. We obtain an overestimation of the dust related properties (SFR, L_{dust}, and M_{dust}). We found this overestimation strongly dependent on the redshift, due to the lack of information in the UV and MIR rest-frame wavelenghts. In Fig. 1 we show the MS relation obtained for the LSST-like sample with the relation obtained from the full UV-FIR SED fitting. At low redshift, the LSST estimation fails to probe low-SFR objects, and this leads to a clear division between the respective MS relations, which overlap at higher redshifts, however. The stellar masses are instead well estimated with LSST, leading to comparable results between the two runs, because their estimation mostly rely on optical data.

To provide a valid method of estimation of the physical properties using LSST observations with SED fitting we tried different methods to correct the observed overestimation. The first approach assumes usage of auxiliary data (e.g. MIR *Spitzer* bands, FIR *Herschel* Spire bands). By adding the rest-frame near infrared (IR) measurements the attenuation of the old stellar population would be better constrained, while the mid- and far- IR mainly constrain the dust emission from the star-forming regions and the SFR hidden by dust. Riccio (2021) showed that adding mid-IR or ultraviolet observations, along with LSST bands, the overestimation is completely corrected.

The second approach is to use a recently explored relation between attenuation in far-UV (A_{FUV}) and $log_{10}(M_{star})$ over a wide mass range ($9 \leq log_{10}(M_{star}) \leq 12$) (Buat 2009, Bogdanoska 2020). This method can be applied using LSST observations only, as the M_{star} employed to retrieve the A_{FUV} is well estimated by LSST. The procedure adopted is as follows: (1) from the LSST data we estimate the M_{star}, (2) from the real multi wavelength (UV-FIR) data we obtain the A_{FUV}-M_{star} relation, (3) we use the relation with the M_{star} estimated with LSST to retrieve the $A_{FUV\ LSST}$, (4) we use $A_{FUV\ LSST}$ as a prior of the new LSST CIGALE run to estimate the physical parameters. Fig. 2 shows the full correction of the estimated SFR employing this method. This result is particularly interesting because proves that knowing the A_{FUV}–M_{star} relation for a given sample of galaxies, it is possible to estimate SFR without the IR counterpart, making it a powerful tool to use with LSST.

Acknowledgements

GR acknowledges support from the Polish National Science Centre (UMO-2018/30/E/ST9/00082, and UMO-2018/30/M/ST9/00757). Authors are grateful for the support from Polish Ministry of Science and Higher Education through a grant DIR/WK/2018/12.

References

Walcher, J., Groves, B., Budavári, T., & Dale, D., *et al.* 2011, *Ap&SS*, 331, 1
Conroy, C., 2013, *ARA&A*, 51, 393
Kennicutt, J., Robert C., *et al.* 1998, *A&A*, 36, 189
Bogdanoska, J., Burgarella, D., *et al.* 2020, *MNRAS*, 496, 4, 5341
Buat, V., *et al.* 2009, *A&A*, 507, 2, 693
Eales, S., *et al.* 2010, *PASP*, 122, 891, 499
Ivezic, Z., *et al.* 2019, *ApJ*, 873, 2, 111
Laigle, C., *et al.* 2016, *ApJS*, 224, 2, 24
Oliver, S., *et al.* 2000, *MNRAS*, 316, 4, 749
Oliver, S. J., *et al.* 2012, *MNRAS*, 424, 3, 1614
Riccio, G., *et al.* 2021, *A&A* 653, A107
Schreiber, C., *et al.* 2021, *A&A* 575, A74
Shirley, R., *et al.* 2021, *MNRAS* 507, 1, 129
Speagle, J. S., Steinhardt, C. L., Capak, P. L., & Silverman, J. D., *ApJS*, 214, 15

Resolving the Rise and Fall of Star Formation in Galaxies
Proceedings IAU Symposium No. 373, 2023
T. Wong & W.-T. Kim, eds.
doi:10.1017/S1743921322004732

Break type and interactions from ultra-deep optical imaging of isolated galaxies

P. M. Sánchez-Alarcón[1,2†], J. Román[1,2,3,4], J. H. Knapen[1,2], L. Verdes-Montenegro[3], and S. Comerón[2,1]

[1]Instituto de Astrofsica de Canarias, E-38205, La Laguna, Tenerife, Spain

[2]Departamento de Astrofsica, Universidad de La Laguna, E-38206, La Laguna, Tenerife, Spain

[3]Instituto de Astrofsica de Andaluca (CSIC), Granada, Spain

[4]Kapteyn Astronomical Institute, University of Groningen, Groningen, the Netherlands

Abstract. In the standard cosmological model of galaxy evolution, mergers and interactions play a fundamental role in shaping galaxies. Galaxies that are currently isolated are thus interesting, allowing us to identify how internal or external processes impact galactic structure. However, current observational limits may be obscuring crucial information in the low-mass or low-brightness regime. We use the AMIGA catalog of isolated galaxies to explore the impact of different factors on the structure of these galaxies. In particular, we study the type of disk break based on the degree of isolation and the presence of interactions which are only detectable in the ultra-low surface brightness regime. We present the first results of an extensive observational campaign of ultra-deep optical imaging targeting a sample of 25 low-redshift ($z < 0.035$) isolated galaxies. The nominal surface brightness limits achieved are comparable to those to be obtained in the 10-year LSST coadds ($\mu_{r,\mathrm{lim}} \gtrsim 29.5\,\mathrm{mag\,arcsec^{-2}}$; 3σ; 10" × 10"). We find that isolated galaxies have a considerably higher fraction of purely exponential disk profiles and a lower presence of up-bending breaks than field or cluster galaxies. Our extreme imaging depth allows us to detect the presence of previously unreported interactions with minor companions in some of the galaxies in our sample (~40% of the galaxies show signs of interaction). The results of our work fit with the general framework of galactic structure in which up-bending breaks (Type III) would be produced by mergers and down-bending breaks (Type II) due to a threshold in star formation that would tend to become single exponential disk (Type I) in case of cessation or decrease of star formation.

Keywords. galaxies: evolution - galaxies: photometry - galaxies: spiral - galaxies: structure

---

1. Results and Implications

Our work is based on the Analysis of the interstellar Medium of Isolated Galaxies (AMIGA) sample (Verdes-Montenegro et al. 2005). The latest revision of the AMIGA catalog was carried out by Argudo-Fernández et al. (2013), a work on which we based our selection of the sample. We used the INT, VST, Jeanne-Rich Telescope, and HSC-SSP archival data (see Snchez-Alarcn et al. 2023, in prep.). Through careful processing of the data to preserve the low surface brightness features, the surface brightness limits achieved are of $\mu_{r,\mathrm{lim}} \gtrsim 29.5\,\mathrm{mag\,arcsec^{-2}}$, ($3\sigma$; 10" × 10") following the nominal depth description by Román et al. (2020). In Fig. 1, we show the galaxy CIG 340 comparing with SDSS and Legacy Survey data. While the morphology of CIG 340 appears similar in SDSS and the Legacy Survey, in our data we detect a clear tidal stream to the south of

†email: pmsa.astro@gmail.com

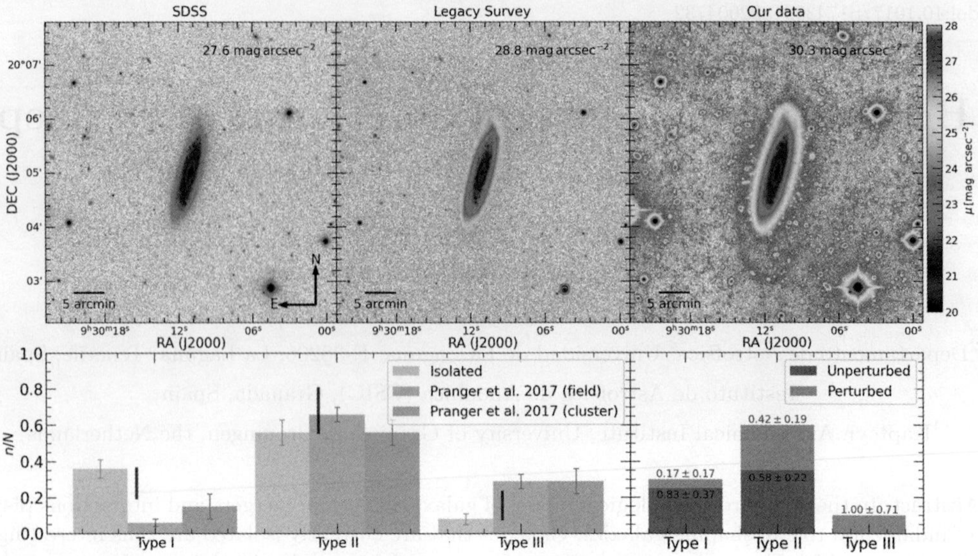

Figure 1. Upper panel: Comparison between the Sloan Digital Sky Survey (left), the Legacy Survey (middle) and our data for the CIG 340 (IC 2487) galaxy. The upper right value indicates the surface brightness limits (3σ, $10'' \times 10''$). Our work shows that the HI asymmetries detected in CIG 340 (Scott et al. 2014) are most likely due to an interaction with a satellite that we detected for the first time in the form of a tidal stream. Lower left panel: Normalized distribution on the frequency of disk break type for isolated galaxies (left, blue color), and field galaxies (middle-green) and cluster galaxies (right-red) according to work by Pranger et al. (2017). Lower right panel: Frequency of perturbed and unperturbed isolated galaxies for each disk break type.

CIG 340 and diffuse light appearing in the direction transversal to the disk. This shows the considerable leap in detection power from previously existing data, and the usefulness of our observations to reveal the presence of minor interactions.

We identify and classify the disk breaks found in the profiles for our sample of 25 isolated galaxies. We assign each galaxy a surface brightness profile Type: Type I $\equiv$ single exponential, Type II $\equiv$ down-bending, and Type III $\equiv$ up-bending. We also classify galaxies according to the presence of interactions, as perturbed and unperturbed. The strong presence of cirrus in galaxies does not allow us to assign a degree of perturbation. We compare the break type of our sample of isolated galaxies to galaxies in other environments, including in high-density (cluster) and low-density (field) environments from Pranger et al. (2017). In total, among our 25 galaxies, we identify 9 Type I with no significant break, 14 Type II, and 2 Type III, corresponding to frequencies of 36% 5%, 56% 7%, and 8% 3%. We estimate the uncertainties using a 95% confidence interval from the sample variance assuming Poisson statistics (as done in Pranger et al. 2017). We find a statistically significant difference between the type distribution of breaks in isolated galaxies as compared to field galaxies or in clusters reported in previous studies. In particular, we find that isolated galaxies have a considerably higher fraction of purely exponential disk, Type I, and a lower presence of Type III than field or cluster galaxies (see Fig. 1, bottom left). We find that 40% of the isolated galaxies in our sample show the presence of interactions, 30% show an asymmetric halo and 15% show some tidal current. Bottom right panel of Fig. 1 we show the fraction of perturbed and unperturbed galaxies for each type of break. We find that for Type I breaks, 6 ($83\% \pm 37\%$) galaxies are unperturbed and 1 ($17\% \pm 17\%$) is perturbed; for Type II, 8 ($58\% \pm 22\%$) are unperturbed and 5 ($42\% \pm 19\%$) perturbed. The two galaxies with Type III are perturbed ($100\% \pm 71\%$).

2. Conclusions

Type III profiles can be produced by mergers of galaxies (e.g., Pfeffer et al. 2022, and references therein). The low number of Type III in our sample of isolated galaxies would be in agreement with this statement. Additionally, the only two Type III found in our sample (CIG 33 and CIG 96) show disturbed morphology with a puffed-up external halo in their outer regions, compatible with a recent major merger. Their appearing isolated would therefore be the consequence of at least one pair of galaxies having merged to produce an isolated galaxy. Type II are widely agreed to be the result of disks truncated by the star formation threshold (e.g., Tang et al. 2020; Pfeffer et al. 2022), cessation or decrease in star formation, through stellar migration effects (Sánchez-Blázquez et al. 2009), would tend to homogenize the stellar populations and produce a Type I (e.g., Pfeffer et al. 2022). We do not have star formation measurements in our sample but an indirect hint that could indicate that this is the case is the considerably higher fraction of perturbed Type II (42%±19%) than Type I (17%±17%) galaxies. While the rate of star formation may depend on various circumstances such as the rate of pristine gas inflow, it is known that satellite interaction are capable of triggering star formation (e.g., Mesa et al. 2021), which would be in accordance with our findings. The high fraction of Type I can be explained by the low ratio of perturbations detected in these galaxies. Additionally, the specific star formation rate in isolated galaxies tends to be lower than in higher-density environments (Melnyk et al. 2015). These arguments would explain the high rate of Type I in isolated galaxies, as unperturbed galaxies, evolving slowly with a low star formation rate. The results of our work thus fit within the general framework of galactic structure.

Acknowledgements

We acknowledge support from the State Research Agency (AEI-MCINN) of the Spanish Ministry of Science and Innovation under the grants "The structure and evolution of galaxies and their central regions" with reference PID2019-105602GB-I00/10.13039/501100011033, and "Thick discs, relics of the infancy of galaxies" with reference PID2020-113213GA-I00, and from the ACIISI, Consejería de Economía, Conocimiento y Empleo del Gobierno de Canarias and the European Regional Development Fund (ERDF) under grant with reference PROID2021010044. LVM acknowledges financial support from the State Agency for Research of the Spanish Ministry of Science, Innovation and Universities through the "Center of Excellence Severo Ochoa" awarded to the IAA (SEV-2017-0709), grant RTI2018-096228-B-C31 and PID2021-123930OB-C21 (MCIU/AEI/FEDER,UE), and IAA4SKA (Ref. R18-RT-3082).

References

Argudo-Fernández, M., Verley, S., Bergond, G., et al. 2013, A&A, 560, A9.
Gutiérrez, L., Erwin, P., Aladro, R., et al. 2011, AJ, 142, 145.
Melnyk, O., Karachentseva, V., & Karachentsev, I. 2015, MNRAS, 451, 1482.
Mesa, V., Alonso, S., Coldwell, G., et al. 2021, MNRAS, 501, 1046.
Pfeffer, J. L., Bekki, K., Forbes, D. A., et al. 2022, MNRAS, 509, 261.
Pohlen, M.; Trujillo, I. 2006, A&A, 454, 759P
Pranger, F., Trujillo, I., Kelvin, L. S., et al. 2017, MNRAS, 467, 2127.
Román, J., Trujillo, I., & Montes, M. 2020, A&A, 644, A42.
Sánchez-Blázquez, P.; Courty, S.; Gibson, B. K.; Brook, C. B. 2009, MNRAS, 398, 591
Scott, T. C., Sengupta, C., Verdes Montenegro, L., et al. 2014, A&A, 567, A56.
Tang, Y., Chen, Q., Zhang, H.-X., et al. 2020, AJ, 897, 79.
Watkins, A. E., Laine, J., Comerón, S., et al. 2019, A&A, 625, A36.
Verdes-Montenegro, L., Sulentic, J., Lisenfeld, U., et al. 2005, A&A, 436, 443.

Resolving the Rise and Fall of Star Formation in Galaxies
Proceedings IAU Symposium No. 373, 2023
T. Wong & W.-T. Kim, eds.
doi:10.1017/S1743921322003751

Spatial variation of the star formation history in the disc of M31 galaxy using evolved stars

Maryam Torki[1], Mahdieh Navabi[2] and Atefeh Javadi[1]

[1]School of Astronomy, Institute for Research in Fundamental Sciences (IPM), Tehran, 19568-36613, Iran
email: maryamtorki84@gmail.com

[2]Physics Department, University of Surrey, Guildford, GU2 7XH, United Kingdon

Abstract. The rate of star formation (SFR) is one of the important quantities that helps to study galaxies' evolutionary path. In fact, measuring the SFR during the life of the Universe shows us how galaxies have acquired their metallicity and star mass. In this regard, the galaxies of the Local Group give us a great opportunity to study the connection between different stellar populations and galaxy evolution. In this paper, we use the Long-Period variable stars to estimate the radial star formation in the disc of the M31 galaxy. These stars are powerful instruments to achieve this goal. They reach their peak luminosity and coldest state at the final point of their evolution. Also, there is a directly related between their mass and luminosity, so using stellar evolution theoretical models, we construct the mass function and hence the star formation history (SFH). In the disc of M31, we see an increase in the rate of star formation and a decrease in the age of stars in the outer parts. These results predict the inside-out growth well.

Keywords. stars: evolution – stars: AGB and LPVs – stars: luminosity function, mass function – galaxies: stellar content – galaxies: structure – galaxies: evolution – galaxies: formation – galaxies: individual: M31

1. Introduction

In the Local Group, the M31 galaxy is the best target for detailed observations of its metallicity, star formation rate (SFR), and stellar population. As a result, this galaxy presents an excellent opportunity to study the detailed evolutionary history of a large spiral galaxy, both because it is the closest galaxy and because it has been undergoing star-forming activity recently. The radial gradient distribution of abundance in spiral galaxy disks shows the chemical evolution of those discs. These galaxies are formed from the inside-out, so metallicity and age are expected to have negative slopes. Galaxy growth inward-outward causes the metallicity gradient, while old stars at the center cause the age gradient. According to recent Hubble Space Telescope studies of M33 and NGC300, older populations are more concentrated in the center, and younger stars are found in larger radii (Williams et al. 2009; Gogarten *et al.* 2010).

2. Data

In this research, we took advantage of the catalogue of long period variable stars (LPVs) in the M31 from Mould *et al.* 2004 to estimate its star formation history (Torki et al. 2019, 2023). Infrared photometry of the M31 has been obtained by Kitt Peak National

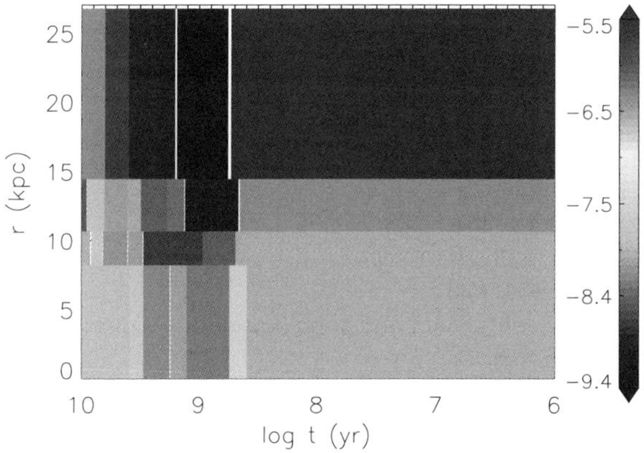

Figure 1. SFH at different radial distances from the centre of M31, in the galaxy (de-projected) plane. The colour bar values are on a logarithmic scale and represent the fraction of the total stellar mass as it formed each year. All radius bin contains the same stars.

Observatory (KPNO) and their results included photometry of almost 2000 variables, most of which were asymptotic giant branch stars (AGBs).

3. Method

Our method for reconstructing star formation history(SFH) is based on using LPVs which previously have successfully applied to other local group galaxies such as M33 (Javadi *et al.* 2011a,b; 2017), Magellanic Clouds (Rezaeikh et al. 2014), NGC147 and NGC185 (Hamedani Golshan *et al.* 2017), IC1613 (Hashemi et al. 2019), Andromeda VII (Navabi *et al.* 2020; 2021) and Andromeda I (Saremi *et al.* 2019; 2020; 2021). In this technique, we used Padova stellar evolutionary models (Marigo *et al.* 2017). In order to obtain SFH, we assumed metallicity was constant over time and obtained the stars' mass, age, and pulsation duration.

SFH is described by the SFR, which provides information about how much gas has become a star each year in the past. SFR, ξ (in $M_\odot$ yr^{-1}) is a function of time and estimated by the equation below (Javadi et al. 2011):

$$\xi(t) = \frac{dn'(t)}{\delta t} \frac{\int_{\min}^{\max} f_{\rm IMF}(m) m \, dm}{\int_{m(t)}^{m(t+dt)} f_{\rm IMF}(m) \, dm} \tag{3.1}$$

4. Results

The variation of SFH as a function of radial distance from the center of M31 is shown in Figure 1. All radius bins contain the same number of variable stars. As can be seen, the old epoch of star formation is more robust in the inner parts $0 < r < 9$ kpc of the disc, and as it moves towards the outer parts, it weakens and finally disappears at the outermost part. On the other hand, the recent star formation and the accumulation of young stars increase as it moves outward and this positive trend is seen across the entire disc for stars younger than log t (yr)=8.6. In addition, star formation in all parts has an almost identical rate at intermediate ages ($1 < \log t < 3$ Gyr ago). Our results are consistent with the result derived by (Sick et al. 2014). They used Bayesian modeling of the optical-infrared spectral energy distribution to estimate stellar mass profiles in the M31 galaxy. Their results also provide evidence of the inside-out disk formation.

References

Gogarten et al. 2010, APJ, 712, 854
Hamedani Golshan, R. et al. 2017, MNRAS, 466, 1764
Hashemi, S.A. et al. 2019, MNRAS, 483, 4751
Javadi, A. et al. 2011a, ASPC, 445, 497
Javadi, A. et al. 2011b, MNRAS, 414, 3394
Javadi, A. et al. 2017, MNRAS, 464, 2103
Marigo, P. et al., 2017, ApJ, 835, 19
Mould, J., et al. 2004, ApJ, 154, 623
Navabi, M. et al. 2020, Proc.conf. Stars and their Variability Observed from Space, 383, 385
Navabi, M. et al. 2021, ApJ, 910, 127
Rezaeikh, S. et al. 2014, MNRAS, 445, 2214
Saremi, E. et al. 2019, Proceedings of IAU Symposium, 339, 336
Saremi, E. et al. 2020, ApJ, 894, 135
Saremi, E. et al. 2021, ApJ, 923, 164
Sick et al. 2014, Proceedings of IAU Symposium, 311, 82
Torki, M. et al. 2019, Proceedings of IAU Symposium, 343, 512
Torki, M. et al. 2023, IAU Symposium, 362, 353
Williams, B. F. et al. 2009, APJ, 695, L15

Chapter 5:
Regulation of Star Formation and the Evolution of Galaxies

Resolving the Rise and Fall of Star Formation in Galaxies
Proceedings IAU Symposium No. 373, 2023
T. Wong & W.-T. Kim, eds.
doi:10.1017/S1743921323000133

Feedback models in galaxy simulations and probing their impact by cosmological hydrodynamic simulations

Kentaro Nagamine[1,2,3]

[1]Theoretical Astrophysics Group, Department of Earth and Space Science, Graduate School of Science, Osaka University
1-1 Machikaneyama-cho, Toyonaka, Osaka, 560-0043, Japan

[2]Department of Physics & Astronomy, University of Nevada, Las Vegas
4505 S. Maryland Pkwy, Las Vegas, NV 89154-4002, USA

[3]Kavli IPMU, The University of Tokyo, 5-1-5 Kashiwanoha, Kashiwa, Chiba, 277-8583, Japan
email: kn@astro-osaka.jp

Abstract. Feedback effects by supernovae (SNe) and active galactic nuclei (AGNs) are believed to be essential for galaxy evolution and shaping present-day galaxies, but their exact mechanisms on galactic scales and their impact on CGM/IGM are not well understood yet. In galaxy formation simulations, it is still challenging to resolve sub-parsec scales, and we need to implement subgrid models to account for the physics on small scales. In this article, we summarize some of the efforts to build more physically based feedback models, discuss about pushing the resolution to its limits in galaxy simulations, testing galaxy formation codes under the AGORA code comparison project, and how to probe the impact of feedback using cosmological hydrodynamic simulations via Lyα absorption and CGM/IGM tomography technique. We also discuss our future directions of research in this field and how we make progress by comparing our simulations with observations.

Keywords. cosmology: theory, galaxies: formation, galaxies: evolution, galaxies: ISM, galaxies: high-redshift, hydrodynamics

1. Introduction

It is widely accepted that feedback processes by SNe and AGNs regulate galaxy formation and evolution. The standard lore is that the AGN feedback suppresses star formation in massive galaxies at lower redshift, and the SN feedback in low-mass galaxies at higher redshift, hence causing the peak in the stellar-to-halo mass relation (Behroozi et al. 2013). In particular, many interesting physical processes take place in high-redshift galaxies as we summarize in a schematic diagram below (Fig. 1), and therefore they are good laboratories for testing feedback models. Low-metallicity gas falls into the dark matter halo via narrow, cold streams, providing abundant fuel for star formation (Kereš et al. 2005; Wright et al. 2021). At lower redshifts of $z \lesssim 2$, gas accretion switches to the hot accretion mode, and prominent cold flows are no longer observed in simulations (Faucher-Giguère et al. 2011; Nelson et al. 2016). The cosmic star formation rate density (SFRD) has a broad peak at $z \simeq 3-5$ (e.g., Nagamine et al. 2000; Nagamine et al. 2004, 2006; Kistler et al. 2009; Madau & Dickinson 2014), and the rising SFRD at high-redshift is driven by gravitational instability where dark matter halos and galaxies actively form (Schaye et al. 2010). Different emission lines such as Lyα, [C II], [O III] are expected

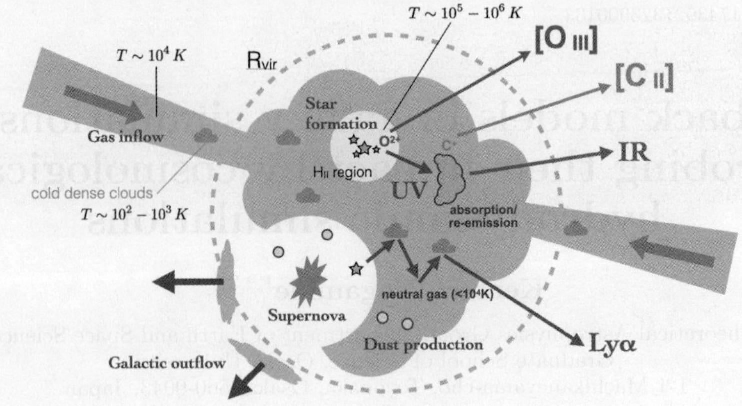

Figure 1. Schematic figure describing various physical processes in high-redshift galaxies. The gas flows into the dark matter halo via cold streams of $T \sim 10^4$ K, and later condenses into dense clouds with $T \sim 10^2 - 10^3$ K via radiative cooling. Once the stars form, the massive stars ionizes the ISM around them and produce H II regions. The UV radiation from massive stars impinges on the surface of ISM and produce photo-dissociation region (PDR), where infrared emission lines such as [C II] are emitted as a reprocessed radiation. These emission from high-z galaxies have been computed using cosmological zoom-in hydrodynamic simulations (e.g., Arata et al. 2019; Katz et al. 2022). Prominent Lyα emissions are also observed from high-z galaxies. The massive stars die after several Myrs, and the gas is ejected as galactic outflows.

from high-z galaxies, and many of those lines have recently been observed by ALMA (e.g., Smit et al. 2018; Hashimoto et al. 2018, 2019), suggesting very early onset of star formation at $z \sim 15$.

One of the important physical parameters to determine the radiative properties of high-z galaxies is the escape fraction $f_{\rm esc}$ of ionizing and ultra-violet (UV) photons. However it is difficult to estimate $f_{\rm esc}$ observationally and only a handful of rough measurements have been made. Therefore, it would be desirable to predict $f_{\rm esc}$ for different types of galaxies directly from hydrodynamic simulations of galaxy formation (e.g., Cen 2003; Razoumov & Sommer-Larsen 2006; Gnedin et al. 2008; Wise & Cen 2009; Yajima et al. 2017). It requires accurate computations of ISM structure and hence a high-resolution to predict $f_{\rm esc}$. Understanding the details of $f_{\rm esc}$ is also quite important for figuring out whether the reionization of the universe favor the 'Early' or 'Late' scenarios (e.g., Finkelstein et al. 2019; Naidu et al. 2020).

2. Feedback models in galaxy formation simulations

In the early cosmological hydrodynamic simulations, SN feedback was modeled with simple injection of thermal energy on large scales (> kpc) (Cen & Ostriker 1992; Katz 1992; Katz et al. 1996; Cen & Ostriker 1999). However, the injected thermal energy would be radiated away quickly in low-resolution simulation because the Sedov-Taylor phase of each SN (or collective superbubble) cannot be resolved in detail, i.e., the well-known overcooling problem. Therefore, effective models of SN feedback need to be developed, and several strategies have been taken in galaxy formation simulations; for example, (i) ignoring and bypassing the physics on unresolved scales; (ii) scaling up the energy dynamics to a resolvable scale by considering cumulative energies; or (iii) modeling the physics on unresolved scales via subgrid models.

As examples of method (i), the 'delayed cooling' model ignores the cooling for a short period of time after the SN event to make thermal feedback effective (Thacker & Couchman 2000; Stinson et al. 2006). The constant velocity wind model

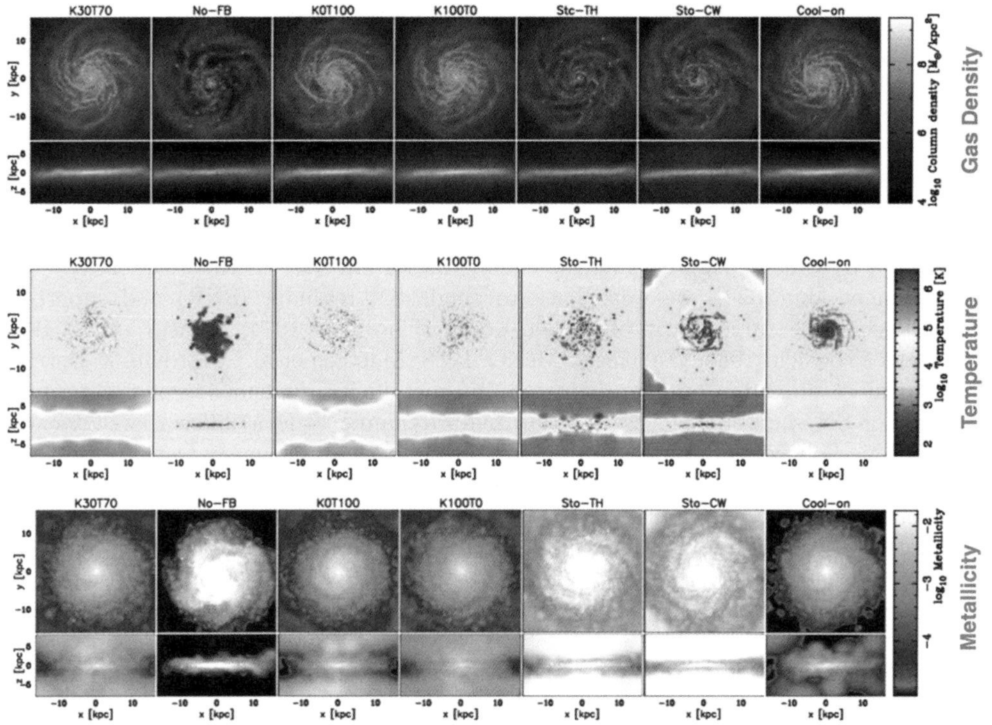

Figure 2. An example of testing SN feedback models in isolated galaxy simulations, showing projected gas density, mass-weighted temperature, and metallicity from top to bottom rows, respectively (Shimizu et al. 2019). In each row, the top panels show face-on view, and the lower panel shows the edge-on view.

of Springel & Hernquist (2003) stochastically kicks gas particles by turning off the hydro forces until the wind particles get out of the galaxies in a smoothed particle hydrodynamics (SPH) code. As an example of method (ii), 'stochastic thermal feedback' model increases the temperature of neighboring fluid elements by ΔT (Kay et al. 2003; Dalla Vecchia & Schaye 2012), so that the subsequent evolution of hot bubble can be solved by a hydro solver with efficient thermal feedback. But the choice of ΔT remains ad hoc and uncertain. As an example of method (iii), 'multiphase ISM model' treats single SPH particle as a multiphase gas, and the energy exchange between hot and cold phase is treated by a subgrid equilibrium model (Yepes et al. 1997; Springel & Hernquist 2003; Keller et al. 2014). Another model is injecting the terminal momentum of single SN explosion based on Sedov–Taylor solution (Kimm & Cen 2014; Hopkins et al. 2018). See Nagamine (2018) and Oku et al. (2022) for more discussions of different feedback treatments.

In Shimizu et al. (2019), we combined the delayed cooling model and kinetic feedback using the Sedov-Taylor self-similar solution in `GADGET3-Osaka` code. We show an example of testing different SN feedback models in Fig. 2. Here in the fiducial model (K30T70), 30% of the SN energy was injected as kinetic energy, and 70% as thermal energy following earlier work (e.g., Chevalier 1974; Durier & Dalla Vecchia 2012). The cooling was temporarily shut off for the neighbor particles which received the thermal feedback energy. One can see a very clumpy, cold, dense gas distribution in the face-on gas distribution image of 'No-feedback' (No-FB) run, with no hot gas above the disk. If the fraction of thermal energy injection is increased, the spiral arms become more fluffy compared to the

Fiducial run. The 'Stochastic thermal' (Stc-TH) and 'Stochastic constant wind velocity' (Sto-CW) runs also have some clumpy knots in the disk, but the feedback effects above the disk is very strong with hot outflowing gas, hence over-enriching the CGM. If the cooling is turned on ('Cool-on' run), the heated gas by the feedback quickly cools back down and will not get out of the disk. Therefore the Cool-on run shows similar disk feature as the No-FB run. The Osaka feedback model by Shimizu et al. (2019) provided useful comparison of various SN feedback treatment, succeeded in the self-regulation of star formation and naturally produced galactic outflow, however, it did contain some unphysical treatment (e.g., turning the cooling off for effective thermal feedback).

In Oku et al. (2022), we went back to single SN remnant (SNR) and superbubble following the same spirit of earlier works (Chevalier 1974; Weaver et al. 1977; Tomisaka & Ikeuchi 1986; Ostriker & McKee 1988; Martizzi et al. 2015; Kim & Ostriker 2015; Kim et al. 2017), and investigated the metallicity dependence of the terminal moment of SN shell. Using the Eulerian hydrodynamic code `ATHENA++`, we extended the analytic solution of the SNR shell-formation time by (Kim & Ostriker 2015) to include the effect of metallicity, and obtained the analytic solution of the superbubble shell-formation time. We found a universal scaling relations for the time evolution of momentum and radius for a superbubble, when the momentum and time are scaled by those at the shell-formation time. We then developed a SN feedback model based on the `ATHENA++` simulation results utilizing Voronoi tessellation around each star particle, and implemented it into the `GADGET3-Osaka` SPH code. We examined the mass/energy/metal loading factors and found that our stochastic thermal feedback model produced galactic outflow that carries metals above the galactic plane but with a weak suppression of star formation. Additional mechanical feedback further suppressed star formation and brought the simulation results into better agreement with the observations of the Kennicutt–Schmidt relation within the uncertainties of observed data. We argued that both thermal and mechanical feedback are necessary for the SN feedback model of galaxy evolution, especially in SPH simulations when an individual SN bubble is unresolved.

Some simulations have succeeded to push the resolution limit by focusing on low-mass galaxies. For example, Hu (2019) used the GADGET-3 SPH simulation to study SN feedback in a dwarf galaxy in a dark matter halo of virial mass $M_{\rm vir} = 10^{10}\,M_\odot$, and achieved a resolution of $m_{\rm gas} = 1\,M_\odot$, and 0.3 pc for gravitational softening length and SPH smoothing length. Even with this resolution, some assumptions are still necessary (e.g. fraction of energy given as kinetic energy, and how many SPH particles receive the feedback energy). Nevertheless, they were able to simulate the formation of superbubbles of a few hundred parsecs in size, and how they break out of the galactic disk. Ma et al. (2020) simulated formation of dwarf galaxies in a halo of $M_{\rm vir} = 3.7 \times 10^{10}\,M_\odot$ in reionization epoch using cosmological zoom-in simulation (FIRE-2 GIZMO), with a mass resolution of $m_{\rm gas} = 100\,M_\odot$ and a spatial resolution of $\sim$pc. They examined where the stars form in these high-redshift dwarf galaxies with respect to the superbubble walls, and found a large spatio-temporal variation in escape fraction of ionizing photons (cf. Wise & Cen 2009; Kimm & Cen 2014). As the mass resolution of simulations improve and approach 'star-by-star' simulation, one needs to devise a way to stochastically sample the IMF for the star formation model (Ploeckinger et al. 2014; Hu 2019; Hirai et al. 2021), and its impact needs to be studied further for feedback as well.

3. The AGORA code comparison project

One of the ways to test galaxy formation codes is to perform a code comparison project such as the Santa Barbara cluster comparison project (Frenk et al. 1999), Aquila project (Scannapieco et al. 2012), nIFTy project (Knebe et al. 2015), and the AGORA project (Kim et al. 2014, 2016; Roca-Fàbrega et al. 2021). For example, the Santa Barbara project highlighted different results from various hydrodynamic schemes and spurious entropy generation in SPH codes. This had a profound impact and led to the development of new SPH schemes that can resolve shocks better, such as the density-independent scheme (Saitoh & Makino 2013) or a more general formulation based on Lagrangian-based derivation (Springel & Hernquist 2002; Hopkins 2013). The Aquila comparison project found large code-to-code variations in galactic properties at $z = 0$ such as the stellar mass, size, morphology and gas content. Their conclusion was that, due to different feedback prescriptions, the simulations could not yet uniquely predict galactic properties even when the dark matter halo assembly history were the same.

To make a more controlled comparison environment, the Assembling Galaxies ofResolved Anatomy (AGORA) project (†) started with a more rigorous calibration steps with common star formation recipe and the same Grackle cooling module (Smith et al. 2017) among all participating codes. With these common astrophysics setup, they were able to show more consistent behaviors between various codes (both SPH and AMR), and Kim et al. (2016) concluded that modern high-resolution galaxy formation simulations are more sensitive to input physics (e.g. feedback prescription) than to intrinsic differences in numerical schemes. Roca-Fàbrega et al. (2021) extended the comparison to cosmological zoom-in hydro simulations, and seven contemporary astrophysical simulation codes (ART-I, ENZO, RAMSES, CHANGA, GADGET-3, GEAR, and GIZMO) were compared. The comparison involved four methodical calibration steps, starting from a simple adiabatic run without cooling and star formation, then gradually adding steps to include cooling/heating and star formation. At the last fourth step, each code was asked to roughly reproduce the stellar mass of $10^9\,M_\odot$ at $z = 4$ in a halo which will become $10^{12}\,M_\odot$ at $z = 0$, with code-dependent SN feedback recipes. With a physical resolution of $\lesssim 100$ pc at $z = 4$, all codes *roughly* agree with each other on gas and stellar properties, but there were also interesting differences in the CGM temperature and chemical enrichment due to differences in feedback treatments (see Fig. 3). Their results also highlight the importance of further refining and constraining SN and AGN feedback models via comparison to various observations.

4. Probing the impact of feedback by cosmological simulations

It is also desirable to use large-scale cosmological hydro simulations to probe feedback in addition to isolated galaxies and zoom-in simulations. With large simulation box size and large galaxy samples, one can examine galaxy statistics such as galaxy stellar mass/luminosity functions or stellar-to-halo-mass ratio (see the contribution by R. Somerville to this proceedings). In addition to galaxies, we would also like to probe the distribution of diffuse baryons via absorption and emission lines. For example, the distribution of neutral hydrogen (HI) probed by the Lyα forest (e.g., Weymann et al. 1981; Cowie et al. 1995; Rauch 1998) reflects the strength of UV background radiation field as well as the local ionizing radiation, and we expect that the impact of feedback is imprinted in CGM/IGM (Cen et al. 1994; Hernquist et al. 1996; Miralda-Escudé et al. 1996; Zhang et al. 1997, 1998; Theuns et al. 2002; Cen et al. 2005; Kollmeier et al. 2006). The Lyα forest is one of the most powerful probes of cosmology, and it has been used to constrain cosmological parameters and matter power spectrum (Weinberg et al. 1998; Croft et al. 1998; McDonald et al. 2006; Iršič et al. 2017), as well as the mass of warm dark matter particles or neutrinos (e.g. Viel et al. 2005, 2013; Palanque-Delabrouille et al. 2015). This subject is also related to the so-called 'Missing baryon problem' (Cen & Ostriker 1999; Nicastro et al. 2005; Shull et al. 2012), where we still struggle to observationally account for the entire cosmic baryons.

† `https://sites.google.com/site/santacruzcomparisonproject/`

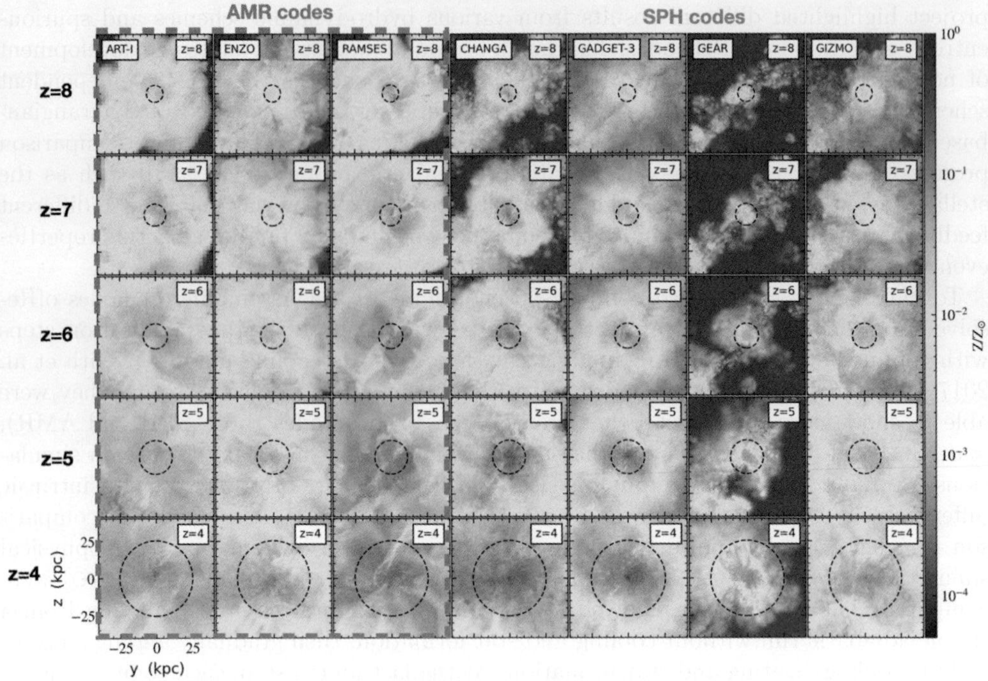

Figure 3. Comparison of projected (density-square-weighted) metallicity of seven different code from $z = 8$ to $z = 4$. Adapted from Fig. 18 of (Roca-Fàbrega et al. 2021).

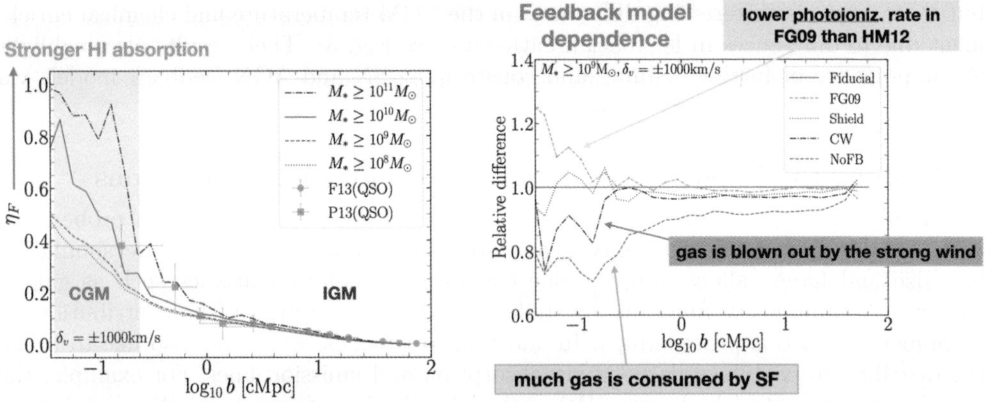

Figure 4. *Left:* Lyα flux contrast as a function of impact parameter from nearby galaxies. The data points are from Font-Ribera et al. (2013, orange filled circle; F13) and Prochaska et al. (2013, orange filled square; P13). *Right:* Relative difference in the flux contrast from the Fiducial model in runs with different feedback and UVB treatments. Both figures are from Nagamine et al. (2021).

The Lyα forest technique has recently been upgraded to IGM tomography, whereby one can generate 3D contour maps of HI density using more numerous star-forming galaxies as background sources than just using quasars sight-lines. A few groups have already demonstrated that this is possible (e.g., Lee et al. 2014, 2018; Cai et al. 2016; Mukae et al. 2020). Massive protoclusters have been discovered at $z = 2 - 3$ by the Lyα tomographic technique (Lee et al. 2016; Cai et al. 2017). In a more general sense, one can derive the correlation between galaxy overdensity and HI overdensity (Mukae et al. 2017; Liang et al. 2021). The scientific goals of IGM tomography are: (i) to characterize the cosmic web at $z > 2$, (ii) to study the association between galaxies/AGNs and HI gas, and (iii) to identify protoclusters and voids in an *unbiased* manner.

As a pathfinder to the IGM tomography studies by the Subaru PFS (Takada et al. 2014; Greene et al. 2022) and upcoming observations by the JWST/TMT/ELT, Nagamine et al. (2021) investigated the impact of feedback on basic Lyα forest statistics by creating a light-cone data set at $z = 2 - 3$ and generating a mock Lyα forest data. They used five cosmological hydro simulations (`GADGET3-Osaka` code) with different models of feedback and UVB treatment (comoving boxsize $L_{\rm box}$=147.6 Mpc, particle number $N = 2 \times 512^3$), and examined the 1D flux probability distribution function, 1D flux power spectrum, flux contrast vs. impact parameter from galaxies, and HI–galaxy cross-correlation. The flux contrast is defined as $\eta_F \equiv -\delta_F = 1 - \frac{F}{\langle F \rangle}$, where F is the transmitted flux ($F = e^{-\tau}$), and $\langle F \rangle$ is the average effective Lyα optical depth adjusted to the observed value (Becker & Bolton 2013). Higher η_F in the vicinity of galaxies means stronger absorption, i.e., more HI (left panel of Fig. 4). In other words, they found stronger HI absorption with decreasing impact parameter from galaxies, consistently with earlier simulation results (e.g., Bruscoli et al. 2003; Kollmeier et al. 2003, 2006; Meiksin et al. 2015; Turner et al. 2017; Meiksin et al. 2017; Sorini et al. 2018). Their simulation results showed overall agreement with current observational data, but with some interesting discrepancies of about 30% on small scales that are due to different treatments of feedback and UVB, or varying observational conditions (right panel of Fig. 4). The massive galaxies with $M_\star \geq 10^{10}\,M_\odot$ contribute strongly to the flux contrast signal (left panel of Fig. 4), and the lower-mass galaxies with $M_\star \approx 10^8 - 10^{10}\,M_\odot$ dilute the flux contrast signal from massive galaxies when the signal is averaged over the entire galaxy sample. The variations in η_F on scales of < 1 Mpc can be probed with future IGM tomography surveys with dense background source sampling by JWST/ELT/TMT. On large scales, the average flux contrast smoothly connects to the IGM level, supporting the spherical infall model and concordance Λ cold dark matter model, similarly to the results by Meiksin et al. (2017); Sorini et al. (2018). It is interesting to note that Sorini et al. (2020) found negligible impact of AGN feedback on the flux contrast, and they argued that stellar feedback is the primary driver determining the average physical properties of CGM at $z = 2 - 3$, which needs to be checked in other simulations with AGN feedback in the future (cf. Tillman et al. 2022).

In addition to HI distribution, metal distribution can also be probed by emission and absorption lines. For example, MEGAFLOW project has observed MgII lines in both absorption and emission in galactic wind region of $z \sim 0.7$ galaxy (Zabl et al. 2020, 2021). Nelson et al. (2021) computed the resonantly scattered MG II emission from TNG50 simulation, and it seems that the simulated galaxies have somewhat steeper profile (declining faster with increasing radii) than the observed data points (their Fig. 3). But the currently observed sources are especially bright ones that are visible, so more comparison needs to be performed in the future with lower mass systems.

Interestingly, though, we find a similar trend in other emission lines, such as [C II] (Arata et al. 2020; Fujimoto et al. 2019), and Lyα (Zhang et al. 2020), where the simulated galaxies cannot account for the observed extended emission profiles. These discrepancies may have interesting implications for the feedback efficiencies (e.g., mass-loading factor of metals in galactic outflows; Pizzati et al. 2020), therefore requires further studies to constrain the efficiencies of chemical enrichment in CGM/IGM.

5. Summary

In this article, we reviewed various feedback treatments in galaxy simulations, and discussed our development of physically-based SN feedback models and the tests in the AGORA project using both isolated galaxies and zoom-in cosmological hydrodynamic simulations. We argued that considering both thermal and kinetic modes is important for SN feedback at the current level of resolution ($\gtrsim 10\,\mathrm{pc}$). In our latest work, Oku et al. (2022) showed that the kinetic feedback suppresses star formation while stochastic thermal feedback drives strong metal outflows. Further studies on both small and large scales are important to constrain feedback effects in galaxy evolution and chemical enrichment of CGM/IGM.

It seems that some very high-resolution simulations ($\lesssim$ pc scale) exhibit weaker winds than larger scale simulations that mimic the physics of galactic winds on supergalactic scales (see the contributions by E. Ostriker and C.-G. Kim in this proceedings, and Hu 2019). Although we did not discuss in this article, it will be essential to consider additional physics such as cosmic rays and magnetic fields for more physically plausible models of star formation and feedback (e.g., Hopkins et al. 2022).

It might also be possible to constrain the physics of feedback at larger scales of circumgalactic and intergalactic scales utilizing the Lyα absorption by neutral hydrogen (i.e., the so-called IGM tomography technique) and distribution of metals and dust. For example, Nagamine et al. (2021) have shown that the SN feedback changes the radial distribution of H I gas and the Lyα flux contrast signal at $\sim$30% level. Future comparisons between simulations and the CGM/IGM tomography surveys by WEAVE, MOONS, Subaru PFS, JWST, ELT, and TMT will allow us to further constrain the physics of feedback.

I am grateful to all of my recent collaborators on the research results discussed in this article, including S. Arata, R. Cen, Y. Oku, L. Romano, I. Shimizu, K. Tomida, H. Yajima, and everyone in the AGORA project.

References

Arata, S., Yajima, H., Nagamine, K., Abe, M., & Khochfar, S. 2020, *MNRAS*, 498, 5541
Arata, S., Yajima, H., Nagamine, K., Li, Y., & Khochfar, S. 2019, *MNRAS*, 488, 2629
Becker, G. D., & Bolton, J. S. 2013, *MNRAS*, 436, 1023
Behroozi, P. S., Wechsler, R. H., & Conroy, C. 2013, *ApJ*, 770, 57
Bruscoli, M., Ferrara, A., Marri, S., et al. 2003, *MNRAS*, 343, L41
Cai, Z., Fan, X., Peirani, S., et al. 2016, *ApJ*, 833, 135
Cai, Z., Fan, X., Bian, F., et al. 2017, *ApJ*, 839, 131
Cen, R. 2003, *ApJ*, 591, 12
Cen, R., Miralda-Escudé, J., Ostriker, J. P., & Rauch, M. 1994, *ApJL*, 437, L9
Cen, R., Nagamine, K., & Ostriker, J. P. 2005, *ApJ*, 635, 86
Cen, R., & Ostriker, J. P. 1992, *ApJL*, 399, L113
—. 1999, *ApJ*, 514, 1
Chevalier, R. A. 1974, *ApJ*, 188, 501
Cowie, L. L., Songaila, A., Kim, T.-S., & Hu, E. M. 1995, AJ, 109, 1522
Croft, R. A. C., Weinberg, D. H., Katz, N., & Hernquist, L. 1998, *ApJ*, 495, 44
Dalla Vecchia, C., & Schaye, J. 2012, *MNRAS*, 426, 140
Durier, F., & Dalla Vecchia, C. 2012, *MNRAS*, 419, 465
Faucher-Giguère, C.-A., Kereš, D., & Ma, C.-P. 2011, *MNRAS*, 417, 2982
Finkelstein, S. L., D'Aloisio, A., Paardekooper, J.-P., et al. 2019, *ApJ*, 879, 36
Font-Ribera, A., Arnau, E., Miralda-Escudé, J., et al. 2013, *JCAP*, 5, 018
Frenk, C. S., White, S. D. M., Bode, P., et al. 1999, *ApJ*, 525, 554
Fujimoto, S., Ouchi, M., Ferrara, A., et al. 2019, *ApJ*, 887, 107
Gnedin, N. Y., Kravtsov, A. V., & Chen, H. 2008, *ApJ*, 672, 765
Greene, J., Bezanson, R., Ouchi, M., Silverman, J., & the PFS Galaxy Evolution Working Group. 2022, arXiv e-prints, arXiv:2206.14908
Hashimoto, T., Laporte, N., Mawatari, K., et al. 2018, *Nature*, 557, 392
Hashimoto, T., Inoue, A. K., Mawatari, K., et al. 2019, *PASJ*, 71, 71

Hernquist, L., Katz, N., Weinberg, D. H., & Miralda-Escudé, J. 1996, *ApJL*, 457, L51
Hirai, Y., Fujii, M. S., & Saitoh, T. R. 2021, *PASJ*, 73, 1036
Hopkins, P. F. 2013, *MNRAS*, 428, 2840
Hopkins, P. F., Butsky, I. S., Panopoulou, G. V., et al. 2022, *MNRAS*, 516, 3470
Hopkins, P. F., Wetzel, A., Kereš, D., et al. 2018, *MNRAS*, 477, 1578
Hu, C.-Y. 2019, *MNRAS*, 483, 3363
Iršič, V., Viel, M., Berg, T. A. M., et al. 2017, *MNRAS*, 466, 4332
Katz, H., Rosdahl, J., Kimm, T., et al. 2022, *MNRAS*, 510, 5603
Katz, N. 1992, *ApJ*, 391, 502
Katz, N., Weinberg, D. H., & Hernquist, L. 1996, *ApJS*, 105, 19
Kay, S. T., Thomas, P. A., & Theuns, T. 2003, *MNRAS*, 343, 608
Keller, B. W., Wadsley, J., Benincasa, S. M., & Couchman, H. M. P. 2014, *MNRAS*, 442, 3013
Kereš, D., Katz, N., Weinberg, D. H., & Davé, R. 2005, *MNRAS*, 363, 2
Kim, C.-G., & Ostriker, E. C. 2015, *ApJ*, 802, 99
Kim, C.-G., Ostriker, E. C., & Raileanu, R. 2017, *ApJ*, 834, 25
Kim, J.-H., Abel, T., Agertz, O., et al. 2014, *ApJS*, 210, 14
Kim, J.-h., Agertz, O., Teyssier, R., et al. 2016, *ApJ*, 833, 202
Kimm, T., & Cen, R. 2014, *ApJ*, 788, 121
Kistler, M. D., Yüksel, H., Beacom, J. F., Hopkins, A. M., & Wyithe, J. S. B. 2009, *ApJL*, 705, L104
Knebe, A., Pearce, F. R., Thomas, P. A., et al. 2015, *MNRAS*, 451, 4029
Kollmeier, J. A., Miralda-Escudé, J., Cen, R., & Ostriker, J. P. 2006, *ApJ*, 638, 52
Kollmeier, J. A., Weinberg, D. H., Davé, R., & Katz, N. 2003, *ApJ*, 594, 75
Lee, K.-G., Hennawi, J. F., Stark, C., et al. 2014, *ApJL*, 795, L12
Lee, K.-G., Hennawi, J. F., White, M., et al. 2016, *ApJ*, 817, 160
Lee, K.-G., Krolewski, A., White, M., et al. 2018, *ApJS*, 237, 31
Liang, Y., Kashikawa, N., Cai, Z., et al. 2021, *ApJ*, 907, 3
Ma, X., Quataert, E., Wetzel, A., et al. 2020, arXiv e-prints, arXiv:2003.05945
Madau, P., & Dickinson, M. 2014, *ARA&A*, 52, 415
Martizzi, D., Faucher-Giguère, C.-A., & Quataert, E. 2015, *MNRAS*, 450, 504
McDonald, P., Seljak, U., Burles, S., et al. 2006, *ApJS*, 163, 80
Meiksin, A., Bolton, J. S., & Puchwein, E. 2017, *MNRAS*, 468, 1893
Meiksin, A., Bolton, J. S., & Tittley, E. R. 2015, *MNRAS*, 453, 899
Miralda-Escudé, J., Cen, R., Ostriker, J. P., & Rauch, M. 1996, *ApJ*, 471, 582
Mukae, S., Ouchi, M., Kakiichi, K., et al. 2017, *ApJ*, 835, 281
Mukae, S., Ouchi, M., Cai, Z., et al. 2020, *ApJ*, 896, 45
Nagamine, K. 2018, The Encyclopedia of Cosmology. Volume 2: Numerical Simulations in Cosmology, Vol. 2 (World Scientific Publishing), doi:10.1142/9496-vol2
Nagamine, K., Cen, R., Hernquist, L., Ostriker, J. P., & Springel, V. 2004, *ApJ*, 610, 45
Nagamine, K., Cen, R., & Ostriker, J. P. 2000, *ApJ*, 541, 25
Nagamine, K., Ostriker, J. P., Fukugita, M., & Cen, R. 2006, *ApJ*, 653, 881
Nagamine, K., Shimizu, I., Fujita, K., et al. 2021, *ApJ*, 914, 66
Naidu, R. P., Tacchella, S., Mason, C. A., et al. 2020, *ApJ*, 892, 109
Nelson, D., Byrohl, C., Peroux, C., Rubin, K. H. R., & Burchett, J. N. 2021, *MNRAS*, 507, 4445
Nelson, D., Genel, S., Pillepich, A., et al. 2016, *MNRAS*, 460, 2881
Nicastro, F., Mathur, S., Elvis, M., et al. 2005, *Nature*, 433, 495
Oku, Y., Tomida, K., Nagamine, K., Shimizu, I., & Cen, R. 2022, *ApJS*, 262, 9
Ostriker, J. P., & McKee, C. F. 1988, Reviews of Modern Physics, 60, 1
Palanque-Delabrouille, N., Yèche, C., Lesgourgues, J., et al. 2015, *JCAP*, 2015, 045
Pizzati, E., Ferrara, A., Pallottini, A., et al. 2020, *MNRAS*, 495, 160
Ploeckinger, S., Hensler, G., Recchi, S., Mitchell, N., & Kroupa, P. 2014, *MNRAS*, 437, 3980
Prochaska, J. X., Hennawi, J. F., Lee, K.-G., et al. 2013, *ApJ*, 776, 136
Rauch, M. 1998, *ARA&A*, 36, 267
Razoumov, A. O., & Sommer-Larsen, J. 2006, *ApJL*, 651, L89

Roca-Fàbrega, S., Kim, J.-H., Hausammann, L., et al. 2021, *ApJ*, 917, 64
Saitoh, T. R., & Makino, J. 2013, *ApJ*, 768, 44
Scannapieco, C., Wadepuhl, M., Parry, O. H., et al. 2012, *MNRAS*, 423, 1726
Schaye, J., Dalla Vecchia, C., Booth, C. M., et al. 2010, *MNRAS*, 402, 1536
Shimizu, I., Todoroki, K., Yajima, H., & Nagamine, K. 2019, *MNRAS*, 484, 2632
Shull, J. M., Smith, B. D., & Danforth, C. W. 2012, *ApJ*, 759, 23
Smit, R., Bouwens, R. J., Carniani, S., et al. 2018, *Nature*, 553, 178
Smith, B. D., Bryan, G. L., Glover, S. C. O., et al. 2017, *MNRAS*, 466, 2217
Sorini, D., Davé, R., & Anglés-Alcázar, D. 2020, *MNRAS*, 499, 2760
Sorini, D., Oñorbe, J., Hennawi, J. F., & Lukić, Z. 2018, *ApJ*, 859, 125
Springel, V., & Hernquist, L. 2002, *MNRAS*, 333, 649
—. 2003, *MNRAS*, 339, 289
Stinson, G., Seth, A., Katz, N., et al. 2006, *MNRAS*, 373, 1074
Takada, M., Ellis, R. S., Chiba, M., et al. 2014, *PASJ*, 66, R1
Thacker, R. J., & Couchman, H. M. P. 2000, *ApJ*, 545, 728
Theuns, T., Viel, M., Kay, S., et al. 2002, *ApJL*, 578, L5
Tillman, M. T., Burkhart, B., Tonnesen, S., et al. 2022, arXiv e-prints, arXiv:2210.02467
Tomisaka, K., & Ikeuchi, S. 1986, *PASJ*, 38, 697
Turner, M. L., Schaye, J., Crain, R. A., et al. 2017, *MNRAS*, 471, 690
Viel, M., Becker, G. D., Bolton, J. S., & Haehnelt, M. G. 2013, *Phys. Rev. D*, 88, 043502
Viel, M., Lesgourgues, J., Haehnelt, M. G., Matarrese, S., & Riotto, A. 2005, *Phys. Rev. D*, 71, 063534
Weaver, R., McCray, R., Castor, J., Shapiro, P., & Moore, R. 1977, *ApJ*, 218, 377
Weinberg, D. H., Burles, S., Croft, R. A. C., et al. 1998, ArXiv Astrophysics e-prints, astro-ph/9810142
Weymann, R. J., Carswell, R. F., & Smith, M. G. 1981, *ARA&A*, 19, 41
Wise, J. H., & Cen, R. 2009, *ApJ*, 693, 984
Wright, R. J., Lagos, C. d. P., Power, C., & Correa, C. A. 2021, *MNRAS*, 504, 5702
Yajima, H., Nagamine, K., Zhu, Q., Khochfar, S., & Dalla Vecchia, C. 2017, *ApJ*, 846, 30
Yepes, G., Kates, R., Khokhlov, A., & Klypin, A. 1997, *MNRAS*, 284, 235
Zabl, J., Bouché, N. F., Schroetter, I., et al. 2020, *MNRAS*, 492, 4576
Zabl, J., Bouché, N. F., Wisotzki, L., et al. 2021, *MNRAS*, 507, 4294
Zhang, H., Ouchi, M., Itoh, R., et al. 2020, *ApJ*, 891, 177
Zhang, Y., Anninos, P., Norman, M. L., & Meiksin, A. 1997, *ApJ*, 485, 496
Zhang, Y., Meiksin, A., Anninos, P., & Norman, M. L. 1998, *ApJ*, 495, 63

Resolving the Rise and Fall of Star Formation in Galaxies
Proceedings IAU Symposium No. 373, 2023
T. Wong & W.-T. Kim, eds.
doi:10.1017/S174392132200391X

Cloud-scale Analyses on Molecular Gas in Simulated and Observed Galaxy Mergers

Hao He

McMaster University
1280 Main St W, Hamilton, ON L8S 4L8, CAN

Abstract. We employ the Feedback In Realistic Environments (FIRE-2) physics model to study how the properties of giant molecular clouds (GMCs) evolve during galaxy mergers. Due to the rarity of mergers in the local Universe, samples of nearby merging galaxies suitable for studies of individual GMCs are limited. Idealized simulations provide us with a new window to study GMC evolution during a merger, and assist in interpreting observations. We conduct a pixel-by-pixel analysis of the simulated molecular gas properties in both undisturbed control galaxies and galaxy mergers. The simulated GMC-pixels follow a similar trend in a diagram of velocity dispersion (σ_v) versus gas surface density ($\Sigma_{\rm mol}$) as observed in normal spiral galaxies in the Physics at High Angular resolution in Nearby GalaxieS (PHANGS) survey. For simulated mergers, we see a significant increase in both the $\Sigma_{\rm mol}$ and σ_v for GMC-pixels by a factor of 5 – 10, which put these pixels to be above the trend of PHANGS galaxies in the σ_v vs $\Sigma_{\rm mol}$ diagram. This deviation indicates that GMCs in the simulated merger are more gravitationally unbound and have higher virial parameter ($\alpha_{\rm vir}$) of 10 – 100, which is much larger than that of simulated control galaxies. Furthermore, we find that the increase in $\alpha_{\rm vir}$ generally happens at the same time as the increase in global star formation rate (SFR), which suggests feedback is playing a role in dispersing the gas. The correspondence between high $\alpha_{\rm vir}$ and SFR also suggests some other physical mechanisms besides self-gravity are helping the GMCs in starburst mergers to collapse and form stars.

Keywords. ISM: clouds, ISM: kinematics and dynamics, ISM: structure, galaxies: interactions, galaxies: starburst, galaxies: star formation

1. Introduction

Giant molecular clouds (GMCs) are the birth places of stars. Hence, to understand the links between molecular gas and star formation rate (SFR) in galaxies, it is essential to study the physical properties of molecular clouds in a broad range of environments.High-resolution CO observations have successfully characterized GMCs in the Milky Way and nearby galaxies. In particular, the recently completed PHANGS-ALMA survey (Leroy et al. 2021) has expanded these observations across a complete sample of nearby spiral galaxies, providing direct measurements of molecular gas surface density $\Sigma_{\rm mol}$, velocity dispersion σ_v and radius R, which are key quantities for determining the physical state of GMCs. However, GMCs in galaxy mergers are rarely studied. On the observational side, the scarcity of nearby mergers means that we have only a handful of systems observed with GMC resolution (Brunetti et al. 2021; Brunetti 2022). These studies show that GMCs in mergers have significantly higher gas surface densities and are less gravitationally bound compared to GMCs in normal spirals. However, it is difficult to draw statistically robust conclusions on how GMC properties evolve across various merging stages based on these limited observational samples. Using a comprehensive library of

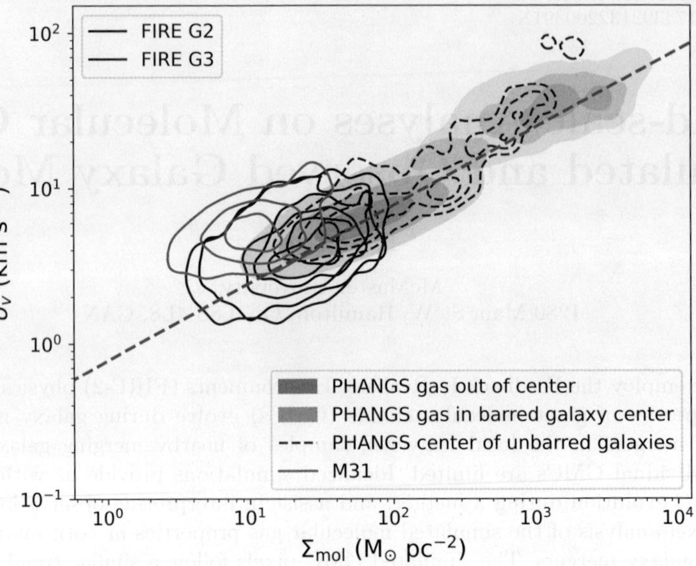

Figure 1. Velocity dispersion versus gas surface density for the G2 (black solid line) and G3 (brown solid line) simulated galaxies with inclination angle of 30 degrees compared to the PHANGS galaxy sample. The contour is mass weighted and set to include 20%, 50% and 80% of the data. The density contours of PHANGS galaxies (Sun et al. 2020) show the distribution of measurements in galaxy disks (blue shaded contours), the centers of barred galaxies (Salmon shaded contours) and the centers of unbarred galaxies (brown dashed contours) with a resolution of 90 pc. The red dashed line marks the position with median values of $\alpha_{\rm vir}$ for PHANGS galaxies of 3.1 (Sun et al. 2020). We also show the data for M31 (green solid contour) at 120 pc resolution from Sun et al. (2018). $\Sigma_{\rm mol}$ and σ_v for the FIRE-2 spiral galaxies follow a similar trend to the PHANGS galaxies but lie at the lower end of the plot.

idealized galaxy merger simulations based on the FIRE-2 physics model, Moreno et al. (2019) show that SFR enhancement is accompanied by an increase in the cold dense gas reservoir. This suite thus provides us with the ideal tool to properly examine GMC evolution along the entire merging sequence.

2. GMCs in the simulated isolated galaxies

To test if the simulation successfully reproduces observed GMCs, Figure 1 shows the well-known correlation between σ_v and $\Sigma_{\rm mol}$ for isolated simulated galaxies and PHANGS-ALMA spiral galaxies. The left panel shows σ_v versus $\Sigma_{\rm mol}$ contours for G2 and G3 galaxies at an inclination angle of 30 degrees, compared with that of observed galaxies, while the right contour shows the contour for G3 galaxies with different inclination angles. The two simulated galaxies, G2 and G3, exhibit similar properties (black and dark red solid contours) and generally lie on the trend followed by the PHANGS galaxies when viewed at an inclination angle of 30 degrees. We also plot the red dashed line indicating GMCs with constant virial parameter $\alpha_{\rm vir}$ of 3.1. We can see both our simulated galaxies and observed PHANGS galaxies follow the trend of the constant $\alpha_{\rm vir}$, which gives the relation of $\sigma_v^2 \propto \Sigma_{\rm mol}$. For this aspect, the simulations reproduce GMCs similar to the observations. However, we can see that the two galaxies lie at the low surface-density end of the PHANGS distribution and thus their gas properties are more similar to those of M31 than to a typical PHANGS galaxy.

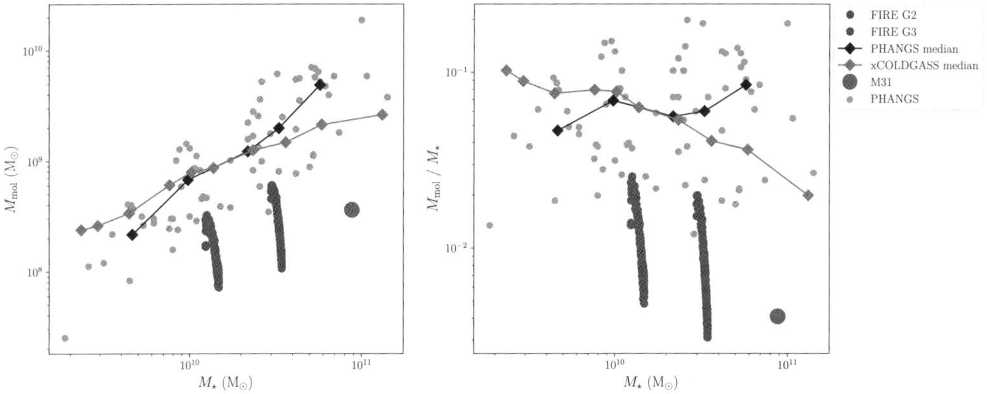

Figure 2. (Left) $M_{\rm mol}$ versus $M_\star$ for PHANGS galaxies (salmon dots; Leroy et al. 2021), M 31 (green filled circle, Nieten et al. 2006) and the G2 (red points) and G3 (blue points) simulated galaxies. Note that the G2 and G3 simulated galaxies lie significantly below the star-forming main sequence defined by the xCOLDGASS sample. (Right) $f_{\rm mol}$ versus $M_\star$ for the same galaxies. The molecular gas fractions of G2 and G3 are significantly lower than those of the PHANGS spiral galaxies.

In interpreting the lower $\Sigma_{\rm mol}$ values seen in Fig. 1, one possibility is that there may not be as much gas available to form high surface density clouds in the two simulated galaxies compared to the PHANGS galaxies. Fig. 2 compares the global molecular gas masses, $M_{\rm mol}$, and molecular gas fractions, $f_{\rm mol} = M_{\rm mol} / M_\star$, for the FIRE-2 mergers with those of the PHANGS galaxies from (Sun et al. 2020). We also show the median value of $M_{\rm mol}$ and $f_{\rm gas}$ in each $M_\star$ bin for the PHANGS galaxies, as well as the weighted median of $M_\star$ and $f_{\rm mol}$ for galaxies in xCOLDGASS sample (Saintonge et al. 2017). The two median values are quite close to each other for galaxies with $M_\star$ of $10^{9.5}$ – 10^{11} $M_\odot$, although the PHANGS galaxies seem to deviate somewhat from the xCOLDGASS sample in the highest and lowest mass bins. In contrast, the G2 and G3 galaxies both have $f_{\rm mol} \sim 3$ times lower than typical PHANGS or xCOLDGASS galaxies of the same stellar mass. Therefore, the small global $f_{\rm mol}$ may be the cause for producing the low $\Sigma_{\rm mol}$ values seen in the simulated galaxies.

3. Merging galaxies

We performed a similar σ_v versus $\Sigma_{\rm mol}$ analysis for our suite of galaxy merger simulations. Since we are particularly interested in how the starburst activities influence GMC properties, we focus on the period right before and after then second passage where you can see the largest contrast in SFR. We show the two example snapshots before and after the second passage in Fig. 3. As we can see, the simulated merger still has the similar $\Sigma_{\rm mol}$ and σ_v as the isolated galaxies right before the start of the second perihelion passage. Then the molecular gas quickly transitions to a more turbulent state with much higher σ_v after the second passage along with dramatic increase in global SFR, as shown in the snapshot at 2.66 Gyr. As shown in the zeroth moment map, the G2&G3 merger at this time still shows two separate nuclei; a similar stage as our observed mergers, the Antennae and NGC 3256. At this time, the σ_v versus $\Sigma_{\rm mol}$ contours for the simulated merger lies above the trend seen for the PHANGS galaxies, similar to NGC 3256. In contrast, it is different from the Antennae, which still lies along the trend of PHANGS galaxies. The larger deviation from the PHANGS trend means higher $\alpha_{\rm vir}$.

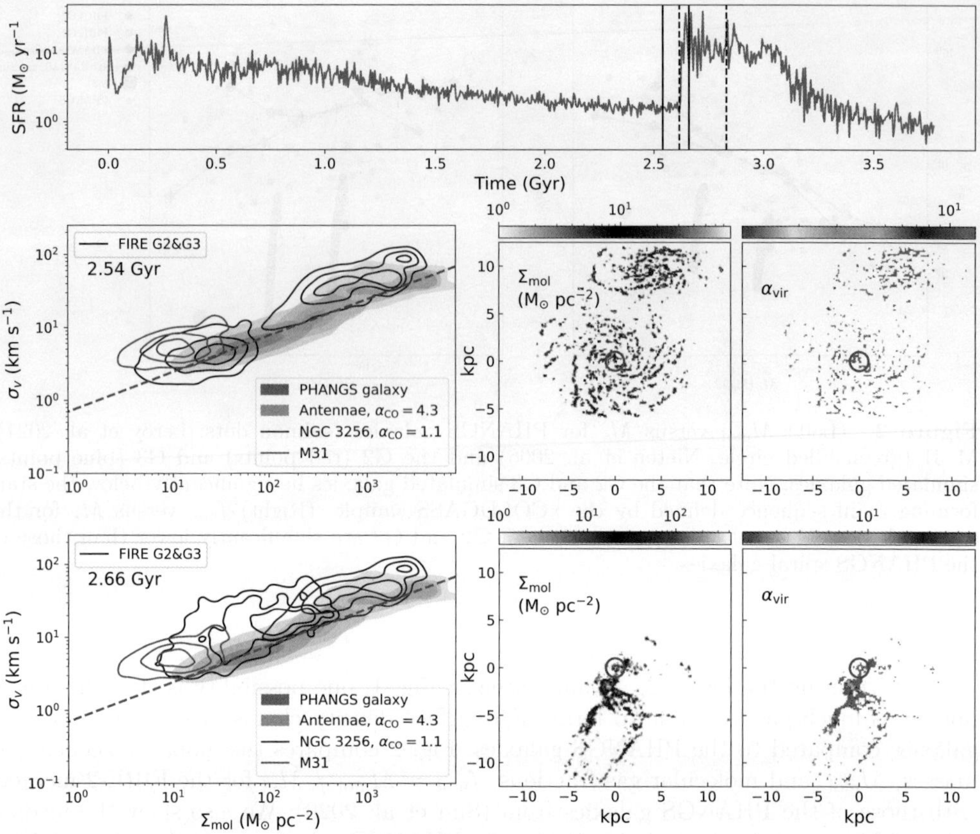

Figure 3. Two of the snapshots for the G2&G3 merger with 'e2' orbit with viewing angle of 'v0'. The top panel shows the SFR history for this merging suite. The two solid vertical lines indicate the time for each snapshot. The two dashed lines indicate the times at the start of second merging and the final coalesce of two nuclei. For the bottom snapshots, the left panel shows the σ_v versus $\Sigma_{\rm mol}$ mass weighted contour with the same setting as Fig.1. We also show the density contours for the PHANGS galaxies (filled blue region), NGC 3256 (blue contours) and the Antennae (orange shaded region). For NGC 3256, $\Sigma_{\rm mol}$ is calculated using the ULIRG $\alpha_{\rm CO}$ of 1.1 $M_\odot$ (K km s^{-1} pc^2)$^{-1}$. For the Antennae, the gas surface density is calculated using the Milky Way $\alpha_{\rm CO}$ of 4.3 $M_\odot$ (K km s^{-1} pc^2)$^{-1}$. We also show the $\Sigma_{\rm mol}$ and $\alpha_{\rm vir}$ map for the two snapshots. The interactive version of the animation is available at `https://htmlpreview.github.io/?https://github.com/heh15/merger_animations/blob/main/G2G3_e2_v0.html`.

We note that different $\alpha_{\rm CO}$ choices will affect the position of the contour. If we choose ULIRG $\alpha_{\rm CO}$ instead of the Milky Way value, the Antennae would have $\alpha_{\rm vir}$ similar to that of NGC 3256 and our G2&G3 merger. The uncertainty in the correct $\alpha_{\rm CO}$ value to use makes it difficult to interpret the data for the Antennae in this context. For our further analysis, we will adopt the Milky Way $\alpha_{\rm CO}$ for the Antennae.

3.1. *The evolution of virial parameter $\alpha_{\rm vir}$*

During the second passage, we see that the σ_v vs $\Sigma_{\rm mol}$ distribution for our simulated mergers lies above the trend observed for the PHANGS galaxies. A higher σ_v for a given $\Sigma_{\rm mol}$ means the GMCs in these mergers are more turbulent and less gravitationally bound

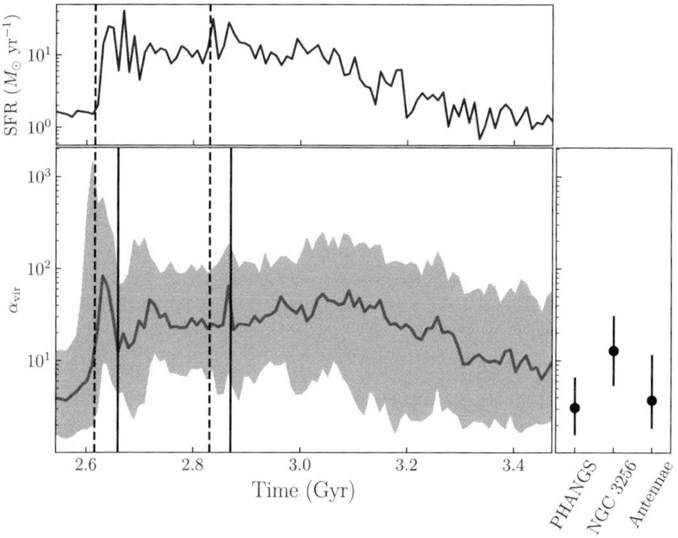

Figure 4. The variation of α_{vir} versus time for the G2&G3 mergers in (left) the e2 orbit and (right) the e1 orbit viewed from 'v0' angle during the final coalescence. (**Left**) the red line is the mass weighted median for α_{vir} from the simulation. The orange shaded region includes data within the 16th and 84th quantile of α_{vir} values. The dashed lines correspond to the start of the second passage and the final coalescence of the two nuclei. The two solid lines correspond to the merger times shown in Fig. 3. The upper panel shows SFR vs time for the second coalescence and the right panel shows the 16th, 50th and 84th quantile of α_{vir} for PHANGS, NGC 3256 and the Antennae from the observations. In calculating α_{vir}, we use the U/LIRG α_{CO} for NGC 3256 and the Milky Way value for PHANGS and the Antennae. (**Right**) Same plot for G2&G3 merger in the 'e1' orbit during the final coalescence. The 'e1' orbit has a smaller impact parameter than the 'e2' orbit.

than in normal spiral galaxies. We adopt the same approach as in observation to calculate α_{vir} for fixed-scale GMC pixels as (Sun et al. 2018)

$$\alpha_{vir} = \frac{9 \ln 2}{2\pi G} \frac{\sigma_v^2}{\Sigma_{mol} R} = 5.77 \left(\frac{\sigma_v}{\text{km } s^{-1}} \right)^2 \left(\frac{\Sigma_{mol}}{\text{M}_{\odot}} \right)^{-1} \left(\frac{R}{40 \text{ pc}} \right)^{-1} \tag{1}$$

where R is the radius of GMCs. In Sun et al. (2018), R is set to be the radius of the beam in the image, as each beam is treated as an independent GMC. For the simulation data, we do not have a telecope beam to convolve with, so we set R to be half the size of each pixel as each pixel can be treated as an independent GMC. We note that the pixel in simulation is equivalent to the beam in observation since both are independent measurement units. With R a constant, α_{vir} depends only on σ_v and Σ_{mol}. The higher σ_v at a similar Σ_{mol} thus implies that α_{vir} values for GMCs in simulated mergers are higher than the values for PHANGS or simulated isolated galaxies. Higher values for α_{vir} are also found for NGC 3256 (Brunetti et al. 2021) and the Antennae (Brunetti 2022). Fig. 4 shows α_{vir} as a function of time during the period near the second pericentric passage for the merger simulation with "e2" and "e1" orbit and viewed from "v0" angle. We can see that α_{vir} stays low before the second passage and suddenly rises after the passage along with a sudden increase in SFR. The peak of median α_{vir} can reach ~ 100. We note that both the Antennae and NGC 3256 are at the very start of their second passages. In this stage, α_{vir} is quite time-sensitive and it is hard to match the exact same stage between the simulated and observed galaxies. Therefore, it is possible that both NGC 3256 and Antennae are caught at a specific merger stage with a lower $\alpha_{\rm vir}$ (although in the case of NGC 3256, still enhanced relative to PHANGS galaxies). In comparison, $\alpha_{\rm vir}$ in the simulations is relatively stable in the post merger stage. This stability suggests that a comparison between simulations and observations of post merger galaxies could be a useful next step. In addition, post mergers have rather simple morphology, which simplifies the task of making quantitative comparisons.

4. Future work

In the future, we would like to expand our comparison to more observed and simulated mergers. From the observational side, we need larger samples of galaxy mergers spanning different evolutionary stages in order to understand how GMCs evolve throughout the merging. In addition, it is easier to compare the observations with simulations at post-merger stages since the morphology is simpler and easier to quantify. In the ALMA archive, we have already had at least 10 U/LIRGs with GMC resolution CO 2-1 observations. We can utilize these archival data to build a more complete sample of GMCs in mergers at different stages. From the simulation side, we can see that the initial set-up could affect our interpretation on the simulation results. Therefore, it is necessary for us to compare with simulations that better match the observed galaxies. For the two observed mergers we have, the Antennae has been widely studied and matched by non-cosmological simulations (e.g. Renaud et al. 2019a; Li et al. 2022). Besides comparing with these non-cosmological simulations, we could also compare observation with some latest cosmological simulations, such as FIREBox (Feldmann et al. 2022), that includes local mergers.

References

Brunetti, N., Wilson, C.D., Sliwa, K. 2021, MNRAS, 500, 4730
Brunetti, N. 2022, thesis
Feldmann, R., Quataert, E., Faucher-Gigu'ere, C.-A., et al. 2022, 2022arXiv220515325F
Leroy, A. K., Hughes, A., Schruba, A., et al. 2016, ApJ, 831, 16
Leroy, A. K., Schinnerer, E., Hughes, A., et al. 2021, ApJS, 257, 43
Li, H., Vogelsberger, M., Bryan, G. L., et al. 2022, MNRAS, 514, 265
Moreno, J., Torrey, P. Ellison, S., 2019, MNRAS, 485, 1320
Nieten, C., Neininger, N., Gu'elin, M., et al. 2006, A&A, 453, 459
Renaud, F., Bournaud, F., Agertz, O., et al. 2019a, A&A, 625, A65
Saintonge, A., Catinella, B., Tacconi, L. J., et al. 2017, APJS, 233, 220
Sun, J., Leroy, A. K., Schruba, A., et al. 2018, ApJ, 860, 172
Sun, J., Leroy, A. K., Schinnerer, E., et al. 2020, ApJL, 901, L8

Resolving the Rise and Fall of Star Formation in Galaxies
Proceedings IAU Symposium No. 373, 2023
T. Wong & W.-T. Kim, eds.
doi:10.1017/S1743921322003738

Variation of the molecular cloud lifecycle across the nearby galaxy population

Jaeyeon Kim[1], Mélanie Chevance[1,2], J. M. Diederik Kruijssen[2] and Adam K. Leroy[3,4]

[1]Zentrum für Astronomie der Universität Heidelberg, Institut für Theoretische Astrophysik, Albert-Ueberle-Str. 2, 69120 Heidelberg

[2]Cosmic Origins Of Life (COOL) Research DAO, coolresearch.io

[3]Department of Astronomy, The Ohio State University, 140 West 18th Avenue, Columbus, OH 43210, USA

[4]Center for Cosmology and Astroparticle Physics, 191 West Woodruff Avenue, Columbus, OH 43210, USA

Abstract. The processes of star formation and feedback take place on the scales of giant molecular clouds (GMCs; $\sim$100 pc) within galaxies and play a major role in governing galaxy evolution. By applying a robust statistical method to PHANGS observations, we systematically measure the evolutionary timeline from molecular clouds to exposed young stellar regions, across an unprecedented sample of 54 galaxies. These timescales depend on galaxy environment, revealing the role of galactic-scale dynamical processes in the small-scale GMC evolution. Furthermore, in the 5 nearest galaxies of our sample, we have refined the GMC timeline further and established the duration of the heavily obscured phase, using 24 μm emission. These results represent a major first step towards a comprehensive picture of cloud assembly and feedback, which will be extended to 19 more galaxies with our ongoing JWST Large Program.

Keywords. stars: formation – ISM: clouds – ISM: structure – galaxies: ISM – galaxies: star formation

1. Introduction

Giant molecular clouds (GMCs) are primary sites for star formation. The properties of the clouds are shaped by the large-scale environment of their host galaxies, directly linking the initial conditions of star formation to galactic-scale properties (e.g. Sun et al. 2022). In turn, the energy, momentum and metals deposited by stellar feedback drive the continuous evolution of the interstellar medium (ISM) in general (e.g. Chevance et al. 2022). The characterisation of the evolutionary time-scales from GMC assembly to star formation, and to young stellar regions devoid of cold gas provides important insights into which physical mechanisms regulate this multi-scale cycle, and is therefore crucial to understanding the evolution of galaxies.

Growing evidence from cloud-scale observations is revealing spatial offsets between molecular gas and young stellar regions, and points towards a view of GMCs as transient objects that are dispersed quickly by violent feedback from young massive stars (e.g. Kawamura et al. 2009; Kruijssen et al. 2019; Chevance et al. 2020). This contradicts a conventional view where GMCs are considered to represent quasi-equilibrium structures that survive over a large fraction of a galactic rotation period (e.g. Scoville & Hersh 1979; Koda et al. 2009).

Kruijssen & Longmore (2014) and Kruijssen et al. (2018) developed a statistical method, called the 'uncertainty principle for star formation', which translates the observed spatial decorrelation between gas and young stellar regions into their underlying evolutionary timeline, spanning from cloud assembly to subsequent star formation and cloud dispersal, and finally to young stellar regions free of molecular gas. Applying this method to CO and Hα observations, several works have determined the evolutionary timeline between quiescent molecular gas to exposed young stellar regions for about a dozen of nearby galaxies (Kruijssen et al. 2019; Chevance et al. 2020; Zabel et al. 2020; Kim et al. 2021; Lu et al. 2022; Ward et al. 2022). So far, these time-scale measurements have been limited to a small number of galaxies due to the lack of CO imaging of star-forming discs at cloud-scale resolution and the fact that our method requires us to resolve at least the separation length between independent star-forming regions (100–300 pc). These previous studies did not allow us to identify the key environmental factors and cloud properties (e.g. total, gas, molecular gas surface densities and masses) responsible for setting these time-scales.

In Kim et al. (2022a), we greatly increase the number of galaxies to 54 galaxies, for which the evolutionary sequence between GMCs and young stellar regions has been determined. We capitalise on our CO observations from PHANGS–ALMA (Leroy et al. 2021) and a narrow-band Hα survey by A. Razza et al. (in prep.; PHANGS–Hα). By applying the statistical method described above to these galaxies, we systematically obtain the evolutionary sequence of GMCs from a quiescent molecular cloud phase to a feedback dispersal phase, and finally to a gas-free HII region phase. This statistically representative PHANGS sample covers a large range of galactic properties ($\sim$2 dex in stellar mass) and diverse spiral arm morphologies. This enables us to quantitatively determine the correlations between the small-scale evolutionary cycle of molecular clouds and galactic-scale environmental properties.

2. Evolutionary timeline from molecular gas to young stars in 54 PHANGS galaxies

Here we present the results obtained from the application of our statistical method (Kruijssen & Longmore 2014; Kruijssen et al. 2018) to CO and Hα observations as tracers of molecular gas and star formation rate (SFR) for 54 galaxies in the PHANGS sample. We measure the cloud lifetime ($t_{\rm CO}$), the phase during which both molecular gas and SFR tracers overlap, the feedback time-scale ($t_{\rm fb}$), and the characteristic separation length between independent regions evolving from gas to stars (λ). We also derive the integrated star formation efficiency per cloud lifecycle ($\epsilon_{\rm sf} = t_{\rm CO}\Sigma_{\rm SFR}/\Sigma_{\rm H_2}^{\rm compact}$, where $\Sigma_{\rm SFR}$ and $\Sigma_{\rm H_2}^{\rm compact}$ are the area-weighted average surface densities of SFR and compact molecular gas across the field of view, respectively). In Figure 1, we show the distributions of our main measurements across the galaxy sample.

Across all galaxies in our sample, the range of measured cloud lifetimes is 5 –30 Myr, with an average of 16 Myr and a 16–84 percentile range of 11–22 Myr. The range of our measurements of $t_{\rm CO}$ corresponds to 1–3 times the average crossing time-scale of massive GMCs in PHANGS–ALMA (Sun et al. 2022), which suggests that clouds are transient objects that disperse within a small multiple of their dynamical time-scale. For 12 galaxies, we do not sufficiently resolve the separation between independent regions and therefore we are only able to obtain upper limits on $t_{\rm fb}$. Without these galaxies, the range of feedback time-scale becomes 1.3–5.4 Myr, constituting 10–30% of the cloud lifetime, with an average and a standard deviation of 3.2 ± 1.1 Myr. This time-scale represents the time it takes for emerging massive stars (visible in Hα) to disrupt the surrounding molecular gas (either by kinetic dispersal or by the photodissociation of CO molecules).

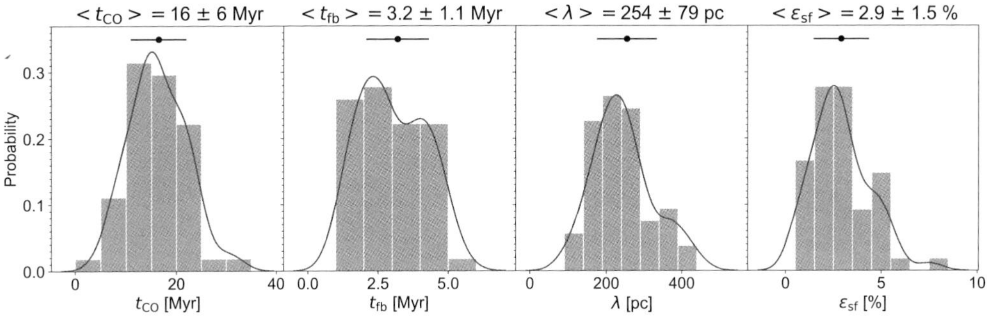

Figure 1. Distributions of the main physical quantities describing the molecular cloud lifecycle in 54 PHANGS galaxies. From left to right, we show a Gaussian kernel density estimate of the cloud lifetime ($t_{\rm CO}$), the feedback time-scale ($t_{\rm fb}$), the average separation length between independent star-forming regions (λ), and the integrated star formation efficiency ($\epsilon_{\rm sf}$). For each histogram, the mean (black dot) and 16-84 percentile range (error bar) are indicated at the top of the panel. Taken from Kim et al. (2022a).

The overall measured range of molecular cloud lifetimes and the feedback time-scales are consistent with that found in previous studies, those using cloud classification methods (e.g. Engargiola et al. 2003; Kawamura et al. 2009), and those using the same statistical method (Kruijssen et al. 2019; Chevance et al. 2020; Kim et al. 2021). This short feedback timescale implies that supernova explosions (4–20 Myr after star formation) take place in a medium that is already heavily pre-processed by early feedback.

For the 46 galaxies where we sufficiently resolve independent regions to get reliable measurements on λ, we find that λ ranges from 100 pc to 400 pc, with an average and standard deviation of 254 ± 79 pc. This is similar to the total cold gas disc thickness (100-300 pc; Heyer & Dame 2015; Patra 2020; Yim et al. 2020), as well as the range of values found in previous application of the same method to relatively nearby and well-resolved galaxies (100-250 pc; Kruijssen et al. 2019; Chevance et al. 2020; Kim et al. 2021). From the similarity of λ to the gas disc scale height, Kruijssen et al. (2019) have suggested that this characteristic length scale might be set when feedback-driven bubbles break out of the galactic disc and depressurise.

We measure a star formation efficiency per GMC of $0.7 - 7.5\%$ across our galaxy sample, illustrating that rapid stellar feedback strongly restricts the conversion of gas to stars in these clouds by limiting their lifetime. Our previous measurements of $\epsilon_{\rm sf}$ (Kruijssen et al. 2019; Chevance et al. 2020; Kim et al. 2021) also fall within this range of values.

3. Relations with global galaxy and average cloud properties

In Figure 2, we show several statistically significant correlations identified between our measurements and galaxy properties. In brief, $t_{\rm CO}$ shows positive correlations with quantities related to galaxy mass (e.g. stellar mass $M_{*,\rm global}$), as well as with the molecular gas surface density ($\Sigma_{\rm H_2}$). These correlations could be caused by higher gas pressures at higher galaxy masses or higher gas surface densities, which may extend the cloud lifetime. Additionally, the higher pressures allow us to also detect the extended envelopes of the GMCs, unlike in low-surface density (atomic gas-dominated) environments where CO is only tracing the densest centres of the GMCs. Finally, in low-mass (low-metallicity) and less dense environments, the molecular gas may be CO-dark, as these environments require higher gas column density in order to shield CO molecules from being dissociated. All of these mechanisms would make the CO-bright GMC lifetime increase with galaxy mass or gas surface density.

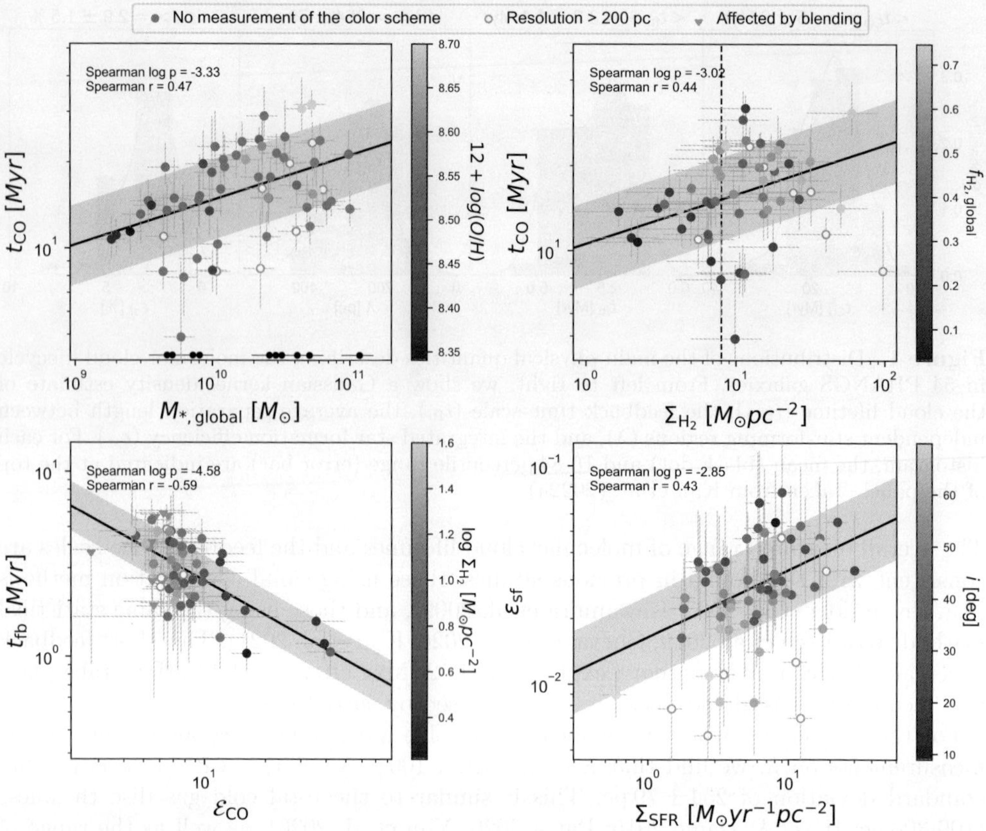

Figure 2. Examples of four statistically significant correlations between our measurements and galactic properties. In the upper panels, the measured cloud lifetime ($t_{\rm CO}$) is shown as a function of galaxy stellar mass ($M_{*,\rm global}$; left) and molecular gas surface density ($\Sigma_{\rm H_2}$; right), with points colour-coded by metallicity ($12 + \log{\rm (O/H)}$) and molecular gas fraction ($f_{\rm H_2,global}$), respectively. The density proposed by Chevance et al. (2020) to divide between internal dynamics and galactic dynamics-driven GMC lifetimes is indicated by the dashed line at $8\,M_\odot{\rm pc}^{-2}$. In the bottom left panel, the measured feedback time-scale ($t_{\rm fb}$) is shown as a function of the surface density contrast between CO emission peaks and the galactic average ($\varepsilon_{\rm CO}$), colour-coded by the molecular gas surface density ($\Sigma_{\rm H_2}$). The bottom right panel shows the integrated star formation efficiency ($\epsilon_{\rm sf}$) as a function of SFR surface density ($\Sigma_{\rm SFR}$), where the points are colour coded by galaxy inclination (i). Galaxies without measurements corresponding to each colour bar scheme are denoted in magenta. Gray circles are the galaxies with resolution worse than 200 pc. Gray triangles indicate upper limits for galaxies suffering from blending of sources (see Kim et al. 2022a). All gray points are excluded from the correlations. For individual data points, 1σ uncertainties are shown. In each panel, the best-fitting linear regression (solid line), 1σ scatter of the data around the fit (shaded area), and the Spearman p-values and correlation coefficients are indicated. Masses of the ten galaxies excluded from our sample due to insufficient emission peaks are also shown (black circles, top left panel). Taken from Kim et al. (2022a).

The feedback time-scale $t_{\rm fb}$ also shows correlations with quantities related to galaxy mass, most likely because $t_{\rm CO}$ and $t_{\rm fb}$ are correlated. In addition, $t_{\rm fb}$ also shows an interesting relation with $\varepsilon_{\rm CO}$, which is the surface density contrast measured in a CO map between emission peaks and the galactic average. We find that $t_{\rm fb}$ is shorter with increasing $\varepsilon_{\rm CO}$ (i.e. towards sharper emission peaks). This can be physically understood as a direct consequence of stellar feedback, with CO emission becoming undetected faster after

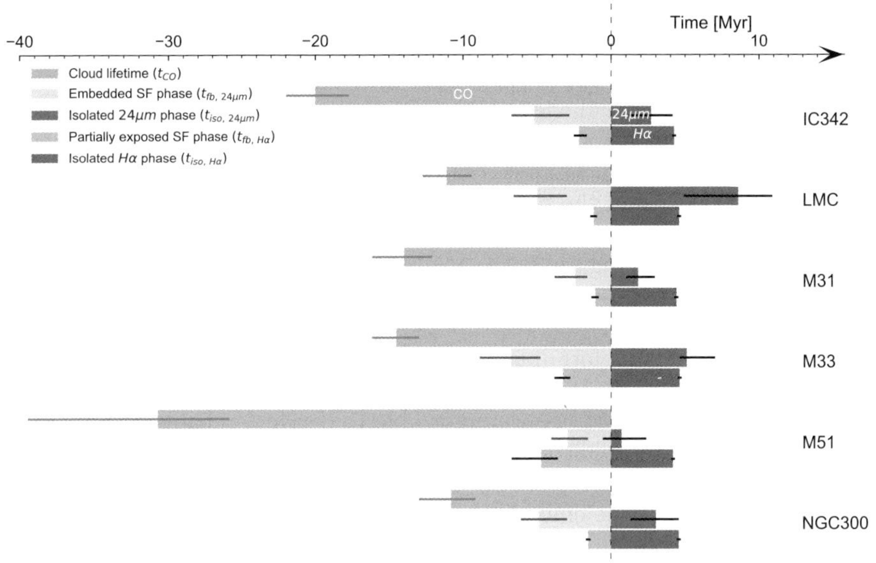

Figure 3. Timelines describing the evolution of molecular clouds from an inert gas phase, to an embedded star-forming phase, and then finally to exposed young stellar regions. The time during which CO is visible ($t_{\rm CO}$) is indicated in orange, the time during which $24\,\mu$m and Hα are visible without CO are shown respectively in dark blue and dark purple. The timescales associated to the feedback phase, during which both CO and SFR tracer emissions are observed co-spatially are shown in light blue (for $24\,\mu$m) and light purple (for Hα). The corresponding 1σ error bars are also indicated. We note that for M51 the feedback timescale constrained using $24\,\mu$m ($t_{\rm fb,\,24\,\mu m}$) is likely to be biased by deconvolution artifacts (see Kim et al. 2021 for details). Taken from Kim et al. (2021).

the onset of massive star formation when the surrounding medium is sparser, facilitating a rapid dispersal of the gas.

The star formation efficiency $\epsilon_{\rm sf}$ shows a correlation with the SFR surface density. This can be at least partly understood by the fact that, naively, a higher SFR would naturally lead more stars formed within a cloud in a given time, and therefore to a higher integrated star formation efficiency per cloud. However, other factors can also contribute to this trend such as the tight correlation between $t_{\rm CO}$ and $\Sigma_{\rm H_2}$, as well as the dependence of extinction correction on inclination, as indicated by the data point colours in Figure 2 showing a trend between i and $\Sigma_{\rm SFR}$.

4. Duration of the early phase of massive star formation in five nearby galaxies

While Hα is typically used as a SFR tracer, the earliest phases of star formation are heavily embedded and invisible in Hα due to the extinction from the surrounding dense gas and dust. Therefore, the duration of these early phases of star formation and the time that clouds spend being truly inert are still poorly constrained, and so is the time needed for the feedback from these embedded stars to pierce through the natal cloud and enable the detection of Hα.

The $24\,\mu$m emission in the mid-infrared observed by the Multiband Imaging Photometer aboard the Spitzer Space Telescope provides an unbiased tracer of embedded massive star formation (Kennicutt & Evans 2012). This mid-infrared emission originates from stochastically heated small dust grains that do not require ionizing photons to be excited but do empirically correlate with tracers of massive star formation and so can be used as a tracer of the presence of large amounts of dust reprocessed photospheric light from massive stars.

In Kim et al. (2021), we apply the same statistical method (Kruijssen et al. 2018) to six nearby star-forming galaxies using CO, *Spitzer* 24 μm, and Hα emission maps, tracing molecular clouds, embedded star formation, and exposed star formation, respectively. The small size of this sample results from the limited resolution of the *Spitzer* 24 μm observations (6") and the requirement that the observations need to resolve each galaxy into its distinctive units of star formation (e.g. GMCs and HII regions, typically separated by $\sim$ 100 pc). In Figure 3, we show an illustration of the evolutionary timelines of molecular clouds and star-forming regions in the six galaxies included in our analysis. GMCs initially emit only in CO, then in 24 μm after the onset of star formation and finally in Hα when the star-forming regions become (partially) exposed. This provides systematic constraints on the duration of the embedded phase of star formation for five of these six galaxies. For M 51, we were not able to derive a reliable result, because the deconvolution algorithms applied to the 24 μm map (Dumas et al. 2011) did not lead to a sufficient data quality to successfully perform this measurement. For the remaining five galaxies, we find that the embedded phase of massive star formation, defined as the time for which CO and 24 μm overlap ($t_{\mathrm{fb24}\,\mu\mathrm{m}}$), to last for 2$-$7 Myr, or 20$-$50% of the cloud lifetime. This indicates that clouds are devoid of massive stars for 50$-$80% of their lifetime. The first half of the embedded phase (1$-$4 Myr) is heavily obscured and invisible in Hα emission.

5. Conclusion

We have quantified the evolutionary lifecycle of GMC formation, evolution, and dispersal across 54 nearby disc galaxies, which is the largest and most statistically complete sample to date. We have demonstrated that this lifecycle depends on the large-scale galactic environment. We have identified strong correlations between GMC evolutionary time-scales and the host galaxy properties (e.g. stellar mass, molecular gas surface density), revealing the role of galactic-scale processes in the small-scale GMC lifecycle. In the 5 most nearby galaxies of our sample, we were able to refine this timeline further and constrain the heavily obscured phase using the *Spitzer* 24 μm observations. In the future, we plan to further extend and refine this evolutionary timeline, for a subset of our parent galaxy sample, by including other observations at different wavelengths: ionised emission lines from MUSE, mid-infrared from the *James Webb Space Telescope*, and HI from the VLA and Meerkat. This will allow us to determine the time-scales of all the successive phases of the gas that participates in star formation, identify their dependences on the large-scale environment, and derive constraints on the physical mechanisms that regulate GMC evolution and star formation in galaxies.

Acknowledgements.

JK gratefully acknowledges funding from the Deutsche Forschungsgemeinschaft (DFG, German Research Foundation) through the DFG Sachbeihilfe (grant number KR4801/2-1). MC gratefully acknowledges funding from the DFG through an Emmy Noether Research Group (grant number CH2137/1-1). COOL Research DAO is a Decentralised Autonomous Organisation supporting research in astrophysics aimed at uncovering our cosmic origins. JMDK gratefully acknowledges funding from the DFG through an Emmy Noether Research Group (grant number KR4801/1-1), as well as from the European Research Council (ERC) under the European Union's Horizon 2020 research and innovation programme via the ERC Starting Grant MUSTANG (grant agreement number 714907).

References

Chevance M., Kruijssen J. M. D., Hygate A. P. S., *et al.*, 2020, *MNRAS*, 493, 2872
Chevance M., Krumholz, Mark R. and McLeod, Anna F., *et al.*, 2022, arXiv:2203.09570

Dumas G., Schinnerer E., Tabatabaei F. S., *et al.*, 2011, *AJ*, 141, 41
Engargiola G., Plambeck R. L., Rosolowsky E., Blitz L., 2003, *ApJS*, 49, 343
Heyer M., Dame T. M., 2015, *ARAA*, 53, 583
Kawamura A., Mizuno1 Y., Minamidani T., *et al.*, 2009, *ApJS*, 184, 1
Kennicutt, R. C., & Evans, N. J. 2012, *ARAA*, 50, 531
Kim J., Chevance M., Kruijssen J. M. D., *et al.*, 2021, *MNRAS*, 504, 487
Kim J., Chevance M., Kruijssen J. M. D., *et al.*, 2022a, *MNRAS*, 516, 3006
Koda, J. and Scoville, N. and Sawada, T., *et al.*, 2009, *ApJL*, 700, 132
Kruijssen J. M. D., Longmore S. N., 2014, *MNRAS*, 439, 3239
Kruijssen J. M. D., Schruba A., Chevance M., *et al.*, 2019, *Nature*, 569, 519
Kruijssen J. M. D., Schruba A., Hygate A. P. S., *et al.*, 2018, *MNRAS*, 479, 1866
Krumholz M. R., 2014, *Phys. Rep.*, 539, 49
Leroy A. K., Schinnerer E., Hughes A., *et al.*, 2021, *ApJS*, 257, 43
Lu, A. and Boyce, H. and Haggard, D., *et al.*, 2022, *MNRAS*, 514, 5035
Patra N. N., 2020, *A&A*, 638, A66
Scannapieco, C. and Wadepuhl, M. and Parry, O. H., *et al.*, 2012, *MNRAS*, 423, 1726
Scoville N. Z., Hersh K., 1979, *ApJ*,229, 578
Sun J., Leroy A. K., Rosolowsky E., *et al.* 2022, *AJ*, 164, 43
Ward, J. L. and Kruijssen, J. M. D. and Chevance, M., *et al.* 2022, *MNRAS*, 516, 4025
Yim K., Wong T., Rand R. J., Schinnerer E., 2020, *MNRAS*, 494, 4558
Zabel, N. and Davis, T. A. and Sarzi, M., *et al.* 2020, *MNRAS*, 496, 2155

Resolving the Rise and Fall of Star Formation in Galaxies
Proceedings IAU Symposium No. 373, 2023
T. Wong & W.-T. Kim, eds.
doi:10.1017/S1743921322003908

Star Formation History of Two Fields in the Halo of NGC 5128

Sima T. Aghdam[1], Atefeh Javadi[1], Seyedazim Hashemi[2,3], Jacco Th. van Loon[4], Habib Khosroshahi[1], Roya H. Golshan[5], Elham Saremi[1,6,7] and Maryam Saberi[8]

[1]School of Astronomy, Institute for Research in Fundamental Sciences (IPM), Tehran, 19568-36613, Iran
email: sima.t.ahdam@gmail.com

[2]Department of Physics and Astronomy, University of California Riverside, CA 92521, The USA

[3]Department of Physics, Sharif University of Technology, Tehran, 11155-9161, Iran

[4]Lennard-Jones Laboratories, Keele University, ST5 5BG, UK

[5]Physikalisches Institut der Universität zu Köln, Zülpicher Str. 77, 50937, Köln, Germany

[6]Instituto de Astrofisica de Canarias, C/ Vìa Làctea s/n, 38205 La Laguna, Tenerife, Spain

[7]Departamento de Astrofisica, Universidad de La Laguna, 38205 La Laguna, Tenerife, Spain

[8]Rosseland Centre for Solar Physics, University of Oslo, P.O. Box 1029, Blindern, NO-0315, Oslo, Norway

Abstract. NGC 5128 galaxy is a giant elliptical galaxy located in the Centaurus group of galaxies at 3.8 Mpc. We aim to study the star formation history (SFH) of two different fields of the galaxy. The northeastern field (Field 1) is located at a distance of 18.8 kpc, while the southern field (Field 2) is at 9.9 kpc. We use a photometric method that is based on identifying long period variable (LPV) stars and asymptotic giant branch (AGB) stars, as they are strong tracers of star formation and galaxy evolution due to their luminosity and variability; 395 LPVs in Field 1 and 671 LPVs in Field 2 have been identified. These two fields present similar SFHs, although the SF rate of Field 2 is more enhanced. We find that the galaxy has three major star formation episodes t $\sim$ 800 Myr ago, t $\sim$ 3.2 Gyr ago, and t $\sim$ 10 Gyr ago, where t is look-back time. The rate of star formation at $\sim$ 800 Myr ago agrees with previous studies suggesting that the galaxy experienced a merger around that time. Furthermore, NGC 5128 has experienced a lower star formation rate in its recent history which could have been driven by jet-induction star formation and multiple outbursts of AGN activity in this galaxy, as well as a minor merger around 400 Myr ago.

Keywords. stars: AGB and LPV – stars: formation – galaxies: jets – galaxies: nuclei – galaxies: halos – galaxies: evolution – galaxies: star formation – galaxies: individual: NGC 5128

---

1. Introduction

There is no giant elliptical galaxy (GE) in the local group (LG), so to study the various features of GE galaxies, we must look at the nearest elliptical galaxy in other groups. NGC 5128 (or Centaurus A), with a distance of 3.8 Mpc ($\mu = 27.87 \pm 0.16$ mag; Rejkuba et al. 2004a, and $E(B-V) = 0.15 \pm 0.05$ mag; Rejkuba et al. 2001) provides for us a distinctive opportunity to study the nearest GE galaxy (Harris et al. 1999;

Charmandaris et al. 2000; Rejkuba 2004b; Rejkuba et al. 2005), located in the Centaurus group of galaxies (Karachentsev 2005).

We aim to find the SFH of two small fields in the halo of NGC 5128 using long period variable stars (LPVs) to better understand the relation between the SFH of the halo and likely recent merger. LPV stars are evolved asymptotic giant branch stars (AGBs) (e.g. Fraser et al. 2005, 2008; Soszyński et al. 2009) which are identified according to high luminosity (1000 to $60,000\ L_{\odot}$), as well as their variability ($\sim$ 100 to 1300 days). In addition, red supergiant (RSG) stars are massive stars with a mass up to $\sim 30\ \mathrm{M}_{\odot}$, known as a tracer of recent star formation (about $10-30$ Myr) in the galaxy with a strong radial pulsation (Vassiliadis & Wood 1993; Javadi et al. 2011b; van Loon et al. 1999, 2005).

2. Data

We used the data that was obtained by the ISAAC near-IR imaging spectrometer at the ESO Paranal UT1 Antu 8.2m Telescope (VLT) in two different fields in the halo of the galaxy that were published by Rejkuba et al. (2003), which have been separated into two fields in the halo of the galaxy and identified as Field 1 and Field 2 by Rejkuba et al. (2001) and Rejkuba et al. (2003). Field 1 is on the distinguished north-eastern part of the halo, at a distance of $\sim$ 18.8 kpc from the center of the galaxy whereas Field 2 is located at a distance of $\sim$ 9.9 kpc from the center (Rejkuba et al. 2001). The detected evolved stars to investigate SFH are the LPV stars with periods longer than 70 days (from Rejkuba et al. 2003). Therefore, 395 and 671 LPVs are selected in Field 1 and Field 2, respectively.

3. Method

We used a method to study the SFH based on LPVs and it was developed by Javadi et al. (2011b) which was explained in the previous papers in details (M33: Javadi et al. 2011b, 2011c, 2016, 2017; LMC & SMC: Rezaei et al. 2014; NGC 147 & NGC 185: Hamedani Golshan et al. 2017; IC 1613: Hashemi et al. 2019; Andromeda VII: Navabi et al. 2021; Andromeda I: Saremi et al. 2021). In the first step, we should obtain mass, age, and pulsation duration of LPVs (the duration that stars are in the LPV phase) by applying Padova evolutionary models (Marigo et al. 2017) for various constant metallicities. Furthermore, for studying the star formation history of the galaxy, we consider $Z = 0.039$ (Woodley et al. 2010; Rejkuba et al. 2011). By assuming all mentioned parameters, the SFRs for different bins with specified intervals in age and mass as follows:

$$\xi(t) = \frac{dn'(t)}{\delta t} \frac{\int_{\min}^{\max} f_{\mathrm{IMF}}(m) m\ dm}{\int_{m(t)}^{m(t+dt)} f_{\mathrm{IMF}}(m)\ dm} \tag{3.1}$$

where m is mass, $f_{\mathrm{IMF}}(m)$ is Kroupa initial mass function (IMF) (Kroupa 2001), dn' is the observed LPVs in each bin, and δt is the pulsation duration.

The statistical error bars for each bin come from the Poisson distribution:

$$\sigma_{\xi(t)} = \frac{\sqrt{N}}{N} \xi(t) \tag{3.2}$$

where N is the number of stars in each age bin.

Stellar evolution in the AGB phase causes that LPV stars inject dust into the interstellar medium (ISM). Therefore, the light from the LPVs will be detected with a fainter magnitude (Javadi et al. 2011b; van Loon et al. 1999, 2005). To correct the circumstellar extinction, we should do a de-reddened process by plotting the color-magnitude diagram and overlaying the theoretical isochrones by Marigo et al. (2017).

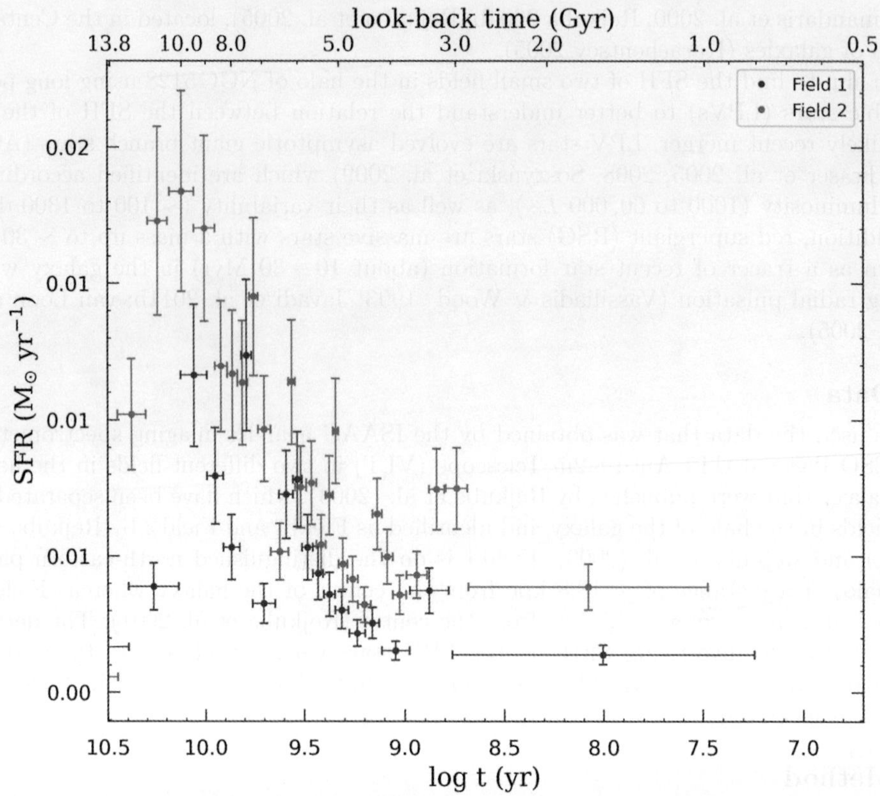

Figure 1. The SFH of Field 1 and Field 2 for Z = 0.039. *The blue* represents the SFH of Field 1, while *the red* one indicates the SFH of Field 2.

Furthermore, since NGC 5128 is a distant galaxy, it is unlikely to detect all of the LPVs in the galaxy. According to Rejkuba et al. (2003) study, they simulated variable stars by using 3 crucial parameters which are the mean magnitude of each star, period, and amplitude. By considering their simulation, we applied the detection probability of each star to our results. It should be noted that it is a key point in the presented method due to its dependency on the number of detected LPVs.

4. Results and Discussion

By considering metallicity Z = 0.039 and applying the probability functions, the SFHs of Field 1 and Field 2 are presented in Fig. 1. Despite being located in very different parts of the galaxy, these two fields have very similar SFHs, as well as the higher rates of star formation for Field 2. There is a similar supply of gas to stars during their evolution. In each field, the consistent patterns of SFRs is seen that are comparable for Field 1 and Field 2 in increasing in $\log t(\mathrm{yr}) \sim 8.9$ (t ∼ 800 Myr), $\log t(\mathrm{yr}) \sim 9.5$ (t ∼ 3.2 Gyr), and $\log t(\mathrm{yr}) \sim 10$ (t ∼ 10 Gyr) where *t* is look-back time.

Additionally, an increase in star formation in Field 1 and Field 2 around 800 Myr ago is consistent with the idea of recent merger of NGC 5128 with a small gas-rich galaxy (Israel 1998; Fassett & Graham 2000; Mould et al. 2000; Rejkuba et al. 2001). After $\log t(\mathrm{yr}) \sim 8.5$, the rate of recent star formation stays constant according to the recent merger which happened around 400 Myr ago (Peng et al. 2002). Furthermore, NGC 5128 has experienced AGN activity based on the central supermassive black hole (SMBH). There is an acceptable scenario leading to starbursts and AGN activity which is the interaction of gas-rich galaxies during mergers (Mo, Van den Bosch & White 2010).

References

Charmandaris V., Combes F., van der Hulst J. M., 2000, A&A, 356, L1

Fassett C. I., & Graham J. A., 2000, ApJ, 538, 594

Fraser O. J., Hawley S. L., Cook K. H., Keller S. C., 2005, AJ, 129, 768

Fraser O. J., Hawley S. L., Cook K. H., 2008, Astron. J, 136, 1242

Hamedani Golshan R., Javadi A., van Loon J. Th., et al. 2017, MNRAS, 466, 1764

Hashemi S. A., Javadi A., van Loon J. Th., 2019, MNRAS, 483, 4751

Harris G. L. H., Harris W. E., Poole G. B., 1999, Astron. J, 117, 855

Israel F., 1998, A&A Rev. 8, 237

Javadi A., van Loon J. Th., Mirtorabi M. T., 2011a, MNRAS, 411, 263

Javadi A., van Loon J. Th., Mirtorabi M. T., 2011b, MNRAS, 414, 3394

Javadi A., van Loon J. Th., Mirtorabi M. T., 2011c, in: Why Galaxies Care About AGB Stars II, eds. F. Kerschbaum, T. Lebzelter & R. F. Wing, ASPC, 445, 497

Javadi A., van Loon J. Th., Khosroshahi H., Mirtorabi M. T., 2013, MNRAS, 432, 2824

Javadi A., van Loon J. Th., Khosroshahi H., 2016, *Memorie della Società Astronomica Italiana*, 87, 278

Javadi A., van Loon J. Th., Khosroshahi H., et al., 2017, MNRAS, 464, 2103

Karachentsev I. D., 2005, Astron. J, 129, 178

Kroupa P., 2001, MNRAS, 322, 231

Marigo P., Girardi L., Bressan A., et al. 2017, ApJ, 835, 77

Mo H., Van den Bosch F., White S. D. M., 2010, *Galaxy formation and evolution. Cambridge University Press*

Mould J. R., Ridgewell A., Gallagher J. S., et al. 2000, ApJ, 536, 266

Navabi M., Saremi E., Javadi A., et al. 2021, ApJ, 910, 127

Peng E. W., Ford H. C., Freeman K. C., White R. L., 2002, Astron. J, 124, 3144

Rejkuba M., Minniti D., Silva D., Bedding T., 2001, A&A, 379, 781

Rejkuba M., Minniti D., Silva D., 2003, A&A, 406, 75

Rejkuba M., Greggio L., Zoccali M., 2004, A&A, 415, 915

Rejkuba M., 2004, A&A, 413, 903

Rejkuba M., Greggio L., Harris W. E., et al. 2005, ApJ, 631, 262

Rejkuba M., Harris W. E., Greggio L., Harris G. L. H., 2011, A&A, 526, A123

Rezaei Kh S., Javadi A., Khosroshahi H., van Loon J. Th., 2014, MNRAS, 445, 2214

Saremi E., Javadi A., Navabi M., et al. 2021, ApJ, 923, 164

Soszyński I., Udalski A., Szymański M. K., et al. 2009, ACTA, 59, 239

van Loon J. Th., Groenewegen M. A. T., de Koter A., 1999, A&A, 351, 559

van Loon J. Th., Cioni M.-R. L., Zijlstra A. A., Loup C., 2005, A&A, 438, 273

Vassiliadis E., & Wood P. R., 1993, ApJ, 413, 641

Woodley K. A., Harris W. E., Puzia T. H., et al. 2010, ApJ, 708, 1335

Resolving the Rise and Fall of Star Formation in Galaxies
Proceedings IAU Symposium No. 373, 2023
T. Wong & W.-T. Kim, eds.
doi:10.1017/S1743921322003933

The volumetric star formation law in nearby galaxies

C. Bacchini[1,2,3], F. Fraternali[2], G. Pezzulli[2], G. Iorio[4], A. Marasco[1] and C. Nipoti[3]

[1]INAF - Astronomical Observatory of Padova, Vicolo dell'Osservatorio 5, IT-35122, Padova, Italy
email: cecilia.bacchini@inaf.it

[2]Kapteyn Astronomical Institute, University of Groningen, Landleven 12, 9747 AD Groningen, The Netherlands

[3]Department of Physics and Astronomy, University of Bologna, via Gobetti 93/2, I-40129, Bologna, Italy

[4]Department of Physics and Astronomy, University of Padova, Vicolo dell'Osservatorio 3, IT-35122, Padova, Italy

Abstract. Star formation laws are empirical relations between the cold gas (HI+H_2) content of a galaxy and its star formation rate (SFR), being crucial for any model of galaxy formation and evolution. A well known example of such laws is the Schmidt-Kennicutt law, which is based on the projected surface densities. However, it has been long unclear whether a more fundamental relation exists between the intrinsic volume densities. By assuming the vertical hydrostatic equilibrium, we infer radial profiles for the thickness of gaseous discs in a sample of 23 local galaxies, and use these measurements to convert the observed surface densities of the gas and the SFR into the de-projected volume densities. We find a tight correlation linking these quantities, that we call the volumetric star formation law. This relation and its properties have crucial implications for our understanding of the physics of star formation.

Keywords. Star formation laws, ISM structure, spiral and dwarf galaxies, Milky Way.

1. Introduction

The characterisation of the relation between star formation and gas properties is crucial to understand how galaxies build their stellar mass with time. The most used and famous star formation is the Schmidt-Kennicutt (SK) law (Schmidt 1959; Kennicutt and Evans 2012), which is a power-law relation with slope $N \approx 1.4$ linking the surface densities of the star formation rate (SFR, $\Sigma_{\rm SFR}$) and the total cold gas (HI+H_2, $\Sigma_{\rm gas}$), namely $\Sigma_{\rm SFR} \propto \Sigma_{\rm gas}^N$. The *global* SK law involves the surface densities averaged over the whole star-forming disc, while the *local* SK law is built using either azimuthally-averaged radial profiles or pixel-by-pixel measurements. Both the global and the local relations show a break at about $\Sigma_{\rm gas} \approx 10\ M_\odot {\rm pc}^{-2}$ (e.g. Bigiel et al. 2010). This break is typically ascribed to the presence of a density threshold for star formation, meaning that the conversion of gas into stars becomes highly inefficient in low-density regions of galaxies. Here, we present a new *volumetric* star formation (VSF) law involving the volume densities instead of the surface densities (Bacchini et al. 2019a,b, 2020). First, we describe the crucial role of the disc thickness in the derivation of the volume densities. Then, we show that the

VSF law has no break and holds for both the high- and low-density regions of galaxies. Finally, we discuss the physical interpretation of our results.

2. The flaring of gas discs in galaxies

Our approach considers two fundamental properties of gas discs in galaxies, which are the thickness and the flaring. The gas pressure is expected to balance the gravitation pull toward the galaxy midplane (vertical hydrostatic equilibrium, hereafter VHE, Olling 1996). Since gravity weakens with increasing galactocentric distance, the gas layer becomes thicker moving from the inner to the outer regions of galaxies (flaring). This expectation is corroborated by observational works finding indications of the gas disc flaring in real galaxies (e.g. Yim et al. 2011; Marasco et al. 2017). Unfortunately, measuring the disc thickness from observations is possible only in very favourable cases, such as the Milky Way (MW) and nearby edge-on galaxies (although often with large uncertainties). Thus, we calculated the gas disc thickness relying on the assumption of the VHE.

We collected a sample of nearby galaxies including 10 spirals and 12 dwarf irregulars, and the MW. This sample allows us to probe not only the inner regions of spirals, which are high-density, molecular-gas rich, and high-metallicity environments, but also the spiral outskirts and the dwarf galaxies, which are low-density, atomic gas-dominated, and low-metallicity regions. Our goal is to convert the azimuthally-averaged radial profiles of the observed surface densities (Σ_i), which are typically used to derive the (local) SK law, into the midplane volume densities (ρ_i) by dividing for the scale height (h_i) of the disc: $\rho_i(R) = \Sigma_i(R)/(\sqrt{2\pi} h_i(R))$, where R is the galactocentric distance, and i is component considered (HI, H_2, or SFR). Our approach based on the VHE requires to measure the gas velocity dispersion and to calculate the gravitation potential of the galaxies. We accurately measured the radial profile of the gas velocity dispersion by modelling HI and CO emission line datacubes using the software ^{3D}BAROLO Di Teodoro and Fraternali (2015), which is based on the tilted-ring model. The galactic potential is derived using a mass model made of a stellar disc and bulge, a dark matter halo, and a gaseous disc, which is fitted to the galaxy rotation curve. For each galaxy, we derived the potential via numerical integration using a mass model obtained though the rotation curve decomposition. We then used the Python tool `Galpynamics` (Iorio 2018) to calculate the scale height for both the atomic and the molecular gas discs. For all the galaxies, we find a significant flaring for both the gas discs. The last ingredient that we need is the SFR scale height, which is unfortunately not known a priori. Hence, as a proxy, we used a linear combination of the scale heights of the atomic gas and molecular gas discs $h_{\rm SFR}(R) = h_{\rm HI}(R) f_{\rm HI}(R) + h_{\rm H_2}(R) f_{\rm H_2}(R)$, where the gas fractions $f_{\rm HI} \equiv \Sigma_{\rm HI}/\Sigma_{\rm gas}$ and $f_{\rm H_2} \equiv \Sigma_{\rm H_2}/\Sigma_{\rm gas}$ are used as weights. This prescription is based on the idea that the cold gas and young stars share the same vertical distribution. We compared this assumption for $h_{\rm SFR}$ with the scale height of classical Cepheids in the MW (age$\lesssim$ 200 Myr, Bono et al. 2005), finding an excellent agreement.

3. The volumetric star formation law

Having derived the volume densities for both the gas (HI and H_2) and the SFR, we obtained the VSF law. Figure 1 compares the SK law and the VSF law. In the left panel, some galaxies tend to follow the SK law in the high-surface density regime, but most of the galaxies depart from the relation in the low-surface density regime, resulting in a large scatter ($\sigma_\perp \approx 0.3$ dex). The points falling out of the SK law are typically those associated with the outskirts of spirals and dwarf galaxies, where the flaring is the most prominent. The situation is very different when we look at the right panel in Fig. 1

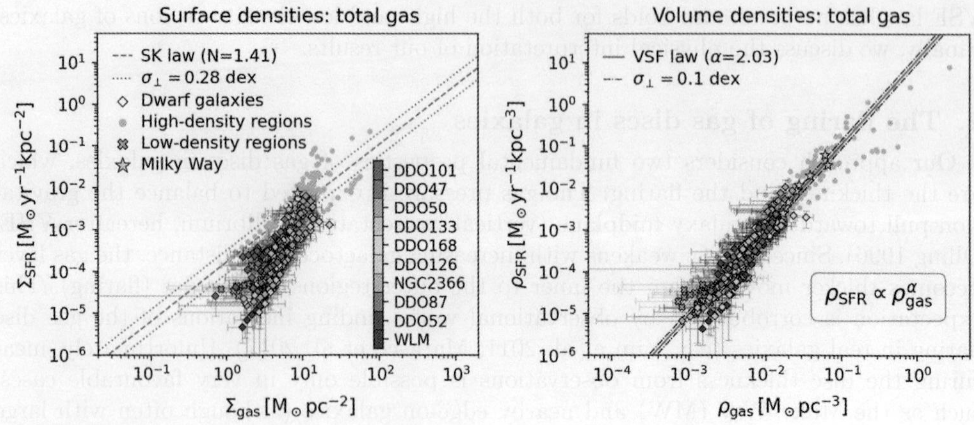

Figure 1. KS law (left) and VSF law (right) respectively based on the surface densities and the volume densities of nearby galaxies. For each galaxy, the azimuthally-averaged radial profiles of the cold gas ($HI+H_2$) and SFR densities are used. Diamonds are for dwarfs, which are individually tracked as indicated by the colour-bar. Lightblue points and green crosses respectively show the regions within and beyond the stellar disc of spirals. The yellow stars are for the MW. $\sigma_\perp$ is the orthogonal intrinsic scatter of the relations.

showing the volume densities. All the galaxies follow the same tight correlation, that is the VSF law. The best-fit relation to the volume densities is

$$\log\left(\frac{\rho_{\rm SFR}}{{\rm M}_\odot{\rm yr}^{-1}{\rm kpc}^{-3}}\right) = (2.03\pm0.03)\log\left(\frac{\rho_{\rm gas}}{{\rm M}_\odot{\rm pc}^{-3}}\right) + (1.10\pm0.01)\,. \tag{3.1}$$

Three compelling points arise from these results. First, the VSF law has a smaller scatter ($\sigma_\perp \approx 0.1$ dex) and spans a wider dynamical range in terms of both galaxy masses and gas densities than the SK law and other surface-based relations (e.g. spatially resolved main-sequence). These important improvements suggest that the VSF law is more fundamental than surface-based relations. The second point results from the absence of a break in the VSF law. This key difference with respect to the SK law indicates that there is no volume density threshold for star formation, at least for $\rho_{\rm gas} \gtrsim 10^{-3} M_\odot {\rm pc}^{-3}$. Hence, the break seen in the SK law is due to the projection effects of the disc flaring, rather than to a real drop in the star formation efficiency. Lastly, the slope of the VSF law disfavours the gravitational collapse as the main mechanism regulating star formation. Let us write the SFR volume density as $\rho_{\rm SFR} \propto \rho_{\rm gas}/\tau_{\rm sf}$, with $\tau_{\rm sf}$ being the timescale of star formation. The quadratic slope of the VSF law (Eq. 3.1) implies $\tau_{\rm sf} \propto \rho_{\rm gas}^{-1}$, which is inconsistent with the gravitational free-fall time but compatible with the timescales of the gas cooling and the H_2 formation, two important processes for star formation.

To conclude, the take-home messages from this proceeding are two. First, the flaring of gas discs in galaxies is significant, so assuming that they are thin or have a constant thickness is not a good approximation. Second, the VSF law (Eq. 3.1) is a new star formation law with smaller scatter and wider validity that the surface-based SK law, being likely more fundamental.

References

Bacchini, C., Fraternali, F., Iorio, G., & Pezzulli, G. 2019,a *A&A*, 622a, A64.
Bacchini, C., Fraternali, F., Pezzulli, G., & Marasco, A. 2020, *A&A*, 644, A125.
Bacchini, C., Fraternali, F., Pezzulli, G., Marasco, A., Iorio, G., & Nipoti, C. 2019,b *A&A*, 632b, A127.

Bigiel, F., Leroy, A., Walter, F., Blitz, L., Brinks, E., de Blok, W. J. G., & Madore, B. 2010, *AJ*, 140, 1194–1213.
Bono, G., Marconi, M., Cassisi, S., Caputo, F., Gieren, W., & Pietrzynski, G. 2005, *ApJ*, 621, 966–977.
Di Teodoro, E. M. & Fraternali, F. 2015, *MNRAS*, 451, 3021–3033.
Iorio, G. 2018,. PhD thesis, PhD Thesis, University of Bologna, (2018).
Kennicutt, R. C. & Evans, N. J. 2012, *ARA&A*, 50, 531–608.
Marasco, A., Fraternali, F., van der Hulst, J. M., & Oosterloo, T. 2017, *A&A*, 607, A106.
Olling, R. P. 1996, *AJ*, 112, 457.
Schmidt, M. 1959, *ApJ*, 129, 243.
Yim, K., Wong, T., Howk, J. C., & van der Hulst, J. M. 2011, *AJ*, 141, 48.

Resolving the Rise and Fall of Star Formation in Galaxies
Proceedings IAU Symposium No. 373, 2023
T. Wong & W.-T. Kim, eds.
doi:10.1017/S1743921323000145

The relation between GC systems and SMBH in spiral galaxies: The link to the $M_\bullet - M_*$ correlation

Rosa A. González-Lópezlira
Instituto de Radioastronomia y Astrofisica, UNAM, Campus Morelia

Abstract. We explore the relationship between globular cluster total number, $N_{\rm GC}$, and central black hole mass, $M_\bullet$, in spiral galaxies. Including cosmic scatter, log $M_\bullet \propto (1.64 \pm 0.24)$ log $N_{\rm GC}$. Whereas in ellipticals the correlation is linear [log $M_\bullet \propto (1.02 \pm 0.10)$ log $N_{\rm GC}$], and hence could be due to statistical convergence through mergers, this mechanism cannot explain the much steeper correlation in spirals. Additionally, we derive total stellar galaxy mass, M_*, from its two-slope correlation with $N_{\rm GC}$ (Hudson et al. 2014).In the $M_\bullet$ versus M_* parameter space, *with* M_* *derived from* $N_{\rm GC}$, $M_\bullet \propto (1.48 \pm 0.18)$ log M_* for ellipticals, and $M_\bullet \propto (1.21 \pm 0.16)$ log M_* for spirals. The observed agreement between ellipticals and spirals may imply that black holes and galaxies co-evolve through "calm" accretion, AGN feedback and other secular processes.

Keywords. black hole physics, galaxies: spiral, galaxies: formation, galaxies: evolution, galaxies: star clusters: general, globular clusters: general

---

1. Introduction

It is widely accepted that all massive galaxies contain a supermassive black hole (SMBH). In spheroidal systems, the masses of the SMBH, $M_\bullet$, correlate with other properties of their host galaxies: the bulge luminosity (the $M_\bullet$–$L_{\rm bulge}$ relation, e.g., Kormendy 1993; Kormendy & Richstone 1995), the bulge mass (the $M_\bullet$–$M_{\rm bulge}$ relation, e.g., Dressler 1989; Magorrian et al. 1998), the bulge stellar velocity dispersion (the $M_\bullet$–σ_* relation, e.g., Ferrarese & Merrit 2000; Gebhardt et al. 2000), the dark matter halo mass (Spitler & Forbes 2009). Today, the $M_\bullet$–σ_* relation is known to have a ≈ 0.6 dex scatter and depend on the galaxy's merger history (Bogdán et al. 2018; Sahu et al. 2019b), and the $M_\bullet$–$M_{\rm bulge}$ relation depends on the morphology of the galaxy (Savorgnan et al. 2016; van den Bosch 2016; Sahu et al. 2019a).

An intriguing correlation, given the extremely disparate scales, is the one between $M_\bullet$ and the total number of globular clusters ($N_{\rm GC}$; Burkert & Tremaine 2010; Harris & Harris 2011; Harris et al. 2014). The correlation can be expressed as $N_{\rm GC} \propto M_\bullet^{1.02\pm0.10}$, and spans over 3 orders of magnitude. This somewhat surprising finding has been since intensely explored, as reviewed by Kormendy & Ho 2013. Possible causal links have been proposed, e.g., feedback by the jets of AGN (e.g., Silk & Rees 1998; Fabian 2012) and cannibalization of GCs by black holes (e.g., Capuzzo-Dolcetta & Donnarumma 2001; Gnedin et al. 2014). Since it is roughly linear for ellipticals, it has been argued that the $M_\bullet - N_{\rm GC}$ correlation could be due to statistical convergence through merging (Peng 2007; Jahnke & Macciò 2011).

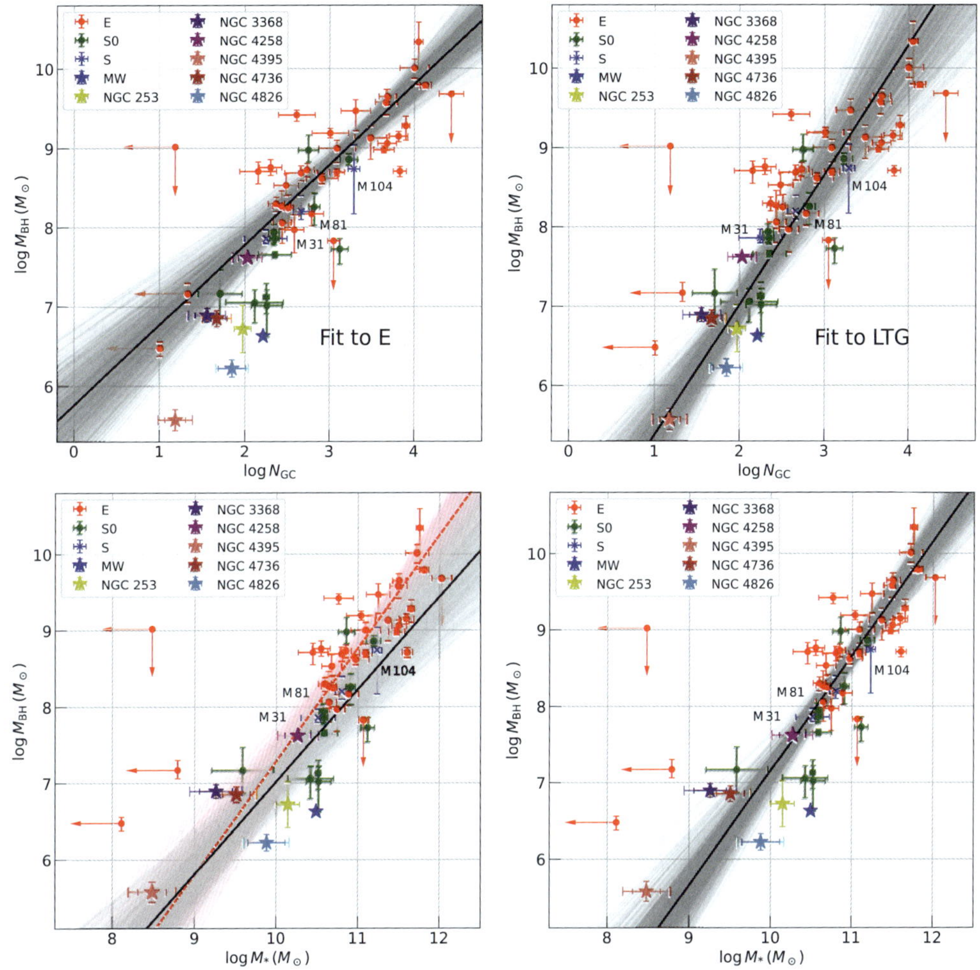

Figure 1. ***Top left:*** Bivariate Correlated Errors and Intrinsic Scatter (BCES; Akritas & Bershady 1996) fit to log $M_\bullet$ versus log $N_{\rm GC}$ of elliptical galaxies, with extra cosmic scatter in both variables. *Solid black lines:* best fits; *translucent gray lines:* range of possible solutions; 5% of the lines belong to fits separated by 2σ or more from the best values. Solid (*red*) and open (*green*) circles represent, respectively, elliptical and lenticular galaxies in the sample of Harris et al. 2014, reclassified as per Sahu et al. 2019b. ***Top right:*** BCES fit to log $M_\bullet$ versus log $N_{\rm GC}$ of LTG. Points and symbols as in left panel. The points in the left and right panels are the same, but only ellipticals are fit in the left, and only spirals in the right. ***Bottom left*** and ***bottom right:*** BCES fits to log $M_\bullet$ versus log M_* with extra cosmic scatter in both variables. *Bottom left, dashed red line:* best fit to elliptical galaxies; *translucent pink lines:* range of possible solutions for elliptical galaxies; *solid black line:* best fit to spiral galaxies; *translucent silver lines:* range of possible solutions for LTG. *Bottom right:* simultaneous fit to ellipticals and LTG. *Solid black lines:* best fit; *translucent gray lines:* range of possible solutions. Points in both panels are the same; ellipticals and LTG are fit separately in the left, simultaneously in the right.

2. The $M_\bullet$ versus $N_{\rm GC}$ correlation

Until relatively recently, there were only 5 spiral galaxies with precise measurements of both $N_{\rm GC}$ and $M_\bullet$: the Milky Way (MW, Sbc), M 104 (Sa), M 81 (Sab), M 31 (Sb), and NGC 253 (Sc). All of them, with the notable exception of the MW, fell right on the

$N_{GC} - M_\bullet$ correlation for ellipticals. Our Galaxy has a black hole that is about 1 order of magnitude lighter than expected from its N_{GC} and the correlation for ellipticals.

We have determined N_{GC} for five additional galaxies with precise black hole measurements (McConnell & Ma 2013): NGC 3368, NGC 4395, NGC 4258, NGC 4736, and NGC 4826. Hence, we have doubled the previous existing sample. Additionally, we have corrected the position of NGC 253, whose number of GC (92±21; Olsen et al. 2004) had been misread by Harris et al. 2014, on the $M_\bullet$ versus N_{GC} parameter space.

For most of these objects, there are strong arguments against major mergers (pseudobulges, low density environments, rotating GC systems). While log $M_\bullet \propto (1.02 \pm 0.10)$ log N_{GC} for ellipticals (Figure 1, *top left*), log $M_\bullet \propto (1.64 \pm 0.24)$ log N_{GC} for late type galaxies (LTG; Figure 1, *top right*), in both cases including cosmic scatter in both variables. Hence, the correlation for LTG is much steeper than and not consistent with the linear correlation for ellipticals, for which statistical convergence through mergers may be important.

3. The $M_\bullet$ versus M_* correlation: secular co-evolution

Finally, we have fitted $M_\bullet$ versus total galaxy stellar mass, M_*, for both ellipticals and LTG. **We derive M_* from N_{GC}.** The relation between log N_{GC} and M_* is a broken power-law, with log $M_* = 2.11 \times$ log N_{GC} + 5.99, for lower galaxy masses, and log $M_* = 0.69 \times$ log N_{GC} + 8.96, for higher galaxy masses, with the inflection point at log $M_*/M_\odot \sim 10.4$, log $N_{GC} \sim 2.1$ (Hudson et al. 2014). Far from mirroring the discrepant slopes of the $M_\bullet$ versus N_{GC} fits, the correlations between log $M_\bullet$ and log M_* for ellipticals and LTG are quite similar, and with a flatter slope for LTG. Once again including cosmic scatter, we obtain $M_\bullet \propto (1.48 \pm 0.18)$ log M_* for ellipticals (Figure 1 *bottom left, red-dashed* and *pink translucent lines*), and $M_\bullet \propto (1.21 \pm 0.16)$ log M_* for LTG (Figure 1, *bottom left, solid black* and *gray translucent lines*). The fits have a small offset of $\sim$ 0.2-0.3 dex at log $M_*/M_\odot = 10$. The simultaneous fit for ellipticals and LTG is shown in Figure 1, *bottom right.* These fits are in line with the results of Reines & Volonteri 2015 and Simmons et al. 2017, and argue for a mostly non-merger driven $M_\bullet$ growth, and galaxy-black hole co-evolution through secular processes, like "calm" accretion and feedback (Simmons et al. 2017; Martin et al. 2018).

References

Akritas, M. G. & Bershady, M. A. 1996, *ApJ*, 470, 706

Bogdán, Á., Lovisari, L., Volonteri, M., et al. 2018, *ApJ*, 852, 131

Burkert, A., & Tremaine, S. 2010, *ApJ*, 720, 516

Capuzzo-Dolcetta, R., & Donnarumma, I. 2001, *MNRAS* 328, 645

Dressler, A. 1989, in IAU Symp. 134, Active Galactic Nuclei, ed. D. E. Osterbrock & J. S. Miller (Dordrecht: Kluwer), 217

Fabian, A. C. 2012, *ARA&A*, 50, 455

Ferrarese, L., & Merritt, D. 2000, *ApJ*, 539, L9

Gebhardt, K., et al. 2000, *ApJ*, 539, L13

Gnedin, O. Y., Ostriker, J. P., & Tremaine, S. 2014, *ApJ*, 785, 71

Harris, G. L. H., & Harris, W. E. 2011, *MNRAS*, 410, 2347

Harris, G. L. H., Poole, G. B., & Harris, W. E. 2014, *MNRAS*, 438, 2117

Hudson, M. J., Harris, G. L., & Harris, W. E. 2014, *ApJL*, 787, L5.

Jahnke, K., & Macciò, A. V. 2011, *ApJ*, 734, 92

Kormendy, J. 1993, *Galactic Bulges*, 153, 209

Kormendy, J. & Ho, L. C. 2013, *ARA&A*, 51, 511

Kormendy, J., & Richstone, D. 1995, *ARA&A*, 33, 581

Magorrian, J., et al. 1998, *AJ*, 115, 2285

Martin, G., Kaviraj, S., Volonteri, M., et al. 2018, *MNRAS*, 476, 2801

McConnell, N. J. & Ma, C.-P. 2013, *ApJ*, 764, 184
Olsen, K. A. G., Miller, B. W., Suntzeff, N. B., et al. 2004, *AJ*, 127, 2674
Peng, C. Y. 2007, *ApJ*, 671, 1098
Reines, A. E. & Volonteri, M. 2015, *ApJ*, 813, 82
Sahu, N., Graham, A. W., & Davis, B. L. 2019a, *ApJ*, 876, 155
—— 2019b, *ApJ*, 887, 10
Savorgnan, G. A. D., Graham, A. W., Marconi, A., et al. 2016, *ApJ*, 817, 21
Silk, J., & Rees, M. J. 1998, *A&A*, 331, L1
Simmons, B. D., Smethurst, R. J., & Lintott, C. 2017, *MNRAS*, 470, 1559
Spitler, L. R., & Forbes, D. A. 2009, *MNRAS*, 392, L1
van den Bosch, R. C. E. 2016, *ApJ*, 831, 134

Resolving the Rise and Fall of Star Formation in Galaxies
Proceedings IAU Symposium No. 373, 2023
T. Wong & W.-T. Kim, eds.
doi:10.1017/S1743921323000157

Star Formation History in the Illustris TNG Simulation

András Péter Joó[1], Bendegúz Koncz[1], Sandor Pinter[2] and L. Viktor Tóth[1]

[1]Dept. of Astronomy, Eötvös Loránd University

[2]Dept. of Natural Science, University of Public Service, Hungary

Abstract. We processed the catalogue data for all snapshots of the Illustris TNG100 cosmological simulation and collected every calculated property of the galaxies formed at different redshifts. With this dataset we can statistically analyze parameters for galaxy samples at given redshifts, as well as trace sample parameters over the entire time range of the simulation. Focusing first on star formation rate (SFR) and metallicity, we see the cosmic star formation history with the mean maximum at around $z \approx 1.6$ and the reionization bump at around $z \approx 5$, while metallicity increases. For a sample of strongly star-forming galaxies with SFR > 10 $M_{\odot}yr^{-1}$ we found different characteristics compared to the whole sample. The mean metallicity of highly star-forming galaxies is higher and changes less, and the mean SFR has its maximum at around the reionization bump.

Keywords. software: simulations, galaxies: star formation, galaxies: starburst, galaxies: high-redshift

1. Introduction

IllustrisTNG (Springel et al. (2018), Marinacci et al. (2018), Nelson et al. (2018), Pillepich et al. (2018), Naiman et al. (2018)) is a suite of large volume, cosmological, gravo-magneto-hydrodynamical simulations including a comprehensive model for galaxy formation. Each TNG simulation self-consistently solves for the coupled evolution of dark matter, cosmic gas, luminous stars, and supermassive blackholes from redshift $z = 127$ to 0 and generates 100 resulting snapshots from $z = 20$ to 0. We used the TNG100 run for analysis, the main high-resolution run including the full TNG physics model, which has the size of 110.73 Mpc^3 and contains more than 10 billion resolution elements.

2. Data processing and results

We processed the catalogue data for all snapshots of the TNG100 run to statistically analyze the properties of galaxies at each redshift. We collected every property from the catalogue files for all the subhalos found by the subfind algorithm, so that we have the subhalo-level properties for all the galaxies exported for every snapshot. With this dataset it becomes possible not only to statistically analyze parameters for galaxy samples at given redshifts, but to easily trace sample parameters over the entire time range of the simulation.

Here we are focusing on how the star formation rate (SFR) and metallicity (Ƶ) of galaxies evolved through time (redshift z). The SFR is calculated for every subhalo from the star formation processes. The metallicity is calculated as Ƶ = M_Z/M_{tot} of all gas cells bound to the subhalo, where Z is any element above He. The star formation rate

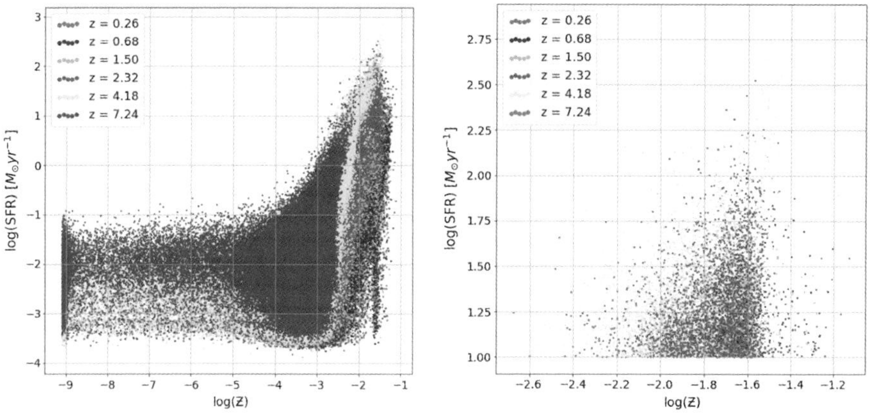

Figure 1. The relation between star formation rate (SFR) and metallicity (Z) for individual galaxies at different redshifts in the TNG100 simulation. Left: all galaxies in the snapshots. The whole population spreads towards lower SFR at low metallicities and higher SFR at higher metallicities, while new higher metallicities appear. Right: galaxies with SFR > 10 $M_\odot yr^{-1}$. Strongly star-forming galaxies are located at high metallicities, and tend towards lower SFR and higher metallicity.

density (SFRD) is calculated as the sum of SFRs for all galaxies in a sample divided by the simulation volume $1000 Mpc^3$. The mean star formation rate (<SFR>) is the average of the SFRs for all galaxies in a sample.

Besides investigating the whole galaxy sample in each redshift, we also separated galaxies with SFR > 10 $M_\odot yr^{-1}$ as a crude sample for strongly star-forming galaxies.

Figure 1 shows parameters of individual galaxies in six snapshots from $z = 7.24$ to $z = 0.26$. The diagram on the left includes all galaxies, while the one on the right contains only strongly star-forming galaxies with SFR > 10 $M_\odot yr^{-1}$. The whole population spreads towards lower SFR at low metallicities and higher SFR at higher metallicities, while new higher metallicities appear. Strongly star-forming galaxies are located at high metallicities for all the six redshifts, and tend toward lower SFR and higher metallicity.

Figure 2 shows the change of star formation rate density (SFRD) and mean metallicity through cosmic time, the left diagram shows data for all galaxies, while the right diagram shows data only for strongly star-forming galaxies with SFR > 10 $M_\odot yr^{-1}$. The cosmic trend of star formation rate density can be seen, in accordance with previous results from the Illustris simulation (eg. Bignone et al. (2017)). We have our highest data point at $z = 2.73$. Strongly star-forming galaxies show similar trend with a noticeable bump from $z \approx 6$ to 4, a period after the reionization epoch (Bauer et al. (2015), Thélie et al. (2022)).

We can observe this bump in detail looking at the mean SFR on Figure 3, on the left for all galaxies, and on the right for galaxies with SFR > 10 $M_\odot yr^{-1}$. Strongly star-forming galaxies have their peak <SFR> after the reionization epoch, from $z \approx 6$ to 4, and a relapse can also be seen around $z \approx 7$.

3. Summary and discussion

We investigated how the star forming rate (SFR) and metallicity of galaxies evolve through time (redshift z) in the Illustris TNG100 simulation. We found that the galaxy population spreads towards lower SFR at low metallicities and high SFR at high metallicities, while strongly star-forming galaxies with SFR > 10 $M_\odot yr^{-1}$ tend towards lower SFR and higher metallicity. The star formation density shows the cosmic trend in accordance with previous results from the Illustris simulation (eg. Bignone et al. (2017)) with

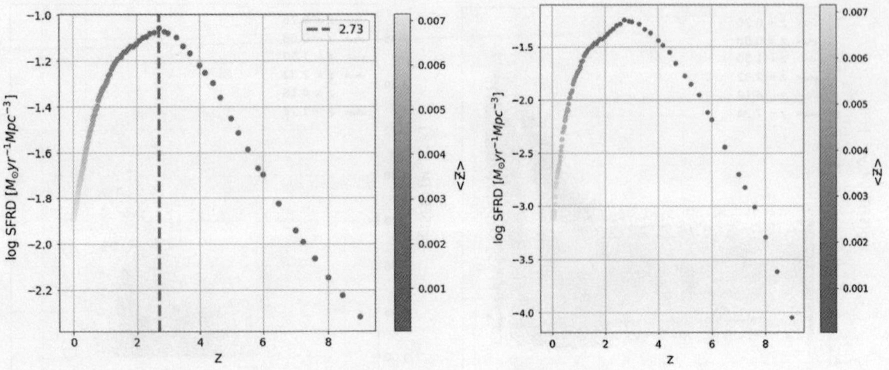

Figure 2. The change of the star formation rate density (SFRD) and mean metallicity (Ƶ) with redshift (z) in the TNG100 simulation. Left: for all galaxies in the snapshots. The cosmic trend of SFRD can be seen with the highest data point at z = 2.73. Right: for strongly star-forming galaxies with SFR > 10 $M_{\odot} yr^{-1}$. Similar trend with a bump after the reionization epoch between z ≈ 6 to 4.

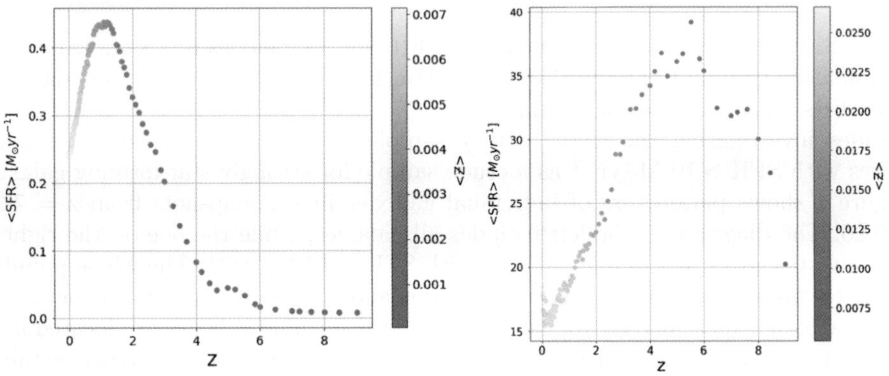

Figure 3. The change of the mean star formation rate (<SFR>) and mean metallicity (Ƶ) with redshift (z) in the TNG100 simulation. Left: for all galaxies in the snapshots. A bump can be seen after the reionization epoch from z ≈ 6 to 4. Right: for strongly star-forming galaxies with SFR > 10 $M_{\odot} yr^{-1}$. <SFR> peaks after the reionization epoch from z ≈ 6 to 4.

a noticeable bump after the reionization epoch (Bauer et al. (2015), Thélie et al. (2022)) from z ≈ 6 to 4 for strongly star-forming galaxies, that can be further observed looking at how the mean SFR changes with redshift. Strongly star-forming galaxies have their peak mean SFR in the z ≈ 6 to 4 period, and also a relapse can be seen around z ≈ 7.

Current models suggest that strongly star-forming environments produce more Long Gamma Ray Bursts (LGRBs). With the use of our dataset we plan to validate these models by assembling possible LGRB host galaxy samples and compare those with results from observational data (for eg. Rácz et al. (2018), Tóth et al. (2019), Horvath et al. (2022)).

Acknowledgements

Simulation data is provided by The TNG Collaboration: `https://www.tng-project.org/`. This work was supported by the Hungarian TKP2021-NVA-16 grant.

Supplementary material

To view supplementary material for this article, please visit https://doi.org/10.1017/S1743921323000157.

References

Springel et al. 2018. MNRAS, 475, 676. doi:10.1093/mnras/stx3304
Marinacci et al. 2018. MNRAS, 480, 5113. doi:10.1093/mnras/sty2206
Nelson et al. 2018. MNRAS, 475, 624. doi: 10.1093/mnras/stx3040
Pillepich et al. 2018. MNRAS, 475, 648. doi: 10.1093/mnras/stx3112
Naiman et al. 2018. MNRAS, 477, 1206. doi: 10.1093/mnras/sty618
Bignone et al. 2017. MNRAS, 469, 4921. doi: 10.1093/mnras/stx1132
Bauer et al. 2015. MNRAS, 453, 3593. doi: 10.1093/mnras/stv1893
Thélie et al. 2022. A&A, 658, A139 doi: 10.1051/0004-6361/202141748
Rácz et al. 2018. AN, 339:347–351. doi.: 10.1002/asna.201813503
Tóth et al. 2019. MNRAS, 486, 4823. doi.: 10.1093/mnras/stz1188
Horvath et al. 2022. Universe, 8(4), 221. doi: 10.3390/universe8040221

Resolving the Rise and Fall of Star Formation in Galaxies
Proceedings IAU Symposium No. 373, 2023
T. Wong & W.-T. Kim, eds.
doi:10.1017/S1743921322004811

Diverse Star Formation History in Presence of Cold-Mode Gas Accretion: From Solar Neighborhood to Distant Galaxies

Masafumi Noguchi

Astronomical Institute, Tohoku University,
6-3, Aramaki, Aoba-ku, Sendai, Miyagi, 980-8578, Japan
email: noguchi@astr.tohoku.ac.jp

Abstract. Existence of cold-mode gas accretion along with the hot-mode accretion of the shock-heated gas can explain the bimodality in the elemental abundance of the Milky Way disk stars as well as the mass-dependence of galaxy morphology represented by mass ratios of thin disks, thick disks, and bulges.

Keywords. Galaxy: abundances, galaxies: formation, galaxies: structure

1. Introduction

Cosmological simulations suggest the importance of the cold gas accretion onto galaxies (cold-mode) as a prime driver of galaxy formation, in addition to the cooling flow of the shock-heated halo gas (hot-mode accretion) (e.g. Dekel & Birnboim 2006). We build a simple model of galaxy evolution including cosmological gas accretion (both cold and hot modes) and suggest that the signature of the cold-mode accretion can be found in various forms. The model includes, in addition to the gas accretion, the star formation from the interstellar medium (ISM), the enrichment of ISM by type II and type Ia supernovae (SNe), and the ejection of ISM by type II SNe.

2. Chemical bimodality in the Milky Way

The stars in the Milky Way disk exhibit bimodal distribution in abundance ratios [α/Fe] and [Fe/H] (e.g. Haywood et al. 2016). The model predicts two distinct episodes of star formation (SF) resulting from the transition from the cold-mode to the hot-mode, thereby explaining the observation (Fig. 1).

Initial cold-mode accretion, proceeding with the free-fall time, produces stars enriched by α-elements synthesized by Type II SNe but relatively free from iron because of delay of Type Ia SNe. These stars are identified with the thick disk stars. When the accretion is switched to the hot mode, SF slows down because the shock-heated gas takes the radiative cooling time which is longer than the free-fall time to arrive at the disk. When the cooling flow arrives at the disk, the second SF starts, which produces stars enriched also by iron due to the action of Type Ia SNe. These stars are regarded as the thin disk stars.

3. Morphology of disk galaxies

Cosmological simulations suggest the existence of three regimes for the halo gas accretion demarcated by the shock mass and the stream mass (Fig. 2). In Domain F, the virial shock does not develop and the unheated gas accretes with the free-fall time (cold-mode).

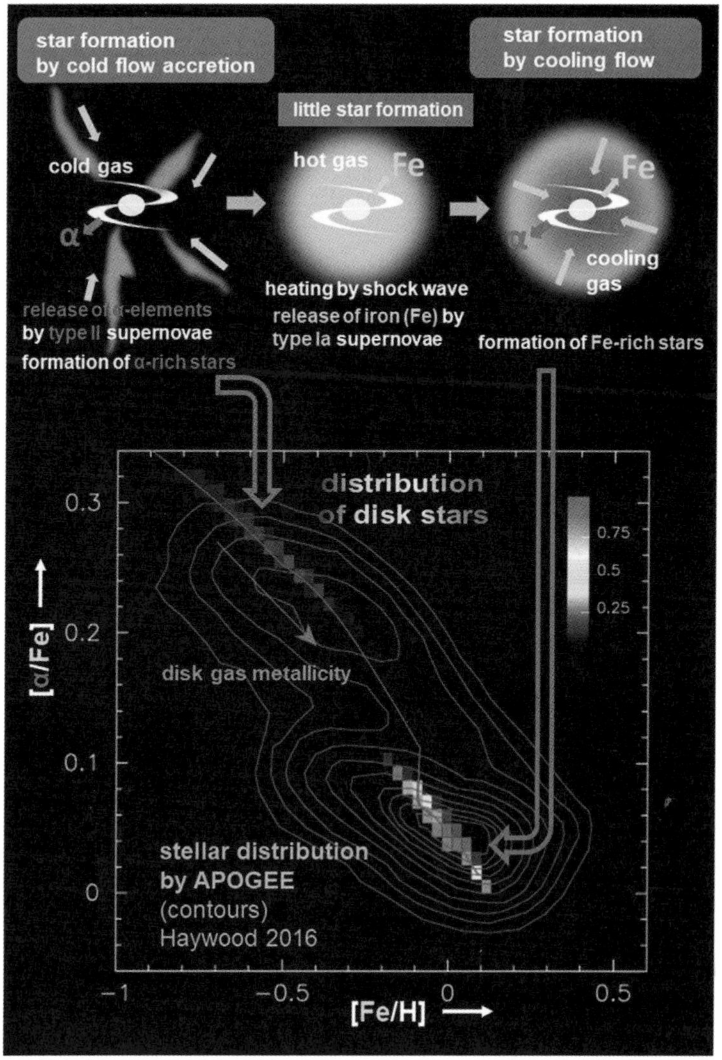

Figure 1. Mechanism for creating chemical bimodality in the Milky Way disk.

In Domain G, the halo gas is totally shock-heated leading to cooling flow (hot-mode). In Domain H, the cold-mode driven by inflowing gas streams and the hot-mode coexist.

If we assume the following correspondence, Domain F = thick disk formation, Domain G = thin disk formation, Domain H (hot gas) = thin disk formation, Domain H (cold gas) = bulge formation, the model produces increasing (decreasing) mass fraction of bulges (thick disks) with the galaxy stellar mass, in agreement with the observation (Fig. 3).

4. Conclusions

The galaxy evolution model including the cold-mode gas accretion explains the chemical bimodality observed in Milky Way disk as a result of transition from the cold to hot modes. When applied to disk galaxies in general, it reproduces the morphological variation along the galaxy mass represented by the mass fraction of thin disks, thick disks, and bulges.

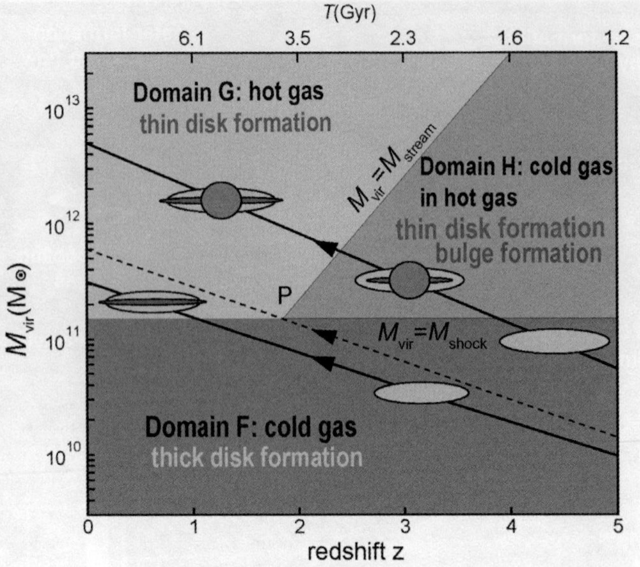

Figure 2. Physical state of the halo gas and the representative evolution of high and low mass galaxies.

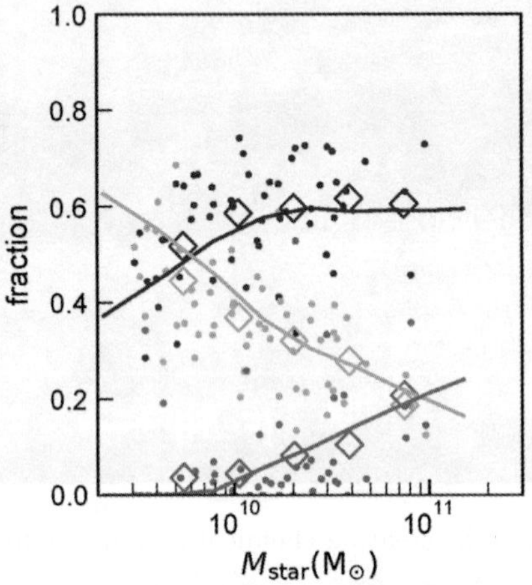

Figure 3. Morphological breakdown of stellar content. Blue: thin disk, orange: thick disk, and red: bulge. Dots represent the observation of Comeron et al. (2014).

References

Comeron, S. et al. 2014, *Astron. Astrophys.*, 571, A58

Dekel, A., & Birnboim, Y. 2006, *Mon. Not. Roy. Astr. Soc.*, 368, 2

Haywood, M. et al. 2016, *Astron. Astrophys.*, 589, A66

Noguchi, M. 2018, *Nature*, 559, 585

Noguchi, M. 2020, *Mon. Not. Roy. Astr. Soc.*, 494, L37

Resolving the Rise and Fall of Star Formation in Galaxies
Proceedings IAU Symposium No. 373, 2023
T. Wong & W.-T. Kim, eds.
doi:10.1017/S1743921322004823

Anatomy of Galactic Star Formation: Roles of Different Modes of Gas Accretion, Feedback, and Recycling

Masafumi Noguchi

Astronomical Institute, Tohoku University,
6-3, Aramaki, Aoba-ku, Sendai, Miyagi, 980-8578, Japan
email: noguchi@astr.tohoku.ac.jp

Abstract. Existence of the cold-mode gas accretion along with the hot-mode accretion can explain the diversity in the galactic star formation history across galaxy mass. We examine the role of various physical processes in producing the observed diversity.

Keywords. galaxies: evolution, galaxies: high-redshift, galaxies: starburst

1. Introduction

Recent observations reveal a rich variety in galactic star formation history (SFH) across galaxy mass. To understand the origin of this diversity and elucidate roles of various baryonic processes, we build a simple model of galaxy evolution including cosmological gas accretion, star formation, ejection of interstellar medium (ISM) by supernovae, recycling of the ejected ISM, and preventive feedback that hinders gas from accreting to the galactic disk (Fig. 1). Cosmological simulations suggest that the property of the halo gas depends on the halo mass and redshift (Fig. 2 'fiducial scheme'). In each Domain, the gas accretes onto the disk with either the free-fall time ($t_{\rm ff}$) (cold mode) or the radiative cooling time ($t_{\rm cool}$) (hot mode) as indicated. Evolution of a series of models (indicated by gray trajectories) is calculated under these accretions. The shock-heating scheme assumes heating to virial temperature for all halos.

2. Star Formation History (SFH)

As shown in Fig. 3, the fiducial scheme predicts that more massive halos reach the peak star formation rate (SFR) earlier and the decrease after that is more prominent, in qualitative agreement with the observation. On the other hand, SFHs in the shock-heating scheme depend little on the halo mass.

3. Main-sequence Evolution

Fig. 4 shows the evolution of the SFR-$M_\star$ relation (Main sequence). The fiducial scheme produces stronger 'turn-over' (decreasing slope) at high masses with time in agreement with the observation, whereas the shock-heating scheme shows an opposite trend.

4. Effect of Feedback and Recycling

We consider the reference model in which only gas accretion of the fiducial scheme is included. Then we add to this accretion-only model, ejective feedback, ejective feedback + recycling, and preventive feedback separately to examine the role of each process.

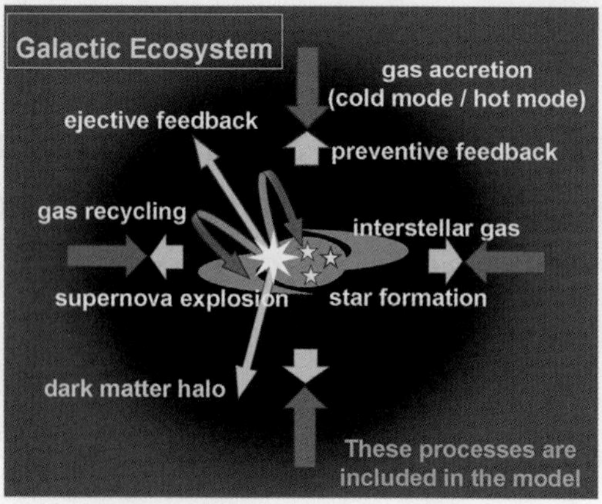

Figure 1. Physical processes included in the model.

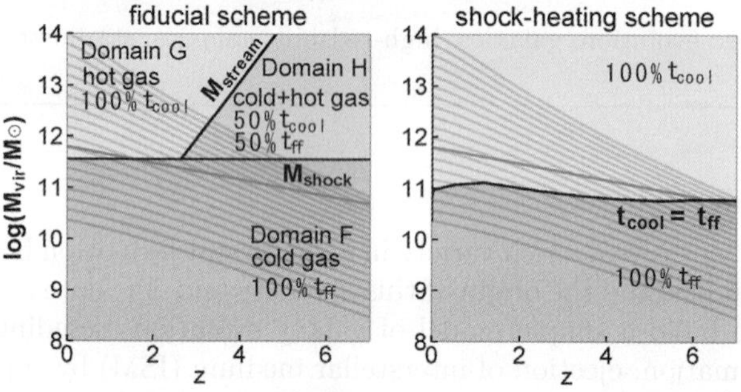

Figure 2. Physical state of the halo gas. Preventive feedback operates below the orange line.

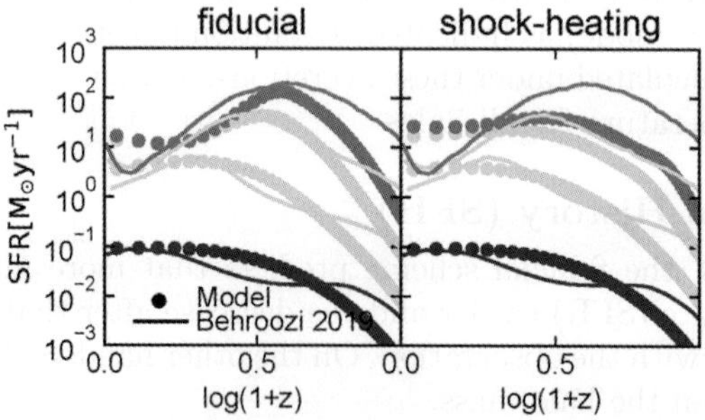

Figure 3. Dots: Model star formation history for the present halo mass of 10^{11} (blue), 10^{12} (green), 10^{13} (yellow), and $10^{14}M_\odot$ (red), respectively. Lines: Empirical analyses by Behroozi et al. (2019).

Fig. 5 shows that the ejective or preventive feedback does not change the ratio of peak and minimum SFRs and the redshift for peak SFR ($z_{\rm peak}$) while gas recycling moves $z_{\rm peak}$ for low mass galaxies to much recent epoch and turns their SFHs to monotonically rising ones (i.e., $\mathrm{SFR_{peak}/SFR_{min}} \sim 1$). This also makes low-mass galaxies younger. These changes improve the agreement with the observation.

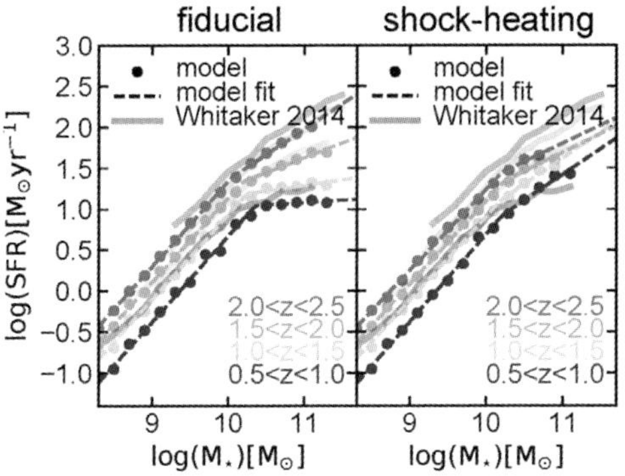

Figure 4. Redshift evolution of the star-forming galaxy main sequence. Broad lines represent the analysis of Whitaker et al. (2014).

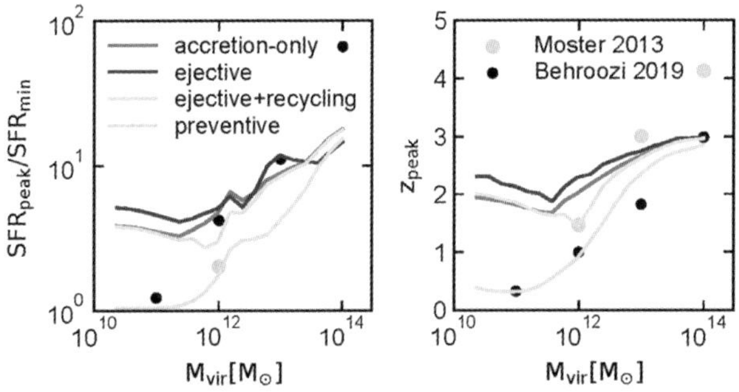

Figure 5. Ratio of the peak and minimum star formation rates and the redshift for the peak SFR. Effect of each added process on the accretion-only model is illustrated.

5. Conclusions

The transition from the cold-mode to hot-mode gas accretion constitutes the basis of star formation history of massive galaxies. Ejective and preventive feedback help reduce the stellar mass of low-mass galaxies to the observed level but the recycling of ejected ISM is required to make these galaxies young enough to match the observation.

References

Behroozi, P. et al. 2019, *Mon. Not. Roy. Astr. Soc.*, 488, 3143
Moster, B.P., Naab, T., & White, S.D.M. 2013, *Mon. Not. Roy. Astr. Soc.*, 428, 3121
Noguchi, M. 2022, submitted to *Mon. Not. Roy. Astr. Soc.*
Whitaker, K.E. et al. 2014, *ApJ*, 795, 104

Resolving the Rise and Fall of Star Formation in Galaxies
Proceedings IAU Symposium No. 373, 2023
T. Wong & W.-T. Kim, eds.
doi:10.1017/S1743921322003969

The Extragalactic Database for Galaxy Evolution (EDGE)

Tony Wong[1], Yixian Cao[2], Yufeng Luo[3], Alberto D. Bolatto[4], Sebastián F. Sánchez[5], and the EDGE-CALIFA Collaboration

[1]Astronomy Department, University of Illinois, Urbana, IL 61801, USA, wongt@illinois.edu

[2]Max-Planck-Institut für extraterrestrische Physik, D-85748 Garching, Germany

[3]Dept. of Physics and Astronomy, University of Wyoming, Laramie, WY 82071, USA

[4]Dept. of Astronomy, University of Maryland, College Park, MD 20742, USA

[5]Instituto de Astronomía, Universidad Nacional Autónoma de México, 04510 México, D.F., Mexico

Abstract. The EDGE-CALIFA collaboration has obtained resolved IFU and CO spectroscopy for 126 nearby galaxies, selected from the CALIFA main sample. We have assembled the spatially resolved products of the survey into Astropy-compatible pixel tables that reduce the oversampling in the original FITS images and facilitate comparison of pixels across different images. By joining these pixel tables to lower dimensional tables that provide profiles, integrated spectra, or global properties, it is possible to investigate the dependence of star formation rate on both local and global conditions. In this short video talk I provide an overview of EDGE, with examples of the use of the database to conduct analysis and generate plots.

Keywords. Galaxy evolution, Interstellar molecules, Star formation, Galaxy photometry

1. Introduction

The EDGE-CALIFA Survey has observed 126 galaxies from the CALIFA sample (Sánchez et al. 2016a) in CO(1–0) emission with the CARMA interferometer (Bolatto et al. 2017) and at optical wavelengths (3745–7500Å) with the PPAK-PMAS integral field spectrograph (IFS) on the 3.5m Calar Alto telescope. In this video talk we have provided an overview of the Python-based `edge_pydb` database which provides access to the sampled image data, as well as integrated properties and 1-D profiles. The images include moment maps derived from the CO(1–0) cubes as well as analysis products (e.g. intensities of strong emission lines) obtained from the IFS cubes using the Pipe3D analysis suite (Sánchez et al. 2016b). We also provide a number of higher-level analysis products, such as the star formation rate surface density based on extinction-corrected Hα emission.

2. Data Access and Examples

The database consists of a number of Astropy (Astropy Collaboration et al. 2022) compatible tables and a Python package (`edge_pydb`) which assists with exploration of the data as well as documenting the process by which the database was constructed. A number of sample iPython notebooks are provided to help users get started with the package. The Python package is freely distributed on Github and PyPI, and is largely based on Astropy. It includes the image data for one galaxy, NGC 4047. The full set

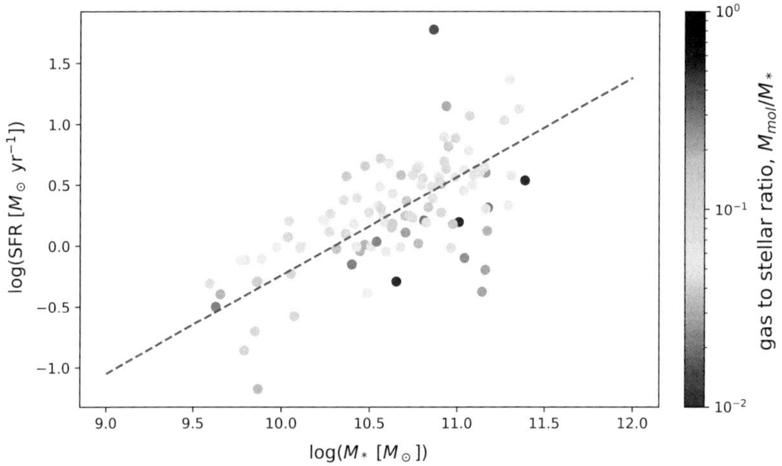

Figure 1. Integrated SFR vs. stellar mass for the EDGE-CALIFA galaxies observed with CARMA. Points are color coded by the ratio of molecular gas to stellar mass. The fiducial locus of the main sequence from Cano-Díaz et al. (2016) is shown as the dashed red line. To first order, the vertical offset from the main sequence is determined by molecular gas content.

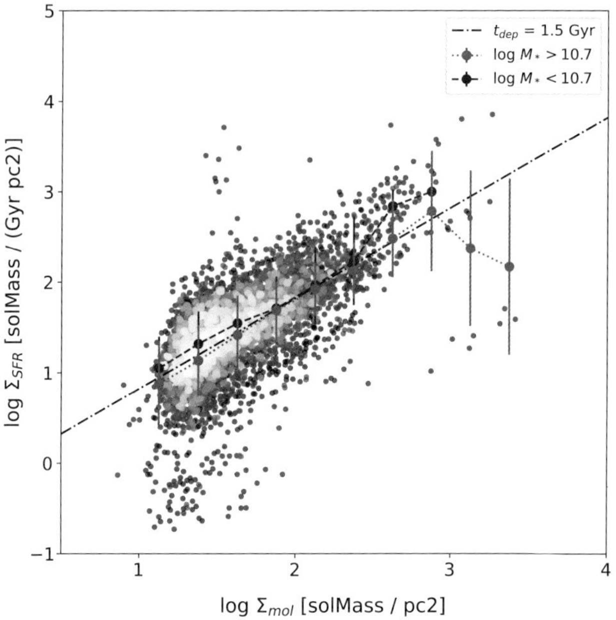

Figure 2. SFR surface density (based on extinction-corrected Hα emission) vs. molecular gas surface density for individual pixels within the EDGE database. Solid blue and green points represent binned averages for the pixels in low mass and high mass galaxies respectively. Overall the data show consistency with a common depletion time of 1.5 Gyr across the sample.

of tables for the 125 galaxies observed with CARMA (omitting one galaxy where the CALIFA data were of poor quality) is expected to be released in 2023 and will be discoverable through the Github site. We encourage interested users to contact us before the release to test out the database for their science.

The database includes sample scripts for plotting global properties of the sample, displaying scatter plots of the mass-metallicity and Kennicutt-Schmidt relations, and converting the tabulated pixel data back into images for visualizing the different data columns.

3. Science Applications

The EDGE-CALIFA data have already been used in a number of publications exploring the links between molecular gas and star formation across the galaxy population (e.g., Bolatto et al. 2017; Utomo et al. 2017; Colombo et al. 2018; Dey et al. 2019). Figure 1 shows an example of plotting the global data tables, producing a main-sequence plot color coded by molecular to stellar mass ratio. Figure 2 shows an example of plotting the image data, showing that the resolved Kennicutt-Schmidt (rKS) relation appears largely insensitive to the mass of the host galaxy in the parameter space covered by the survey. The code to generate these and other plots are included in the sample notebooks on Github.

The EDGE data architecture is easily extended to other surveys and will be used in future EDGE-CALIFA programs that extend the CARMA-observed sample using CO data from other millimeter-wave observatories (e.g., APEX, ALMA, GBT, LMT). Feedback from users is encouraged to facilitate future improvements in the database.

Acknowledgments. We acknowledge support from the National Science Foundation through AAG grants 1616199 and 2009849.

References

Astropy Collaboration, Price-Whelan, A. M., Lim, P. L., et al. 2022, ApJ, 935, 167.
Bolatto, A. D., Wong, T., Utomo, D., et al. 2017, ApJ, 846, 159.
Cano-Díaz, M., Sánchez, S. F., Zibetti, S., et al. 2016, ApJ, 821, L26.
Colombo, D., Kalinova, V., Utomo, D., et al. 2018, MNRAS, 475, 1791.
Dey, B., Rosolowsky, E., Cao, Y., et al. 2019, MNRAS, 488, 1926.
Sánchez, S. F., García-Benito, R., Zibetti, S., et al. 2016a, A&A, 594, A36.
Sánchez, S. F., Pérez, E., Sánchez-Blázquez, P., et al. 2016b, RMxAA, 52, 171
Utomo, D., Bolatto, A. D., Wong, T., et al. 2017, ApJ, 849, 26.

Author index